Nanofluids

Preparation, Applications, and Simulation Methods

Nanofluids

Preparation, Applications, and Simulation Methods

Edited by

Mohammad Mehdi Rashidi

Institute of Fundamental and Frontier Sciences, University of Electronic Science and Technology of China, Chengdu, P.R. China

Sahar Zinatloo-Ajabshir

Department of Chemical Engineering, University of Bonab, Bonab, Iran

ELSEVIER

Elsevier
Radarweg 29, PO Box 211, 1000 AE Amsterdam, Netherlands
125 London Wall, London EC2Y 5AS, United Kingdom
50 Hampshire Street, 5th Floor, Cambridge, MA 02139, United States

ISBN: 978-0-443-13625-2

For Information on all Elsevier publications
visit our website at https://www.elsevier.com/books-and-journals

Publisher: Matthew Deans
Acquisitions Editor: Gwen Jones
Editorial Project Manager: John Leonard
Production Project Manager: Fizza Fathima
Cover Designer: Greg Harris

Typeset by MPS Limited, Chennai, India

Contents

List of contributors

Munawwar Ali Abbas
Department of Mathematics, University of Baltistan, Skardu, Pakistan

Zeeshan Ali
School of Engineering, Monash University Malaysia, Bandar Sunway, Selangor, Malaysia

Mahnaz Amiri
Neuroscience Research Center, Institute of Neuropharmacology, Kerman University of Medical Science, Kerman, Iran

Muhammad Hossein Ashoub
Stem Cells and Regenerative Medicine Innovation Center, Kerman University of Medical Sciences, Kerman, Iran

Emmanuel O. Atofarati
Department of Mechanical and Aeronautical Engineering, University of Pretoria, Hatfield, Gauteng, South Africa

M.M. Bhatti
Material Science Innovation and Modelling (MaSIM) Research Focus Area, North-West University (Mafikeng Campus), Mmabatho, South Africa

Molka El Maghraoui
Faculty of Science of Tunis, University of Tunis El Manar, Tunis, Tunisia

Mahdiyeh Esmaeili-Zare
Institute of Nano Science and Nano Technology, University of Kashan, Kashan, Iran

Zahra Fakharshakeri
Institute of Nano Science and Nano Technology, University of Kashan, Kashan, Iran

Amin Moradi Hasan-Abad
Autoimmune Diseases Research Center, Shahid Beheshti Hospital, Kashan University of Medical Sciences, Kashan, Iran

Kamyar Hosseini
Department of Mathematics, Near East University TRNC, Mersin, Turkey

Ghazal Khajouei
Neuroscience Research Center, Institute of Neuropharmacology, Kerman University of Medical Science, Kerman, Iran

Morteza Khoshvaght-Aliabadi
Department of Chemical Engineering, Shahrood Branch, Islamic Azad University, Shahrud, Iran

Yuan Ma
Department of Mechanical Engineering, The Hong Kong Polytechnic University, Kowloon, Hong Kong, P.R. China

Maryam Masjedi-Arani
Institute of Nano Science and Nano Technology, University of Kashan, Kashan, Iran

Josua P. Meyer
Department of Mechanical and Aeronautical Engineering, University of Pretoria, Hatfield, Gauteng, South Africa; Department of Mechanical and Mechatronic Engineering, Stellenbosch University, Stellenbosch, Western Cape, South Africa

Raed Muslim Mhaibes
Department of Biochemistry, College of Medicine, Misan University, Misan, Iraq

Amir Hossein Mohammadi
Department of Biochemistry, School of Medicine, AJA University of Medical Sciences, Tehran, Iran

Rasul Mohebbi
School of Engineering, Damghan University, Damghan, Iran

Sharifpur Mohsen
Department of Mechanical and Aeronautical Engineering, University of Pretoria, Hatfield, Gauteng, South Africa; School of Mechanical, Industrial and Aeronautical Engineering, University of the Witwatersrand, Gauteng, South Africa; Department of Medical Research, China Medical University Hospital, China Medical University Taichung, Taichung, Heilongjiang, Taiwan

Kimiya Mosaddeghi
School of Medicine, Kashan University of Medical Sciences, Kashan, Iran; Student Research Committee, Kashan University of Medical Sciences, Kashan, Iran

Saima Muhammad
Department of Mathematics, University of Baltistan, Skardu, Pakistan

Mohammad Olfati
Mechanical Engineering Department, Faculty of Engineering, Razi University, Kermanshah, Iran

Faranak Rabiei
Department of Mathematics, Texas A&M University Kingsville, Kingsville, TX, United States; School of Engineering, Monash University Malaysia, Bandar Sunway, Selangor, Malaysia

Mohammad Mehdi Rashidi
Institute of Fundamental and Frontier Sciences, University of Electronic Science and Technology of China, Chengdu, P.R. China

Razieh Razavi
Department of Chemistry, Faculty of Science, University of Jiroft, Jiroft, Iran

Hossein Safardoust-Hojaghan
Institute of Nano Science and Nano Technology, University of Kashan, Kashan, Iran

Aida Salimi
Department of Mechanical Engineering, West Tehran Branch, Islamic Azad University, Tehran, Iran

Ali Sobhani-Nasab
Physiology Research Center, Institute for Basic Sciences, Kashan University of Medical Sciences, Kashan, Iran

Sahar Zinatloo-Ajabshir
Department of Chemical Engineering, University of Bonab, Bonab, Iran

About the editors

Prof. Mohammad Mehdi Rashidi

Mohammad Mehdi Rashidi is a Professor of mechanical engineering, at the Institute of Fundamental and Frontier Sciences, at the University of Electronic Science and Technology of China, in Chengdu, China. He completed his PhD in mechanical engineering from Tarbiat Modares University, Iran, in 2002 and then went on to pursue his professional career at Bu Ali Sina University in Iran until 2014. He then took up the position of full Professor at Bu Ali Sina University and moved to Tongji University in Shanghai until the end of 2016. Following a short tenure at the University of Birmingham, UK, he returned to Tongji University in December 2018.

He has held numerous invited positions at international universities, including the University of Sherbrooke in Canada; the University of the Witwatersrand in Johannesburg, South Africa; the Universite Paris Ouest in France, and King Mongkut's University of Technology in Bangkok, Thailand. A commended Highly Cited Researcher for 2018, 2019, and 2020, he has published over 345 journal papers indexed in Scopus and Web of Science (h-index = 72), and 50 conference papers. He is an Associate Editor for more than five international journals and is an Active Member of the editorial boards for more than 45 national or international journals.

He is an Honorary Fellow of the Australian Institute of High Energetic Materials and is an Active Member of the South African Institution of Mechanical Engineering and the Iranian Society of Mechanical Engineers. His research interests include energy, energy generation, conversion and storage engineering, entropy, heat transfer, mechanical engineering, nonlinear analysis, and thermodynamics.

His research details are available on the following links:
ORCID ID: 0000−0002−6309−8688, https://orcid.org/0000−0002−6309−8688
Mendeley, https://www.mendeley.com/profiles/mohammad-mehdi-rashidi/
KUDOS, https://growkudos.com/hub/63551/publications
Researcher ID: P-2692−2014, http://www.researcherid.com/rid/P-2692−2014
Scopus Author ID: 57189276752
https://www.scopus.com/authid/detail.uri?authorId = 57189276752
Publons ID: 336837, https://publons.com/author/336837/mohammad-mehdi-rashidi#profile
LinkedIn, https://www.linkedin.com/in/mohammad-mehdi-rashidi-09167831/

Dr. Sahar Zinatloo-Ajabshir, PhD

Sahar Zinatloo-Ajabshir is an Associate Professor of nanoscience and nanotechnology. She obtained her MSc in nanoscience and nanotechnology from the University of Tehran, in 2010. Then after completing her PhD in nanoscience and nanotechnology in 2015 from the University of Kashan, she went on to join the group of Prof. Masoud Salavati-Niasari as a postdoc researcher. She became a Faculty Member of the University of Bonab, in 2017.

Her research interests include the development of different types of nanoscale materials, especially the preparation of ceramic nanomaterials and investigation of their applications in various fields such as hydrogen storage, photocatalysis, environmental remediation, and drug delivery systems. During her career, she has published several scientific papers in high-impact factor journals, as well as book chapters, books, and conference proceedings. She is also an active referee on the editorial board for several international journals.

Preface

Recently, nanofluids have obtained the attention of researchers and the scientific community owing to their unique properties and significant applications in several fields such as engineering, solar energy, nanomedicine, space technology, and renewable energies. Nanofluids, which contain ultrafine particles in the nanometer range, have been developed and investigated as a novel generation of heat transfer fluids. Among their substantial advantages is greater thermal conductivity, but the stability of nanofluid should be examined, more pumping power is another issue for them. It has been reported that the stability and dispersion of nanosized particles in the host fluid can greatly affect the performance of nanofluids in a variety of fields.

In this book, we focus on the recent advances related to preparation methods, applications, and numerical methods of nanofluids. Different approaches to preparing some remarkable families of nanofluids such as aluminum oxide-based nanofluids, CuO/Cu-based nanofluids, carbon nanotubes/graphene-based nanofluids, ZnO-based nanofluids, Fe_3O_4-based nanofluids, and SiO_2-based nanofluids are presented along with their advantages and disadvantages as well as their applications in various fields in the form of separate chapters of the book. This book is, therefore, intended as a valuable reference source for researchers and students and will cover the content related to the preparation, applications, and numerical methods for nanofluids.

Finally, we like to thank all the scientists who contributed to the book chapters and all the people who, in different ways, supported us in this work. We hope that this book will be a reference for those directly involved in advanced rare earth–based ceramic nanomaterials and also will be able to stimulate the creativity of newcomers working in different fields pushing them to find new fascinating applications for advanced rare earth–based ceramic nanomaterials.

Mohammad Mehdi Rashidi
Sahar Zinatloo-Ajabshir

CHAPTER

Nanofluids at a glance

1

Sahar Zinatloo-Ajabshir[1] and Mohammad Mehdi Rashidi[2]

[1]Department of Chemical Engineering, University of Bonab, Bonab, Iran

[2]Institute of Fundamental and Frontier Sciences, University of Electronic Science and Technology of China, Chengdu, P.R. China

1.1 Nanofluids

In recent years, with the development of efficient solutions based on nanotechnology, humanity has witnessed improvements in energy systems and also an increase in the efficiency of conversion processes, which has been accompanied by extensive positive economic and environmental effects [1−6]. The use of solutions based on nanotechnology has been reported in various fields, including medical applications or drug delivery, IT technology, clean energy production or storage, environmental science and engineering, materials engineering, solving environmental problems, and food safety and packaging [7−12]. Among these, the design and development of nanofluids is one of the remarkable fields of nanotechnology. Many efforts have been made to synthesize nanostructured materials by controlling their size and morphology because the features and performance of these materials are highly dependent on the preparation method [13−18]. Nanoscale materials include components in the atomic or molecular scale in the range of 1−100 nm, which exhibit special or better features and performance compared to bulk materials [19−22]. Different kinds of nanomaterials have been prepared and investigated in various approaches, including carbon-based nanocompounds such as graphene oxide, single-walled and multiwalled carbon nanotubes, metal oxides, semiconductors and also metals, and nanocomposites [23−25]. The research community has paid specific attention to nanofluids owing to their unique thermal conductivity and great heat transfer efficiency [26,27]. In 2013, for the first time, a colloidal mixture comprising base fluid and nanosized particles was synthesized and studied [28]. Nanofluids, which contain ultrafine particles in the nanometer range, have been developed and investigated as a novel generation of heat transfer fluids. Among their substantial advantages are greater stability, decreased viscosity, better thermal conductivity, and more pumping power [29,30]. The usage of nanofluids has been reported in diverse fields, such as biomedicine, transportation, energy production microelectronics [31,32]. In general, nanofluids include various nanocompounds dispersed in a base fluid like

Nanofluids. DOI: https://doi.org/10.1016/B978-0-443-13625-2.00001-2

oil, water, and ethylene glycol [33,34]. It has been reported that the stability and dispersion of nanocompounds in the host fluid can greatly affect the efficiency of nanofluids in a variety of fields.

Nowadays, using nanofluids is a promising tool for increasing heat-transfer devices' performance. In this book, a comprehensive study of new analytical and numerical simulation methods for nanofluids is carried out. In fact, the governing equations of nanofluid flow are Navier–Stokes equations, that is, conservation of mass (continuity equations), conservation of momentums, and conservation of energy. A nanofluid consists of a base fluid and nanoparticle/particles that, in reality, have two phases (liquid and solid phase). As the used solid phase nanoparticle/nanoparticles can have a higher thermal conductivity, by adding them to the base fluid, the overall thermal conductivity will be increased. Now let us come back to the governing equations, for analytical/numerical simulation of nanofluid flow, we have to solve the conservation of species. One simpler method can be used for analytical/numerical simulation, in this method, we can compute the viscosity and thermal conductivity of nanofluid using some formulas, in this approach, there is no need to solve the conservation of species. Definitely, the accuracy of this approach is less than considering adding the conservation of species to the Navier–Stokes equations.

For numerical simulation of nanofluids, three approaches could be used: macroscopic, mesoscopic, and microscopic. For the macroscopic approach, finite difference method, finite volume method (FVM), finite element method (FEM), and compact method can be utilized. Lattice Boltzmann method (LBM) is a mesoscopic approach that is widely used for the numerical simulation of nanofluids. Molecular dynamics method (MDM) is a good and powerful approach for the numerical simulation of nanofluid on a microscopic scale. Every approach has its own pros and cons. For instance, MDM has a better accuracy with respect to FVM, FEM, and LBM but its computational time is much higher than these methods. Recently some researchers have using multiscale computational methods for optimal solutions of nanofluid problems for achieving higher accuracy with less computational cost.

In addition, in this book, contemporary topics related to nanofluids, such as Cu-based nanofluids and their utilization within renewable energy applications, along with an examination and exploration of semianalytical solutions pertaining to the time-fractional Navier–Stokes equation, have been thoroughly presented and studied with enough details. Utilizing nanofluid in thermal, renewable energy systems and other engineering systems to enhance their efficiency is expanding continuously with each passing day. A nanofluid comprises particles with dimensions as small as 100 nm in suspension. As mentioned before, every nanofluid contains two main parts, base fluid, which can be water, ethylene glycol, engine oil, and decene, and another important part is nanoparticle/nanoparticles, some popular nanoparticles are as follows: copper, copper oxide, alumina oxide, silicon dioxide (SiO_2), and zinc oxide (ZnO).

A wide range of nanomaterials have been investigated as the dispersed phase including carbon nanotubes, metals (Cu, Ag, Fe), metal oxides (Al_2O_3, CuO,

ZrO_2, $CoFe_2O_4$, TiO_2), graphene, and so on. Pioneering studies in the 1990s revealed that by adding just a few percent of nanoparticles to popular base fluids such as water or ethylene glycol, the overall thermal conductivity will be increased significantly, and consequently the heat transfer efficiency will increase.

This discovery sparked tremendous interest in nanofluids to harness their properties for applications from heating and cooling systems, energy generation and management, and also in using thermal storage. The advantageous heat transfer properties have positioned nanofluids for numerous thermal management applications including transportation, microelectronics, HVAC systems, solar collectors, and nuclear reactors. Experimental studies have tested nanofluids for convective heat transfer enhancement and improved efficiency in heat exchangers and cooling devices. Nanoparticle additives have also enabled innovative smart fluids with tunable viscosity, magnetically-driven motions, electrorheological behavior, and nanoparticle self-thermophoresis effects. Biomedical applications such as magnetic hyperthermia cancer treatment and targeted drug delivery are also emerging. Indeed, it is crucial to note that the utilization of nanofluid helps to increase the heat transfer efficiency, but utilizing nanofluid in systems should be under control and even sometimes using nanofluid will damage the system (such as pump vanes, particularly when nanofluids are employed in solar collectors). In fact, the stability of nanofluid is a very important issue in practical applications and it must be considered as a very important parameter, otherwise using nanofluid will cause some serious problems in the thermal systems.

At present, many experimental investigations are being conducted in the field of heat transfer and fluid flow for using nanofluids in engineering systems. These experiments encompass various aspects, including simulating nanofluid usage, exploring recent advancements in preparation techniques, and examining applications such as Cu-based nanofluids. Additionally, researchers have turned to numerical, analytical, and semianalytical solutions such as the homotopy analysis method (HAM) and optimal homotopy asymptotic method (OHAM). It worth mentioning that if the governing equations of the considered problem can be converted to a similarity problem, one can easily implement advanced analytical methods such as homotopy perturbation method (HPM), HAM, Spectral HAM (SHAM), variational iteration method (VIM) and Adomian Decomposition Method (ADM).

In chapter eight of this book, which is dedicated to using nanofluids in engineering systems, the implementation of Cu-based nanofluids for increasing heat transfer performance is studied with enough detail. Using new types of nanofluids is developing fast, in fact, this method is a practical and relatively easy approach for increasing heat transfer performance. For instance, zirconium dioxide and copper (Cu), within a base fluid composed of ethylene glycol (EG), have been comprehensively examined by different research groups. The perception of hybrid nanofluids and their benefits compared to single-component nanofluids *is explored through an analysis of existing literature. Additionally, their potential* *applications in renewable energy sources have been ass*essed. Furthermore, certain research investigations have examined the impact of employing a MgO-CuO/water nanofluid on the energy-related performance of an evacuated heat pipe solar system, encompassing flat collectors, solar stills, pipes, and related components. Recent developments in the preparation methods and applications of Cu-based nanofluids such as heat exchangers, solar collectors, electro-cooling systems, medical applications, and nanocomposite materials have propelled these innovative materials into a broad spectrum of industries, promising improved efficiency, performance, and versatility across various technological domains.

Nanofluids and their applications in renewable energy are studied in another chapter. Recently, many researchers are implementing nanotechnology in renewable energy fields, such as solar still, heat transfer in solar collectors, improved heat dissipation in photovoltaic (PV) cells, industrial desalination, thermal energy enhancement, geometrical energy enhancement, biofuel production, biomass processing, and so on. For instance, (1) adding nanoparticles to conventional heat transfer fluids enhances solar thermal collector efficiency by increasing absorption of radiation and improving conductive/convective heat transfer, (2) reported performance benefits include increased collector thermal efficiency and higher operating temperatures allowing collection of higher-grade thermal energy, (3) Cooling solar PV panels using nanofluid-based heat removal systems improves solar cell efficiency and power output and nanofluids demonstrate promising potential as advanced heat transfer fluids for renewable energy technologies, but further research into sustainability and real-world efficacy is still required.

In another chapter of this book, an investigation was conducted on the semianalytical solution for the time-fractional Navier–Stokes equations employing Caputo fractional derivatives, with a particular focus on the utilization of the advanced version of HAM. In this method, the most important variable of HAM will be optimized for obtaining the most accurate solution. It is important to mention that despite this method having a better accuracy with respect to HAM, sometimes obtaining the optimal values could be a little difficult. This method is known as OHAM. For solving nanofluid flow problems, computational fluid dynamics modeling can be categorized into two important classifications: single-phase and two-phase simulation. Most recent simulations of nanofluid flow can be considered as a: single-phase approach because in these simulations, some equations were used for obtaining viscosity and thermal conductivity of the nanofluid (base fluid + nanoparticle/particles), and there is no conservation of species equation to be solved. In two two-phase simulations, one should solve the conservation of species equation. In fact, for OHAM, both methods can be selected.

One important advantage of using OHAM for solving nanofluid problems is that the solution will be obtained in a type of polynomial with all parameters that can change the solution, in other words, if one can implement OHAM for the solution, by changing any parameter, the user can obtain solution just from the polynomial. For instance, if, in the numerical method, one parameter is changed, the solution should be done again, but for OHAM, it is just enough to replace the new variable in the polynomial to obtain the results. The reader may ask why we have considered OHAM for our solution method. Many analytical methods can be used for solving these kinds of problems: perturbation method (PM), HPM, ADM, VIM, and differential transform method (DTM). Every method has its advantages and disadvantages, but in general conclusion, HAM and DTM are more robust and powerful methods for solving highly nonlinear problems. In terms of convergence and size of solution polynomial, DTM is more powerful than HAM, but it has some difficulties in solving boundary value problems. Therefore, for obtaining the analytical solution of nanofluid flow, we have used an advanced version of HAM, called OHAM. For more generalization, we used the fractional form of the governing equation. Fractional equations have many applications in science and engineering. The OHAM provides a way to find an approximate analytical solution to these fractional differential equations. It works by constructing a homotopy between the original equation and an initial approximation. OHAM involves expanding the solution in a series and finding optimal values for a more accurate solution. This results in a semianalytical solution containing a rapidly converging series term. OHAM provides several advantages over purely numerical techniques, including more efficient computation and physical insight into fractional dynamics. Overall, it is a powerful method for handling the mathematical complexities of fractional calculus applied to fluid flow problems.

Recent advances related to numerical simulation methods and applications of nanofluid in heat and mass transfer systems are investigated in another chapter of this book. The improvement of heat transfer can enhance the thermal and overall performance of the system, leading to more compact and cost-effective thermal equipment while saving energy and space.

In the last chapter of this book, applications of binary nanofluids including emerging issues are presented, the authors have presented new aspects of using binary nanofluids in engineering and science problems. Nanofluids are used widely for increasing heat transfer, one new application of nanofluids is using them for cooling by jet impingement. The same mechanism of heat transfer can increase the cooling performance of conventional fluid. Both experimental and numerical simulations are carried out to analyze this cooling method. The effect of the different physical parameters was considered in this regard.

This book focuses on the recent advances related to preparation methods, applications, and numerical methods of nanofluids. Different approaches to preparing some remarkable families of nanofluids such as aluminum oxide-based nanofluids, CuO/Cu-based nanofluids, carbon nanotubes/graphene-based nanofluids, ZnO-based nanofluids, Fe_3O_4-based nanofluids, and SiO_2-based nanofluids are presented along with their advantages and disadvantages as well as their applications in various fields in the form of separate chapters of the book. This book is, therefore, intended as a valuable reference source for researchers and students and will cover the content related to the preparation, applications, and numerical methods for nanofluids.

References

[1] T. Ambreen, M.-H. Kim, Influence of particle size on the effective thermal conductivity of nanofluids: a critical review, Appl. Energy 264 (2020)114684.

[2] Z. Said, M.A. Sohail, A.K. Pandey, P. Sharma, A. Waqas, W.-H. Chen, et al., Nanotechnology-integrated phase change material and nanofluids for solar applications as a potential approach for clean energy strategies: progress, challenges, and opportunities, J. Clean. Prod. 416 (2023)137736.

[3] M. Samadi Kazemi, A. Sobhani, $CuMn_2O_4$/chitosan micro/nanocomposite: green synthesis, methylene blue removal, and study of kinetic adsorption, adsorption isotherm experiments, mechanism and adsorbent capacity, Arab. J. Chem. 16 (2023) 104754.

[4] S. Zinatloo-Ajabshir, M.H. Esfahani, C.A. Marjerrison, J. Greedan, M. Behzad, Enhanced electrochemical hydrogen storage performance of lanthanum zirconium oxide ceramic microstructures synthesized by a simple approach, Ceram. Int. 49 (2023) 37415–37422.

[5] S. Alinavaz, M. Ghiyasiyan-Arani, E.A. Dawi, M. Salavati-Niasari, Sonochemical synthesis, characterization and investigation of the electrochemical hydrogen storage properties of $Ho_3Fe_5O_{12}/Mg_2Al(OH)_7$ nanocomposites, J. Energy Storage 66 (2023) 107369.

[6] A. Chachvalvutikul, T. Luangwanta, S. Kaowphong, Double Z-scheme $FeVO_4/Bi_4O_5Br_2$/BiOBr ternary heterojunction photocatalyst for simultaneous photocatalytic removal of hexavalent chromium and rhodamine B, J. Colloid Interface Sci. 603 (2021) 738–757.

[7] S.F. Ahmed, J.C. Debnath, F. Mehejabin, N. Islam, R. Tripura, M. Mofijur, et al., Utilization of nanomaterials in accelerating the production process of sustainable biofuels, Sustain. Energy Technol. Assess. 55 (2023) 102894.

[8] J.F. Li, W. Lu, Y.B. Zeng, Z.P. Luo, Simultaneous enhancement of latent heat and thermal conductivity of docosane-based phase change material in the presence of spongy graphene, Sol. Energy Mater. Sol. Cell 128 (2014) 48–51.

[9] A. Fahmi, S.R. Zavaragh, M.R. Hanafi, H. Rahimpour, S. Zinatloo-Ajabshir, A. Asghari, Facile preparation, characterization, and investigation of mechanical strength of Starchy NaCl-binder as a lightweight construction material, Sci. Rep. 13 (2023) 19042.

[10] A. Sobhani, Hydrothermal synthesis of $CuMn_2O_4$/CuO nanocomposite without capping agent and study its photocatalytic activity for elimination of dye pollution, Int. J. Hydrog. Energy 47 (2022) 20138–20152.

[11] S. Zinatloo-Ajabshir, H. Mahmoudi-Moghaddam, M. Amiri, H. Akbari Javar, A facile green fabrication of sponge-like $Pr_2Ce_2O_7$ nanostructure using sucrose for electrochemical quantification of anti-Parkinson drug pramipexole, Microchem. J. 195 (2023) 109480.

[12] Z. Heydariyan, R. Monsef, E.A. Dawi, M. Salavati-Niasari, $EuMnO_3/EuMn_2O_5$/MWCNT nanocomposites: insights into synthesis and application as potential materials for development of hydrogen storage capacity, Fuel 351 (2023) 128885.

[13] F. Ghadami, M. Valian, F. Atoof, A. Ryadh, A. Khoobi, M. Bagher Miranzadeh, M. Salavati-Niasari, Synthesis of a novel $HoFeO_3/Ho_3Fe_5O_{12}$ nanoadsorbent for efficient absorption of ciprofloxacin from liquid-phase: modeling and optimization using response surface methodology, J. Ind. Eng. Chem. 129 (2023) 143–157.

[14] A. Sobhani, S. Alinavaz, Co-precipitation synthesis of $CuMn_2O_4$/CuMnO nanocomposites without capping agent and investigation of their applications for removing pollutants from wastewater, Biomass Convers. Biorefinery (2023). Available from: https://doi.org/10.1007/s13399-022-03732-2.

[15] M. Mousavi-Kamazani, F. Azizi, Facile sonochemical synthesis of Cu doped CeO_2 nanostructures as a novel dual-functional photocatalytic adsorbent, Ultrason. Sonochem. 58 (2019) 104695.

[16] S. Hamzeh, H. Mahmoudi-Moghaddam, S. Zinatloo-Ajabshir, M. Amiri, S.A. Razavi Nasab, Eco-friendly synthesis of mesoporous praseodymium oxide nanoparticles for highly efficient electrochemical sensing of carmoisine in food samples, Food Chem. 433 (2024) 137363.

[17] L. Wang, B. Cheng, L. Zhang, J. Yu, In situ irradiated XPS investigation on S-scheme TiO_2@$ZnIn_2S_4$ photocatalyst for efficient photocatalytic CO_2 reduction, Small 17 (2021) 2103447.

[18] X. Zhang, Z. Cheng, P. Deng, L. Zhang, Y. Hou, $NiSe_2$/$Cd_{0.5}Zn_{0.5}S$ as a type-II heterojunction photocatalyst for enhanced photocatalytic hydrogen evolution, Int. J. Hydrog. Energy 46 (2021) 15389−15397.

[19] X. Wang, J. Zhuang, Q. Peng, Y. Li, A general strategy for nanocrystal synthesis, Nature 437 (2005) 121−124.

[20] A.P. Alivisatos, Semiconductor clusters, nanocrystals, and quantum dots, Science 271 (1996) 933−937.

[21] F. Soofivand, F. Mohandes, M. Salavati-Niasari, Silver chromate and silver dichromate nanostructures: sonochemical synthesis, characterization, and photocatalytic properties, Mater. Res. Bull. 48 (2013) 2084−2094.

[22] A. Sobhani, $CuMn_2O_4$/Mn_2O_3 micro composites: sol-gel synthesis in the presence of sucrose and investigation of their photocatalytic properties, Arab. J. Chem. 16 (2023) 105201.

[23] K. Scida, P.W. Stege, G. Haby, G.A. Messina, C.D. García, Recent applications of carbon-based nanomaterials in analytical chemistry: critical review, Anal. Chim. Acta 691 (2011) 6−17.

[24] S. Zinatloo-Ajabshir, Advanced Rare Earth-Based Ceramic Nanomaterials Elsevier Series on Advanced Ceramic Materials, Elsevier, Paperback ISBN: 9780323899574, 2022.

[25] N.S.K. Gowthaman, M. Amalraj, S. Kesavan, K. Rajalakshmi, S. Shankar, B. Sinduja, et al., Zero-, one- and two-dimensional carbon nanomaterials as low-cost catalysts in optical and electrochemical sensing of biomolecules and environmental pollutants, Microchem. J. 194 (2023) 109291.

[26] D.K. Devendiran, V.A. Amirtham, A review on preparation, characterization, properties and applications of nanofluids, Renew. Sustain. Energy Rev. 60 (2016) 21−40.

[27] A. Shirole, M. Wagh, V. Kulkarni, Thermal performance comparison of parabolic trough collector (PTC) using various nanofluids, Int. J. Renew. Energy Dev. 10 (2021) 875−889.

[28] B.P. McGrail, P.K. Thallapally, J. Blanchard, S.K. Nune, J.J. Jenks, L.X. Dang, Metal-organic heat carrier nanofluids, Nano Energy 2 (2013) 845−855.

[29] S. Margoum, C. El Fouas, B. Hajji, S. Aneli, M.G. Tina, A. Gagliano, Modelling and performances assessment of a nanofluids-based PV/T hybrid collector, Energy Sources Part A: Recovery Util. Environ. Eff. 45 (2023) 3070−3086.

[30] Y. Wu, M. Ibrahim, A. Althobaiti, Y. Cao, M. Dahari, D. Nguyen, et al., Employing numerical method for evaluating the heat transfer rate of a hot tube by nanofluid natural convection, Case Stud. Therm. Eng. 35 (2022) 102006.
[31] R. Saidur, K.Y. Leong, H.A. Mohammed, A review on applications and challenges of nanofluids, Renew. Sustain. Energy Rev. 15 (2011) 1646–1668.
[32] A. Dev, R. Kumar, R.P. Saini, Experimental evaluation of performance of a hybrid solar photovoltaic (PV/T) panel integrated with effective cooling solutions with water base nanofluids and phase change materials, Energy Sources Part A: Recovery Util. Environ. Eff. 44 (2022) 7287–7302.
[33] N. Priyanka, S. Kumar, A. Kumar, Effect of mono/hybrid nanofluids and passive techniques on thermal performance of parabolic trough solar collector: a review, Energy Sources Part A: Recovery Util. Environ. Eff. 45 (2023) 1686–1709.
[34] M.I. Pryazhnikov, A.V. Minakov, V.Y. Rudyak, D.V. Guzei, Thermal conductivity measurements of nanofluids, Int. J. Heat. Mass. Transf. 104 (2017) 1275–1282.

CHAPTER

CuO/Cu-based nanofluids

2

Mahdiyeh Esmaeili-Zare

Institute of Nano Science and Nano Technology, University of Kashan, Kashan, Iran

2.1 General information

The management and research of atoms and molecules at the nanoscale to the production of useful materials is called nanoscience and nanotechnology. Nanotechnology functions with the border between the domain of special atoms and molecules, where quantum mechanics rules, and the macroworld, in which bulk attributes of materials rise from the composed treatment of a large number of atoms and molecules. At the nanotechnology level, there are usually nanomaterials with a size of less than 100 nm [1–5].

A kind of nanotechnology that combines a two-phase system, following which solids (nanomaterials) are suspended in liquids (base fluid), is called as nanofluid. Nanomaterials are a type of engineered materials having nanoscale measurements. Nanomaterials include rods, particles, rods, fibers, tubes, and sheets. The physical specifications of liquids contain viscosity, density, thermal diffusivity, thermal conductivity, and specific heat capacity. Nanomaterials are generically diffused in liquids at low concentrations (up to 9%) to create nanofluids. Higher quantities of nanoparticles in liquid might cause precipitation because of their low stability [6–11].

Nanofluids include types of metallic and nonmetallic nanofluids [12–14]. Metallic nanofluids are synthesized using placing nanoparticles taken from metals such as nickel, copper [15], aluminum, and so forth. Nonmetallic nanofluids are synthesized using adding nanoparticles of nonmetals such as the metal oxides (ZrO_2 [16–19], CuO [20–24], TiO_2 [25–27], Al_2O_3 [28–30], and SiO_2 [31,32]), different allotropes of carbon (graphene [33,34], carbon nanotube, MWCNT [35–37]), and etcetera suspended in water, ethanol, oil . In recent years, nanofluids have been obtaining excellent notice because of their potential application, not only due to their big gravity in uses such as oil recovery, medicine, biology, and biomedicine but also as an increased thermophysical heat transfer fluid [38–43]. In this chapter, we focus on the recent advances related to preparation methods and applications of CuO/Cu-based nanofluids.

Nanofluids. DOI: https://doi.org/10.1016/B978-0-443-13625-2.00002-4

2.2 CuO/Cu nanoparticle properties

Researchers pay special attention to the construction and applications of metal and metal oxide nanoparticles because of their new chemical and physical features [44–51]. The Cu/CuO-NPs (copper and copper oxide nanoparticles) as metallic nanoparticles are used more than other metallic NPs because of their special physiochemical and other features such as magnetic [52], catalytic [53], optical [54], electrical [55], thermal [56] and antimicrobial activity [57]. Different characteristics such as morphology, production method, and size are effective on nanoparticle properties [58,59]. Copper oxide nanoparticles have excellent chemical and physical features, so these nanoparticles are one of the key transition metal oxides for application in different fields [60]. Copper and oxygen, which appertain to the d and p block, respectively, are two basic elements of copper oxide. Copper oxide has a direct band gap of 1.74 eV at room temperature, so it is a p-type semiconductor. Furthermore, its crystal structure is monoclinic. CuO nanoparticles have a high surface-to-volume proportion, so compared to their bulk peers, they indicate high selectivity and catalytic activity [61–63]. Copper oxide nanoparticles have applications in different fields such as biosensors [64], photocatalytic degradation [65], gas sensors [66], photoelectrochemical water splitting [67], and superconductors [68]. Nanoparticle features such as size, composition, and shape are effective in the interlinkage between living cells and nanoparticles [60,69–71]. Copper oxide nanoparticles are synthesized using various methods such as sonochemical method [72,73], sol-gel method [74], electrochemical method [75], high-temperature combustion [76], single-step solid state reaction technique [77,78], colloidal thermal [79], and hydrothermal process [80].

Copper nanoparticles obtained much regard because of their useful properties such as their great solderability, good thermal and electrical conductivity, high melting point, low material cost, little electrochemical migration action, and so on [81]. According to these features, copper nanoparticles are used in conductive inks [82] and cooling liquid for electronic devices [83]. Besides, copper nanoparticles can be used in many fields in optical systems and nonlinear optical materials, such as optical switches or photochromic glasses because of plasmon surface resonance [84]. Various processes are utilized for the preparation of Cu nanoparticles such as microemulsion [85], metal vapor synthesis [86], ultrasonic [87], vapor deposition [88,89], reducing flame synthesis [90,91], thermal reduction [92], and chemical reduction [93,94].

2.3 Procurement techniques for nanofluids

A colloidal suspension that has nanometer-sized particles (1–100 nm) dispersed in the base fluids, containing engine oil, ethylene glycol, water, and so on, is called nanofluide. Different forces such as gravitational energy, floating energy, and electrostatic repellent energy lead to instability and sediment formation in nanofluids. The stability of any nanofluid depended on van der Waal adsorbent energy and gravitational energy as opposing energy. The stability of a nanofluid is one of the most important characteristics that affect the efficiency of nanofluids for different applications. In nanofluids, nanoparticles tend to hit each other and create agglomeration nanoparticles that lead to big cluster configurations. Therefore, this process leads to the instability of nanofluids due to the high surface action of nanoparticles. To achieve stability in nanofluids, this tendency to aggregate must be removed. The reason for the deposition of nanoparticles is the existence of an absorbent energy of van der Waal that overcomes the electrostatic repulsive energy [9,95–97]. The beneficial performance of nanofluids can be limited due to the deposition of nanoparticles in the microfluidic system. Also, the deposition of particles may result in enhanced viscosity, and reduced heat features.

The stability of nanofluids is evaluated using various methods as follows:

- DLS (dynamic light scattering) [98–100],
- 3 ω technique [101–104],
- precipitation and centrifugation [105,106],
- evaluation of spectral absorbance and transmittance [107–109],
- zeta potential evaluation [110,111], and
- TEM (transmission electron microscopy) [112–114].

Nanofluids are classified into four specified categories due to nanoparticle type used to prepare nanofluids: (1) carbon-based nanofluid, (2) metal-based nanofluid (copper, aluminum, iron, gold, and silver) [115–118], (3) metal oxide nanofluid (CuO, Al_2O_3, TiO_2, and Fe_3O_4) [119–125], and (4) hybrid/combined metal based nanofluid [126–129]. In order to prepare nanofluids, these nanoparticles are hung in methanol, kerosene, water, ethylene glycol, oil, and so on, as basefluids.

2.3.1 Metal based nanofluid

To synthesize metal-based nanofluids, metallic nanoparticles such as copper, silver, gold, and so on should be suspended in base fluids. Qiang et al. [130] built an experimental set to study convective thermal transfer and current specifications of the nanofluid in a tube. They calculated convective thermal transfer factor and attrition factor of copper/water nanofluid for the streamline and turbulent current.

2.3.2 Metal oxide based nanofluid

Metal oxide-based nanofluids, such as Fe_2O_3, Fe_3O_4, TiO_2, CuO, ZnO, and so on, are the most widely used type of nanofluids due to their low cost of preparation compared to metallic nanoparticles.

2.3.3 Carbon based nanofluid

The use of carbon-based nanofluids instead of base fluid leads to an increase in heat properties. Nevertheless, broad commercial usage is restricted due to its high cost. Most articles on carbon-based nanofluids (graphene oxide, graphene, carbon nanotube, and so on) have evaluated their heat conductivity, applications, and electrical conductivity [131–133].

2.3.4 Hybrid/combined metal based nanofluid

Hybrid nanofluid is a novel kind of nanofluid in which more than one kind of nanoparticle is mixed into a nanofluid leading to great heat features compared to nanofluids constructed of one kind of nanoparticles. Tong et al. [134] synthesized Fe_3O_4 and MWCNT/Fe_3O_4 nanofluids. They found that Fe_3O_4/MWCNT hybrid nanofluid shows a great potential to modify the thermal transfer yield compared to Fe_3O_4 nanofluid at a similar weight concentration. Munkhbayar et al. [135] in their research achieved the conclusion that the heat conductivity of Ag/nanoparticle-based nanofluids is increased using the addition of little amounts of MWCNTs. Yarmand et al. [136] synthesized the graphene nanoplatelets–silver nanofluids diffusing the nanocomposite into water. They found that heat conductivity and thermal transfer yield of graphene nanoplatelets–silver nanofluids enhanced.

There are two principal methods in order to prepare nanofluids, including the single-step method and the two-step method [106,137–147].

2.4 Single-step method

The single-step process is a technique that straight mixes the produced nanoparticles with synthesized nanofluid. In this process, some procedures are kept away from such as storage, dispersion, drying, and transportation of nanostructures. The single-step process is also divided into two chemical and physical methods. Submerged arc, laser ablation, and vapor deposition methods are introduced as physical processes. Simultaneously, chemical processes employ chemical reactions to make nanofluids. The benefits of the single-step manufacturing technique are that the collection of the nanoparticles is decreased and it is more money-saving.

The laser ablation in liquid (LAL) method is a greater fitting process for the preparation of NPs through ablating a solid target in a liquid medium (Fig. 2.1). In this method, the target is warmed up to high temperature using a laser beam; as a result, target atom plasma is produced or thawed droplets can be excreted from warmed target. The constructed plasma develops adiabatically, which induces a hit wave subject to the limitation of solution and ionization of the liquid molecules occurring at the plasma liquid junction. The target atoms' nucleation

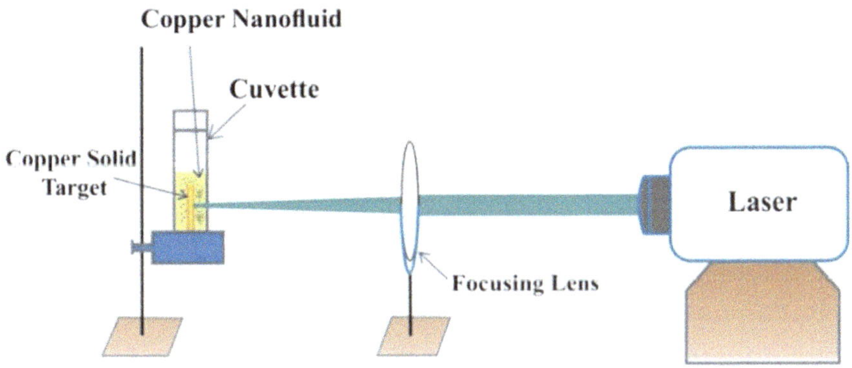

FIGURE 2.1

Schematic of setup for laser ablation synthesis in liquid (LAL) for CuO-NFs [148].

Reprinted with permission from M. Rafique, M.S. Rafique, S.H. Butt, A. Afzal, U. Kalsoom, Laser nature dependence on enhancement of optical and thermal properties of copper oxide nanofluids, Appl. Surf. Sci. 483 (2019) 187–193.

occurs throughout the plasma concentration and the liquid types have the capability to reach the lattice of goal atoms. Afterward, nuclei adhere together to create NPs and nanosize droplets could put out and grain as NPs.

LAL is utilized for synthesis of semiconductor, metal, and ceramic nanostructures [149–154]. This method has benefits compared to other methods, including easiness of the method, freeing from chemical precursors, chemical purity because of lack of toxic materials, which generates chemical pollution-free nanoparticles, no requirement for vacuum and high temperature, and so on [155,156]. Furthermore, attaint of laser parameters, such as pulse duration [157,158], wavelength [159,160], and laser pulse energy [161,162], can lead to control of shape, composition, and size of NPs. Also, the optical and structural features of the prepared materials can be affected by liquid medium features such as liquid temperature [163,164] and type of solvent [165–167]. Moreover, it is a single-step process for the synthesis of copper/copper oxide nanoparticles without needing elevated temperate and any chemical reducer that opens a novel vista in biological usages.

Azadi et al. produced copper/copper oxide NPs through laser ablation of Cu target in H_2O_2 solvent with the Ce:Nd:YAG (pulsed cerium doped Nd:YAG) (Fig. 2.2) laser with pulses of 1064 nm [168].

They considered the effects of the synthesis parameters such as laser pulse energy and H_2O concentration on the structural and optical features of the fabricated NPs. They found that the nanoparticle band gap enhances from 2.19 to 2.34 eV reducing pulse energy and also reducing water concentration up to 5 vol. %. Table 2.1 shows that enhancing pulse energy has led to a rise in the crystallite size of laser-ablated NPs.

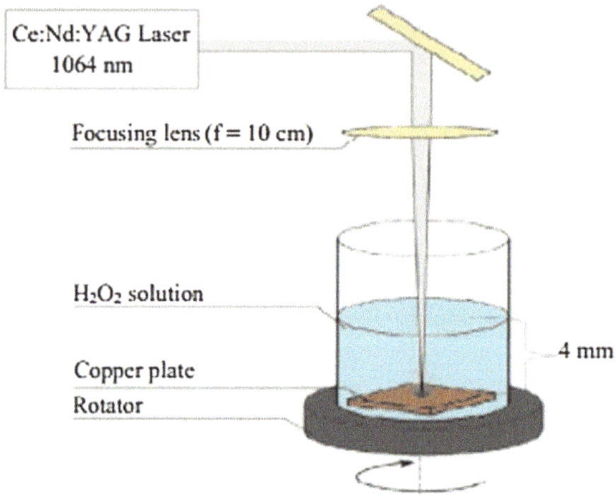

FIGURE 2.2

Laser ablation setup.

Reprinted with permission from H. Azadi, H.D. Aghdam, R. Malekfar, S.M. Bellah, Effects of energy and hydrogen peroxide concentration on structural and optical properties of CuO nanosheets prepared by pulsed laser ablation, Results Phys. 15 (2019) 102610.

Table 2.1 Structural and optical features of laser ablated nanoparticles in different H_2O_2 concentration or different laser pulse energy.

H_2O_2 con. (vol.%)	Pulse energy (mJ)	2θ (°)	Mean crystallite size (nm)	Interplaner distance (Å)	Band gap (eV)
10	100	37.58	3.1	2.777	2.19
10	70	37.6	2.9	2.775	2.21
10	40	37.5	2.8	2.782	2.22
5	40	37.53	3.3	2.78	2.34
1	40	37.5	2.1	2.782	2.26

Reprinted with permission from H. Azadi, H.D. Aghdam, R. Malekfar, S.M. Bellah, Effects of energy and hydrogen peroxide concentration on structural and optical properties of CuO nanosheets prepared by pulsed laser ablation, Results Phys. 15 (2019) 102610.

Rafique et al. synthesized successfully copper/copper oxide nanofluids through a continuous wave laser and pulsed laser in deionized water [148]. The results showed that the CuO NPs produced using continuous wave laser and pulsed laser, with average sizes of 25 and 10 nm, indicating a 29% and 41% increase in thermal conductivity, respectively. Figs. 2.3 and 2.4 show the AFM images of Cu/CuO nanoparticles synthesized using continuous wave laser and pulsed laser, respectively. Both of them indicate that spherical nanoparticles with size range of

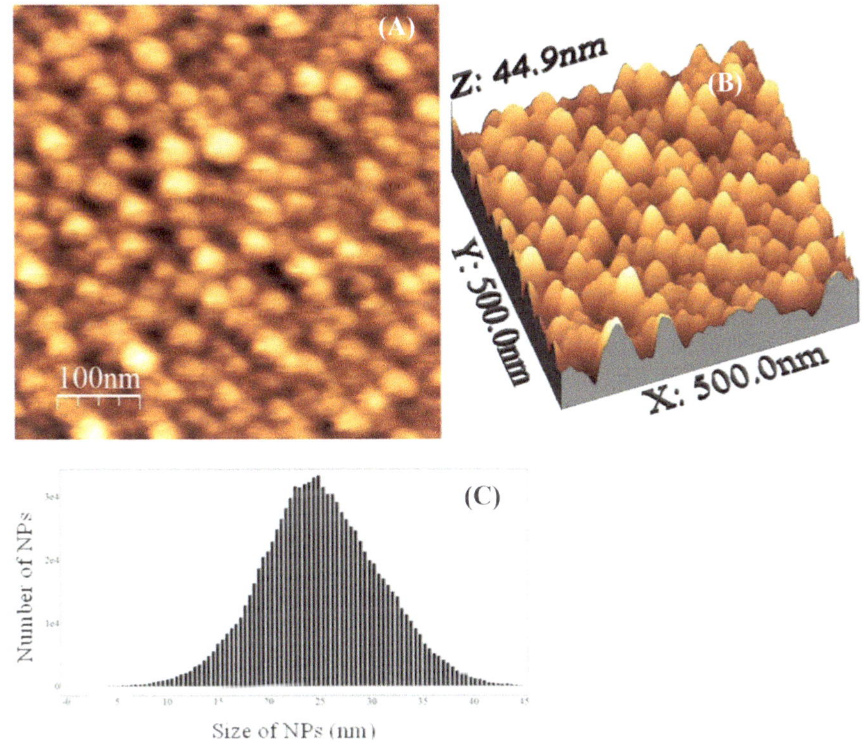

FIGURE 2.3

Optical band gap of CuO-NPs; graph (A) is the optical band gap of Ag-NPs produced by CW laser and (B) is the optical band gap of CuO-NPs produced by pulsed laser.

Reprinted with permission from M. Rafique, M.S. Rafique, S.H. Butt, A. Afzal, U. Kalsoom, Laser nature dependence on enhancement of optical and thermal properties of copper oxide nanofluids, Appl. Surf. Sci. 483 (2019) 187–193.

5–45 nm (via Continuous Wave laser) and 2–24 nm (via pulsed laser) have been synthesized.

Kao et al. synthesized CuO brake nanofluide by the arc-submerged nanoparticle synthesis system (ASNSS) [169] (See Fig. 2.5). In this method, they vaporized a clean Cu rod as the electrode in dielectric liquid and then quickly put it out to form nanocrystalline Cu powder.

In other research, Lo et al. produced CuO nanofluid using a submerged arc nanoparticle synthesis system via a Cu electrode [170]. This group synthesized CuO nanofluid at the dielectric liquid temperature of 2°C, current of 8.5–10 A, pulse duration of 12 μs, and voltage of 220 V.

Zhu et al. developed a single-step process for the synthesis of Cu nanofluids reducing copper sulfate pentahydrate with sodium hypophosphite monohydrate in

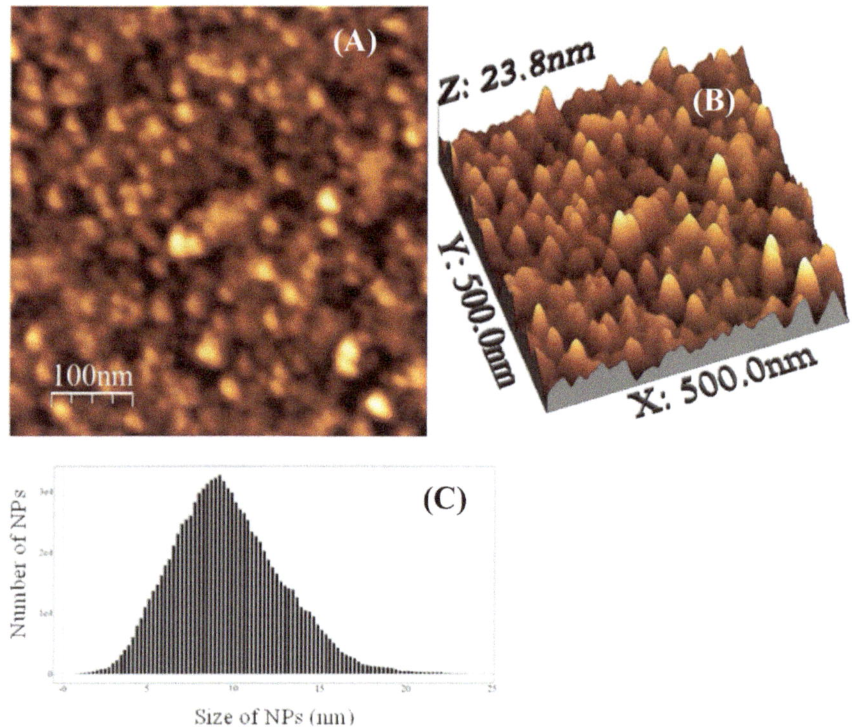

FIGURE 2.4

FTIR spectra of synthesized CuO-NPs; graph (A) is a spectrum of CuO-NPs produced by pulsed laser and (B) is a absorption spectrum of CuO-NPs produced by CW diode laser.

Reprinted with permission from M. Rafique, M.S. Rafique, S.H. Butt, A. Afzal, U. Kalsoom, Laser nature dependence on enhancement of optical and thermal properties of copper oxide nanofluids, Appl. Surf. Sci. 483 (2019) 187–193.

ethylene glycol under microwave irradiation [171]. Suspended copper nanofluids with high stability and good morphology were created.

The usage of gas-phase methods with low pressure for the manufacture of nanofluids is according to the nanoparticle nucleation procedures introduced from atomic metal flows into the vacuum/liquid junction [168]. Produced atoms may return from the surface to the gas phase and they can so spread upon the surface or interpenetrate the bulk. Choukourov et al. applied the magnetron sputtering as a one-step procedure to the synthesis of metal/polymer nanofluides using a filler (22 nm-sized copper nanoparticles) and a liquid phase (polyethylene glycol) [172]. In this method, copper nanoparticles as target materials are synthesized using dc magnetron sputtering that is kept in a distinct vacuum chamber (gas aggregation cluster source) (Fig. 2.6) [170,171,173–177]. Ar is made cold, and its pressure is selected to derive the activities of voluntary nucleation in the

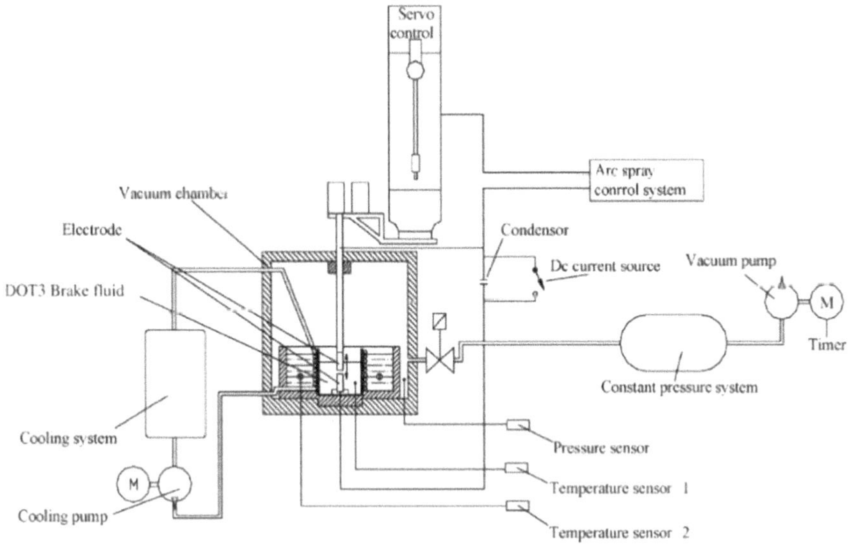

FIGURE 2.5

The arc-submerged nanoparticle synthesis system (ASNSS).

Reprinted with permission from M. Kao, C. Lo, T. Tsung, Y. Wu, C. Jwo, H. Lin, Copper-oxide brake nanofluid manufactured using arc-submerged nanoparticle synthesis system, J. Alloy. Compd. 434 (2007) 672–674.

vapor of atomic metals, followed by the construction and formation of metal nanoparticles.

2.5 Two-step method

The most common process for the synthesis of nanofluids is the two-step method. In this process, nanofibers, nanoparticles, nanotubes, and other nanomaterials are synthesized using chemical methods (for example, sol-gel [178,179], coprecipitation [180–182], ultrasonic agitation [183–186], and so on), or physical methods (such as spray pyrolysis [187,188], ball milling [169,189], aggressive magnetic force agitation, and so on), will then be used to spread the nanosized powder into a base fluid. One of the most principal characteristics of nanofluids is their enhanced thermal features, so they have obtained critical attention. This rise was because of the supplement of nanoparticles in the basic fluid. Different factors such as nanoparticle size, agglomeration, value of pH, and so on can lead to this increase.

The ultrasonic irradiation method is one of the most favorite and fruitful two-step process for preparation of the long-stable nanofluids with homogeneous dispersion. Different parameters such as ultrasonic power, ultrasonic time, and

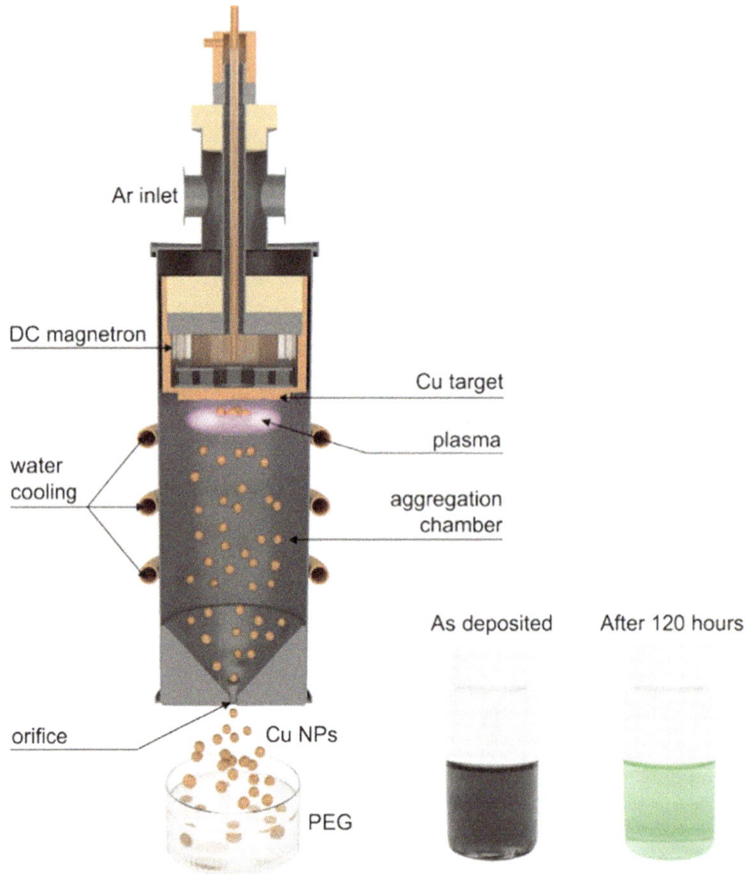

FIGURE 2.6

Scheme of the dc magnetron-based gas aggregation cluster source used for the synthesis of Cu NPs by sputtering a Cu target and subsequent deposition of the NPs onto liquid PEG. The vials show the same nanofluid prepared with the dc of 150 mA, deposition time of 15 min, and a flux of 550 NPs/μm^2 min (the net concentration is 50 μg/mL or 10^{12} NPs/mL) and photographed immediately after the deposition and after the 120-h storage.

Reprinted with permission from A. Choukourov, D. Nikitin, P. Pleskunov, R. Tafiichuk, K. Biliak, M. Protsak, K. Kishenina, J. Hanuš, M. Dopita, M. Cieslar, Residual-and linker-free metal/polymer nanofluids prepared by direct deposition of magnetron-sputtered Cu nanoparticles into liquid PEG, J. Mol. Liq. 336 (2021) 116319.

ultrasonic type are effective parameters that affect the dispersion of nanofluids. The two-step method is the most cost-efficient procedure for large-scale nanofluid production due to the use of synthesized nanopowders at commercial production levels. Nanoparticles have a large surface area, so they tend to aggregate. Surfactants are a typical method for correcting the resistance of nanoparticles in fluids.

Rashin et al. synthesized copper oxide–ethylene glycol nanofluids with different concentrations by ultrasonically facilitated two-step process [184]. In this study, the prepared copper oxide NPs using the coprecipitation method. Then, prepared copper oxide NPs were dispersed in ethylene glycol under ultrasonic irradiation for 3 hours to obtain homogenous suspension sans any phase divorce and precipitation. They found that ultrasonic irradiation leads to an increase in the stability of the suspension. Nanofluids with different concentrations (0.08, 0.16, 0.24, 0.32, 0.4, and 0.48 vol.%) were obtained sans the addition of any surfactant (Shown in Fig. 2.7).

Copper oxide NPs were prepared via a wet chemical technique by Agarwal et al. [24]. In this work, two copper salt precursors were used for the synthesis of CuO NPs ($Cu(CH_3COO)_2$ and $CuSO_4$). They found that copper oxide NPs prepared from $CuSO_4$ have smaller particle sizes and more regular shapes in comparison to CuO NPs prepared from $Cu(CH_3COO)_2$. Copper oxide NPs prepared from $CuSO_4$ precursor were utilized for the synthesis of nanofluids in pure water, ethylene glycol, and machine oil base fluids by two-step method. Fig. 2.7 exhibits TEM images of CuO NPs prepared from two various salt forerunners. Figs. 2.8A and B shows average particle sizes 67 and 54 nm for prepared samples from Cu $(CH_3COO)_2$ and $CuSO_4$, individually. A summary of prepared nanofluids in different fluids is listed in Table 2.2.

2.6 Applications of CuO/Cu nanofluids

2.6.1 Thermal transfer

Various kinds of heat exchangers containing microchannel type, compact, shell, plate, and tube type, and more of the same are being broadly applied in the heavy trade, processing trade, etc.

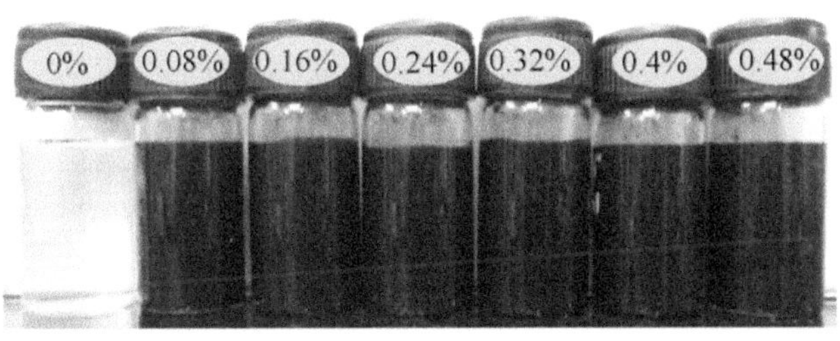

FIGURE 2.7

Copper oxide–ethylene glycol nanofluids of different concentrations.

Reprinted with permission from M.N. Rashin, J. Hemalatha, A novel ultrasonic approach to determine thermal conductivity in CuO–ethylene glycol nanofluids, J. Mol. Liq., 197 (2014) 257–262.

FIGURE 2.8

SEM images of CuO NPs synthesized from (A) Copper acetate and (B) Copper sulfate.

Reprinted with permission from R. Agarwal, K. Verma, N.K. Agrawal, R.K. Duchaniya, R. Singh, Synthesis, characterization, thermal conductivity and sensitivity of CuO nanofluids, Appl. Therm. Eng. 102 (2016) 1024–1036.

One of the most important activities in different industries is heat processes. Therefore, researchers use additive materials such as nanoparticles to improve the heat efficiency of processes. Boiling is one of the most significant thermal transition modes because of its broadly applied application in various industries such as nuclear and petroleum procedures so cooling of great power electronics ingredients [205–207]. Boiling is a 2-phase event and generates bubbles, which transition great speeds of energy from the heated area into the working fluid. Hence, higher thermal transition speed and heat efficiency are advantages of boiling compared to the one-phase thermal transition. Using expanded fluids such as nanofluids is one approach to improve heat performance. Nanofluids have a good heat conductivity that has an outstanding effect on the thermal transition between the fluid and warmed wall, so nanofluids have displayed better heat conductivity than water, a conventional working fluid [145,208].

Table 2.2 Summary of prepared CuO and Cu nanofluides.

Nanoparticles	Base fluids	References
CuO	Water	[190]
CuO	Water	[191]
CuO	Water	[192]
Cu	Water	[193]
CuO	Ethylene glycol	[194]
Cu	Ethylene glycol	[195]
CuO	Ethylene glycol and water	[196]
CuO	Water	[197]
Cu		
CuO	Oil	[198]
CuO	Deionized water	[199]
CuO	Oleic acid	[200]
Cu	R113	[201]
Cu	Ethylene glycol	[202]
Cu	Deionized water	[203]
Cu	Transformer oil	[204]

The effect of nanoparticles on the flow boiling of copper oxide cooler nanofluid was studied by Peng et al. [209]. They found that the flow boiling thermal transition coefficient of copper oxide/R113 was better than R113 cooler by 29.7%.

In other research, an experimental survey was conducted by Boudouh et al. on the convective boiling thermal transition of less concentration of copper/water nanofluids in vertical microchannels [210]. This group opined that two benefits, especially nanofluid and microchannel, could remedy boiling thermal transition.

Also, Sarafraz et al. and Sun et al. [211,212] studied the effect of the addition of nanoparticles on the thermal transfer in flow boiling, separately.

An experimental and theoretical investigation of thermal transition in compiled convection and subchillied flow boiling of copper oxide/water nanofluids was performed by Zangeneh et al. [213]. The influences of different parameters such as nanoparticle concentration, thermal current, and fluid flow speed on the thermal transition efficiency of the device were considered. Fig. 2.9 displays a schematic view of the experimental setup, which includes different tools noted in the graphic.

Fazeli et al. [214] utilized a hybrid nanofluid (multiwalled carbon nanotubes-copper oxide) that was synthesized using a novel stability technique to prepare the cooling fluid in the soldered plate thermal exchanger. They used four nonionic surfactants to investigate the stability of nanofluids. In this work, the results of water and hybrid nanofluid on heat features in the brazed disk thermal exchanger were investigated.

Asokan et al. [215] prepared CuO, Al_2O_3-CuO and Al_2O_3 hybrid nanofluids with low concentration. The results indicated an increase in heat conductivity of

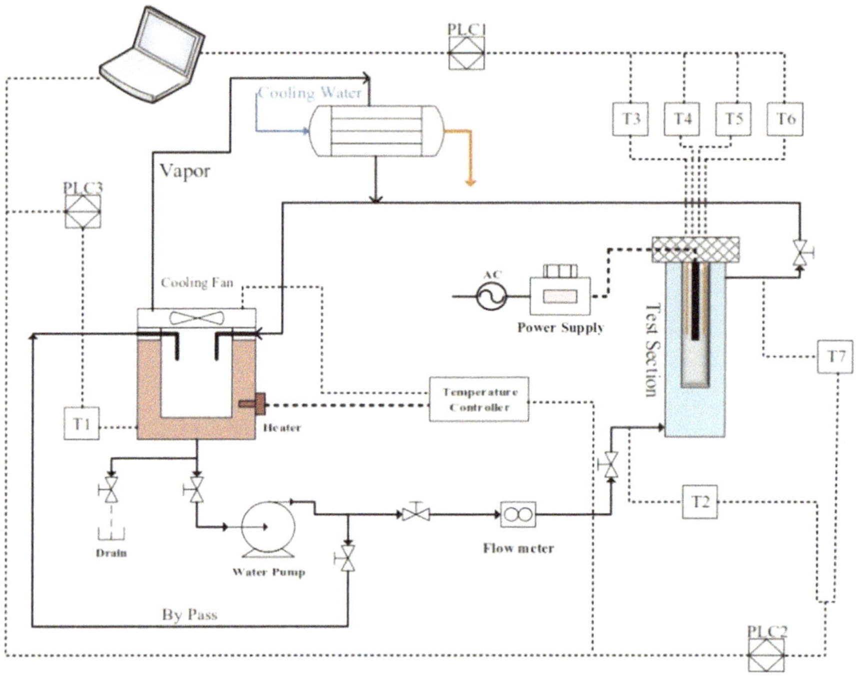

FIGURE 2.9

A simple schematic diagram of the experimental setup.

Reprinted with permission from A. Zangeneh, S. Peyghambarzadeh, A. Bayat, A. Vatani, Application of general multilevel factorial design approach in forced convection and subcooled flow boiling heat transfer to CuO/water nanofluids, J. Mol. Liq., 313 (2020) 113502.

Al_2O_3-CuO hybrid nanofluids about 2.3% and 3.6% compared to CuO and Al_2O_3 nanofluids, respectively.

Du et al. [216] used an experimental geothermal heat exchanger set to check the effect of copper oxide/water nanofluide on the thermal transition enhancement of geothermal heat exchanger. In this research, they found that using copper oxide/water nanofluids as thermal transition fluid, the thermal transition speeds and pumping capacitance utilization of the geothermal heat exchanger system had an enhancement of 39.84% and 16.6%, respectively.

Lately, traditional conductive fluids are gently being replaced using nanofluids because of the rising requirement for green construction and environmental procedures [217–220]. Chaudhari et al. [221] introduced the experimental research of Al_2O_3 and copper oxide/water nanofluids at various volume percentage concentrations such as 0.05%, 0.15%, 0.3%, 0.5%, and 1% for thermal transition application. They reached the conclusion that the superior thermal transition fluid is gained at optimum values of 0.3% and 0.15% volume concentration of Al_2O_3 and copper oxide nanofluide, respectively, for investigation of heat conductivity.

Navas et al. [222] synthesized the copper and nickel nanofluids using a two-step technique. At first, they prepared copper and nickel nanoparticles and then dispersed them into the thermal transition base fluid such as a mixture of diphenyl oxide and biphenyl. The results showed that copper nanofluids improve the heat features compared to the basic fluid. The thermal transition coefficient using copper nanofluid was 11% higher than the base fluid.

2.6.2 Solar panel

Nowadays global energy tension has enhanced due to the rapid increase in world population, industrial development, and the fast expansion of society. We witness that access to fossil fuels is reducing as their use is increasing. Today, a great choice to replace fossil fuels is the use of renewable vigor produced from innate resources such as wind, solar, and geothermic energy [223–225]. One of the most significant renewable sources of vigor to generate electricity and therm is solar energy. Increasing the thermal transition process in solar energy devices is one of the most significant features to attain a preferable efficiency of these systems with conjunct patterns. To solve this issue, the researchers can use working fluids with raised thermophysical attributes. One of the efficient ways is to change the working fluid with nanofluids as new tactics to modify the thermal transition specifications of the fluid [208,226–228]. Lately, many researchers have become attracted to the use of nanofluids in solar systems because of their better attributes over the formal fluids. The merits of applying nanofluids in solar vigor apparatuses are listed below [27,229–242]:

- The large surface area to volume ratio of nanoparticles because of their small particle size enhances the thermal capability of the nanofluids and also the solar energy absorption.
- The nanofluids have superior optical properties than the base fluid. They display great absorption in the solar spectrum area and also little emission in the infrared spectrum area.
- Nanoparticles in nanofluids lead to their high heat conductivity compared to the basis fluid.
- The small size of nanofluids compared to micro-sized particles prevents precipitation and blocking of pumps and pipes, which is a fruitful feature in many solar usages.
- Nanofluids decrease the full cost of solar energy systems because of the reduction of the needed thermal transition area of heat devices.

Chen et al. [243] synthesized a constant two-ingredient nanofluids including copper oxide and antimony-doped tin oxide nanoparticles. They used a wet chemical approach to synthesize copper oxide/water nanofluids [244]. Also, antimony-doped tin oxide/water nanofluids were constructed using dispersing antimony-doped tin oxide nanoparticles into water. Finally, two-component (antimony doped tin oxide-copper oxide/water) nanofluids were obtained as follows: sodium citrate was added to the as-prepared copper oxide/water nanofluids. Then, the mixture was added to antimony-doped tin oxide/water nanofluids and stirred for two hours. They showed that these synthesized 2-component nanofluids have strong absorption in both near-infrared and visible areas, due to supplementary optical absorption of the antimony-doped tin oxide and copper oxide nanoparticles. The outcomes indicated that the solar heat usage efficiency of the 2-component nanofluids was about 92.5% compared to 81.3% for antimony-doped tin oxide nanofluids and 80.7% for copper oxide nanofluids.

Kanti et al. [245] used fly ash/Cu (80:20 V/V%) hybrid nanofluids as absorber fluid for application in solar energy, which can be used to increase the yield of solar systems. They utilized these nanofluids in solar systems with the laminar and turbulent current. The results showed that a concentration of 4 vol.% of hybrid nanofluid has the most increase in the dynamic viscosity and heat conductivity properties of hybrid nanofluid with 45.18% and 49.8% at temperatures of 30°C and 60°C, respectively. These groups used their experimental conclusions to count the Mouromtseff numeral and thermal transition yield for solar power utilization.

Sundar et al. [246] applied the copper/water nanofluids as an absorber fluid to obtain the thermal from solar irradiation to the flat disk collector. The outcomes indicated an increase in the yield of the solar collector to 58% using nanofluid with a volume concentration of 0.3%.

Sint et al. [247] theoretically analyzed the performance of the flat sheet solar gatherer by copper oxide/water nanofluid as a working fluid. They considered the performance of a homemade solar water thermal system in a city in Myanmar using a mathematical pattern. The consequences showed that the performance of homemade solar water thermal systems can improved by up to 5% using copper oxide/water nanofluid as a working fluid.

Zamzamian et al. [248] used a single-step technique to synthesize Cu nanofluids. In this method, copper sulfate pentahydrate was reduced using sodium hypophosphite monohydrate in ethylene glycol solvent. The findings showed that the yield of the collector could be improved using enhancing the weight percentage of the nanoparticle. Further, copper/ethylene glycol nanofluid with 0.3 wt.% at 1.5 L/min showed the lowest and the highest eliminated energy and absorbed energy parameters, respectively.

Alrowaili et al. [249] studied the performance of an emptied tube solar gatherer using uni-copper oxide and copper oxide/copper/water nanofluids as the operant fluid. They received to conclusions as follows:

1. The resistance of the nanofluids synthesized by ultrasound was investigated using zeta potential studies.
2. The heat conductivity of the hybrid nanofluid was enhanced by 21%.
3. The collector surface was decreased by up to 38% using a hybrid nanofluid.
4. The thermal elimination factor was obtained to be 0.894.

Akilu et al. [250] prepared a hybrid nanofluid (ceramic copper oxide/carbon nanoparticles in an 80:20 (weight percentage) mixture) using an ultrasonic-aided wet mixing method. In this work, the effect of nanoparticles on the increase of specific heat, heat conductivity, and viscosity was examined in the confine of volume concentration of 0.5%–2% in the 303.15°K to 353.15°K temperature confine. The outcomes showed that the prepared SiO_2-copper oxide/carbon hybrid nanoparticles increased the thermo-physical attributes of the basis fluid combination, which is superior to using SiO_2 merely. Therefore, SiO_2-copper oxide/carbon hybrid nanofluids due to their higher heat conductivity, less viscosity, and special thermal capacity are a lot noteworthy medium for solar heat energy transportation in comparison to single-part SiO_2 nanofluids. Table 2.3 shows the thermo-physical features increase of hybrid nanofluids.

Karammi et al. [257] reported an archetype of copper oxide nanofluid-based direct absorption solar collector that was made for application in the homemade solar water heater. Copper oxide nanofluids were synthesized using polyvinylpyrrolidone with the 0.25:1 (polyvinylpyrrolidone: CuO) weight ratio [258]. In this research, the thermo-physical and optical features of prepared CuO nanofluids with various volume ratios in ethyleneglycol and water combination (30%:70%) were investigated. The outcomes indicated that the collector performance has improved by 9%–17% compared to the basis fluid using nanofluids.

Amalraj et al. [259] considered the performance of solar cooling using copper oxide and Al_2O_3 nanofluids. They found that the solar panels with fluid (water), without fluid, with copper oxide/water, with Al_2O_3/water, and Al_2O_3-CuO/water

Table 2.3 Summary of thermophysical property enhancement of hybrid nanofluids.

Authors	Nanoadditive	Base fluid	Maximum thermal conductivity enhancement (%)	Maximum viscosity enhancement (%)
Suresh et al. [251]	Al_2O_3/Cu	W	12.1	205
Hemmat Esfe [252]	Cu/TiO_2	W/EG 60:40	40	—
Madhesh et al. [253]	Cu/TiO_2	W	19.3	—
Nabil et al. [254]	TiO_2/SiO_2	W/EG 60:40	22.8	62.5
Nikkam et al. [255]	SiC/SiO_2	W/EG 50:50	20	14
Chopkar et al. [256]	Al/Cu	EG	30	—

Reprinted with permission from S. Akilu, A.T. Baheta, M.A.M. Said, A.A. Minea, K. Sharma, Properties of glycerol and ethylene glycol mixture based SiO_2-CuO/C hybrid nanofluid for enhanced solar energy transport, Sol. Energy Mater. Sol. Cell 179 (2018) 118–128.

as coolants show the efficiency 10%, 12%, 13%, 16%, and 18% and fill factor about 0.8, 0.55, 0.6, 0.82, and 0.95, respectively. This displays that copper oxide and Al_2O_3 will be hopeful applicants with water as a base fluid for solar panel requests.

2.6.3 Heat sink

The last technological need in the engineering industries is related to the preparation of thermal transition fluids with high efficiency. Thermal transfer has an important role in some industrial sections such as chemical processing plants, electronics, transport, and power production. Cooling of electronic components is one of the most important thermal transition scopes. A prosperous high-efficiency electronic system needs to manage thermal dissipation from electronic components. Air-cooled thermal sinks have very small heat conductivity and thermal carrying content. Therefore, they have a very small ability to eliminate thermal from electronic constituents. An electronic cooling device can be efficient and compact using liquids instead of air, due to their great heat conductivity compared to air. Nanofluids are a new group of composites that consist of nanoparticles dispersed in the base fluids of thermal transition. The efficient heat conductivity of the nanofluids is enhanced using the arrival of NPs with superior heat conductivity amounts than the basis fluids.

Selvakumar et al. [260] investigated the influence of copper oxide/water nanofluids prepared by using the ultrasonic method (volume ratios of 0.1% and 0.2%), in the narrow channeled Cu thermal sink under steady thermal flux conditions. In this study, they used a thin block of channeled Cu water with an overall size of $55 \times 55 \times 19$ mm. The results showed that the junction temperature of the water block was decreased using copper oxide/water nanofluids as base fluid.

Krishna et al. [261] synthesized CuO/water, MWCNT/water nanofluids, and MWCNT-CuO/water hybrid nanofluids. The results indicated an increase in Nusselt number of MWCNT-CuO/water hybrid nanofluids about 4.68% and 12.64% compared to CuO/water and MWCNT/water nanofluids, respectively.

Rimbault et al. [262] prepared copper oxide nanofluids with three various ratios of particle volume, specifically 0.24%, 1.03%, and 4.5%. They investigated the effect of the hydraulic and heat fields of copper oxide/water nanofluids with a particle size of 29 nm, flowing under both quiescent and turbulent internal conditions of a rectangular microchannel heat sink. They achieved this conclusion that copper oxide/water nanofluids with volume ratios of 1.03% and 0.24% showed the best and middle thermal transition developments and the average and highest energetic efficiency factors compared to water, respectively. But 4.5% volume ratio copper/water nanofluid displayed the worst thermal transition developments and energetic efficiency factor.

In other research, Ghasemi et al. [263] simulated the slow compelled convection thermal transition with nanofluid as coolant in the rectangular and spherical mini-channel thermal sinks and examined the effect of different volume ratios of copper oxide nanoparticles on the thermal transition and pressure fall features. They found that the thermal transition of a heat sink using nanofluids is better than a heat sink with water as a coolant. Also, the results showed that at the similar Reynold number and cross-section region, the thermal sink with rectangular channels has less heat persistence compared to circular channels.

2.7 Conclusion and outlook

A composition of solid-liquid materials including particles, tubes, roads, or fibers with nanometer size is called nanofluid. This group of materials has many advantages for use in the heat transfer system. Also, the stability of nanofluids has been improved due to the addition of nanometer solid particles to fluids. Researchers pay special attention to the construction and applications of metal and metal oxide nanoparticles because of their new chemical and physical features. CuO nanoparticles have a high surface-to-volume proportion, so compared to their bulk peers, they indicate high selectivity and catalytic activity. Copper oxide nanoparticles have applications in different fields. Also, copper nanoparticles obtained much regard because of their useful properties such as their great solderability, good thermal and electrical conductivity, high melting point, low material cost, and little electrochemical migration action. Nanofluids are classified into 4 specified categories due to nanoparticle type used to prepare nanofluids: (1) carbon-based nanofluid, (2) metal-based nanofluid (copper, aluminum, iron, gold, and silver), (3) metal oxide nanofluid (CuO, Al_2O_3, TiO_2, and Fe_3O_4), and (4) hybrid/combined metal based nanofluid. There are two principal methods in order to prepare nanofluids, including the single-step method and the two-step method. In this chapter, we considered the application of nanofluids in several fields such as thermal transfer, solar panel, and heat sink.

References

[1] A.T. Alshareeda, M.N. Khatijah, B.S. Al-Sowayan, Nanotechnology: a revolutionary approach to prevent breast cancer recurrence, Asian J. Surg. 46 (2022) 13–17.

[2] Q. Ding, Z. Li, Y. Yang, G. Guo, F. Luo, Z. Chen, et al., Preparation and therapeutic application of docetaxel-loaded poly (d, l-lactide) nanofibers in preventing breast cancer recurrence, Drug Deliv. 23 (2016) 2677–2685.

[3] E. Ekrami, M. Pouresmaieli, E. sadat Hashemiyoon, N. Noorbakhsh, M. Mahmoudifard, Nanotechnology: a sustainable solution for heavy metals remediation, Environ. Nanotechnol. Monit. Manage. 18 (2022) 100718.

[4] R.-J. Li, X. Ying, Y. Zhang, R.-J. Ju, X.-X. Wang, H.-J. Yao, et al., All-trans retinoic acid stealth liposomes prevent the relapse of breast cancer arising from the cancer stem cells, J. Controlled Rel. 149 (2011) 281–291.

[5] A.B. Shirode, D.J. Bharali, S. Nallanthighal, J.K. Coon, S.A. Mousa, R. Reliene, Nanoencapsulation of pomegranate bioactive compounds for breast cancer chemoprevention, Int. J. Nanomed. 10 (2015) 475.

[6] S.U. Ilyas, R. Pendyala, N. Marneni, Stability of nanofluids, Engineering Applications of Nanotechnology, Springer, 2017, pp. 1–31.

[7] S. Chakraborty, P.K. Panigrahi, Stability of nanofluid: a review, Appl. Therm. Eng. 174 (2020) 115259.

[8] S. Mukherjee, S. Paria, Preparation and stability of nanofluids—a review, IOSR J. Mech. Civil. Eng. 9 (2013) 63–69.

[9] H. Farzaneh, A. Behzadmehr, M. Yaghoubi, A. Samimi, S. Sarvari, Stability of nanofluids: molecular dynamic approach and experimental study, Energy Convers. Manage. 111 (2016) 1–14.

[10] W. Yu, H. Xie, A review on nanofluids: preparation, stability mechanisms, and applications, J. Nanomater. 2012 (2012) 1687–4110.

[11] V. Fuskele, R. Sarviya, Recent developments in nanoparticles synthesis, preparation and stability of nanofluids, Mater. Today: Proc. 4 (2017) 4049–4060.

[12] A. Ramadhan, W. Azmi, R. Mamat, K. Hamid, S. Norsakinah, Investigation on stability of tri-hybrid nanofluids in water-ethylene glycol mixture, IOP Conference Series: Materials Science and Engineering, IOP Publishing, 2019, p. 012068.

[13] O.A. Alawi, N.A.C. Sidik, H.W. Xian, T.H. Kean, S.N. Kazi, Thermal conductivity and viscosity models of metallic oxides nanofluids, Int. J. Heat. Mass. Transf. 116 (2018) 1314–1325.

[14] A. Akash, S. Abraham, A. Pattamatta, S.K. Das, Experimental assessment of the thermo-hydraulic performance of automobile radiator with metallic and nonmetallic nanofluids, Heat. Transf. Eng. 41 (2020) 235–251.

[15] Y. Xuan, Q. Li, Investigation on convective heat transfer and flow features of nanofluids, J. Heat. Transf. 125 (2003) 151–155.

[16] J. Traciak, J. Sobczak, J.P. Vallejo, L. Lugo, J. Fal, G. Żyła, Experimental study on the density, surface tension and electrical properties of ZrO_2–EG nanofluids, Phys. Chem. Liq. 61 (2022) 1–11.

[17] A. Hernández Battez, J. Viesca, R. González, A. García, T. Reddyhoff, A. Higuera-Garrido, Effect of shear rate, temperature, and particle concentration on the rheological properties of ZnO and ZrO_2 nanofluids, Tribol. Trans. 57 (2014) 489–495.

[18] M. Chopkar, A. Das, I. Manna, P. Das, Pool boiling heat transfer characteristics of ZrO2–water nanofluids from a flat surface in a pool, Heat. Mass. Transf. 44 (2008) 999–1004.

[19] M. Mohammadi, M. Akbari, Z. Fakhroueian, A. Bahramian, R. Azin, S. Arya, Inhibition of asphaltene precipitation by TiO_2, SiO_2, and ZrO_2 nanofluids, Energy Fuels 25 (2011) 3150–3156.

[20] H.H. Balla, S. Abdullah, W. MohdFaizal, R. Zulkifli, K. Sopian, Numerical study of the enhancement of heat transfer for hybrid CuO-Cu nanofluids flowing in a circular pipe, J. Oleo Sci. 62 (2013) 533–539.

[21] G. Kumaresan, S. Venkatachalapathy, L.G. Asirvatham, Experimental investigation on enhancement in thermal characteristics of sintered wick heat pipe using CuO nanofluids, Int. J. Heat. Mass. Transf. 72 (2014) 507–516.

[22] L.-P. Zhou, B.-X. Wang, X.-F. Peng, X.-Z. Du, Y.-P. Yang, On the specific heat capacity of CuO nanofluid, Adv. Mech. Eng. 2 (2010) 172085.

[23] L.S. Sundar, M.H. Farooky, S.N. Sarada, M. Singh, Experimental thermal conductivity of ethylene glycol and water mixture based low volume concentration of Al_2O_3 and CuO nanofluids, Int. Commun. Heat. Mass. Transf. 41 (2013) 41–46.
[24] R. Agarwal, K. Verma, N.K. Agrawal, R.K. Duchaniya, R. Singh, Synthesis, characterization, thermal conductivity and sensitivity of CuO nanofluids, Appl. Therm. Eng. 102 (2016) 1024–1036.
[25] K.A. Hamid, W. Azmi, R. Mamat, K. Sharma, Experimental investigation on heat transfer performance of TiO_2 nanofluids in water–ethylene glycol mixture, Int. Commun. Heat. Mass. Transf. 73 (2016) 16–24.
[26] H.M. Ali, H. Babar, T.R. Shah, M.U. Sajid, M.A. Qasim, S. Javed, Preparation techniques of TiO_2 nanofluids and challenges: a review, Appl. Sci. 8 (2018) 587.
[27] L. Yang, K. Du, A comprehensive review on heat transfer characteristics of TiO_2 nanofluids, Int. J. Heat. Mass. Transf. 108 (2017) 11–31.
[28] R. Mostafizur, R. Saidur, A.A. Aziz, M. Bhuiyan, Thermophysical properties of methanol based Al_2O_3 nanofluids, Int. J. Heat. Mass. Transf. 85 (2015) 414–419.
[29] R. Agarwal, K. Verma, N.K. Agrawal, R. Singh, Sensitivity of thermal conductivity for Al_2O_3 nanofluids, Exp. Therm. Fluid Sci. 80 (2017) 19–26.
[30] M. Zawrah, R. Khattab, L. Girgis, H. El Daidamony, R.E. Abdel Aziz, Stability and electrical conductivity of water-base Al_2O_3 nanofluids for different applications, HBRC J. 12 (2016) 227–234.
[31] G.M. Moldoveanu, G. Huminic, A.A. Minea, A. Huminic, Experimental study on thermal conductivity of stabilized Al_2O_3 and SiO_2 nanofluids and their hybrid, Int. J. Heat. Mass. Transf. 127 (2018) 450–457.
[32] W. Azmi, K. Sharma, P. Sarma, R. Mamat, G. Najafi, Heat transfer and friction factor of water based TiO_2 and SiO_2 nanofluids under turbulent flow in a tube, Int. Commun. Heat. Mass. Transf. 59 (2014) 30–38.
[33] H. Akhavan-Zanjani, M. Saffar-Avval, M. Mansourkiaei, F. Sharif, M. Ahadi, Experimental investigation of laminar forced convective heat transfer of Graphene–water nanofluid inside a circular tube, Int. J. Therm. Sci. 100 (2016) 316–323.
[34] M. Mehrali, E. Sadeghinezhad, M.A. Rosen, A.R. Akhiani, S.T. Latibari, M. Mehrali, et al., Heat transfer and entropy generation for laminar forced convection flow of graphene nanoplatelets nanofluids in a horizontal tube, Int. Commun. Heat. Mass. Transf. 66 (2015) 23–31.
[35] R. Walvekar, M.K. Siddiqui, S. Ong, A.F. Ismail, Application of CNT nanofluids in a turbulent flow heat exchanger, J. Exp. Nanosci. 11 (2016) 1–17.
[36] Y. Ding, H. Alias, D. Wen, R.A. Williams, Heat transfer of aqueous suspensions of carbon nanotubes (CNT nanofluids), Int. J. Heat. Mass. Transf. 49 (2006) 240–250.
[37] A.T. Utomo, E.B. Haghighi, A.I. Zavareh, M. Ghanbarpourgeravi, H. Poth, R. Khodabandeh, et al., The effect of nanoparticles on laminar heat transfer in a horizontal tube, Int. J. Heat. Mass. Transf. 69 (2014) 77–91.
[38] K. Solangi, S. Kazi, M. Luhur, A. Badarudin, A. Amiri, R. Sadri, et al., A comprehensive review of thermo-physical properties and convective heat transfer to nanofluids, Energy 89 (2015) 1065–1086.
[39] K. Jafari, M.H. Fatemi, P. Estellé, Deep eutectic solvents (DESs): a short overview of the thermophysical properties and current use as base fluid for heat transfer nanofluids, J. Mol. Liq. 321 (2021) 114752.

[40] E. Joonaki, S. Ghanaatian, The application of nanofluids for enhanced oil recovery: effects on interfacial tension and coreflooding process, Pet. Sci. Technol. 32 (2014) 2599–2607.

[41] L. Hendraningrat, B. Engeset, S. Suwarno, O. Torsæter, Improved oil recovery by nanofluids flooding: an experimental study, SPE Kuwait International Petroleum Conference and Exhibition, OnePetro, 2012.

[42] B. Yuan, R.G. Moghanloo, W. Wang, Using nanofluids to control fines migration for oil recovery: nanofluids co-injection or nanofluids pre-flush? A comprehensive answer, Fuel 215 (2018) 474–483.

[43] L.J. Habeeb, H.S. Majdi, Nanofluids and computational applications in medicine and biology, Applications of Nanobiotechnology, IntechOpen, 2019.

[44] M. Esmaeili-Zare, M. Behpour, CIS/CdS/ZnO/ZnO: Al modified photocathode for enhanced photoelectrochemical behavior under visible irradiation: effects of pH and concentration of electrolyte solution, Int. J. Hydrog. Energy 45 (2020) 8273–8281.

[45] M. Esmaeili-Zare, M. Behpour, Influence of deposition parameters on surface morphology and application of $CuInS_2$ thin films in solar cell and photocatalysis, Int. J. Hydrog. Energy 45 (2020) 16169–16182.

[46] R. Hayward, D. Saville, I.A. Aksay, Electrophoretic assembly of colloidal crystals with optically tunable micropatterns, Nature 404 (2000) 56–59.

[47] A.N. Shipway, E. Katz, I. Willner, Nanoparticle arrays on surfaces for electronic, optical, and sensor applications, ChemPhysChem 1 (2000) 18–52.

[48] S.J. Park, A.A. Lazarides, C.A. Mirkin, R.L. Letsinger, Directed assembly of periodic materials from protein and oligonucleotide-modified nanoparticle building blocks, Angew. Chem. Int. Ed. 40 (2001) 2909–2912.

[49] A.V. Kabashin, M. Meunier, C. Kingston, J.H. Luong, Fabrication and characterization of gold nanoparticles by femtosecond laser ablation in an aqueous solution of cyclodextrins, J. Phys. Chem. B 107 (2003) 4527–4531.

[50] E. Braun, Y. Eichen, U. Sivan, G. Ben-Yoseph, DNA-templated assembly and electrode attachment of a conducting silver wire, Nature 391 (1998) 775–778.

[51] Y. Volokitin, Jd Sinzig, L. De Jongh, G. Schmid, M. Vargaftik, I. Moiseevi, Quantum-size effects in the thermodynamic properties of metallic nanoparticles, Nature 384 (1996) 621–623.

[52] A. Ghasemi, E. Ghasemi, E. Paimozd, Influence of copper cations on the magnetic properties of NiCuZn ferrite nanoparticles, J. Magnetism Magnetic Mater. 323 (2011) 1541–1545.

[53] K. Judai, S. Numao, J. Nishijo, N. Nishi, In situ preparation and catalytic activation of copper nanoparticles from acetylide molecules, J. Mol. Catal. A: Chem. 347 (2011) 28–33.

[54] T.M.D. Dang, T.T.T. Le, E. Fribourg-Blanc, M.C. Dang, Synthesis and optical properties of copper nanoparticles prepared by a chemical reduction method, Adv. Nat. Sci.: Nanosci. Nanotechnol. 2 (2011) 015009.

[55] Q.-l Zhang, Z.-M. Yang, B.-j Ding, X.-z Lan, Y.-j Guo, Preparation of copper nanoparticles by chemical reduction method using potassium borohydride, Trans. Nonferrous Met. Soc. China 20 (2010) s240–s244.

[56] W. Yu, H. Xie, L. Chen, Y. Li, Investigation on the thermal transport properties of ethylene glycol-based nanofluids containing copper nanoparticles, Powder Technol. 197 (2010) 218–221.

[57] J. Ramyadevi, K. Jeyasubramanian, A. Marikani, G. Rajakumar, A.A. Rahuman, Synthesis and antimicrobial activity of copper nanoparticles, Mater. Lett. 71 (2012) 114–116.
[58] R.A. Zargar, M.M. Hassan, N. Boora, I. Ahmed, S. Ahmed, Ku Nissa, et al., A comparative study of micro-and nano-ZnO films fabricated by sol-gel syringe spray method, Int. J. Ceram. Eng. Sci. 2 (2020) 169–176.
[59] T. Hemalatha, S. Akilandeswari, Synthesis and characterization of hmta assisted cuo nanoparticles with its potential antibacterial application, Int. J. Recent. Sci. Res. 6 (2015) 7502–7507.
[60] H.S. Devi, T.D. Singh, Synthesis of copper oxide nanoparticles by a novel method and its application in the degradation of methyl orange, Adv. Electron. Electr. Eng. 4 (2014) 83–88.
[61] M.J. Guajardo-Pacheco, J. Morales-Sánchez, J. González-Hernández, F. Ruiz, Synthesis of copper nanoparticles using soybeans as a chelant agent, Mater. Lett. 64 (2010) 1361–1364.
[62] J. Singh, G. Kaur, M. Rawat, A brief review on synthesis and characterization of copper oxide nanoparticles and its applications, J. Bioelectron. Nanotechnol. 1 (2016) 1–9.
[63] Y. Li, B. Tan, Y. Wu, Mesoporous Co_3O_4 nanowire arrays for lithium ion batteries with high capacity and rate capability, Nano Lett. 8 (2008) 265–270.
[64] A. Sharma, R.K. Dutta, A. Roychowdhury, D. Das, A. Goyal, A. Kapoor, Cobalt doped CuO nanoparticles as a highly efficient heterogeneous catalyst for reduction of 4-nitrophenol to 4-aminophenol, Appl. Catal. A: Gen. 543 (2017) 257–265.
[65] F. Martinez-Gutierrez, P.L. Olive, A. Banuelos, E. Orrantia, N. Nino, E.M. Sanchez, et al., Synthesis, characterization, and evaluation of antimicrobial and cytotoxic effect of silver and titanium nanoparticles, Nanomed.: Nanotechnol. Biol. Med. 6 (2010) 681–688.
[66] I. Nam, N.D. Kim, G.-P. Kim, J. Park, J. Yi, Amorphous Mn oxide-ordered mesoporous carbon hybrids as a high performance electrode material for supercapacitors, J. Nanosci. Nanotechnol. 12 (2012) 5704–5708.
[67] C.-Y. Chiang, K. Aroh, N. Franson, V.R. Satsangi, S. Dass, S. Ehrman, Copper oxide nanoparticle made by flame spray pyrolysis for photoelectrochemical water splitting—Part II. Photoelectrochemical study, Int. J. Hydrog. Energy 36 (2011) 15519–15526.
[68] S.B. Rizvi, S. Ghaderi, M. Keshtgar, A.M. Seifalian, Semiconductor quantum dots as fluorescent probes for in vitro and in vivo bio-molecular and cellular imaging, Nano Rev. 1 (2010) 5161.
[69] M. Shah, A. Qurashi, Novel surfactant-free synthesis of MgO nanoflakes, J. Alloy. Compd. 482 (2009) 548–551.
[70] M. Hosseinpour, S. Ahmadi, T. Mousavand, M. Outokesh, Production of granulated-copper oxide nanoparticles for catalytic application, J. Mater. Res. 25 (2010) 2025–2034.
[71] M. Silambarasan, S. Manikandan, K. Rajan, Viscosity and thermal conductivity of dispersions of sub-micron TiO_2 particles in water prepared by stirred bead milling and ultrasonication, Int. J. Heat. Mass. Transf. 55 (2012) 7991–8002.
[72] R. Vijaya Kumar, R. Elgamiel, Y. Diamant, A. Gedanken, J. Norwig, Sonochemical preparation and characterization of nanocrystalline copper oxide embedded in poly (vinyl alcohol) and its effect on crystal growth of copper oxide, Langmuir 17 (2001) 1406–1410.

[73] R.V. Kumar, Y. Diamant, A. Gedanken, Sonochemical synthesis and characterization of nanometer-size transition metal oxides from metal acetates, Chem. Mater. 12 (2000) 2301–2305.

[74] A.A. Eliseev, A.V. Lukashin, A.A. Vertegel, L.I. Heifets, A.I. Zhirov, Y.D. Tretyakov, Complexes of Cu (II) with polyvinyl alcohol as precursors for the preparation of CuO/SiO_2 nanocomposites, Mater. Res. Innov. 3 (2000) 308–312.

[75] K. Borgohain, J.B. Singh, M.R. Rao, T. Shripathi, S. Mahamuni, Quantum size effects in CuO nanoparticles, Phys. Rev. B 61 (2000) 11093.

[76] Y. Chang, H.C. Zeng, Controlled synthesis and self-assembly of single-crystalline CuO nanorods and nanoribbons, Cryst. Growth Des. 4 (2004) 397–402.

[77] J. Xu, W. Ji, Z. Shen, S. Tang, X. Ye, D. Jia, et al., Preparation and characterization of CuO nanocrystals, J. Solid State Chem. 147 (1999) 516–519.

[78] N. Akhtar, K. Pathak, Probiotics as a tool to biosynthesize metallic nanoparticles: research reports and patents survey, Recent. Pat. Drug Deliv. Formula. 11 (2017) 5–18.

[79] D.I. Son, C.H. You, T.W. Kim, Structural, optical, and electronic properties of colloidal CuO nanoparticles formed by using a colloid-thermal synthesis process, Appl. Surf. Sci. 255 (2009) 8794–8797.

[80] Y. Zhang, S. Wang, X. Li, L. Chen, Y. Qian, Z. Zhang, CuO shuttle-like nanocrystals synthesized by oriented attachment, J. Cryst. Growth 291 (2006) 196–201.

[81] W. Songping, J. Li, N. Jing, Z. Zhenou, L. Song, Preparation of ultra fine copper–nickel bimetallic powders for conductive thick film, Intermetallics 15 (2007) 1316–1321.

[82] Y. Lee, J.-r Choi, K.J. Lee, N.E. Stott, D. Kim, Large-scale synthesis of copper nanoparticles by chemically controlled reduction for applications of inkjet-printed electronics, Nanotechnology 19 (2008) 415604.

[83] H.-S. Kim, S.R. Dhage, D.-E. Shim, H.T. Hahn, Intense pulsed light sintering of copper nanoink for printed electronics, Appl. Phys. A 97 (2009) 791–798.

[84] Y. Tamgadge, P. Gedam, N. Thakare, S. Talwatkar, A. Sunatkari, G. Muley, Optical limiting in gelatin stabilized Cu-PVP nanocomposite colloidal suspension, AIP Conference Proceedings, AIP Publishing LLC, 2018, p. 030002.

[85] M.P. Pileni, B.W. Ninham, T. Gulik-Krzywicki, J. Tanori, I. Lisiecki, A. Filankembo, Direct relationship between shape and size of template and synthesis of copper metal particles, Adv. Mater. 11 (1999) 1358–1362.

[86] G. Vitulli, M. Bernini, S. Bertozzi, E. Pitzalis, P. Salvadori, S. Coluccia, et al., Nanoscale copper particles derived from solvated Cu atoms in the activation of molecular oxygen, Chem. Mater. 14 (2002) 1183–1186.

[87] R.V. Kumar, Y. Mastai, Y. Diamant, A. Gedanken, Sonochemical synthesis of amorphous Cu and nanocrystalline Cu_2O embedded in a polyaniline matrix, J. Mater. Chem. 11 (2001) 1209–1213.

[88] I.G. Casella, T.R. Cataldi, A. Guerrieri, E. Desimoni, Copper dispersed into polyaniline films as an amperometric sensor in alkaline solutions of amino acids and polyhydric compounds, Anal. Chim. Acta 335 (1996) 217–225.

[89] Z. Liu, Y. Bando, A novel method for preparing copper nanorods and nanowires, Adv. Mater. 15 (2003) 303–305.

[90] N.A. Luechinger, E.K. Athanassiou, W.J. Stark, Graphene-stabilized copper nanoparticles as an air-stable substitute for silver and gold in low-cost ink-jet printable electronics, Nanotechnology 19 (2008) 445201.

[91] E.K. Athanassiou, R.N. Grass, W.J. Stark, Large-scale production of carbon-coated copper nanoparticles for sensor applications, Nanotechnology 17 (2006) 1668.

[92] N.A. Dhas, C.P. Raj, A. Gedanken, Synthesis, characterization, and properties of metallic copper nanoparticles, Chem. Mater. 10 (1998) 1446–1452.

[93] M. Grouchko, A. Kamyshny, K. Ben-Ami, S. Magdassi, Synthesis of copper nanoparticles catalyzed by pre-formed silver nanoparticles, J. Nanopart. Res. 11 (2009) 713–716.

[94] X. Zhang, Z. Liu, Y. Huang, X. Wan, J. Tian, Y. Ma, et al., Synthesis, characterization and nonlinear optical property of graphene-C60 hybrid, J. Nanosci. Nanotechnol. 9 (2009) 5752–5756.

[95] A.I. Khan, A.V. Arasu, A review of influence of nanoparticle synthesis and geometrical parameters on thermophysical properties and stability of nanofluids, Therm. Sci. Eng. Prog. 11 (2019) 334–364.

[96] J. Lee, K. Han, J. Koo, A novel method to evaluate dispersion stability of nanofluids, Int. J. Heat. Mass. Transf. 70 (2014) 421–429.

[97] Y.-j Hwang, J. Lee, C. Lee, Y. Jung, S. Cheong, C. Lee, et al., Stability and thermal conductivity characteristics of nanofluids, Thermochim. Acta 455 (2007) 70–74.

[98] R. Sadeghi, S.G. Etemad, E. Keshavarzi, M. Haghshenasfard, Investigation of alumina nanofluid stability by UV–vis spectrum, Microfluidics Nanofluidics 18 (2015) 1023–1030.

[99] M. Kole, T. Dey, Effect of aggregation on the viscosity of copper oxide–gear oil nanofluids, Int. J. Therm. Sci. 50 (2011) 1741–1747.

[100] S. Bhattacharjee, DLS and zeta potential–what they are and what they are not? J. Controlled Rel. 235 (2016) 337–351.

[101] N. Karthikeyan, J. Philip, B. Raj, Effect of clustering on the thermal conductivity of nanofluids, Mater. Chem. Phys. 109 (2008) 50–55.

[102] R. Prasher, W. Evans, P. Meakin, J. Fish, P. Phelan, P. Keblinski, Effect of aggregation on thermal conduction in colloidal nanofluids, Appl. Phys. Lett. 89 (2006) 143119.

[103] D.-W. Oh, A. Jain, J.K. Eaton, K.E. Goodson, J.S. Lee, Thermal conductivity measurement and sedimentation detection of aluminum oxide nanofluids by using the 3ω method, Int. J. Heat. Fluid Flow. 29 (2008) 1456–1461.

[104] N. Ali, J.A. Teixeira, A. Addali, A review on nanofluids: fabrication, stability, and thermophysical properties, J. Nanomater. 2018 (2018) 1–33.

[105] R.G. Harrison, P.W. Todd, S.R. Rudge, D.P. Petrides, Bioprocess design and economics, Bioseparations Science and Engineering, Oxford University Press, 2015.

[106] A.K. Singh, V.S. Raykar, Microwave synthesis of silver nanofluids with polyvinylpyrrolidone (PVP) and their transport properties, Colloid Polym. Sci. 286 (2008) 1667–1673.

[107] D.F. Swinehart, The beer-lambert law, J. Chem. Educ. 39 (1962) 333.

[108] A. Ghadimi, R. Saidur, H. Metselaar, A review of nanofluid stability properties and characterization in stationary conditions, Int. J. Heat. Mass. Transf. 54 (2011) 4051–4068.

[109] W. Chamsa-Ard, S. Brundavanam, C.C. Fung, D. Fawcett, G. Poinern, Nanofluid types, their synthesis, properties and incorporation in direct solar thermal collectors: a review, Nanomaterials 7 (2017) 131.

[110] K. Cacua, F. Ordoñez, C. Zapata, B. Herrera, E. Pabón, R. Buitrago-Sierra, Surfactant concentration and pH effects on the zeta potential values of alumina

nanofluids to inspect stability, Colloids Surf. A: Physicochem. Eng. Asp. 583 (2019) 123960.
[111] K. Suganthi, K. Rajan, Temperature induced changes in ZnO–water nanofluid: zeta potential, size distribution and viscosity profiles, Int. J. Heat. Mass. Transf. 55 (2012) 7969–7980.
[112] S. Chakraborty, I. Sengupta, I. Sarkar, S.K. Pal, S. Chakraborty, Effect of surfactant on thermo-physical properties and spray cooling heat transfer performance of Cu-Zn-Al LDH nanofluid, Appl. Clay Sci. 168 (2019) 43–55.
[113] S. Chakraborty, I. Sarkar, A. Ashok, I. Sengupta, S.K. Pal, S. Chakraborty, Thermo-physical properties of Cu-Zn-Al LDH nanofluid and its application in spray cooling, Appl. Therm. Eng. 141 (2018) 339–351.
[114] S. Chakraborty, I. Sarkar, D.K. Behera, S.K. Pal, S. Chakraborty, Experimental investigation on the effect of dispersant addition on thermal and rheological characteristics of TiO_2 nanofluid, Powder Technol. 307 (2017) 10–24.
[115] A.D. Manasrah, U.A. Al-Mubaiyedh, T. Laui, R. Ben-Mansour, M.J. Al-Marri, I.W. Almanassra, et al., Heat transfer enhancement of nanofluids using iron nanoparticles decorated carbon nanotubes, Appl. Therm. Eng. 107 (2016) 1008–1018.
[116] C. Tsai, H. Chien, P. Ding, B. Chan, T.Y. Luh, P. Chen, Effect of structural character of gold nanoparticles in nanofluid on heat pipe thermal performance, Mater. Lett. 58 (2004) 1461–1465.
[117] I. Javed, S.W. Baek, K. Waheed, Autoignition and combustion characteristics of heptane droplets with the addition of aluminium nanoparticles at elevated temperatures, Combust. Flame 162 (2015) 191–206.
[118] C. Pang, W. Wu, W. Sheng, H. Zhang, Y.T. Kang, Mass transfer enhancement by binary nanofluids (NH_3/H_2O + Ag nanoparticles) for bubble absorption process, Int. J. Refrig. 35 (2012) 2240–2247.
[119] C.V. Popa, C.T. Nguyen, I. Gherasim, New specific heat data for Al_2O_3 and CuO nanoparticles in suspension in water and Ethylene Glycol, Int. J. Therm. Sci. 111 (2017) 108–115.
[120] B. Bhanvase, M. Sarode, L. Putterwar, K. Abdullah, M. Deosarkar, S. Sonawane, Intensification of convective heat transfer in water/ethylene glycol based nanofluids containing TiO_2 nanoparticles, Chem. Eng. Processing: Process. Intensif. 82 (2014) 123–131.
[121] T. Krishnakumar, A. Sheeba, V. Mahesh, M.J. Prakash, Heat transfer studies on ethylene glycol/water nanofluid containing TiO_2 nanoparticles, Int. J. Refrig. 102 (2019) 55–61.
[122] R.S. Khedkar, A.S. Kiran, S.S. Sonawane, K. Wasewar, S.S. Umre, Thermo–physical characterization of paraffin based Fe_3O_4 nanofluids, Procedia Eng. 51 (2013) 342–346.
[123] M. Bahiraei, M. Berahmand, A. Shahsavar, Irreversibility analysis for flow of a non-Newtonian hybrid nanofluid containing coated CNT/Fe_3O_4 nanoparticles in a minichannel heat exchanger, Appl. Therm. Eng. 125 (2017) 1083–1093.
[124] M.I. Hassan, I.A. Alzarooni, Y. Shatilla, The effect of water-based nanofluid incorporating Al_2O_3 nanoparticles on heat pipe performance, Energy Procedia 75 (2015) 3201–3206.
[125] J.-H. Lee, K.S. Hwang, S.P. Jang, B.H. Lee, J.H. Kim, S.U. Choi, et al., Effective viscosities and thermal conductivities of aqueous nanofluids containing low volume

concentrations of Al_2O_3 nanoparticles, Int. J. Heat. Mass. Transf. 51 (2008) 2651–2656.

[126] N. Gupta, S.M. Gupta, S. Sharma, Preparation of stable metal/COOH-MWCNT hybrid nanofluid, Mater. Today: Proc. 36 (2021) 649–656.

[127] W. Ahmed, S. Kazi, Z. Chowdhury, M.R.B. Johan, S. Mehmood, M.E.M. Soudagar, et al., Heat transfer growth of sonochemically synthesized novel mixed metal oxide ZnO + Al_2O_3 + TiO_2/DW based ternary hybrid nanofluids in a square flow conduit, Renew. Sustain. Energy Rev. 145 (2021) 111025.

[128] I. Tlili, S.M. Seyyedi, A. Dogonchi, M. Hashemi-Tilehnoee, D. Ganji, Analysis of a single-phase natural circulation loop with hybrid-nanofluid, Int. Commun. Heat. Mass. Transf. 112 (2020) 104498.

[129] M.N. Labib, M.J. Nine, H. Afrianto, H. Chung, H. Jeong, Numerical investigation on effect of base fluids and hybrid nanofluid in forced convective heat transfer, Int. J. Therm. Sci. 71 (2013) 163–171.

[130] Q. Li, Y. Xuan, Convective heat transfer and flow characteristics of Cu-water nanofluid, Sci. China Series E: Technol. Sci. 45 (2002) 408–416.

[131] E. Sadeghinezhad, M. Mehrali, R. Saidur, M. Mehrali, S.T. Latibari, A.R. Akhiani, et al., A comprehensive review on graphene nanofluids: recent research, development and applications, Energy Convers. Manag. 111 (2016) 466–487.

[132] A. Naddaf, S.Z. Heris, Experimental study on thermal conductivity and electrical conductivity of diesel oil-based nanofluids of graphene nanoplatelets and carbon nanotubes, Int. Commun. Heat. Mass. Transf. 95 (2018) 116–122.

[133] W. Yu, H. Xie, W. Chen, Experimental investigation on thermal conductivity of nanofluids containing graphene oxide nanosheets, J. Appl. Phys. 107 (2010) 094317.

[134] Y. Tong, T. Boldoo, J.H. Mr, H. Cho, Improvement of photo-thermal energy conversion performance of MWCNT/Fe_3O_4 hybrid nanofluid compared to Fe_3O_4 nanofluid, Energy 196 (2020) 117086.

[135] B. Munkhbayar, M.R. Tanshen, J. Jeoun, H. Chung, H. Jeong, Surfactant-free dispersion of silver nanoparticles into MWCNT-aqueous nanofluids prepared by one-step technique and their thermal characteristics, Ceram. Int. 39 (2013) 6415–6425.

[136] H. Yarmand, S. Gharehkhani, G. Ahmadi, S.F.S. Shirazi, S. Baradaran, E. Montazer, et al., Graphene nanoplatelets–silver hybrid nanofluids for enhanced heat transfer, Energy Convers. Manage. 100 (2015) 419–428.

[137] L. Wang, J. Fan, Nanofluids research: key issues, Nanoscale Res. Lett. 5 (2010) 1241–1252.

[138] W. Yu, H. Xie, L. Chen, Y. Li, Enhancement of thermal conductivity of kerosene-based Fe_3O_4 nanofluids prepared via phase-transfer method, Colloids Surf. A: Physicochem. Eng. Asp. 355 (2010) 109–113.

[139] Y. Chen, X. Wang, Novel phase-transfer preparation of monodisperse silver and gold nanoparticles at room temperature, Mater. Lett. 62 (2008) 2215–2218.

[140] H.T. Zhu, C.Y. Zhang, Y.M. Tang, J.X. Wang, Novel synthesis and thermal conductivity of CuO nanofluid, J. Phys. Chem. C 111 (2007) 1646–1650.

[141] W. Yu, H. Xie, X. Wang, X. Wang, Highly efficient method for preparing homogeneous and stable colloids containing graphene oxide, Nanoscale Res. Lett. 6 (2011) 1–7.

[142] A. Kumar, H. Joshi, R. Pasricha, A. Mandale, M. Sastry, Phase transfer of silver nanoparticles from aqueous to organic solutions using fatty amine molecules, J. Colloid Interface Sci. 264 (2003) 396–401.

[143] H. Bönnemann, S. Botha, B. Bladergroen, V. Linkov, Monodisperse copper-and silver-nanocolloids suitable for heat-conductive fluids, Appl. Organomet. Chem. 19 (2005) 768–773.

[144] C.-H. Lo, T.-T. Tsung, L.-C. Chen, C.-H. Su, H.-M. Lin, Fabrication of copper oxide nanofluid using submerged arc nanoparticle synthesis system (SANSS), J. Nanopart. Res. 7 (2005) 313–320.

[145] J.A. Eastman, S. Choi, S. Li, W. Yu, L. Thompson, Anomalously increased effective thermal conductivities of ethylene glycol-based nanofluids containing copper nanoparticles, Appl. Phys. Lett. 78 (2001) 718–720.

[146] C.-H. Lo, T.-T. Tsung, L.-C. Chen, Shape-controlled synthesis of Cu-based nanofluid using submerged arc nanoparticle synthesis system (SANSS), J. Cryst. Growth 277 (2005) 636–642.

[147] Y. Li, S. Tung, E. Schneider, S. Xi, A review on development of nanofluid preparation and characterization, Powder Technol. 196 (2009) 89–101.

[148] M. Rafique, M.S. Rafique, S.H. Butt, A. Afzal, U. Kalsoom, Laser nature dependence on enhancement of optical and thermal properties of copper oxide nanofluids, Appl. Surf. Sci. 483 (2019) 187–193.

[149] M. Dell'Aglio, V. Mangini, G. Valenza, O. De Pascale, A. De Stradis, G. Natile, et al., Silver and gold nanoparticles produced by pulsed laser ablation in liquid to investigate their interaction with ubiquitin, Appl. Surf. Sci. 374 (2016) 297–304.

[150] A. Simakin, V. Voronov, N. Kirichenko, G. Shafeev, Nanoparticles produced by laser ablation of solids in liquid environment, Appl. Phys. A 79 (2004) 1127–1132.

[151] G. Yang, Laser ablation in liquids: applications in the synthesis of nanocrystals, Prog. Mater. Sci. 52 (2007) 648–698.

[152] T. Sakka, S. Iwanaga, Y.H. Ogata, A. Matsunawa, T. Takemoto, Laser ablation at solid–liquid interfaces: an approach from optical emission spectra, J. Chem. Phys. 112 (2000) 8645–8653.

[153] A.K. Mahmoud, Z. Fadhill, S.I. Al-nassar, F.I. Husein, E. Akman, A. Demir, Synthesis of zirconia nanoparticles in distilled water solution by laser ablation technique, J. Mater. Sci. Eng. B 3 (2013) 364–368.

[154] H. Zeng, X.W. Du, S.C. Singh, S.A. Kulinich, S. Yang, J. He, et al., Nanomaterials via laser ablation/irradiation in liquid: a review, Adv. Funct. Mater. 22 (2012) 1333–1353.

[155] D. Zhang, B. Gokce, S. Barcikowski, Laser synthesis and processing of colloids: fundamentals and applications, Chem. Rev. 117 (2017) 3990–4103.

[156] S.V. Rao, G.K. Podagatlapalli, S. Hamad, Ultrafast laser ablation in liquids for nanomaterials and applications, J. Nanosci. Nanotechnol. 14 (2014) 1364–1388.

[157] D. Riabinina, M. Chaker, J. Margot, Dependence of gold nanoparticle production on pulse duration by laser ablation in liquid media, Nanotechnology 23 (2012) 135603.

[158] E. Giorgetti, M.M. Miranda, S. Caporali, P. Canton, P. Marsili, C. Vergari, et al., TiO_2 nanoparticles obtained by laser ablation in water: influence of pulse energy and duration on the crystalline phase, J. Alloy. Compd. 643 (2015) S75–S79.

[159] N. Patra, K. Akash, S. Shiva, R. Gagrani, H.S.P. Rao, V. Anirudh, et al., Parametric investigations on the influence of nano-second Nd^{3+}: YAG laser wavelength and fluence in synthesizing NiTi nano-particles using liquid assisted laser ablation technique, Appl. Surf. Sci. 366 (2016) 104–111.

[160] G. García Guillén, V.A. Zuniga Ibarra, M.I. Mendivil Palma, B. Krishnan, D. Avellaneda Avellaneda, S. Shaji, Effects of liquid medium and ablation wavelength on the properties of cadmium sulfide nanoparticles formed by pulsed-laser ablation, ChemPhysChem 18 (2017) 1035–1046.

[161] Y. Al-Douri, S. Abdulateef, A.A. Odeh, C. Voon, N. Badi, GaNO colloidal nanoparticles synthesis by nanosecond pulsed laser ablation: laser fluence dependent optical absorption and structural properties, Powder Technol. 320 (2017) 457–461.

[162] Y. Al-Douri, R.A. Al-Samarai, S.A. Abdulateef, A.A. Odeh, N. Badi, C.H. Voon, Nanosecond pulsed laser ablation to synthesize GaO colloidal nanoparticles: optical and structural properties, Optik 178 (2019) 337–342.

[163] E. Solatı, D. Dorranian, Effect of temperature on the characteristics of ZnO nanoparticles produced by laser ablation in water, Bull. Mater. Sci. 39 (2016) 1677–1684.

[164] N.E. Jasbi, D. Dorranian, Dependence of laser ablation produced TiO_2 nanoparticles on the ablation environment temperature, Optical Quantum Electron. 49 (2017) 1–13.

[165] M. Moradi, E. Solati, S. Darvishi, D. Dorranian, Effect of aqueous ablation environment on the characteristics of ZnO nanoparticles produced by laser ablation, J. Clust. Sci. 27 (2016) 127–138.

[166] R. Ganeev, M. Baba, A. Ryasnyansky, M. Suzuki, H. Kuroda, Characterization of optical and nonlinear optical properties of silver nanoparticles prepared by laser ablation in various liquids, Opt. Commun. 240 (2004) 437–448.

[167] M.H. Mahdieh, B. Fattahi, Size properties of colloidal nanoparticles produced by nanosecond pulsed laser ablation and studying the effects of liquid medium and laser fluence, Appl. Surf. Sci. 329 (2015) 47–57.

[168] J. Dupont, J.D. Scholten, On the structural and surface properties of transition-metal nanoparticles in ionic liquids, Chem. Soc. Rev. 39 (2010) 1780–1804.

[169] M. Kao, C. Lo, T. Tsung, Y. Wu, C. Jwo, H. Lin, Copper-oxide brake nanofluid manufactured using arc-submerged nanoparticle synthesis system, J. Alloy. Compd. 434 (2007) 672–674.

[170] J.-G. Mattei, P. Grammatikopoulos, J. Zhao, V. Singh, J. Vernieres, S. Steinhauer, et al., Gas-phase synthesis of trimetallic nanoparticles, Chem. Mater. 31 (2019) 2151–2163.

[171] B. Das, B. Balasubramanian, P. Manchanda, P. Mukherjee, R. Skomski, G.C. Hadjipanayis, et al., Mn_5Si_3 nanoparticles: synthesis and size-induced ferromagnetism, Nano Lett. 16 (2016) 1132–1137.

[172] A. Choukourov, D. Nikitin, P. Pleskunov, R. Tafiichuk, K. Biliak, M. Protsak, et al., Residual-and linker-free metal/polymer nanofluids prepared by direct deposition of magnetron-sputtered Cu nanoparticles into liquid PEG, J. Mol. Liq. 336 (2021) 116319.

[173] B. Balasubramanian, R. Skomski, X. Li, S.R. Valloppilly, J.E. Shield, G.C. Hadjipanayis, et al., Cluster synthesis and direct ordering of rare-earth transition-metal nanomagnets, Nano Lett. 11 (2011) 1747–1752.

[174] I. Barke, H. Hartmann, D. Rupp, L. Flückiger, M. Sauppe, M. Adolph, et al., The 3D-architecture of individual free silver nanoparticles captured by X-ray scattering, Nat. Commun. 6 (2015) 1–7.

[175] G.E. Johnson, R. Colby, J. Laskin, Soft landing of bare nanoparticles with controlled size, composition, and morphology, Nanoscale 7 (2015) 3491–3503.

[176] J. Zhao, E. Baibuz, J. Vernieres, P. Grammatikopoulos, V. Jansson, M. Nagel, et al., Formation mechanism of Fe nanocubes by magnetron sputtering inert gas condensation, ACS Nano 10 (2016) 4684–4694.

[177] J. Vernieres, S. Steinhauer, J. Zhao, A. Chapelle, P. Menini, N. Dufour, et al., Gas phase synthesis of multifunctional Fe-based nanocubes, Adv. Funct. Mater. 27 (2017) 1605328.

[178] J.A. Lopez, F. González, F.A. Bonilla, G. Zambrano, M.E. Gómez, Synthesis and characterization of Fe_3O_4 magnetic nanofluid, Rev. Latinoamericana de. Metalurgia y. Materiales 30 (2010) 60–66.

[179] N.P. Devi, M. Maisnam, Characterizations of sol-gel synthesized and high energy ball milled spinel nanoferrites: MFe_2O_4 (M = Li, Ni, Zn, Mn) for nanofluid preparations, Integr. Ferroelectr. 204 (2020) 133–141.

[180] A. Ghadimi, I.H. Metselaar, The influence of surfactant and ultrasonic processing on improvement of stability, thermal conductivity and viscosity of titania nanofluid, Exp. Therm. Fluid Sci. 51 (2013) 1–9.

[181] I. Sharifi, H. Shokrollahi, S. Amiri, Ferrite-based magnetic nanofluids used in hyperthermia applications, J. Magnetism Magnetic Mater. 324 (2012) 903–915.

[182] T.-W. Lin, H.C. Weng, Electrostatically stabilized nanofluid preparation by chemical co-precipitation and the effect of particle size on nanofluid viscosity, Smart Sci. 6 (2018) 197–204.

[183] K. Nemade, S. Waghuley, A novel approach for enhancement of thermal conductivity of CuO/H_2O based nanofluids, Appl. Therm. Eng. 95 (2016) 271–274.

[184] M.N. Rashin, J. Hemalatha, A novel ultrasonic approach to determine thermal conductivity in CuO–ethylene glycol nanofluids, J. Mol. Liq. 197 (2014) 257–262.

[185] T. Gao, X. Zhang, C. Li, Y. Zhang, M. Yang, D. Jia, et al., Surface morphology evaluation of multi-angle 2D ultrasonic vibration integrated with nanofluid minimum quantity lubrication grinding, J. Manuf. Process. 51 (2020) 44–61.

[186] S. Chakraborty, S.K. Saha, J. Pandey, S. Das, Experimental characterization of concentration of nanofluid by ultrasonic technique, Powder Technol. 210 (2011) 304–307.

[187] B. Munkhbayar, M. Bat-Erdene, B. Ochirkhuyag, D. Sarangerel, B. Battsengel, H. Chung, et al., An experimental study of the planetary ball milling effect on dispersibility and thermal conductivity of MWCNTs-based aqueous nanofluids, Mater. Res. Bull. 47 (2012) 4187–4196.

[188] C.-Y. Chiang, K. Aroh, S.H. Ehrman, Copper oxide nanoparticle made by flame spray pyrolysis for photoelectrochemical water splitting–Part I. CuO nanoparticle preparation, Int. J. Hydrog. Energy 37 (2012) 4871–4879.

[189] L. Chen, W. Yu, H. Xie, Enhanced thermal conductivity of nanofluids containing Ag/MWNT composites, Powder Technol. 231 (2012) 18–20.

[190] S. Lee, S.-S. Choi, S. Li, J. Eastman, Measuring thermal conductivity of fluids containing oxide nanoparticles, J. Heat Transfer. 121 (1999) 280–289.

[191] C.-S. Jwo, T.-P. Teng, H. Chang, A simple model to estimate thermal conductivity of fluid with acicular nanoparticles, J. Alloy. Compd. 434 (2007) 569–571.

[192] M.-H. Chang, H.-S. Liu, C.Y. Tai, Preparation of copper oxide nanoparticles and its application in nanofluid, Powder Technol. 207 (2011) 378–386.

[193] M.-S. Liu, M.C.-C. Lin, C. Tsai, C.-C. Wang, Enhancement of thermal conductivity with Cu for nanofluids using chemical reduction method, Int. J. Heat. Mass. Transf. 49 (2006) 3028–3033.

[194] M.S. Liu, M.C. Lin, I.T. Huang, C.C. Wang, Enhancement of thermal conductivity with CuO for nanofluids, Chem. Eng. Technol. 29 (2006) 72–77.

[195] J. Eastman a, S.U.S. Choi, S. Li, W. Yu, L.J. Thompson, Anomalously increased effective thermal conductivities of ethylene glycolbased nanofluids containing copper nanoparticles, Appl. Phys. Lett. 78 (2001) 718.

[196] R.S. Khedkar, S.S. Sonawane, K.L. Wasewar, Influence of CuO nanoparticles in enhancing the thermal conductivity of water and monoethylene glycol based nanofluids, Int. Commun. Heat. Mass. Transf. 39 (2012) 665–669.

[197] M. Liu, M.C. Lin, C. Wang, Enhancements of thermal conductivities with Cu, CuO, and carbon nanotube nanofluids and application of MWNT/water nanofluid on a water chiller system, Nanoscale Res. Lett. 6 (2011) 1–13.

[198] M. Akhavan-Behabadi, F. Hekmatipour, S. Mirhabibi, B. Sajadi, An empirical study on heat transfer and pressure drop properties of heat transfer oil-copper oxide nanofluid in microfin tubes, Int. Commun. Heat. Mass. Transf. 57 (2014) 150–156.

[199] S. Fotukian, M.N. Esfahany, Experimental study of turbulent convective heat transfer and pressure drop of dilute CuO/water nanofluid inside a circular tube, Int. Commun. Heat. Mass. Transf. 37 (2010) 214–219.

[200] S. Harikrishnan, S. Kalaiselvam, Preparation and thermal characteristics of CuO–oleic acid nanofluids as a phase change material, Thermochim. Acta 533 (2012) 46–55.

[201] H. Peng, G. Ding, H. Hu, Effect of surfactant additives on nucleate pool boiling heat transfer of refrigerant-based nanofluid, Exp. Therm. Fluid Sci. 35 (2011) 960–970.

[202] E. De Robertis, E. Cosme, R. Neves, A.Y. Kuznetsov, A. Campos, S. Landi, et al., Application of the modulated temperature differential scanning calorimetry technique for the determination of the specific heat of copper nanofluids, Appl. Therm. Eng. 41 (2012) 10–17.

[203] R. Kathiravan, R. Kumar, A. Gupta, R. Chandra, Preparation and pool boiling characteristics of silver nanofluids over a flat plate heater, Heat. Transf. Eng. 33 (2012) 69–78.

[204] Y. Xuan, Q. Li, Heat transfer enhancement of nanofluids, Int. J. Heat. Fluid Flow. 21 (2000) 58–64.

[205] J. Wu, J. Zhao, A review of nanofluid heat transfer and critical heat flux enhancement—research gap to engineering application, Prog. Nucl. Energy 66 (2013) 13–24.

[206] Z. Liu, Q. Zhu, Application of aqueous nanofluids in a horizontal mesh heat pipe, Energy Convers. Manage. 52 (2011) 292–300.

[207] W. Escher, T. Brunschwiler, N. Shalkevich, A. Shalkevich, T. Burgi, B. Michel, et al., On the cooling of electronics with nanofluids, J. Heat. Transf. 133 (2011).

[208] S.U. Choi, J.A. Eastman, Enhancing Thermal Conductivity of Fluids with Nanoparticles, Argonne National Lab.(ANL), Argonne, IL, 1995.

[209] H. Peng, G. Ding, W. Jiang, H. Hu, Y. Gao, Heat transfer characteristics of refrigerant-based nanofluid flow boiling inside a horizontal smooth tube, Int. J. Refrig. 32 (2009) 1259–1270.

[210] M. Boudouh, H.L. Gualous, M. De Labachelerie, Local convective boiling heat transfer and pressure drop of nanofluid in narrow rectangular channels, Appl. Therm. Eng. 30 (2010) 2619–2631.

[211] B. Sun, D. Yang, Flow boiling heat transfer characteristics of nano-refrigerants in a horizontal tube, Int. J. Refrig. 38 (2014) 206–214.
[212] M. Sarafraz, F. Hormozi, Convective boiling and particulate fouling of stabilized CuO-ethylene glycol nanofluids inside the annular heat exchanger, Int. Commun. Heat. Mass. Transf. 53 (2014) 116–123.
[213] A. Zangeneh, S. Peyghambarzadeh, A. Bayat, A. Vatani, Application of general multilevel factorial design approach in forced convection and subcooled flow boiling heat transfer to CuO/water nanofluids, J. Mol. Liq. 313 (2020) 113502.
[214] I. Fazeli, M.R.S. Emami, A. Rashidi, Investigation and optimization of the behavior of heat transfer and flow of MWCNT-CuO hybrid nanofluid in a brazed plate heat exchanger using response surface methodology, Int. Commun. Heat. Mass. Transf. 122 (2021) 105175.
[215] N. Asokan, P. Gunnasegaran, V.V. Wanatasanappan, Experimental investigation on the thermal performance of compact heat exchanger and the rheological properties of low concentration mono and hybrid nanofluids containing Al_2O_3 and CuO nanoparticles, Therm. Sci. Eng. Prog. 20 (2020) 100727.
[216] R. Du, D. Jiang, Y. Wang, K.W. Shah, An experimental investigation of CuO/water nanofluid heat transfer in geothermal heat exchanger, Energy Build. 227 (2020) 110402.
[217] H.-J. Kim, K.-J. Seo, K.H. Kang, D.-E. Kim, Nano-lubrication: a review, Int. J. Precis. Eng. Manuf. 17 (2016) 829–841.
[218] J.-H. Wu, B. Phillips, W. Jiang, J. Sanders, J. Zabinski, A. Malshe, Bio-inspired surface engineering and tribology of MoS_2 overcoated cBN–TiN composite coating, Wear 261 (2006) 592–599.
[219] N. Sankar, N. Mathew, C. Sobhan, Molecular dynamics modeling of thermal conductivity enhancement in metal nanoparticle suspensions, Int. Commun. Heat. Mass. Transf. 35 (2008) 867–872.
[220] Y. Ding, D. Wen, Particle migration in a flow of nanoparticle suspensions, Powder Technol. 149 (2005) 84–92.
[221] S. Chaudhari, R. Chakule, P. Talmale, Experimental study of heat transfer characteristics of Al_2O_3 and CuO nanofluids for machining application, Mater. Today: Proc. 18 (2019) 788–797.
[222] J. Navas, A. Sánchez-Coronilla, E.I. Martín, M. Teruel, J.J. Gallardo, T. Aguilar, et al., On the enhancement of heat transfer fluid for concentrating solar power using Cu and Ni nanofluids: an experimental and molecular dynamics study, Nano Energy 27 (2016) 213–224.
[223] A.A. Prasad, R.A. Taylor, M. Kay, Assessment of solar and wind resource synergy in Australia, Appl. Energy 190 (2017) 354–367.
[224] M. Mussard, Solar energy under cold climatic conditions: a review, Renew. Sustain. Energy Rev. 74 (2017) 733–745.
[225] A.-E. Barkaoui, S. Boldyryev, N. Duic, G. Krajacic, Z. Guzović, Appropriate integration of geothermal energy sources by Pinch approach: case study of Croatia, Appl. energy 184 (2016) 1343–1349.
[226] M. Bahiraei, R. Rahmani, A. Yaghoobi, E. Khodabandeh, R. Mashayekhi, M. Amani, Recent research contributions concerning use of nanofluids in heat exchangers: a critical review, Appl. Therm. Eng. 133 (2018) 137–159.

[227] S. Bazri, I.A. Badruddin, M.S. Naghavi, M. Bahiraei, A review of numerical studies on solar collectors integrated with latent heat storage systems employing fins or nanoparticles, Renew. Energy 118 (2018) 761–778.
[228] M. Bahiraei, S. Heshmatian, Electronics cooling with nanofluids: a critical review, Energy Convers. Manage. 172 (2018) 438–456.
[229] S.S. Murshed, C.N. de Castro, Conduction and convection heat transfer characteristics of ethylene glycol based nanofluids—a review, Appl. Energy 184 (2016) 681–695.
[230] O. Arthur, M. Karim, An investigation into the thermophysical and rheological properties of nanofluids for solar thermal applications, Renew. Sustain. Energy Rev. 55 (2016) 739–755.
[231] D.K. Devendiran, V.A. Amirtham, A review on preparation, characterization, properties and applications of nanofluids, Renew. Sustain. Energy Rev. 60 (2016) 21–40.
[232] M.M. Tawfik, Experimental studies of nanofluid thermal conductivity enhancement and applications: a review, Renew. Sustain. Energy Rev. 75 (2017) 1239–1253.
[233] S. Akilu, K. Sharma, A.T. Baheta, R. Mamat, A review of thermophysical properties of water based composite nanofluids, Renew. Sustain. Energy Rev. 66 (2016) 654–678.
[234] S. Ahmad, R. Saidur, I. Mahbubul, F. Al-Sulaiman, Optical properties of various nanofluids used in solar collector: a review, Renew. Sustain. Energy Rev. 73 (2017) 1014–1030.
[235] R. Ganvir, P. Walke, V. Kriplani, Heat transfer characteristics in nanofluid—a review, Renew. Sustain. Energy Rev. 75 (2017) 451–460.
[236] K. Leong, K.K. Ahmad, H.C. Ong, M. Ghazali, A. Baharum, Synthesis and thermal conductivity characteristic of hybrid nanofluids—a review, Renew. Sustain. Energy Rev. 75 (2017) 868–878.
[237] S. Sharma, S.M. Gupta, Preparation and evaluation of stable nanofluids for heat transfer application: a review, Exp. Therm. Fluid Sci. 79 (2016) 202–212.
[238] R.V. Pinto, F.A.S. Fiorelli, Review of the mechanisms responsible for heat transfer enhancement using nanofluids, Appl. Therm. Eng. 108 (2016) 720–739.
[239] L.S. Sundar, K. Sharma, M.K. Singh, A. Sousa, Hybrid nanofluids preparation, thermal properties, heat transfer and friction factor—a review, Renew. Sustain. Energy Rev. 68 (2017) 185–198.
[240] T.B. Gorji, A. Ranjbar, A review on optical properties and application of nanofluids in direct absorption solar collectors (DASCs), Renew. Sustain. Energy Rev. 72 (2017) 10–32.
[241] W. Azmi, K. Sharma, R. Mamat, G. Najafi, M. Mohamad, The enhancement of effective thermal conductivity and effective dynamic viscosity of nanofluids—a review, Renew. Sustain. Energy Rev. 53 (2016) 1046–1058.
[242] M. Gupta, V. Singh, R. Kumar, Z. Said, A review on thermophysical properties of nanofluids and heat transfer applications, Renew. Sustain. Energy Rev. 74 (2017) 638–670.
[243] N. Chen, H. Ma, Y. Li, J. Cheng, C. Zhang, D. Wu, et al., Complementary optical absorption and enhanced solar thermal conversion of CuO-ATO nanofluids, Sol. Energy Mater. Sol. Cell 162 (2017) 83–92.

[244] H. Zhu, D. Han, Z. Meng, D. Wu, C. Zhang, Preparation and thermal conductivity of CuO nanofluid via a wet chemical method, Nanoscale Res. Lett. 6 (2011) 1–6.

[245] P. Kanti, K. Sharma, K.M. Yashawantha, M. Jamei, Z. Said, Properties of water-based fly ash-copper hybrid nanofluid for solar energy applications: application of RBF model, Sol. Energy Mater. Sol. Cell 234 (2022) 111423.

[246] L.S. Sundar, A. Misganaw, M.K. Singh, A.M. Pereira, A.C. Sousa, Efficiency, energy and economic analysis of twisted tape inserts in a thermosyphon solar flat plate collector with Cu nanofluids, Renew. Energy Focus 35 (2020) 10–31.

[247] N.K.C. Sint, I. Choudhury, H.H. Masjuki, H. Aoyama, Theoretical analysis to determine the efficiency of a CuO-water nanofluid based-flat plate solar collector for domestic solar water heating system in Myanmar, Sol. Energy 155 (2017) 608–619.

[248] A. Zamzamian, M. KeyanpourRad, M. KianiNeyestani, M.T. Jamal-Abad, An experimental study on the effect of Cu-synthesized/EG nanofluid on the efficiency of flat-plate solar collectors, Renew. Energy 71 (2014) 658–664.

[249] Z. Alrowaili, M. Ezzeldien, N.M. Shaaalan, E. Hussein, M. Sharafeldin, Investigation of the effect of hybrid CuO-Cu/water nanofluid on the solar thermal energy storage system, J. Energy Storage 50 (2022) 104675.

[250] S. Akilu, A.T. Baheta, M.A.M. Said, A.A. Minea, K. Sharma, Properties of glycerol and ethylene glycol mixture based SiO_2-CuO/C hybrid nanofluid for enhanced solar energy transport, Sol. Energy Mater. Sol. Cell 179 (2018) 118–128.

[251] S. Suresh, K. Venkitaraj, P. Selvakumar, M. Chandrasekar, Synthesis of Al_2O_3–Cu/water hybrid nanofluids using two step method and its thermo physical properties, Colloids Surf. A: Physicochem. Eng. Asp. 388 (2011) 41–48.

[252] M.H. Esfe, S. Wongwises, A. Naderi, A. Asadi, M.R. Safaei, H. Rostamian, et al., Thermal conductivity of Cu/TiO_2–water/EG hybrid nanofluid: experimental data and modeling using artificial neural network and correlation, Int. Commun. Heat. Mass. Transf. 66 (2015) 100–104.

[253] D. Madhesh, R. Parameshwaran, S. Kalaiselvam, Experimental investigation on convective heat transfer and rheological characteristics of Cu–TiO_2 hybrid nanofluids, Exp. Therm. Fluid Sci. 52 (2014) 104–115.

[254] M. Nabil, W. Azmi, K.A. Hamid, R. Mamat, F.Y. Hagos, An experimental study on the thermal conductivity and dynamic viscosity of TiO_2-SiO_2 nanofluids in water: ethylene glycol mixture, Int. Commun. Heat. Mass. Transf. 86 (2017) 181–189.

[255] N. Nikkam, M. Saleemi, M. Behi, R. Khodabandeh, M.S. Toprak, Experimental investigation on the effect of SiO_2 secondary phase on thermophysical properties of SiC nanofluids, Int. Commun. Heat. Mass. Transf. 87 (2017) 164–168.

[256] M. Chopkar, P.K. Das, I. Manna, Synthesis and characterization of nanofluid for advanced heat transfer applications, Scr. Materialia 55 (2006) 549–552.

[257] M. Karami, M. Akhavan-Bahabadi, S. Delfani, M. Raisee, Experimental investigation of CuO nanofluid-based direct absorption solar collector for residential applications, Renew. Sustain. Energy Rev. 52 (2015) 793–801.

[258] M. Sahooli, S. Sabbaghi, M. Shariaty Niassar, Preparation of CuO/water nanofluids using polyvinylpyrolidone and a survey on its stability and thermal conductivity, Int. J. Nanosci. Nanotechnol. 8 (2012) 27–34.
[259] S. Amalraj, P.A. Michael, Synthesis and characterization of Al_2O_3 and CuO nanoparticles into nanofluids for solar panel applications, Results Phys. 15 (2019) 102797.
[260] P. Selvakumar, S. Suresh, Convective performance of CuO/water nanofluid in an electronic heat sink, Exp. Therm. Fluid Sci. 40 (2012) 57–63.
[261] V.M. Krishna, M.S. Kumar, O. Mahesh, P.S. Kumar, Numerical investigation of heat transfer and pressure drop for cooling of microchannel heat sink using MWCNT-CuO Water hybrid nanofluid with different mixture ratio, Mater. Today: Proc. 42 (2021) 969–974.
[262] B. Rimbault, C.T. Nguyen, N. Galanis, Experimental investigation of CuO–water nanofluid flow and heat transfer inside a microchannel heat sink, Int. J. Therm. Sci. 84 (2014) 275–292.
[263] S.E. Ghasemi, A. Ranjbar, M. Hosseini, Numerical study on effect of CuO-water nanofluid on cooling performance of two different cross-sectional heat sinks, Adv. Powder Technol. 28 (2017) 1495–1504.

CHAPTER

Aluminum oxide-based nano-fluids

3

Maryam Masjedi-Arani

Institute of Nano Science and Nano Technology, University of Kashan, Kashan, Iran

3.1 General introduction

For the first time, in 1995, Choi applied the word nanofluids [1]. The nanofluid consists of a suspension of nanosized particles with dimensions of about 1–100 nm in a liquid fluid.

A new investigation has tried to alter the features of background fluids by adding nanosized resources to develop different application performances. Nanoparticles with a high surface-to-volume ratio improve the nanofluids performance. There are three reasons for better implementation of the nanofluids than the background fluids: (1) advanced fundamental features of the nanosamples, (2) the movement of nanostructures in the foundation fluid, and (3) the other sample fluid relations, including the development of an interfacial layer. Size, morphology, concentration, and uniformity of nanoparticles are important factors in the determination of nanofluids properties.

According to the category of utilized nanoparticles, dissimilar nanofluids can be distributed into four assemblies:

- Metal oxide nanofluids (CuO, silica, Al_2O_3, titanium dioxide, zinc oxide, Fe_3O_4, Fe_2O_3, and magnesium oxide) [2–4].
- Metal nanofluids (Au, Ag, Cu, and Fe) [5–7].
- Alloy nanofluids [8–10].
- Carbon-based nanofluids (SiC, TiC, diamond, SWCNT, MWCNT, and graphite) [11–13].

Also, water, refrigerant, ethylene glycol, and engine oil are common fluids that are utilized as foundation liquids.

This chapter discusses about synthesis methods and applications of Al_2O_3-based nanofluids. Also, Al_2O_3 hybrid nanofluids were reported. The usages of Al_2O_3 nanoparticle-based fluids in renewable energy devices, machining industries, cooling of electronics, and different components of automobiles were considered. Despite some widespread research on the use of nanofluids in various sets, there is no recent complete manuscript on the Al_2O_3-based nanofluid applications and synthesis methods. In the current chapter, first, an outline of Al_2O_3

Nanofluids. DOI: https://doi.org/10.1016/B978-0-443-13625-2.00003-6

nanofluids and their preparation approaches is given. Following, the different applications of Al_2O_3 nanofluids and hybrid-based nanofluids were studied.

3.2 Al_2O_3-based nanofluids

Nanoparticles compared to bulk materials have favorable properties owing to their great surface-to-volume ratio. Owing to the intense effect of production approaches on the morphology and dimension of nanoparticles, the preparation path of nanosamples is significant. A variation of probable methods, including hydrothermal [14], sol-gel [15], sonochemical [16,17], solid-state [18], and precipitation [19] exist for the synthesis of uniform nanoparticles. Extensive research has been carried out for synthesis of Al_2O_3 nanostructures *via* different approaches of precipitation [20,21], hydrothermal [22–24], sol-gel [25,26], combustion [27,28], microwave [29,30], mechanochemical [31,32], and wet chemical methods [33] that various morphologies have obtained such as nanoparticles, nanoplates, nanoflakes, and nanorods [34]. Zhu et al. synthesized α-Al_2O_3 nanostructures at low temperatures *via* ball-milling of aluminum hydroxide [35].

Wang et al. fabricated the Al_2O_3 nanoparticles by simple homogeneous precipitation route through aluminum sulfate ($Al_2(SO_4)_3 0.18H_2O$) and aluminum nitrate ($Al(NO_3)_3 0.9H_2O$) [36]. The particle size of nanosamples was controlled with changing of aluminum sulfate:aluminum nitrate molar ratio.

Based on the previously reported studies, adding nanomaterials to the background fluids results in increasing the viscosity, density, and thermal conductivity, while reducing the specific heat [37–39] (Table 3.1).

3.2.1 Al_2O_3-hybrid nanofluids

In hybrid nanofluids, there are two or more nanoparticles simultaneously in a foundation fluid. Hybrid nanofluids have better properties than nanofluids [53,54].

In an experimental education, Taherialekouhi et al. [55] examined the thermal conductivity and stability of H_2O/graphene oxide/Al_2O_3 nanoparticles as a refrigeration hybrid nanofluid at a temperature range of 25°C–50°C and volume fractions of 0.1%–1% (Fig. 3.1). The outcomes designated that growing the temperature and the Al_2O_3 volume fraction improved the thermal conductivity of the nanosized fluid up to 33.9%.

In experimental and calculated research, Kannaiyan et al. [56] studied the thermophysical features of Al_2O_3/CuO nanoparticles dispersed in H_2O/ethylene glycol. This hybrid-based nanofluid presented an improvement of approximately 45% in its thermal conductivity.

In 2019, thermophysical properties and fuzzy models of $Al_2O_3/TiO_2/H_2O$/ethylene glycol hybrid nanofluid were considered [57]. The influence of surfactant,

Table 3.1 Different Al_2O_3 nanofluids.

Nanofluids	Concentration %	Synthesis method	Application	References
Al_2O_3/water/ethylene glycol	0.5%–2%	Two-step	Thermal conductivity	[40]
Al_2O_3/water	0.3%–2%	–	Heat exchanger	[41]
Al_2O_3/water	0%–4%	–	Plate heat exchanger	[42]
Al_2O_3/water/bioglycol	0.5%–2%	Two-step	Thermal and electrical conductivity	[43]
Al_2O_3/water	0%–0.6%	–	Heat transfer	[44]
Al_2O_3/water	0.05%	–	Solar panel	[45]
Al_2O_3/ethylene glycol	5%	–	Thermal conductivity	[46]
Al_2O_3/water	3%–50%	Two-step	Heat transfer	[47]
Al_2O_3/ethylene glycol	0.05%–0.1%	Two-step	Flat plate solar collector	[48]
Al_2O_3/water				
Al_2O_3/oil	0%–0.03%	–	Parabolic trough solar collector	[49]
Al_2O_3/water	4%	–	Cooling of lithium-ion batteries	[50]
Al_2O_3/water/ethylene glycol	0.01%–20%	–	Plate heat exchanger	[51]
Al_2O_3/water/ethylene glycol	–	Two-step	Heat transfer	[52]

temperature, and volume fraction was studied on the stability, density, viscosity, and thermal conductivity of hybrid nanofluid. The density and viscosity decreased through the increment of temperature, and the heat conductivity improved by the intensification in temperature.

Aparna et al. explored the improvement of the thermal transmission rate of Al_2O_3/Ag hybrid nanofluid [58]. Polyvinylpyrrolidone (PVP) was applied to stabilize the Al_2O_3/Ag nanoparticles in H_2O (Fig. 3.2). The results specify that Al_2O_3/Ag hybrid nanofluids exhibit developed thermal conductivity improvement compared to Al_2O_3 nanofluids alone (Table 3.2).

3.2.2 Synthesis methods of nanofluids

Many properties of nanofluids such as thermophysical behavior and stability depend on their synthesis methods [74]. A wide range of nanomaterials such as carbon nanostructures, metal nanoparticles, and metal oxide nanoparticles are

FIGURE 3.1

SEM micrographs of nanoparticles [55].

Reprinted with permission from R. Taherialekouhi, S. Rasouli, A. Khosravi, An experimental study on stability and thermal conductivity of water-graphene oxide/aluminum oxide nanoparticles as a cooling hybrid nanofluid, Int. J. Heat. Mass. Transf. 145 (2019) 118751, Copyright 2020 Elsevier.

utilized for the synthesis of nanofluids. Because of the Van der Waals interactions of nanomaterials, nanofluids are not stable, and their synthesis routes have special importance. To gain stable nanofluids, various methods are utilized [75], such as surfactant addition [76], pH control [77], ultrasonic agitation [78], functionalization [79], magnetic stirring [75], and high-pressure homogenization [80] (Fig. 3.3; Table 3.3).

Many techniques are utilized to evaluate nanofluids durability consisting of (1) sedimentation/centrifugation, (2) zeta potential evaluation, (3) spectral absorbance and transmittance calculation, (4) transmission electron microscopy, and (4) dynamic light scattering.

There are two main methods for the preparation of nanofluids, including one-step and two-step techniques [92]. In the one-step synthesis route, nanomaterials are prepared and circulated in a distinct step in the base fluid, so the agglomeration of nanostructures in single-step route is less, but it is not appropriate for mass production. In summary, one-step production technique of nanofluids involves the in situ creation and distribution of nanoparticles in the foundation fluid simultaneously, without the need for an intermediate particle drying step that limits the probability of particle agglomeration. Also, there is superior control over nanoparticle dimension and form during preparation. The disadvantages of the single-step route for nanofluid synthesis are nanofluid preparation in low quantity, expensive method, and also the unreacted starting materials residual can make problem during experiment.

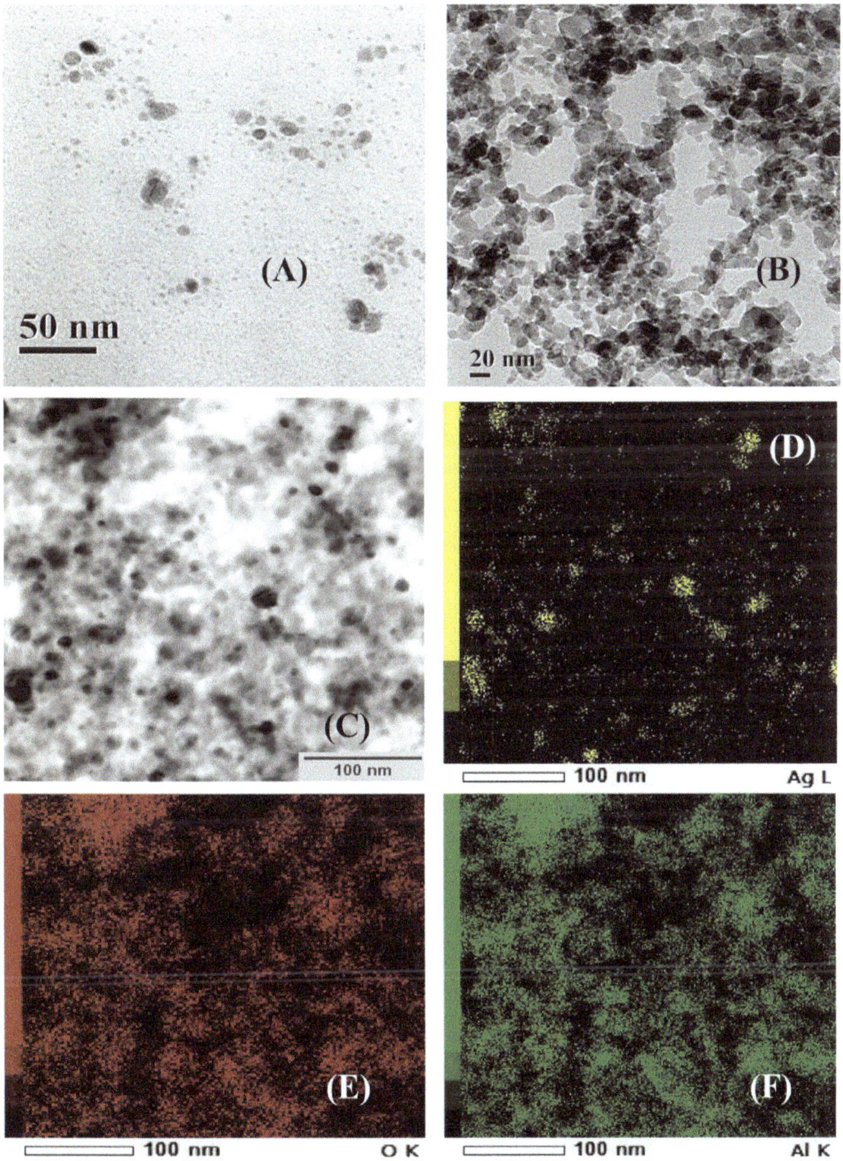

FIGURE 3.2

(A) TEM bright field image of PVP-coated (A) Ag nanoparticles, (B) Al_2O_3 nanoparticles, (C) Al_2O_3/Ag hybrid nanofluid, and (D–F) elements Ag, O, and Al map scan of Al_2O_3/Ag hybrid nanofluid in image (C) [58].

Reprinted with permission from Z. Aparna, M. Michael, S. Pabi, S. Ghosh, Thermal conductivity of aqueous Al_2O_3/Ag hybrid nanofluid at different temperatures and volume concentrations: an experimental investigation and development of new correlation function, Powder Technol. 343 (2019) 714–722, Copyright 2019 Elsevier.

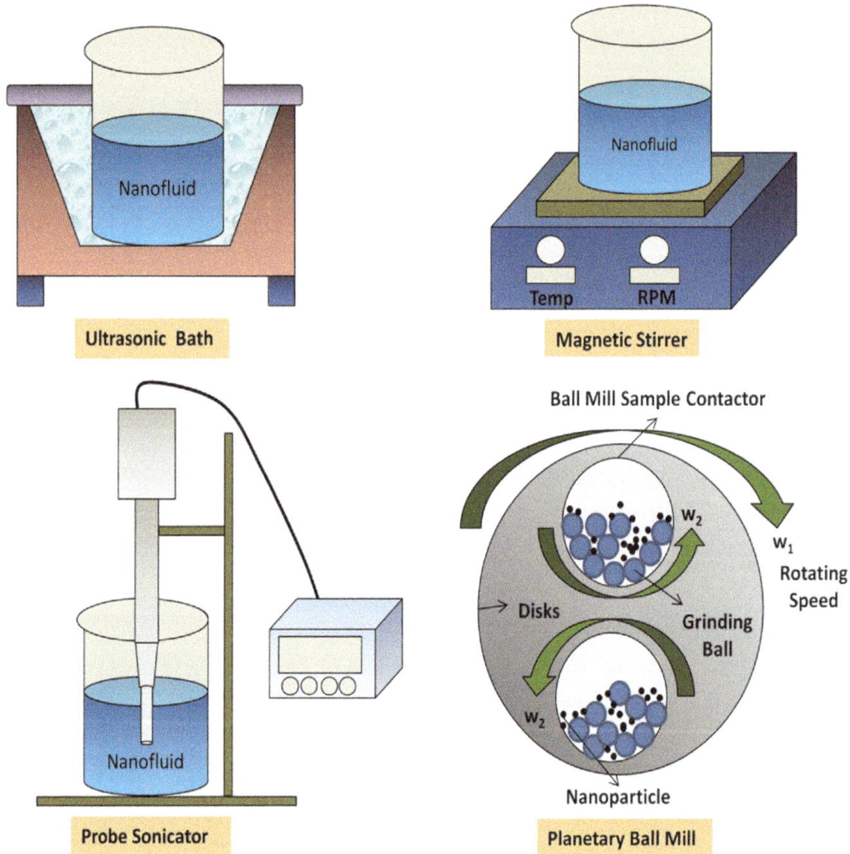

FIGURE 3.3

Schematic of different types of equipment for nanofluid stabilization via mechanical route [75].

Reprinted with permission from S. Chakraborty, P.K. Panigrahi, Stability of nanofluid: a review, Appl. Therm. Eng. 174 (2020) 115259, Copyright 2020 Elsevier.

In the two-step technique, nanoparticles are first created in the system of nanopowder, and at that time, the produced nanomaterials are distributed in fluids by agitation, stirring, or ultrasonication methods [93,94]. Two-step technique is an ideal model for mass fabrication. In brief, the two-step route includes two individual phases of nanoparticle preparation and dispersion in base fluid by means of physical (ultrasonic waves, magnetic stirrer, and great pressure homogenizer) and chemical (steric, electrostatic, and electrosteric) ways. The advantages of the two-step method are production on a large scale and cost-effectiveness. The weaknesses are more agglomeration of synthesized nanoparticles, frequent usage

Table 3.2 Properties and applications of different Al_2O_3 hybrid nanofluids.

Hybrid nanoparticles	Base fluid	Concentration	Application	References
Al_2O_3/Cu	Ethylene glycol	0.125–2.0	Thermal conductivity	[59]
Al_2O_3/CuO	Ethylene glycol/Water	0.05–0.2	Thermophysical properties	[56]
Al_2O_3/CuO	Ethylene glycol/Water	0.5–2.0	Flat plate solar collector	[60]
Al_2O_3/SiO_2	Water	–	–	[61]
Al_2O_3/ZnO	Water	0.1–0.8	Hydrate formation	[62]
Al_2O_3/ graphene oxide	Water	0.1–1.0	Thermal conductivity	[55]
Al_2O_3/TiO_2	Ethylene glycol/Water	0.02–0.1	Thermophysical properties	[63]
$Al_2O_3/TiO_2/SiO_2$	Water	–	Thermophysical properties	[64]
Al_2O_3/Fe_2O_3	Ethylene glycol/Water	–	Viscosity and rheological behavior	[65]
Al_2O_3/Cu	Water	–	Heat transfer	[66]
Al_2O_3/CNT	Water	0.05 –0.1	Thermophysical properties	[67]
Al_2O_3/MWCNT	Thermal oil	0.125–1.5	Cooling fluid	[68]
Al_2O_3/Ag	Water	0.005–0.1	Heat transfer	[38]
Cu/ $Al_2O_3/C_3H_8O_2$	Propylene glycol	–	Solar collector optimization	[69]
$Al_2O_3/ZnO/Fe_3O_4$	Water	–	Solar collector	[70]
Al_2O_3/MWCNT	Engine oil	0.13–1	Tube heat exchanger	[71]
Al_2O_3/AlN	Water	1–6	Heat transfer	[72]
Al_2O_3/TiO_2	Water	2	Photovoltaic performance	[73]

of physical or chemical methods to stabilize, and finally, limited control over particle size and form during production.

Wang et al. synthesized reduced graphene oxide-Al_2O_3 nanocomposite (rGO-Al_2O_3) *via* the hydrothermal method [95]. They utilized rGO-Al_2O_3 nanofluid as a lubricant with superior dispersion stability.

Mahirah Bahari et al. prepared Al_2O_3–Silica/H_2O hybrid nanofluids and considered the effect of different surfactants on their distribution and durability [61]. Aluminum oxide and silica nanoparticles were manufactured by the sol-gel route, and ultrasonic radiation was applied to distribute the nanopowders into the H_2O

Table 3.3 various stabilization methods of Al_2O_3-based nanofluids.

Nanofluids	Stabilization method	Synthesis route	References
Al_2O_3/ethylene glycol/water	Ultrasonic radiation (40 kHz, 490 W)	Two-step	[81]
Al_2O_3/SiO_2/water	Ultrasonic radiation (60 min)	Two-step	[82]
Al_2O_3/TiO_2/water	Different surfactants (CTAB, SDBS, acetic acid, and SDS)	–	[83]
Al_2O_3/water	Surfactant (CTAB and SDBS) pH (HCl and NaOH)	Two-step	[84]
Al_2O_3/water	Different surfactants (CTAB, SDBS, and Tween 60)	Two-step	[85]
Al_2O_3/TiO_2/ therminol-55	Ball mill	Two-step	[86]
Al_2O_3/water	Magnetic field and thermal radiation	Two-step	[87]
MWCNT/Al_2O_3/ ZnO/water	Magnetic field, ultrasonic radiation, and homogenizer	–	[88]
Cu/Al_2O_3/water	No stabilizing agent	Two-step	[89]
Al_2O_3/CuO/water	High-speed mechanical stirring SDBS surfactant Adjusting pH Ultrasonic vibration	Two-step	[90]
Al_2O_3/water	CTAB and SDS surfactant	Two-step	[91]

to create nanofluids. The effect of different anionic (SDS), cationic (CTAB), and nonionic (PVP) surfactants was investigated on the durability, electrical conductivity, and viscosity of the obtained nanofluid. The surfactant type played a significant role in diffusing the nanomaterials and creating stable nanofluids. SDS has a positive effect on the stability of the nanofluids. Also, nanofluids comprising cationic and anionic surfactants had an advanced growth in conductivity.

Tahat et al. fabricated the alumina/CuO hybrid nanostructures distributed in ethylene glycol and H_2O mixture [60]. Al_2O_3/CuO hybrid nanoparticles were synthesized by a thermochemical synthesis method. Al_2O_3/CuO hybrid nanofluids using concentrations of 0.5%–2% were then arranged by mixing the manufactured nanopowders in a combination of H_2O (75%) and ethylene glycol (25%) using the two-step method. Al_2O_3/CuO nanomaterials were stabilized in combined foundation fluid by magnetic stirring and ultrasonic irradiation. The effect of the mentioned hybrid nanomaterial-based fluids on the heat proficiency of plane plate solar gatherers was considered. The results of this paper show that thermal conductivity, density, and viscosity rise with nanopowder concentration. The proficiency of the gatherer was developed by 45% by growing the nanomaterials' weight fraction.

Mahbubul et al. prepared the alumina (Al_2O_3)–water nanofluids of 0.5 vol.% by ultrasonication (up to 5 hours) and stability of 30 days [96]. Al_2O_3-water nanofluids were fabricated *via* a two-step route. The result of the current work shows that increasing ultrasonication time reduced the sedimentation rate and increased stability of nanofluids and thermal conductivity, while viscosity decreased.

Khdher et al. synthesized Al_2O_3 nanofluids by bioglycol with volume concentrations of 0.1–1.0 and investigated their heat conductivity, dispersion durability, and electrical conductivity of them [97]. The heat conductivity enhancement and thermo-electrical conductivity of the obtained Al_2O_3 nanofluids were 17% and 9.5 at 30 C, respectively.

3.3 Applications

The advantages of nanofluids over base liquids convinced scientists to consume various nanofluids in industrialized presentations. The researchers considered the influence of nanofluids in different industries, such as pharmaceuticals, medicine, food, and cosmetics.

3.3.1 Renewable energy systems

The requests for nanomaterial-based liquids in renewable energy arrangements have attracted growing consideration in recent years. Nanofluids can be useful for the improvement of the usage of power firms, means of transportation, electronic tools, bio-photovoltaic sets, waste thermal recovery organizations, thermionic energy creators, and supplementary renewable energy sets [98–100].

3.3.1.1 Solar heat gatherers

The solar gatherer is a heat apparatus that performs solar treatment and creates beneficial thermal fabrication in different temperatures. There are various solar heat accumulators such as uniform plate collectors, linear Fresnel reflectors, gatherers with evacuated cylinders, concentrating solar dishes, multiple parabolic concentrators, and parabolic troughs. The employment of nanomaterial-based liquids in solar heat components is a favorable impression for growing their proficiency [101]. The nanofluids are tested in different collector categories to proliferate the valuable output and to develop heat efficacy. The most significant advantages of applying nanofluids are the growth of the fluid heat conductivity and the heat convection factor between the liquid and the absorber [102].

Kim et al. utilized an Al_2O_3 nanoparticle-based liquid in a U-tube solar gatherer and investigated different concentrations and particle sizes of Al_2O_3 nanofluid [103]. The heat conductivity of the Al_2O_3 nanofluid augmented with a rise in its concentration and decreased in its nanoparticle size (Fig. 3.4). The solar gatherer with 1.0 vol.% alumina nanofluid of 20 nm presented the maximum efficacy, 24.1% greater than that of the solar gatherer with H_2O.

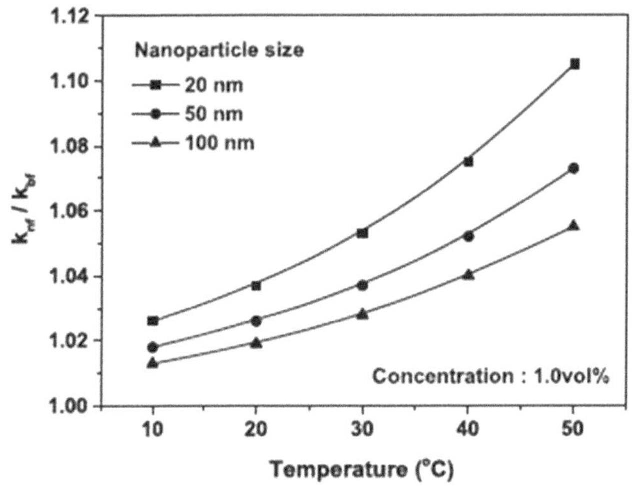

FIGURE 3.4

Thermal conductivity ratio as a function of temperature for different nanoparticle sizes (nanofluid concentration: 1.0 vol.%) [103].

Reprinted with permission from H. Kim, J. Kim, H. Cho, Experimental study on performance improvement of U-tube solar collector depending on nanoparticle size and concentration of Al_2O_3 nanofluid, Energy 118 (2017) 1304–1312, Copyright 2017 Elsevier.

Ghaderian et al. considered the effect of Al_2O_3/H_2O nanofluid on the heat efficacy of an all-glass inert flow evacuated tube solar gatherer [104]. The highest proficiency was obtained to be 58% for 0.06 vol.% of nanofluids with an Al_2O_3 particle size of 40 nm.

Tong et al. investigated the activity of the flat-plate solar gatherer using Al_2O_3/H_2O and CuO/H_2O nanoparticle-based fluid [105]. Alumina nanoparticles (20 nm) and CuO nanopowders (40 nm) were used. The highest efficiencies of the flat plate solar gatherer for Al_2O_3/water and CuO/water fluid nanosystems have been compared. The highest efficiencies of the Al_2O_3 and CuO fluid nanosystems were 77.5% and 73.9%, respectively. The use of the Al_2O_3 fluid nanosystems in the solar system improved the efficiency by 4.9% more than the usage of the CuO nanofluid. When the 1.0 vol.%-Al_2O_3 nanofluid and 0.5 vol.%-CuO nanofluid were utilized, the highest efficacy was amplified by 21.9% and 16.2%, individually, associated with the case of H_2O. Generally, the thermal achievement factor of the alumina nanofluid was greater than that of the CuO fluid nanosystems, and the thermal loss constant of the alumina nanofluid was inferior to that of the CuO nanofluid (Table 3.4).

3.3.1.2 Solar photovoltaic thermic structures

Solar photovoltaic thermic organizations are advantageous in requests where both thermal and electricity creation are required, at the same time. The system should

Table 3.4 different Al_2O_3 nanofluids used in solar collector applications.

Nanofluids	Application	Concentration %	Synthesis route	References
Al_2O_3/CuO/water/thermia oil	Various solar thermal collectors	–	–	[101]
Al_2O_3-diathermic oil	Solar energy systems	0.3, 0.7, and 1.0	Two-step	[106]
Al_2O_3/water	Solar thermal energy	0.02, 0.1, and 0.5	–	[107]
Al_2O_3/water	Solar parabolic collector	0.05–0.5	–	[108]
Al_2O_3/CuO/water/ethylene glycol	Direct absorption solar thermal energy	–	Two-step	[109]
synthetic oil/Cu/Al_2O_3	Finned parabolic trough solar collector	0.02	–	[110]
Al_2O_3/water	Solar flat plate collector	0.1, 0.2, and 0.3	–	[111]
Al_2O_3/graphene oxide/water	Solar energy application	0.1–1	–	[112]
Al_2O_3/water	Solar parabolic trough collector	0.5	–	[113]
Al_2O_3/MWCNT/water	Solar flat plate collector	0.005, 0.01, 0.025, and 0.5	Two-step	[114]
Al_2O_3/water	Solar flat plate collector	0.1–0.3	–	[115]
Al_2O_3/water	U-tube solar collector	1.0, 2.0, 3.0, and 4.0	–	[116]
Al_2O_3/TiO_2/therminol-55	Concentrating solar collectors	0.5	Two-step	[117]
Al_2O_3/soybean oil	Parabolic trough solar collector	–	–	[118]

involve a heat absorber and a photovoltaic cell, with the cells being chilled by a nanoparticle-based fluid. The quantity of solar treatment that is not transformed in a straight line to electricity in the sets is taken in the absorber. The nanofluids play a significant role in this system because the nanofluid can cool successfully the cell with more useful created heat and electricity. Incorporating nanofluids in photovoltaic-thermal systems is noteworthy because any development in thermal abstraction is supplemented by a secondary advantage in electrical proficiency. Therefore, there is extensive research on the usage of nanofluids in these systems.

Hader et al. studied the implementation of hybrid photovoltaic heat sets employing Al_2O_3/H_2O nanofluid [119]. Simulation outcomes display that by growing the concentration of the Al_2O_3 nanoparticles, improved general efficacy is attained but with a consequence of augmented friction factor.

Elminshawy et al. examined the photovoltaic systems utilizing Al_2O_3 nanofluid [120]. Ultrasonic stimulation and the employment of different surfactants are the most proper approaches for decreasing nanoparticle accumulation and growing dispersion durability. The results indicate that electrical yield and electrical and thermal proficiencies of the photovoltaic system of Al_2O_3-nanofluid improved by 10.40%, 10.24%, and 35.79%, respectively, compared to only water. Flowing the 0.7% Al_2O_3-H_2O nanofluid over the C-PV system develops the module's thermal proficiency by 89.47% to H_2O base fluid due to the circulating Al_2O_3- H_2O-nanofluid straight and broader interaction area with the C-PV system's backside (Fig. 3.5). Agreeing to results, rising the nanofluid flow rate can increase cooling effectiveness. The exergy proficiencies of 0.7% Al_2O_3-nanofluid/C-PV and H_2O/C-PV arrangements are 55.78% and 18.10%, respectively.

Zamen et al. considered the use of Al_2O_3/H_2O nanoparticle-based liquid as the coolant in photovoltaic heat sets [121]. Diverse concentrations of Al_2O_3 fluid nanosamples (0.05–0.5 wt.%) were arranged. To increase the durability of created fluids nanosamples, Triton X-100 as a surfactant is utilized. According to the attained consequences, the addition of Al_2O_3 nanoparticles to the pure H_2O can remarkably raise the efficiency of the photovoltaic/thermal apparatus. The efficiency of 126.71% and 7.38% development in terms of heat and electrical efficiency extended by the presentation of 0.5% Al_2O_3 nanofluid related to pure H_2O.

Adun et al. studied the activity of photovoltaic heat sets applying alumina-ZnO-Fe_3O_4/water ternary nanofluid [122]. The alumina-ZnO-Fe_3O_4/H_2O ternary nanofluid has been revealed to have enhanced thermophysical and rheological properties as associated with predictable nanofluids. The ternary nanofluid was prepared via a two-step method. The best electrical and thermal proficiency calculated was 13.8% and 59.4%, respectively (Table 3.5).

3.3.1.3 Geothermal energy

Geothermal energy as a renewable energy source is the heat storage inside the ground that can be utilized in numerous thermic usages, including space warming and chilling, power production, hot water fabrication, region, and greenhouse warming [127]. The nanofluids as thermal transmission liquids are utilized in a closed-loop geothermal exchanger. The nanoparticle-based liquids can improve the heat transmission factors inside the movement and so a more operative heat transfer can be attained. In geothermal energy submissions, the use of nanoliquids can lead to advanced thermal recovery from the earth, higher valuable heat quantities, less length of the borehole, and less expensive investment [128].

Jamshidi and Mosaffa investigated the application of Al_2O_3 nano-liquids in a finned helical geothermic exchanger. They found an improvement of approximately 10% in the thermal transference with a 0.5% volumetric concentration of Al_2O_3 in the H_2O [129].

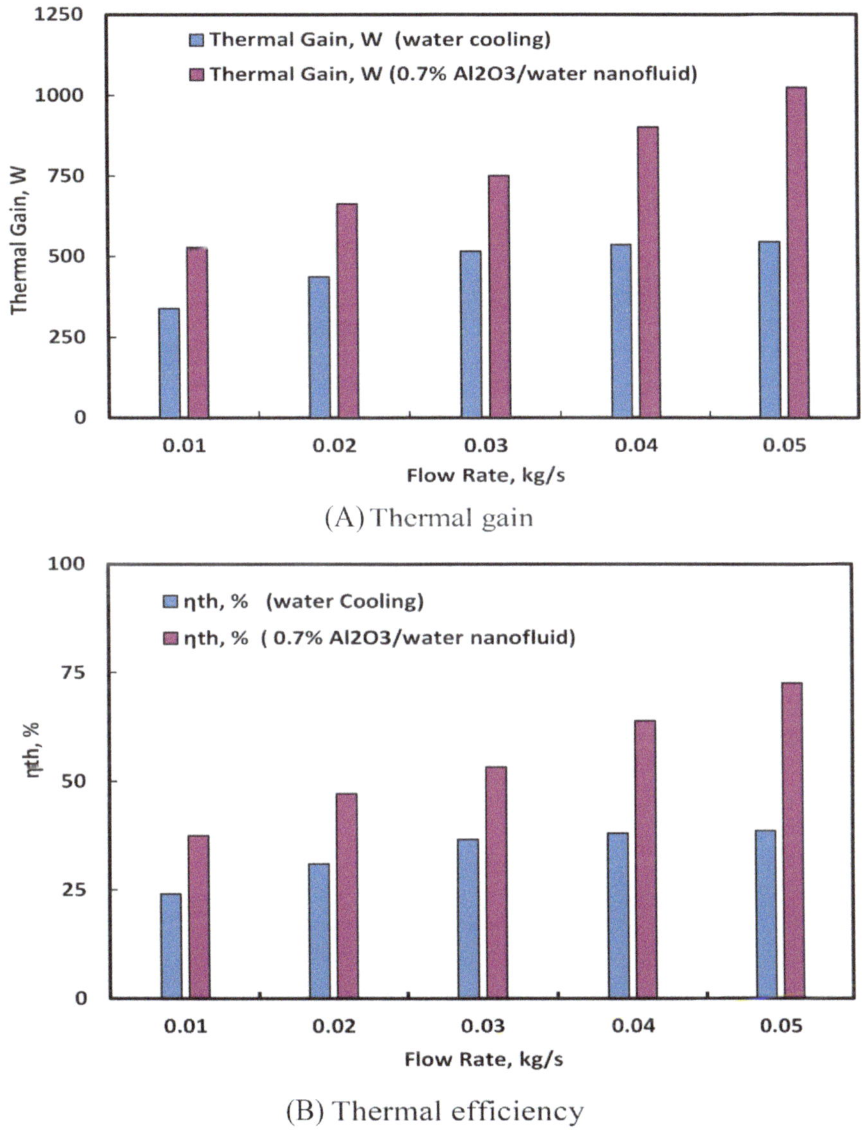

FIGURE 3.5

Thermal performance of C-PV module versus nanofluid mass flow rates presented in; (A) thermal gain and (B) thermal efficiency [120].

Reprinted with permission from N. Elminshawy, A. Elminshawy, A. Osama, M. Bassyouni, M. Arıcı, Experimental performance analysis of enhanced concentrated photovoltaic utilizing various mass flow rates of Al_2O_3-nanofluid: energy, exergy, and exergoeconomic study, Sustain. Energy Technol. Assess. 53 (2022) 102723, Copyright 2022 Elsevier.

Table 3.5 Different Al_2O_3-based nanofluids utilized in photovoltaic-thermal system.

Nanofluid	Concentration %	Synthesis route	References
Al_2O_3/CuO/MWCNT/water	0.5, 1, 2.5, and 5	–	[123]
Al_2O_3/water	–	–	[124]
Cu/Al_2O_3/water	–	Two-step	[125]
Al_2O_3/water	0.2	–	[126]

3.3.1.4 Fuel cells

Due to the shortage of fossil energy sources and environmental pollution, the application of fuel cells has received significant attention, recently. Designing a well-organized chilling organization for fuel cells can improve their efficiency and period, considerably. Considering the economy, thermophysical properties, and stability, Al_2O_3 nanoliquids are the greatest candidates. Zakaria et al. considered the behavior of Al_2O_3/water/ethylene glycol nanofluid in a distinct chilling plate of proton exchange membrane fuel cell [130]. Thermal assignment was developed up to 13.87% for 0.5 vol.% Al_2O_3 as associated with the foundation liquid. The progressive thermic outcomes inferred that Al_2O_3 nanofluid is a favorable choice for future requests in fuel cell thermic controlling.

Zeiny et al. considered the influence of Al_2O_3/water nanoliquids on the cooling polymer electrolyte membrane fuel cells [131]. The result of this research displays the highest decrease of 7.9% in the thermic usefulness and the highest growth of 6% in the pumping power by the Al_2O_3/water nanofluids in its place of H_2O. This weakening is due to the decrease in the operative particular thermal capacity and growth in viscosity of the nanofluids.

The index of uniform temperature (IUT) as a meaning of Reynolds number falls while the molar ratio of the nanoliquid increases, giving an improved thermic routine. However, the IUT as a function of current rate and pumping influence increases as the concentration of the nanoparticle-based fluid increases, resulting in lower thermic activity (Fig. 3.6).

3.3.2 Machining industries

The benefits of using nanoliquids in different apparatus industries are increasing the lifetime of the device, decreasing the distortion produced by thermic tension, and removing the particles from the cutting zone. Nanofluids must have possessions to accomplish their responsibilities consisting of lubrication at low incision speeds, preservation of utilized sections at great slicing speeds, rollerblades, chilling and lubrication of drill edges, grinding rolls, cooling apparatus for repairing, and keeping the instrument from deterioration. A perfect cutting liquid must have extraordinary heat conductivity, and nanofluids have this feature [132,133].

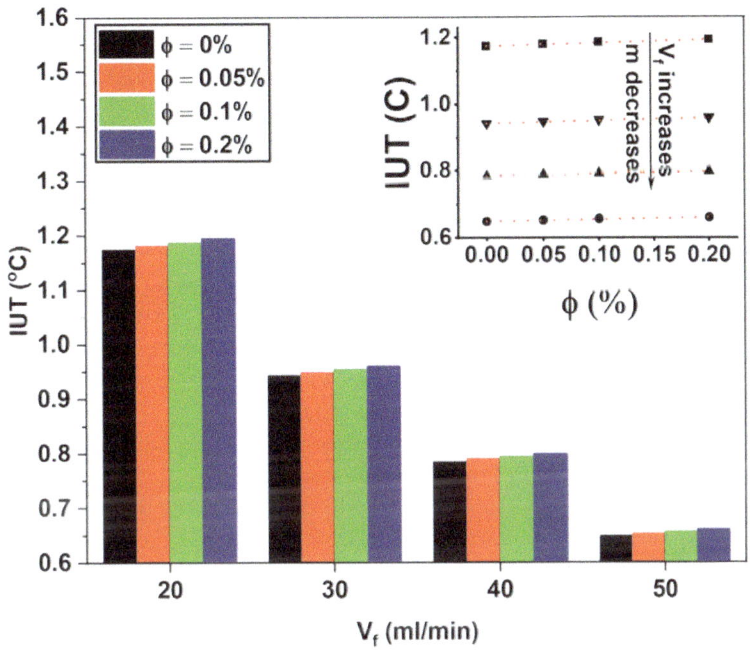

FIGURE 3.6

The index of uniform temperature (IUT) as a function of the flow rate of the coolant and as a function of the nanoparticles concentration (inset). Note that "m" is the slope of the trend lines (the dotted lines) [131].

Reprinted with permission from A. Zeiny, M.A. Al-Baghdadi, W.F. Arear, M.S. Ismail, Al_2O_3–H_2O nanofluids for cooling PEM fuel cells: a critical assessment, Int. J. Hydrog. Energy 47 (2022) 38823–38836, Copyright 2022 Elsevier.

Chaudhari et al. utilized Al_2O_3/CuO/water nanofluids at different volume percentage concentrations for thermic transfer application. The outcome displays that the thermic conductivity proportion is improved by 19.74% and 36.21% for Al_2O_3 and CuO nanoparticle-based fluids at 1 vol.% concentration, respectively. Likewise, the total viscosity percentage is enlarged by 29.77% and 48.71% [134].

3.3.2.1 Turning operations

Extensive research has been reported on the turning process of Al_2O_3 nanofluids. Zaman et al. described the utilization of hybrid alumina/MWCNT nanofluid in turning Ti-6Al-4V alloy based on the least quantity lubrication [135].

Sharma et al. studied the effect of Alumina nanostructure-based slicing liquid in turning of AISI 1040 steel in minimum quantity lubrication [136]. Al_2O_3 nanofluids play an energetic character by preserving and lubricating the tool-work portion boundary and eliminating imperfections from the cutting region. The result of the mentioned research reveals that the performance of Al_2O_3 nanofluid in terms of surface irregularity, instrument wear, slicing force, and piece morphology is well associated with dry devices, wet instruments with conservative cutting liquid, and MQL by predictable slicing fluid.

3.3.2.2 Grinding operations

Nanofluids exhibit developed heat transfer ability and lubricating possessions in machining, particularly in the grinding route. Attrition between rasper grains and work portions in grinding can straight affect grinding power, grinding roll uniform, and work part feature. Consequently, growing smooth effects by nanofluids is an operative technique to expand device enactment. Li et al. evaluated the effect of Al_2O_3/palm oil nanofluid in different concentrations for the external edge grinding of a nickel-based alloy in MQL method [137]. Results demonstrated that the Al_2O_3 nanofluid currents over the grinding wheel and work portion boundary reduce friction, significantly.

Zhang et al. considered lubricating features of minimal quantity lubrication grinding of alumina/SiC hybrid nanoliquid with dissimilar morphology and unit size [138]. According to the obtained data of this work, the greatest dimension proportion of the alumina/SiC mixed nanosized fluid was 30:70 that has the minimum specific standard grinding power (Fig. 3.7). Spherical Al_2O_3 nanoparticle with great filling capacity and suitable chemical durability were scattered by the MQL outlet and followed onto the grinding wheel workpiece boundary and filled the "furrows" on the rasping external edge that can improve the smooth effect. The SiC nanoparticles with nonuniform morphology and several boundaries could not attain the perfect "ball effect." SiC nanoparticles can drudge the work set and decrease the surface irregularity, efficiently (Fig. 3.8).

3.3.2.3 Drilling operations

The nanofluid minimal quantity lubrication procedure advances the device life and superficial features by decreasing the slicing hotness and confirming promising alterations in the drilling region. Pal et al. utilized the alumina/varied vegetable-oil-based cutting fluid in the drilling of AISI 321 stainless steel [139]. Results specified that drilling with Al_2O_3/vegetable-oil-based nanofluid accomplished superior act as associated with the dry, wet, and pure minimal quantity lubrication drilling. Insertion force, rotation, surface irregularity, and drill tip heat around decreased by 44%, 67%, 56%, and 26%, respectively (Fig. 3.9).

Subhedar et al. investigated the conclusion of alumina/slicing oil nanoassisted coolant for MQL drilling of SS 304 [140]. Different volume fractions of 0.3%, 0.8%, and 1% of Al_2O_3 nanoparticles were used in cutting fluid. The result displays that nanoassisted coolants intensify instrument life and decrease surface coarseness. Device life is perceived as the maximum for nanocutting liquid of 1% volume portion and attains the minimum external edge roughness of 2.01 μm at 1000 rpm. By increasing the volume portion of the nanosized particle, the

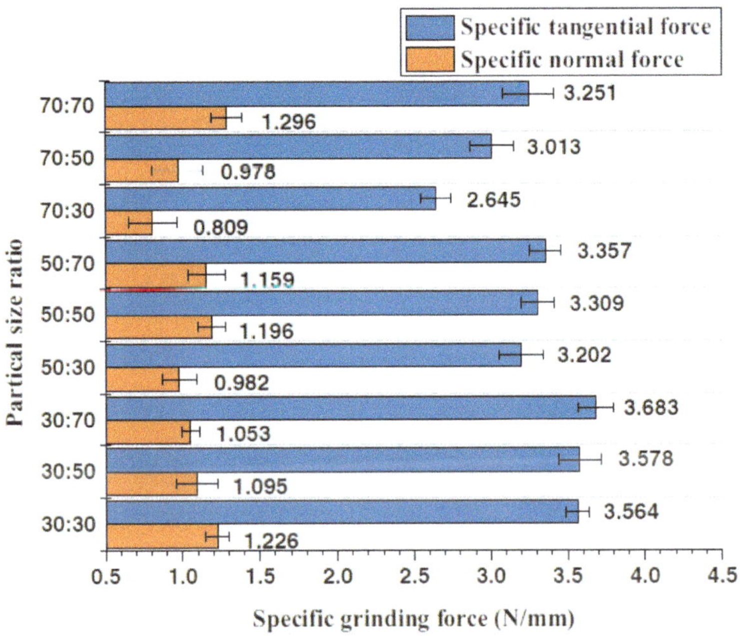

FIGURE 3.7

Specific grinding forces of different Al_2O_3/SiC particle size ratios [138].

Reprinted with permission from X. Zhang, C. Li, Y. Zhang, Y. Wang, B. Li, M. Yang, S. Guo, G. Liu, N. Zhang, Lubricating property of MQL grinding of Al_2O_3/SiC mixed nanofluid with different particle sizes and microtopography analysis by cross-correlation, Precis. Eng. 47 (2017) 532–545, Copyright 2017 Elsevier.

constant of friction falls due to the systematic action of the sphere-shaped nanomaterials.

3.3.2.4 Milling operations

Alumina nanoparticles at different concentrations of 0.5–1.5 vol.% were assessed for powdering of Hastelloy C276 alloy [141]. At Al_2O_3 concentration of 1 vol.% evidently obtained progress by up to 23% and 10% in instrument life, related to 0.5 and 1.5 vol.% concentration, individually.

Jamil et al. premeditated the performance of Al_2O_3/MWCNT nanofluid for milling of Ti-6Al-4V [142]. Results summary showed properly chosen minimal quantity lubrication factors refining the chilling/smooth presentation on condition that oil film and increasing the milling act processes.

3.3.3 Preservation of electronics

Because of the fast progress of machinery, contemporary electronic devices produce a massive amount of hotness that interrupts the ideal routine of the strategies

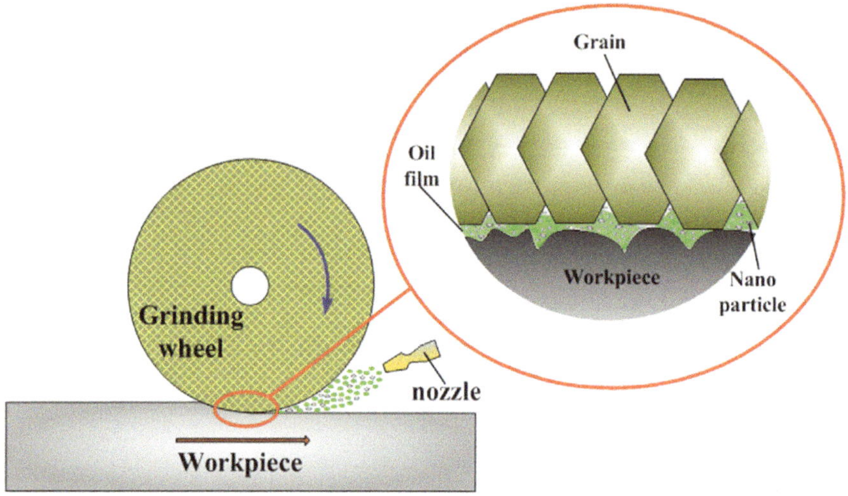

FIGURE 3.8

Antifriction mechanism of Al_2O_3/SiC mixed NPs in the grinding zone [138].

Reprinted with permission from X. Zhang, C. Li, Y. Zhang, Y. Wang, B. Li, M. Yang, S. Guo, G. Liu, N. Zhang, Lubricating property of MQL grinding of Al_2O_3/SiC mixed nanofluid with different particle sizes and microtopography analysis by cross-correlation, Precis. Eng. 47 (2017) 532–545, Copyright 2017 Elsevier.

and decreases the predictable lifecycle. Therefore, a proficient cooling structure is one of the most significant complications in planning automatic apparatuses. There have been many efforts to eliminate great hotness fluidity efficiently such as air freshening, liquid refrigerating, and two-phase chilling; one of the best methods is utilizing nanofluids [143,144].

Nguyen et al. utilized Al_2O_3/water nanofluid for heat transfer improvement in an electronic cooling arrangement [145]. Heat transfer coefficient rises as much as 40% associated with foundation fluid for 6.8% Al_2O_3 nanoparticle capacity.

In other research, Nada et al. used the Al_2O_3/CuO/water hybrid nanofluid for electronic cooling [146]. The existence of the Al_2O_3 and CuO nanoparticles provides a reduction in the thermic opposition and growth in the average Nusselt number compared with pure H_2O for all the liquid channels conformations with better performance of CuO nanoparticle compared with Al_2O_3 nanomaterial.

3.3.4 Oscillating heat pipes

Nanofluids are significantly utilized in oscillating thermic pipes due to increasing the hotness transport capability.

Qu et al. considered the thermic yield of an oscillating thermal pipe with alumina–H_2O nanoparticle-assisted fluids [147]. The obtained results presented

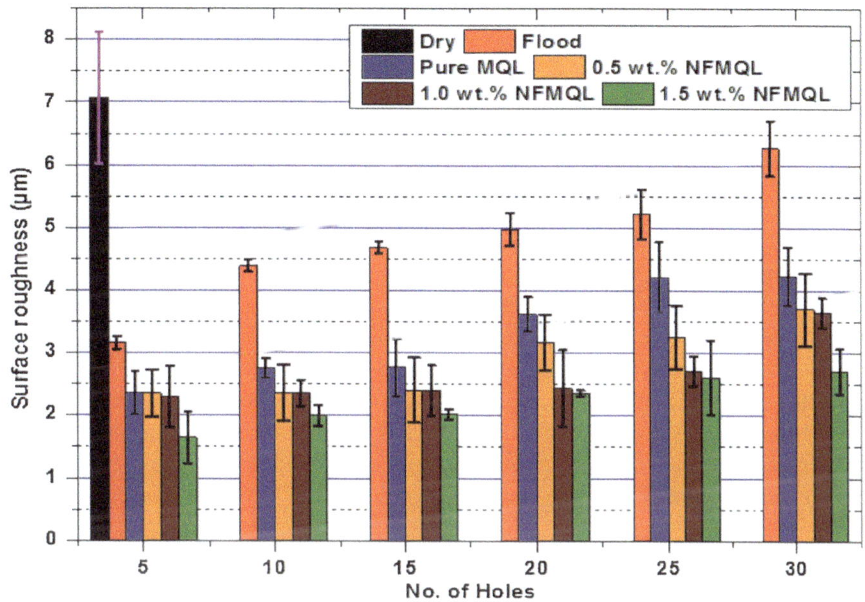

FIGURE 3.9

Variation in average surface roughness values of machined surface under different cooling-lubrication strategies (dry, flood, pure MQL, and nanofluid MQL) [139].

Reprinted with permission from A. Pal, S.S. Chatha, H.S. Sidhu, Performance evaluation of the minimum quantity lubrication with Al_2O_3-mixed vegetable-oil-based cutting fluid in drilling of AISI 321 stainless steel, J. Manuf. Process. 66 (2021) 238–249, Copyright 2021 Elsevier.

that the Al_2O_3 nanofluids considerably enhanced the thermic routine of the oscillating hotness pipes, with an ideal quantity fraction of 0.9 wt.% for the highest temperature transmission development. Paralleled with H_2O, the greatest thermic resistance was reduced by 0.14 C/W.

Tanshen et al. studied the pressure circulation in an oscillating thermic pipe with Al_2O_3/MWCNTs/water nanofluid [148]. Al_2O_3/MWCNTs/water nanofluid at 60% filling proportion attains the maximum variation frequency and the minimum thermic resistance.

3.3.5 Cooling performance of different components of automobile

A decrease in power utilization is probable by refining the act of thermal conversation structures. Surrounded by the efforts for improvement of hotness transmission, the request nanofluids is more perceptible.

Moghaieb et al. utilized the Al_2O_3–water nanofluids for engine cooling [149]. The maximum thermic assignment constant is 78.67% reached at a 1% volume ratio of Al_2O_3 nanosized sample with widths range of (21−37) nm compared with H_2O.

Peyghambarzadeh et al. inspected the outcome of the improved preservation act of vehicle radiator with Al_2O_3/water nanofluid [150]. Diverse doping of Al_2O_3 nanostructures about 0.1−1 vol.% were considered. Using Al_2O_3 nanosized fluid with low doping can improve hotness transmission efficacy by up to 45%.

Salami Tijani et al. applied the Al_2O_3/CuO/H_2O/antirefrigerating nanofluid as a coolant for a machine radiator [151]. This combination is a favorable candidate for cooling car radiators due to its thermophysical properties. The addition of Al_2O_3 and CuO nanosized particle samples to the coolant liquid of the machine radiator improved the hotness transmission routine, with a growth in the thermal conduction, hotness transmission constant, and Nusselt number.

Choi et al. studied the thermal assignment act of a machine with Al_2O_3/ethylene glycol/H_2O nanosized liquids utilized to chill a 100 kW extraordinary power organization [152]. The thermal transfer ratio of the radiator for the 100 kW great power arrangement can be enhanced by Al_2O_3/ethylene glycol/H_2O nanofluids.

3.3.6 Domestic refrigerator

In recent years, extensive research has been done on refrigerant-based nanofluids such as CNT/water [153], TiO_2 [154], CuO [155], and Al_2O_3 [156]. Nanoparticles are one way to attain unlimited energy storage and subsequent emissions falls. The impact of nanosamples on the thermal transference properties of refrigerant-assisted nanosized liquids current boiling in a flat smooth tube, and offered an association for foretelling the hotness transfer act of refrigerant-based nanoliquids was considered [157].

Nikulin et al. reported the investigational research of the pool steaming route by isopropanol/alumina nanofluid [158]. The impact of alumina nanostructures on the thermal transfer factor and bubble leaving width through alcohol boiling was dignified.

Mahbubul et al. examined the influence of Al_2O_3 nanosized structures doping and heat on thermic conduction and viscosity of alumina/R141b nanorefrigerant [159]. The experiment was carried out at 0.5−2 vol.% doping ratio at a hotness of 5°C−20°C with an alumina unit dimension of 13 nm. According to the obtained results, the maximum thermic conductivity and viscosity are 1.626 and 179 times superior to the foundation fluid for 2% doping of Al_2O_3 nanoparticles, correspondingly.

In 2023, Bibin et al. considered the performance of Al_2O_3, TiO_2, and CuO nanoparticles in refrigeration systems [160]. By adding TiO_2 nanoparticle to the R134a cooler, the highest development in the constant of routine about 18.75% was obtained. Also, Al_2O_3/R134a proposals a 27% discount in authority treatment and associated to R134a/CuO, the Al_2O_3/R134a nano-cooler deals energy reserves that are more than five times larger and 80% more active.

3.4 Conclusion and outlook

In summary, Al_2O_3 nanofluids have been the emphasis of attention in contemporary years due to their unique characteristics. In this chapter of the book, the synthesis method and applications of Al_2O_3-based nanofluids are studied. Also, Al_2O_3 hybrid nanofluids were reported. There are two main methods for the preparation of nanofluids, including one-step and two-step techniques that have their own advantages and disadvantages. The utilizations of Al_2O_3 nanoliquids in renewable energy structures, machining industries, cooling of electronics, and different components of automobiles were considered. Renewable energy organizations include solar thermal gatherers, solar photovoltaic-thermal schemes, geothermal energies, and fuel cells. Al_2O_3 nanofluids have many usages in different industries due to favorable heat conductivity. In spite of some complete analysis papers on the use of nanofluids in various systems, there is no recent comprehensive review on the Al_2O_3-based nanoliquids presentations and synthesis methods that in this chapter have been explained about them.

References

[1] S.U. Choi, J.A. Eastman, Enhancing Thermal Conductivity of Fluids with Nanoparticles, Argonne National Lab.(ANL), Argonne, IL, 1995.
[2] K. Suganthi, K. Rajan, Metal oxide nanofluids: review of formulation, thermo-physical properties, mechanisms, and heat transfer performance, Renew. Sustain. Energy Rev. 76 (2017) 226–255.
[3] D. Saidina, M. Abdullah, M. Hussin, Metal oxide nanofluids in electronic cooling: a review, J. Mater. Sci.: Mater. Electron. 31 (2020) 4381–4398.
[4] A.K. Tiwari, N.S. Pandya, H. Shah, Z. Said, Experimental comparison of specific heat capacity of three different metal oxides with MWCNT/water-based hybrid nanofluids: proposing a new correlation, Appl. Nanosci. 13 (2020) 189–199.
[5] H.E. Patel, T. Sundararajan, S.K. Das, An experimental investigation into the thermal conductivity enhancement in oxide and metallic nanofluids, J. Nanopart. Res. 12 (2010) 1015–1031.
[6] K.K. Sarojini, S.V. Manoj, P.K. Singh, T. Pradeep, S.K. Das, Electrical conductivity of ceramic and metallic nanofluids, Colloids Surf. A: Physicochem. Eng. Asp. 417 (2013) 39–46.
[7] M. Muneeshwaran, G. Srinivasan, P. Muthukumar, C.-C. Wang, Role of hybrid-nanofluid in heat transfer enhancement—a review, Int. Commun. Heat. Mass. Transf. 125 (2021) 105341.
[8] C. RamReddy, H.L. Saran, Dual solutions and their stability analysis for inclined magnetohydrodynamics and Joule effects in Ti-alloy nanofluid: flow separation, Proc. Inst. Mech. Eng. Part E: J. Process. Mech. Eng. 236 (2022) 2558–2569.
[9] S. Peyghambarzadeh, S. Hashemabadi, A. Chabi, M. Salimi, Performance of water based CuO and Al_2O_3 nanofluids in a Cu–Be alloy heat sink with rectangular microchannels, Energy Convers. Manage. 86 (2014) 28–38.

[10] H.B. Kulkarni, M.M. Nadakatti, S.C. Kulkarni, R.M. Kulkarni, Investigations on effect of nanofluid based minimum quantity lubrication technique for surface milling of Al7075-T6 aerospace alloy, Mater. Today: Proc. 27 (2020) 251–256.
[11] A. Borode, N. Ahmed, P. Olubambi, A review of solar collectors using carbon-based nanofluids, J. Clean. Prod. 241 (2019) 118311.
[12] A.A. Alrashed, M.S. Gharibdousti, M. Goodarzi, L.R. de Oliveira, M.R. Safaei, E.P. Bandarra Filho, Effects on thermophysical properties of carbon based nanofluids: experimental data, modelling using regression, ANFIS and ANN, Int. J. Heat. Mass. Transf. 125 (2018) 920–932.
[13] A. Joseph, S. Thomas, Energy, exergy and corrosion analysis of direct absorption solar collector employed with ultra-high stable carbon quantum dot nanofluid, Renew. Energy 181 (2022) 725–737.
[14] M. Masjedi-Arani, M. Ghiyasiyan-Arani, M. Salavati-Niasari, Synthesis and characterization of carbon sphere-supported sand-rose like N-GQDs/$NiCo_2S_4$ structures with synergetic effect for development of hydrogen storage capacity, Fuel 312 (2022) 122956.
[15] M. Masjedi, N. Mir, E. Noori, T. Gholami, M. Salavati-Niasari, Effect of Schiff base ligand on the size and the optical properties of TiO_2 nanoparticles, Superlattices Microstructures 62 (2013) 30–38.
[16] M. Masjedi-Arani, M. Salavati-Niasari, A simple sonochemical approach for synthesis and characterization of Zn_2SiO_4 nanostructures, Ultrason. Sonochem. 29 (2016) 226–235.
[17] M. Masjedi-Arani, M. Salavati-Niasari, Cd_2SiO_4/graphene nanocomposite: ultrasonic assisted synthesis, characterization and electrochemical hydrogen storage application, Ultrason. Sonochem. 43 (2018) 136–145.
[18] F. Mazloom, M. Masjedi-Arani, M. Salavati-Niasari, Rapid and solvent-free solid-state synthesis and characterization of $Zn_3V_2O_8$ nanostructures and their phenol red aqueous solution photodegradation, Solid. State Sci. 70 (2017) 101–109.
[19] M. Masjedi-Arani, M. Salavati-Niasari, Effect of carbohydrate sugars as a capping agent on the size and morphology of pure Zn_2SnO_4 nanostructures and their optical properties, Mater. Lett. 174 (2016) 71–74.
[20] Z. Gholizadeh, M. Aliannezhadi, M. Ghominejad, F.S. Tehrani, High specific surface area γ-Al_2O_3 nanoparticles synthesized by facile and low-cost co-precipitation method, Sci. Rep. 13 (2023) 6131.
[21] S. Hassanzadeh-Tabrizi, E. Taheri-Nassaj, Economical synthesis of Al_2O_3 nanopowder using a precipitation method, Mater. Lett. 63 (2009) 2274–2276.
[22] J. Sun, Y. Wang, H. Zou, X. Guo, Z.-j Wang, Ni catalysts supported on nanosheet and nanoplate γ-Al_2O_3 for carbon dioxide methanation, J. Energy Chem. 29 (2019) 3–7.
[23] W.L. Suchanek, Hydrothermal synthesis of alpha alumina (α-Al_2O_3) powders: study of the processing variables and growth mechanisms, J. Am. Ceram. Soc. 93 (2010) 399–412.
[24] W. Wang, J. Zhou, Z. Zhang, J. Yu, W. Cai, Different surfactants-assisted hydrothermal synthesis of hierarchical γ-Al_2O_3 and its adsorption performances for parachlorophenol, Chem. Eng. J. 233 (2013) 168–175.
[25] A.A. Mohammed, Z.T. Khodair, A.A. Khadom, Preparation and investigation of the structural properties of α-Al_2O_3 nanoparticles using the sol-gel method, Chem. Data Collect. 29 (2020) 100531.

[26] J. Li, Y. Pan, C. Xiang, Q. Ge, J. Guo, Low temperature synthesis of ultrafine α-Al_2O_3 powder by a simple aqueous sol−gel process, Ceram. Int. 32 (2006) 587−591.

[27] P. Prashanth, R. Raveendra, R. Hari Krishna, S. Ananda, N. Bhagya, B. Nagabhushana, et al., Synthesis, characterizations, antibacterial and photoluminescence studies of solution combustion-derived α-Al_2O_3 nanoparticles, J. Asian Ceram. Soc. 3 (2015) 345−351.

[28] V. Zhuravlev, V. Bamburov, A. Beketov, L. Perelyaeva, I. Baklanova, O. Sivtsova, et al., Solution combustion synthesis of α-Al_2O_3 using urea, Ceram. Int. 39 (2013) 1379−1384.

[29] Z. Wang, Z. Mai, Y. Wu, Y. Wang, Z. Cao, M. Li, et al., Preparation of spherical α-Al_2O_3 nanoparticles by microwave hydrothermal synthesis and addition of nano-Al seeds, J. Am. Ceram. Soc. 105 (2022) 5585−5597.

[30] K. Laishram, R. Mann, N. Malhan, A novel microwave combustion approach for single step synthesis of α-Al_2O_3 nanopowders, Ceram. Int. 38 (2012) 1703−1706.

[31] T.D. Isfahani, J. Javadpour, A. Khavandi, R. Dinnebier, M. Goodarzi, H.R. Rezaie, Mechanochemical synthesis of alumina nanoparticles: formation mechanism and phase transformation, Powder Technol. 229 (2012) 17−23.

[32] B. Szczęśniak, J. Choma, M. Jaroniec, Facile mechanochemical synthesis of highly mesoporous γ-Al_2O_3 using boehmite, Microporous Mesoporous Mater. 312 (2021) 110792.

[33] R. López-Juárez, N. Razo-Perez, T. Pérez-Juache, O. Hernandez-Cristobal, S.Y. Reyes-López, Synthesis of α-Al_2O_3 from aluminum cans by wet-chemical methods, Results Phys. 11 (2018) 1075−1079.

[34] Z. Wu, Y. Shen, Y. Dong, J. Jiang, Study on the morphology of α-Al_2O_3 precursor prepared by precipitation method, J. Alloy. Compd. 467 (2009) 600−604.

[35] L. Zhu, L. Liu, C. Sun, X. Zhang, L. Zhang, Z. Gao, et al., Low temperature synthesis of polyhedral α-Al_2O_3 nanoparticles through two different modes of planetary ball milling, Ceram. Int. 46 (2020) 28414−28421.

[36] J. Wang, D. Zhao, G. Zhou, C. Zhang, P. Zhang, X. Hou, Synthesis of nano-sized γ-Al_2O_3 with controllable size by simple homogeneous precipitation method, Mater. Lett. 279 (2020) 128476.

[37] K. Bashirnezhad, S. Bazri, M.R. Safaei, M. Goodarzi, M. Dahari, O. Mahian, et al., Viscosity of nanofluids: a review of recent experimental studies, Int. Commun. Heat. Mass. Transf. 73 (2016) 114−123.

[38] M.H. Ahmadi, A. Mirlohi, M.A. Nazari, R. Ghasempour, A review of thermal conductivity of various nanofluids, J. Mol. Liq. 265 (2018) 181−188.

[39] I. Shahrul, I. Mahbubul, S. Khaleduzzaman, R. Saidur, M. Sabri, A comparative review on the specific heat of nanofluids for energy perspective, Renew. Sustain. Energy Rev. 38 (2014) 88−98.

[40] N. Usri, W. Azmi, R. Mamat, K.A. Hamid, G. Najafi, Thermal conductivity enhancement of Al_2O_3 nanofluid in ethylene glycol and water mixture, Energy Procedia 79 (2015) 397−402.

[41] J. Albadr, S. Tayal, M. Alasadi, Heat transfer through heat exchanger using Al_2O_3 nanofluid at different concentrations, Case Stud. Therm. Eng. 1 (2013) 38−44.

[42] S.D. Pandey, V. Nema, Experimental analysis of heat transfer and friction factor of nanofluid as a coolant in a corrugated plate heat exchanger, Exp. Therm. Fluid Sci. 38 (2012) 248−256.

[43] M.K. Abdolbaqi, W. Azmi, R. Mamat, K. Sharma, G. Najafi, Experimental investigation of thermal conductivity and electrical conductivity of BioGlycol–water mixture based Al_2O_3 nanofluid, Appl. Therm. Eng. 102 (2016) 932–941.
[44] R. Ali, A. Shahzad, K. us Saher, Z. Elahi, T. Abbas, The thin film flow of Al_2O_3 nanofluid particle over an unsteady stretching surface, Case Stud. Therm. Eng. 29 (2022) 101695.
[45] A. Ibrahim, M.R. Ramadan, A.E.-M. Khallaf, M. Abdulhamid, A comprehensive study for Al_2O_3 nanofluid cooling effect on the electrical and thermal properties of polycrystalline solar panels in outdoor conditions, Environ. Sci. Pollut. Res. 30 (2023) 106838–106859.
[46] M.A. Taher, S. Siddiqa, M. Kamrujjaman, M.M. Molla, Free convection of temperature-dependent thermal conductivity based ethylene glycol-Al_2O_3 nanofluid in an open cavity with wall heat flux, Int. Commun. Heat. Mass. Transf. 138 (2022) 106379.
[47] M. Ghanbarpour, E.B. Haghigi, R. Khodabandeh, Thermal properties and rheological behavior of water based Al_2O_3 nanofluid as a heat transfer fluid, Exp. Therm. Fluid Sci. 53 (2014) 227–235.
[48] Z. Said, M.H. Sajid, M.A. Alim, R. Saidur, N.A. Rahim, Experimental investigation of the thermophysical properties of AL_2O_3-nanofluid and its effect on a flat plate solar collector, Int. Commun. Heat. Mass. Transf. 48 (2013) 99–107.
[49] M.V. Bozorg, M.H. Doranehgard, K. Hong, Q. Xiong, CFD study of heat transfer and fluid flow in a parabolic trough solar receiver with internal annular porous structure and synthetic oil–Al_2O_3 nanofluid, Renew. Energy 145 (2020) 2598–2614.
[50] M. Mashayekhi, E. Houshfar, M. Ashjaee, Development of hybrid cooling method with PCM and Al_2O_3 nanofluid in aluminium minichannels using heat source model of Li-ion batteries, Appl. Therm. Eng. 178 (2020) 115543.
[51] W. Ajeeb, R.R.T. da Silva, S.S. Murshed, Experimental investigation of heat transfer performance of Al_2O_3 nanofluids in a compact plate heat exchanger, Appl. Therm. Eng. 218 (2023) 119321.
[52] T.J. Choi, M.S. Park, S.H. Kim, S.P. Jang, Experimental study on the effect of nanoparticle migration on the convective heat transfer coefficient of EG/water-based Al_2O_3 nanofluids, Int. J. Heat. Mass. Transf. 169 (2021) 120903.
[53] L. Yang, W. Ji, M. Mao, J.-n Huang, An updated review on the properties, fabrication and application of hybrid-nanofluids along with their environmental effects, J. Clean. Prod. 257 (2020) 120408.
[54] M.H. Esfe, H.R. Raki, M.R.S. Emami, M. Afrand, Viscosity and rheological properties of antifreeze based nanofluid containing hybrid nano-powders of MWCNTs and TiO_2 under different temperature conditions, Powder Technol. 342 (2019) 808–816.
[55] R. Taherialekouhi, S. Rasouli, A. Khosravi, An experimental study on stability and thermal conductivity of water-graphene oxide/aluminum oxide nanoparticles as a cooling hybrid nanofluid, Int. J. Heat. Mass. Transf. 145 (2019) 118751.
[56] S. Kannaiyan, C. Boobalan, A. Umasankaran, A. Ravirajan, S. Sathyan, T. Thomas, Comparison of experimental and calculated thermophysical properties of alumina/cupric oxide hybrid nanofluids, J. Mol. Liq. 244 (2017) 469–477.
[57] Z. Said, M.A. Abdelkareem, H. Rezk, A.M. Nassef, Fuzzy modeling and optimization for experimental thermophysical properties of water and ethylene glycol mixture for Al_2O_3 and TiO_2 based nanofluids, Powder Technol. 353 (2019) 345–358.

[58] Z. Aparna, M. Michael, S. Pabi, S. Ghosh, Thermal conductivity of aqueous Al2O3/Ag hybrid nanofluid at different temperatures and volume concentrations: an experimental investigation and development of new correlation function, Powder Technol. 343 (2019) 714–722.

[59] A. Parsian, M. Akbari, New experimental correlation for the thermal conductivity of ethylene glycol containing Al_2O_3–Cu hybrid nanoparticles, J. Therm. Anal. Calorim. 131 (2018) 1605–1613.

[60] M.S. Tahat, A.C. Benim, Experimental analysis on thermophysical properties of Al_2O_3/CuO hybrid nano fluid with its effects on flat plate solar collector, Defect and Diffusion Forum, Trans Tech Publications, 2017, pp. 148–156.

[61] N.M. Bahari, S.N. Che Mohamed Hussein, N.H. Othman, Synthesis of Al_2O_3–SiO_2/water hybrid nanofluids and effects of surfactant toward dispersion and stability, Particulate Sci. Technol. 39 (2021) 844–858.

[62] H. Kakati, A. Mandal, S. Laik, Promoting effect of Al_2O_3/ZnO-based nanofluids stabilized by SDS surfactant on CH4 + C2H6 + C3H8 hydrate formation, J. Ind. Eng. Chem. 35 (2016) 357–368.

[63] W. Urmi, M. Rahman, W. Hamzah, An experimental investigation on the thermophysical properties of 40% ethylene glycol based TiO_2-Al_2O_3 hybrid nanofluids, Int. Commun. Heat. Mass. Transf. 116 (2020) 104663.

[64] A.A. Minea, Hybrid nanofluids based on Al_2O_3, TiO_2 and SiO_2: numerical evaluation of different approaches, Int. J. Heat. Mass. Transf. 104 (2017) 852–860.

[65] V.V. Wanatasanappan, P.K. Kanti, P. Sharma, N. Husna, M. Abdullah, Viscosity and rheological behavior of Al_2O_3-Fe_2O_3/water-EG based hybrid nanofluid: a new correlation based on mixture ratio, J. Mol. Liq. 375 (2023) 121365.

[66] S. Mehryan, E. Izadpanahi, M. Ghalambaz, A. Chamkha, Mixed convection flow caused by an oscillating cylinder in a square cavity filled with Cu–Al_2O_3/water hybrid nanofluid, J. Therm. Anal. Calorim. 137 (2019) 965–982.

[67] M. Devarajan, N. Parasumanna Krishnamurthy, M. Balasubramanian, B. Ramani, S. Wongwises, K. Abd El-Naby, et al., Thermophysical properties of CNT and CNT/Al_2O_3 hybrid nanofluid, Micro Nano Lett. 13 (2018) 617–621.

[68] A. Asadi, M. Asadi, A. Rezaniakolaei, L.A. Rosendahl, M. Afrand, S. Wongwises, Heat transfer efficiency of Al_2O_3-MWCNT/thermal oil hybrid nanofluid as a cooling fluid in thermal and energy management applications: an experimental and theoretical investigation, Int. J. Heat. Mass. Transf. 117 (2018) 474–486.

[69] S. Salawu, A. Obalalu, S. Okoya, Thermal convection and solar radiation of electromagnetic actuator Cu–Al_2O_3/$C_3H_8O_2$ and Cu–$C_3H_8O_2$ hybrid nanofluids for solar collector optimization, Mater. Today Commun. 33 (2022) 104763.

[70] A. Fattahi, N. Karimi, Numerical simulation of the effects of superhydrophobic coating in an oval cross-sectional solar collector with a wavy absorber filled with water-based Al_2O_3-ZnO-Fe_3O_4 ternary hybrid nanofluid, Sustain. Energy Technol. Assess. 50 (2022) 101881.

[71] A. Heydari, M. Goharimanesh, M.R. Gharib, Dynamic viscosity analysis of hybrid nanofluid MWCNT-Al_2O_3/engine oil using statistical models with evaluating its performance in a double tube heat exchanger, J. Therm. Anal. Calorim. 148 (2022) 8025–8039.

[72] M.I.I. Rabby, M.A.R. Sharif, F. Hossain, Numerical study of laminar convective heat transfer from a corrugated pipe into an Al_2O_3–AlN/H_2O hybrid nanofluid, Case Stud. Therm. Eng. 39 (2022) 102454.

[73] T.K. Murtadha, Effect of using Al_2O_3/TiO_2 hybrid nanofluids on improving the photovoltaic performance, Case Stud. Therm. Eng. 47 (2023) 103112.
[74] F. Mashali, E.M. Languri, J. Davidson, D. Kerns, W. Johnson, K. Nawaz, et al., Thermo-physical properties of diamond nanofluids: a review, Int. J. Heat. Mass. Transf. 129 (2019) 1123–1135.
[75] S. Chakraborty, P.K. Panigrahi, Stability of nanofluid: a review, Appl. Therm. Eng. 174 (2020)115259.
[76] G. Xia, H. Jiang, R. Liu, Y. Zhai, Effects of surfactant on the stability and thermal conductivity of Al_2O_3/de-ionized water nanofluids, Int. J. Therm. Sci. 84 (2014) 118–124.
[77] Z. Said, R. Saidur, M. Sabiha, A. Hepbasli, N. Rahim, Energy and exergy efficiency of a flat plate solar collector using pH treated Al_2O_3 nanofluid, J. Clean. Prod. 112 (2016) 3915–3926.
[78] A. Ghadimi, I.H. Metselaar, The influence of surfactant and ultrasonic processing on improvement of stability, thermal conductivity and viscosity of titania nanofluid, Exp. Therm. Fluid Sci. 51 (2013) 1–9.
[79] X. Yang, Z.-h Liu, A kind of nanofluid consisting of surface-functionalized nanoparticles, Nanoscale Res. Lett. 5 (2010) 1324–1328.
[80] Y. Hwang, J.-K. Lee, J.-K. Lee, Y.-M. Jeong, S.-i Cheong, Y.-C. Ahn, et al., Production and dispersion stability of nanoparticles in nanofluids, Powder Technol. 186 (2008) 145–153.
[81] L.S. Sundar, E.V. Ramana, M.K. Singh, A.C. Sousa, Thermal conductivity and viscosity of stabilized ethylene glycol and water mixture Al_2O_3 nanofluids for heat transfer applications: an experimental study, Int. Commun. Heat. Mass. Transf. 56 (2014) 86–95.
[82] G.M. Moldoveanu, G. Huminic, A.A. Minea, A. Huminic, Experimental study on thermal conductivity of stabilized Al_2O_3 and SiO_2 nanofluids and their hybrid, Int. J. Heat. Mass. Transf. 127 (2018) 450–457.
[83] P.K. Das, A.K. Mallik, A.H. Molla, A.K. Santra, R. Ganguly, A. Saha, et al., Experimental investigation for stability and surface properties of TiO_2 and Al_2O_3 water-based nanofluids, J. Therm. Anal. Calorim. 147 (2022) 5617–5635.
[84] K. Cacua, F. Ordoñez, C. Zapata, B. Herrera, E. Pabón, R. Buitrago-Sierra, Surfactant concentration and pH effects on the zeta potential values of alumina nanofluids to inspect stability, Colloids Surf. A: Physicochem. Eng. Asp. 583 (2019)123960.
[85] J. Ji, X. Yao, J. Gao, W. Lu, W. Wang, D. Chu, Effect of surfactants and pH values on stability of γ-Al_2O_3 nanofluids, Chem. Phys. Lett. 781 (2021) 138996.
[86] O. Gulzar, A. Qayoum, R. Gupta, Experimental study on stability and rheological behaviour of hybrid Al_2O_3-TiO_2 Therminol-55 nanofluids for concentrating solar collectors, Powder Technol. 352 (2019) 436–444.
[87] C.G.N. Ketchate, P.T. Kapen, D. Fokwa, G. Tchuen, Stability analysis of mixed convection in a porous horizontal channel filled with a Newtonian Al_2O_3/water nanofluid in presence of magnetic field and thermal radiation, Chin. J. Phys. 79 (2022) 514–530.
[88] M.H. Esfe, D. Toghraie, S. Esfandeh, S. Alidoust, Measurement of thermal conductivity of triple hybrid water based nanofluid containing MWCNT (10%)-Al_2O_3 (60%)-ZnO (30%) nanoparticles, Colloids Surf. A: Physicochem. Eng. Asp. 647 (2022)129083.

[89] F.R. Siddiqui, C. Tso, K.C. Chan, S.C. Fu, C.Y. Chao, On trade-off for dispersion stability and thermal transport of Cu-Al_2O_3 hybrid nanofluid for various mixing ratios, Int. J. Heat. Mass. Transf. 132 (2019) 1200–1216.
[90] S. Zhang, L. Lu, T. Wen, C. Dong, Turbulent heat transfer and flow analysis of hybrid Al_2O_3-CuO/water nanofluid: an experiment and CFD simulation study, Appl. Therm. Eng. 188 (2021) 116589.
[91] J. Nishit, S. Bekal, Experimental investigation on polymer solar water heater using Al_2O_3 nanofluid for performance improvement, Mater. Today: Proc. 92 (2023) 249–257.
[92] S. Sharma, S.M. Gupta, Preparation and evaluation of stable nanofluids for heat transfer application: a review, Exp. Therm. Fluid Sci. 79 (2016) 202–212.
[93] A. Asadi, F. Pourfattah, I.M. Szilágyi, M. Afrand, G. Żyła, H.S. Ahn, et al., Effect of sonication characteristics on stability, thermophysical properties, and heat transfer of nanofluids: a comprehensive review, Ultrason. Sonochem. 58 (2019)104701.
[94] L.H. Kumar, S. Kazi, H. Masjuki, M. Zubir, A review of recent advances in green nanofluids and their application in thermal systems, Chem. Eng. J. 429 (2022) 132321.
[95] C. Wang, J. Sun, C. Ge, H. Tang, P. Wu, Synthesis, characterization and lubrication performance of reduced graphene oxide-Al_2O_3 nanofluid for strips cold rolling, Colloids Surf. A: Physicochem. Eng. Asp. 637 (2022)128204.
[96] I. Mahbubul, E. Elcioglu, M. Amalina, R. Saidur, Stability, thermophysical properties and performance assessment of alumina–water nanofluid with emphasis on ultrasonication and storage period, Powder Technol. 345 (2019) 668–675.
[97] A.M. Khdher, N.A.C. Sidik, W.A.W. Hamzah, R. Mamat, An experimental determination of thermal conductivity and electrical conductivity of bio glycol based Al_2O_3 nanofluids and development of new correlation, Int. Commun. Heat. Mass. Transf. 73 (2016) 75–83.
[98] M. Sangeetha, S. Manigandan, B. Ashok, K. Brindhadevi, A. Pugazhendhi, Experimental investigation of nanofluid based photovoltaic thermal (PV/T) system for superior electrical efficiency and hydrogen production, Fuel 286 (2021) 119422.
[99] S.H. Park, Y.S. Song, Carbon nanofluid flow based biophotovoltaic cell, Nano Energy 81 (2021) 105624.
[100] A. Olabi, K. Elsaid, E.T. Sayed, M.S. Mahmoud, T. Wilberforce, R.J. Hassiba, et al., Application of nanofluids for enhanced waste heat recovery: a review, Nano Energy 84 (2021)105871.
[101] W. Huang, M. Marefati, Energy, exergy, environmental and economic comparison of various solar thermal systems using water and Thermia Oil B base fluids, and CuO and Al_2O_3 nanofluids, Energy Rep. 6 (2020) 2919–2947.
[102] O. Mahian, E. Bellos, C.N. Markides, R.A. Taylor, A. Alagumalai, L. Yang, et al., Recent advances in using nanofluids in renewable energy systems and the environmental implications of their uptake, Nano Energy 86 (2021) 106069.
[103] H. Kim, J. Kim, H. Cho, Experimental study on performance improvement of U-tube solar collector depending on nanoparticle size and concentration of Al_2O_3 nanofluid, Energy 118 (2017) 1304–1312.
[104] J. Ghaderian, N.A.C. Sidik, An experimental investigation on the effect of Al_2O_3/distilled water nanofluid on the energy efficiency of evacuated tube solar collector, Int. J. Heat. Mass. Transf. 108 (2017) 972–987.

[105] Y. Tong, H. Lee, W. Kang, H. Cho, Energy and exergy comparison of a flat-plate solar collector using water, Al_2O_3 nanofluid, and CuO nanofluid, Appl. Therm. Eng. 159 (2019) 113959.

[106] G. Colangelo, E. Favale, P. Miglietta, M. Milanese, A. de Risi, Thermal conductivity, viscosity and stability of Al_2O_3-diathermic oil nanofluids for solar energy systems, Energy 95 (2016) 124–136.

[107] Y.R. Sekhar, K. Sharma, R.T. Karupparaj, C. Chiranjeevi, Heat transfer enhancement with Al_2O_3 nanofluids and twisted tapes in a pipe for solar thermal applications, Procedia Eng. 64 (2013) 1474–1484.

[108] J. Subramani, P. Nagarajan, S. Wongwises, S. El-Agouz, R. Sathyamurthy, Experimental study on the thermal performance and heat transfer characteristics of solar parabolic trough collector using Al_2O_3 nanofluids, Environ. Prog. Sustain. Energy 37 (2018) 1149–1159.

[109] A. Menbari, A.A. Alemrajabi, Analytical modeling and experimental investigation on optical properties of new class of nanofluids (Al_2O_3–CuO binary nanofluids) for direct absorption solar thermal energy, Optical Mater. 52 (2016) 116–125.

[110] S. Samiezadeh, R. Khodaverdian, M.H. Doranehgard, H. Chehrmonavari, Q. Xiong, CFD simulation of thermal performance of hybrid oil-Cu-Al_2O_3 nanofluid flowing through the porous receiver tube inside a finned parabolic trough solar collector, Sustain. Energy Technol. Assess. 50 (2022) 101888.

[111] L.S. Sundar, Y.T. Sintie, Z. Said, M.K. Singh, V. Punnaiah, A.C. Sousa, Energy, efficiency, economic impact, and heat transfer aspects of solar flat plate collector with Al_2O_3 nanofluids and wire coil with core rod inserts, Sustain. Energy Technol. Assess. 40 (2020) 100772.

[112] P.K. Kanti, P. Sharma, M.P. Maiya, K.V. Sharma, The stability and thermophysical properties of Al_2O_3-graphene oxide hybrid nanofluids for solar energy applications: application of robust autoregressive modern machine learning technique, Sol. Energy Mater. Sol. Cell 253 (2023)112207.

[113] J. Subramani, P. Sevvel, S. Srinivasan, Influence of CNT coating on the efficiency of solar parabolic trough collector using AL_2O_3 nanofluids-a multiple regression approach, Mater. Today: Proc. 45 (2021) 1857–1861.

[114] E. Elshazly, A.A. Abdel-Rehim, I. El-Mahallawi, 4E study of experimental thermal performance enhancement of flat plate solar collectors using MWCNT, Al_2O_3, and hybrid MWCNT/Al_2O_3 nanofluids, Results Eng. 16 (2022) 100723.

[115] N.S. Rajput, D.D. Shukla, D. Rajput, S.K. Sharm, Performance analysis of flat plate solar collector using Al_2O_3/distilled water nanofluid: an experimental investigation, Mater. Today: Proc. 10 (2019) 52–59.

[116] H. Kaya, M. Alkasem, K. Arslan, Effect of nanoparticle shape of Al_2O_3/Pure Water nanofluid on evacuated U-Tube solar collector efficiency, Renew. Energy 162 (2020) 267–284.

[117] O. Gulzar, A. Qayoum, R. Gupta, Experimental study on thermal conductivity of mono and hybrid Al_2O_3–TiO_2 nanofluids for concentrating solar collectors, Int. J. Energy Res. 45 (2021) 4370–4384.

[118] V.P. Kalbande, P.V. Walke, K. Rambhad, Performance of oil-based thermal storage system with parabolic trough solar collector using Al_2O_3 and soybean oil nanofluid, Int. J. Energy Res. 45 (2021) 15338–15359.

[119] M. Hader, W. Al-Kouz, Performance of a hybrid photovoltaic/thermal system utilizing water-Al_2O_3 nanofluid and fins, Int. J. Energy Res. 43 (2019) 219–230.

[120] N. Elminshawy, A. Elminshawy, A. Osama, M. Bassyouni, M. Arıcı, Experimental performance analysis of enhanced concentrated photovoltaic utilizing various mass flow rates of Al_2O_3-nanofluid: energy, exergy, and exergoeconomic study, Sustain. Energy Technol. Assess. 53 (2022) 102723.

[121] M. Zamen, M. Kahani, B. Rostami, M. Bargahi, Application of Al_2O_3/water nanofluid as the coolant in a new design of photovoltaic/thermal system: an experimental study, Energy Sci. Eng. 10 (2022) 4273–4285.

[122] H. Adun, M. Adedeji, T. Ruwa, M. Senol, D. Kavaz, M. Dagbasi, Energy, exergy, economic, environmental (4E) approach to assessing the performance of a photovoltaic-thermal system using a novel ternary nanofluid, Sustain. Energy Technol. Assess. 50 (2022) 101804.

[123] M. Sangeetha, S. Manigandan, M.T. Chaichan, V. Kumar, Progress of MWCNT, Al_2O_3, and CuO with water in enhancing the photovoltaic thermal system, Int. J. Energy Res. 44 (2020) 821–832.

[124] O.M.J. Jasim, S. Selimli, H. Dumrul, S. Yilmaz, Closed-loop aluminium oxide nanofluid cooled photovoltaic thermal collector energy and exergy analysis, an experimental study, J. Energy Storage 50 (2022) 104654.

[125] A.A. Memon, W. Khan, T. Muhammad, Numerical investigation of photovoltaic thermal energy efficiency improvement using the backward step containing Cu-Al_2O_3 hybrid nanofluid, Alex. Eng. J. 75 (2023) 391–406.

[126] A. Aydın, İ. Kayri, H. Aydin, Electrical and thermal performance enhancement of a photovoltaic thermal hybrid system with a novel inner plate-finned collective cooling with different nanofluids, Int. J. Green. Energy 21 (2023) 555–569.

[127] A. Habibollahzade, Z.K. Mehrabadi, C.N. Markides, Comparative thermoeconomic analyses and multi-objective particle swarm optimization of geothermal combined cooling and power systems, Energy Convers. Manage. 234 (2021) 113921.

[128] M. Faizal, A. Bouazza, R.M. Singh, Heat transfer enhancement of geothermal energy piles, Renew. Sustain. Energy Rev. 57 (2016) 16–33.

[129] N. Jamshidi, A. Mosaffa, Investigating the effects of geometric parameters on finned conical helical geothermal heat exchanger and its energy extraction capability, Geothermics 76 (2018) 177–189.

[130] I. Zakaria, W. Mohamed, A.B. Mamat, R. Saidur, W. Azmi, R. Mamat, et al., Experimental investigation of Al_2O_3-water ethylene glycol mixture nanofluid thermal behaviour in a single cooling plate for PEM fuel cell application, Energy Procedia 79 (2015) 252–258.

[131] A. Zeiny, M.A. Al-Baghdadi, W.F. Arear, M.S. Ismail, $Al_2O_3-H_2O$ nanofluids for cooling PEM fuel cells: a critical assessment, Int. J. Hydrog. Energy 47 (2022) 38823–38836.

[132] B. Shen, A.J. Shih, S.C. Tung, Application of nanofluids in minimum quantity lubrication grinding, Tribol. Trans. 51 (2008) 730–737.

[133] A.H. Pordanjani, S. Aghakhani, M. Afrand, M. Sharifpur, J.P. Meyer, H. Xu, et al., Nanofluids: physical phenomena, applications in thermal systems and the environment effects-a critical review, J. Clean. Prod. 320 (2021) 128573.

[134] S. Chaudhari, R. Chakule, P. Talmale, Experimental study of heat transfer characteristics of Al_2O_3 and CuO nanofluids for machining application, Mater. Today: Proc. 18 (2019) 788–797.

[135] P.B. Zaman, M.I.H. Tusar, N.R. Dhar, Selection of appropriate process inputs for turning Ti-6Al-4V alloy under hybrid Al_2O_3-MWCNT nano-fluid based MQL, Adv. Mater. Process. Technol. 8 (2022) 380–400.
[136] A.K. Sharma, R.K. Singh, A.R. Dixit, A.K. Tiwari, Characterization and experimental investigation of Al_2O_3 nanoparticle based cutting fluid in turning of AISI 1040 steel under minimum quantity lubrication (MQL), Mater. Today: Proc. 3 (2016) 1899–1906.
[137] C. Li, H.M. Ali, Experimental evaluation on tribological performance of the wheel/workpiece interface in NMQL grinding with different concentrations of Al_2O_3 nanofluids, Research Anthology on Synthesis, Characterization, and Applications of Nanomaterials, IGI Global, 2021, pp. 1608–1627.
[138] X. Zhang, C. Li, Y. Zhang, Y. Wang, B. Li, M. Yang, et al., Lubricating property of MQL grinding of Al_2O_3/SiC mixed nanofluid with different particle sizes and microtopography analysis by cross-correlation, Precis. Eng. 47 (2017) 532–545.
[139] A. Pal, S.S. Chatha, H.S. Sidhu, Performance evaluation of the minimum quantity lubrication with Al_2O_3-mixed vegetable-oil-based cutting fluid in drilling of AISI 321 stainless steel, J. Manuf. Process. 66 (2021) 238–249.
[140] D.G. Subhedar, Y.S. Patel, B.M. Ramani, G.S. Patange, An experimental investigation on the effect of Al_2O_3/cutting oil based nano coolant for minimum quantity lubrication drilling of SS 304, Clean. Eng. Technol. 3 (2021) 100104.
[141] F. Günan, T. Kıvak, Ç.V. Yıldırım, M. Sarıkaya, Performance evaluation of MQL with AL_2O_3 mixed nanofluids prepared at different concentrations in milling of Hastelloy C276 alloy, J. Mater. Res. Technol. 9 (2020) 10386–10400.
[142] M. Jamil, N. He, W. Zhao, A.M. Khan, R.A. Laghari, Tribology and machinability performance of hybrid Al_2O_3-MWCNTs nanofluids-assisted MQL for milling Ti-6Al-4 V, Int. J. Adv. Manuf. Technol. 119 (2022) 2127–2144.
[143] M. Bahiraei, S. Heshmatian, Electronics cooling with nanofluids: a critical review, Energy Convers. Manage. 172 (2018) 438–456.
[144] K. Aglawe, R. Yadav, S. Thool, Preparation, applications and challenges of nanofluids in electronic cooling: a systematic review, Mater. Today: Proc. 43 (2021) 366–372.
[145] C.T. Nguyen, G. Roy, C. Gauthier, N. Galanis, Heat transfer enhancement using Al_2O_3–water nanofluid for an electronic liquid cooling system, Appl. Therm. Eng. 27 (2007) 1501–1506.
[146] S.A. Nada, R. El-Zoheiry, M. Elsharnoby, O. Osman, Enhancing the thermal performance of different flow configuration minichannel heat sink using Al_2O_3 and CuO-water nanofluids for electronic cooling: an experimental assessment, Int. J. Therm. Sci. 181 (2022)107767.
[147] J. Qu, H.-y Wu, P. Cheng, Thermal performance of an oscillating heat pipe with Al_2O_3–water nanofluids, Int. Commun. Heat. Mass. Transf. 37 (2010) 111–115.
[148] M.R. Tanshen, S. Lee, J. Kim, D. Kang, J. Noh, H. Chung, et al., Pressure distribution inside oscillating heat pipe charged with aqueous Al_2O_3 nanoparticles, MWCNTs and their hybrid, J. Cent. South. Univ. 21 (2014) 2341–2348.
[149] H.S. Moghaieb, H. Abdel-Hamid, M.H. Shedid, A. Helali, Engine cooling using Al_2O_3/water nanofluids, Appl. Therm. Eng. 115 (2017) 152–159.
[150] S. Peyghambarzadeh, S. Hashemabadi, M.S. Jamnani, S. Hoseini, Improving the cooling performance of automobile radiator with Al_2O_3/water nanofluid, Appl. Therm. Eng. 31 (2011) 1833–1838.

[151] A.S. Tijani, A.S. bin Sudirman, Thermos-physical properties and heat transfer characteristics of water/anti-freezing and Al_2O_3/CuO based nanofluid as a coolant for car radiator, Int. J. Heat. Mass. Transf. 118 (2018) 48–57.

[152] T.J. Choi, S.H. Kim, S.P. Jang, D.J. Yang, Y.M. Byeon, Heat transfer enhancement of a radiator with mass-producing nanofluids (EG/water-based Al_2O_3 nanofluids) for cooling a 100 kW high power system, Appl. Therm. Eng. 180 (2020) 115780.

[153] S. Baskar, M. Chandrasekaran, T. Vinod Kumar, P. Vivek, S. Ramasubramanian, Experimental studies on flow and heat transfer characteristics of secondary refrigerant-based CNT nanofluids for cooling applications, Int. J. Ambient. Energy 41 (2020) 285–288.

[154] L. Yang, W. Jiang, X. Chen, K. Du, Dynamic characteristics of an environment-friendly refrigerant: ammonia-water based TiO_2 nanofluids, Int. J. Refrig. 82 (2017) 366–380.

[155] H. Peng, G. Ding, W. Jiang, H. Hu, Y. Gao, Measurement and correlation of frictional pressure drop of refrigerant-based nanofluid flow boiling inside a horizontal smooth tube, Int. J. Refrig. 32 (2009) 1756–1764.

[156] F. Sarrafzadeh Javadi, R. Saidur, Thermodynamic and energy efficiency analysis of a domestic refrigerator using Al_2O_3 nano-refrigerant, Sustainability 13 (2021) 5659.

[157] H. Peng, G. Ding, W. Jiang, H. Hu, Y. Gao, Heat transfer characteristics of refrigerant-based nanofluid flow boiling inside a horizontal smooth tube, Int. J. Refrig. 32 (2009) 1259–1270.

[158] A. Nikulin, O. Khliyeva, N. Lukianov, V. Zhelezny, Y. Semenyuk, Study of pool boiling process for the refrigerant R11, isopropanol and isopropanol/Al_2O_3 nanofluid, Int. J. Heat. Mass. Transf. 118 (2018) 746–757.

[159] I. Mahbubul, R. Saidur, M. Amalina, Influence of particle concentration and temperature on thermal conductivity and viscosity of Al_2O_3/R141b nanorefrigerant, Int. Commun. Heat. Mass. Transf. 43 (2013) 100–104.

[160] B. Bibin, P. Amartya, K. Govil, G. Edison, Performance of refrigeration system using Al_2O_3, CuO and TiO_2 nano-enhanced refrigerants: a review, Mater. Today: Proc. (2023). Available from: https://doi.org/10.1016/j.matpr.2023.01.125.

CHAPTER

ZnO-based nanofluids

4

Amin Moradi Hasan-Abad[1], Amir Hossein Mohammadi[2], Ali Sobhani-Nasab[3] and Kimiya Mosaddeghi[4,5]

[1]*Autoimmune Diseases Research Center, Shahid Beheshti Hospital, Kashan University of Medical Sciences, Kashan, Iran*

[2]*Department of Biochemistry, School of Medicine, AJA University of Medical Sciences, Tehran, Iran*

[3]*Physiology Research Center, Institute for Basic Sciences, Kashan University of Medical Sciences, Kashan, Iran*

[4]*School of Medicine, Kashan University of Medical Sciences, Kashan, Iran*

[5]*Student Research Committee, Kashan University of Medical Sciences, Kashan, Iran*

4.1 General introduction

Nanotechnology is an enabling technology that works with nanometer-sized objects. Nanotechnology is predicted to progress at three levels: materials, electronics, and systems. At the moment, the nanomaterials level is the most advanced, both in scientific knowledge and in commercial applications such as biosensors photocatalyst, supercapacitors, and biomedical [1–13]. Nanoparticles were investigated a decade ago because of their size-dependent physical and chemical features [14–22]. At this point, they have moved into the phase of commercial exploration [23]. Nanofluids (NFs) are a novel class of fluids created by distributing nanometer-sized materials in base fluids (nanoparticles, nanofibers, nanotubes, nanowires, nanorods, nanosheets, or droplets). NFs, in other terms, are nanoscale colloidal solutions containing condensed nanomaterials. They are two-phase systems, with one phase (solid phase) contained within another (liquid phase). When compared to base fluids such as oil or water, NFs have improved thermophysical properties such as thermal conductivity, thermal diffusivity, viscosity, and convective heat transfer coefficients. It has shown a wide range of possible applications [24–34].

NFs have the potential to increase heat transfer in a wide range of industries, including food, medical, electronics, transportation, and heat exchangers [35–38]. When compared to their solid form, the physicochemical properties of NFs changed dramatically [39]. As a result, there is a growing demand for NFs in nanobiomedical fields. Because of their particular biological properties, NFs have recently been employed in antibacterial treatments and pharmaceutical administration [40]. Furthermore, several studies have shown that NFs are used in

Nanofluids. DOI: https://doi.org/10.1016/B978-0-443-13625-2.00004-8

multimodal imaging (MR imaging + PET/optical) or combination systems (imaging + high fever/medication delivery) [41]. Many critical requirements, such as proper size, content, morphology, crystallinity, biocompatibility, nonagglomeration, and stability, must be met for NFs to be used optimally and safely [42]. A substantial proportion of nanoparticles disperse in solutions with a watery basis. The consistency of dispersion is largely dependent on the size, shape, structure, and content of the nanoparticles. To establish stable NFs, several processes must be addressed, including the use of surfactants or other chemical additives, surface engineering, ultrasonic homogenization, physical stirring, and nanomaterial creation [43]. Furthermore, the legal permission and medical safety of the additions must be considered. The great majority of tests on colloid theory, surface science, and material have been conducted. Despite this, the biomedical application of NFs is far from easy and requires additional research [44]. Despite the scarcity of published data on the biomedical uses of NFs, there have been several studies that have usually explored a number of relevant applications in a wide range of fields such as biomedical. Wong et al. briefly mentioned some studies about the diverse applications of NFs [45]. Yu et al. demonstrated that NFs have medicine delivery and antimicrobial properties [46]. Another study indicated the utilization of NFs in medicine. In this review, the biomedical zinc oxide's medicinal and antibacterial properties and applications in drug delivery, particularly nanosuspensions or suspensions containing zinc oxide NPs, were thoroughly investigated and reviewed. In addition, the applications of NFs in a wide range of biomedical developments were examined [44]. The homogenous colloidal dispersion of nanoparticles into the base fluid results in NFs. Researchers have so far created unique NFs for a variety of uses. To examine the properties, various characterization approaches were applied. The following are the findings of the preceding study: Chemical methods for NFs synthesis are less expensive and faster than physical methods. Because nanoparticles are now commercially available on an industrial scale, the two-step technique is currently the most energy-efficient and cost-effective means of producing NFs. NFs characterization is used to examine materials, structure, form, size, composition, and physical and chemical properties. The microscope approach and the spectroscopic method are two well-known characterization techniques. The zeta potential estimates the stability of nanofluid by measuring the potential difference between the fluid layer absorbed on the nanoparticle surface and the bulk phase of the liquid.

4.2 Different methods for creating zinc oxide nanofluids

There are several methods for creating nanofluids. Methods are classified into two types: one-step methods and two-step approaches. Nanoparticles are synthesized and disseminated in fluid with a the one-step method. The one-step method enhances fluid stability and reduces buildup by removing the need for dispersion, drying, transportation, and storage of nanoparticles [47]. The one-step method was employed to achieve the synthesis of a stable NFs with consistent dispersion of nanoparticles [48]. The manufacture of nanofibers was achieved through many methods, including the chemical solution approach, vacuum submerged arc nanoparticles synthesis system (vacuum-SANSS), liquid chemical technique, and physical mist deposition [49]. While the one-step method has certain advantages, it also exhibits several notable limitations, such as its inability to adequately elucidate the impacts of nanoparticles while simultaneously accounting for the influence of impurities. Furthermore, empirical evidence has demonstrated that the one-step method is exclusively applicable in the context of fluids characterized by a low vapor pressure foundation. Consequently, it is unsuitable for large-scale production purposes [48,49]. The initial stage of the two-step method involves the generation of a suspension comprising nanopowder particles in a base fluid. The two-step method is mostly employed for the generation of NFs (Fig. 4.1). Additionally, the utilization of nanoparticles in this context is particularly cost-effective due to their large quantities. Nanoparticles are initially generated in the form of a dry powder by

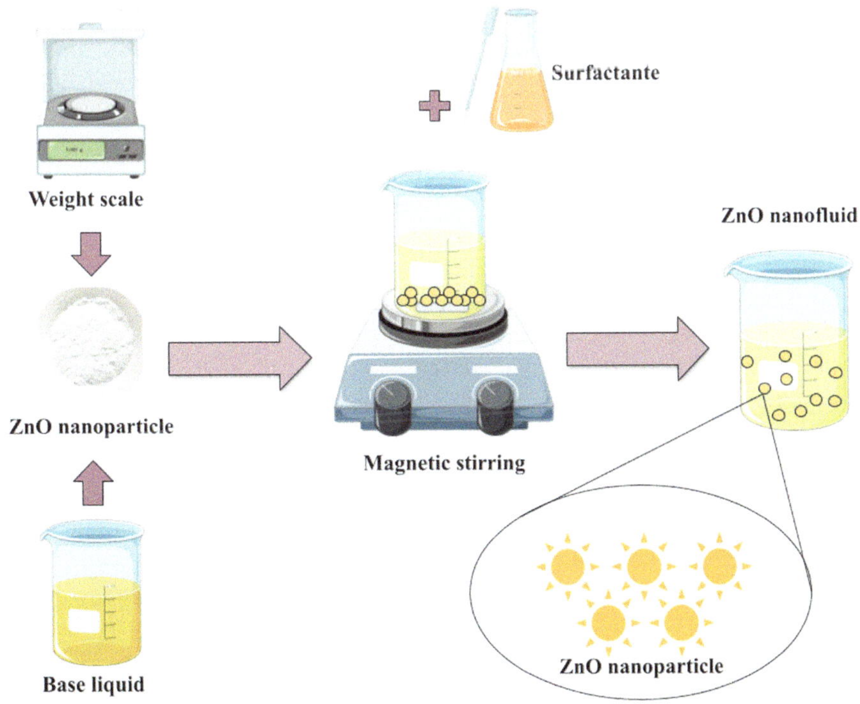

FIGURE 4.1

Nanoparticles are synthesized and disseminated in fluid by two-step method.

the use of physical or chemical methodologies. Subsequently, employing ball milling, homogenization, high shear combination, and ultrasonic agitation/ intense magnetic force, the particles are suspended inside the lower region of the fluid. The second stage is of significant importance in effectively mitigating the accumulation of nanoparticles [49]. The study conducted by Cacua et al. examined the impact of ultrasonic dispersion and disruption on nanoparticles. The researchers additionally examined the impact of temperature on levels of thermal conductivity [49]. The stability of a water/metal oxide nanofiltration (NF) membrane was investigated by Mahbubul et al. through a two-step fabrication process. Initially, the researchers conducted measurements pertaining to heat transmission, followed by measurements of the viscosity and thermal conductivity of the NFs. The investigation conducted by the researchers indicated a negative correlation between the duration of ultrasonic usage and the deposition of nanoparticles [50]. Due to the distinct production processes employed for NFs and NPs, a notable drawback of this technology is the occurrence of particle buildup resulting from inadequate van der Waals forces during the stages of transportation, storage, and drying. Moreover, the presence of surfactants enhances the stability of nanoparticles. The employment of surfactants at elevated temperatures holds significant importance, particularly in applications involving high temperatures [49]. The herbal extract contains a diverse range of compounds, including alkaloids, tryptophan, thymol, terpenoids, gentisic acid, ferulic acid, polyphenols, amino acids, polysaccharides, and flavonoids. These compounds serve various functions, such as chelation, stabilization, and reduction. The aforementioned metabolites exhibit properties such as acting as hydrogen donors, functioning as chelating agents, serving as stabilizing and reducing agents, and acting as oxygen quenchers. These compounds are capable of reducing metal ions present in herbal essence, leading to the formation of nanomaterials with comparable properties. The utilization of herb leaves' essence for the manufacture of metal NFs is a prevalent practice, owing to the presence of vast metabolite resources within them. The manufacture of metal nanofibers (NFs) with the use of herbs serves as a straightforward and convenient process. In a controlled laboratory setting, the amalgamation of metal salts with herbal essences has been conducted, with the duration of the process spanning several hours [51]. The utilization of Fourier-transform infrared spectroscopy (FTIR) and transmission electron microscopy (TEM) has revealed the significant involvement of herbal essences in the reduction and stabilization of nanorods. This can be attributed to the presence of diverse organic components within the essence of the herbs [52]. The latex obtained from *Calotropis procera* was utilized in the production of metal oxide nanofibers, which exhibited an average diameter of 10 nm and a spherical morphology. The aforementioned NFs have been determined to be entirely safe, as evidenced by their low toxicity levels observed only at significantly elevated concentrations. Hence, its application extends to other domains, particularly within the realm of medicine [46,51].

4.3 Strategies for improving nanofluid stability

4.3.1 The surfactants employed in nanofluids

Surfactants are utilized as dispersants in NFs. In the two-step technique, dispersant agents are added to promote NFs stability. Small concentrations of dispersant agents can impact the surface properties of a system. To promote wettability (contact between two substances), dispersant agents with a hydrophilic polar head and a hydrophobic tail are used. Some conductivity between fluids and NPs exists around the two-step system interface, which is the exact optimal spot for a dispersing agent. Surfactants are classified into four major groups based on their head structure: amphoteric surfactants (certain lecithin, betaines), cationic surfactants (may be protonated), anionic surfactants (sulfonates, phosphates, alkyl sulfates, sulfosuccinates, and long-chain fatty acids), and nonionic surfactants (alcohols, polyethylene oxide). The key problem here is selecting suitable dispersants and agents. In the presence of polar solvent NF base fluid, water-soluble surfactants must be employed; otherwise, oil-soluble surfactants must be used. The hydrophilic/lipophilic balance (HLB) determines the solubility of nonionic surfactants. The HLB value exhibits an inverse relationship with oil-soluble surfactants but a direct relationship with water-soluble surfactants. HLB values can be found in a variety of textbooks. Despite the fact that adding surfactant improves NPs dispersion, there are several issues [53]. Heat media transfer is caused by the use of surfactants. Cooling and heating are common operations in heat exchange systems. The surfactant heating causes foam production. Furthermore, molecular attachment of surfactant to the NPs surface increases thermal resistance between the base fluid and NPs, resulting in restrictions in thermal conductivity efficiency.

4.3.2 Surfactant-free surface modification techniques

The use of functional NPs without surfactants is a viable strategy for achieving long-term stability for nanofluids. During the pool boiling process, there was no sedimentation on the heated surface, which was a distinguishing feature of NFs. Using a mechanochemical reaction, Hwang et al. proposed hydrophilic functionalized groups on the surface of nanotubes [54]. Because of their high thermal conductivity, high stability, low viscosity, and lack of media contamination, the suggested NFs can operate as a cooling agent in updated thermal systems. Surfactant-free NFs, including CTNs with single or double walls, were created utilizing a wet mechanochemical approach. The insertion of hydroxyl groups into the surfaces of CTNs was indicated by a zeta potential test and an infrared spectrum. Carbon nanotubes' solvent stability was enhanced by applying standard chemical treatments, resulting in the surficial functionalization of carbon nanotubes [55]. Wepasnick and colleagues summarize carbon nanotube structural alterations in one review [56]. In aqueous solutions, polymethacrylic acid (PMAA) can modify zinc oxide NPs. PMAA carboxyl groups are conjugated to the hydroxyl groups of the zinc oxide NPs surface, resulting in the formation of a poly (zinc methacrylate) complex. PMAA improved the scattering ability of zinc oxide NPs. The crystalline structure of zinc oxide NPs did not change [57].

4.3.3 Mechanisms of nanofluid stability

Dispersed nanoparticles can adhere to one another and form increasing aggregations that deposit due to gravity. Particles will be stable if they do not agglomerate much. The aggregation rate is determined by the possibility of adhesiveness and the frequency of collisions. A colloidal stability theory was developed by Derjaguin-Laudau-Verwey-Overbeek (DLVO) [58,59].

According to DLVO theory, the NP stability in solution is controlled by the double-layered repulsion electric force that exists between particles when they become adjacent due to Brownian motion and the entire attraction of van der Waals force. Only when the attraction force exceeds, the repulsion force do the particles agglomerate and become unstable; otherwise, the suspension remains stable. A sufficient repelling force between particles is required to provide the requisite stability in colloids or NFs. As previously demonstrated, two types of basic mechanisms influence colloid stability. The first type of repulsion is steric repulsion, while the second type is electrostatic repulsion. Polymers, which absorb into the surface of particles, play an important role in achieving more effective steric repulsive stability. PMAA modifies ZnO nanoparticles that are compatible with polar liquids [57].

4.4 The application of zinc oxide nanofluids in heat transfer processes

NFs are composed of various base fluids, including polymer solutions, bio-fluids, oils, ethylene glycols (EGs), water, and other regularly employed fluids. These base fluids are supplemented with suspended NPs that possess diameters within the range of 1–100 nm [60–62]. One of the most important applications of zinc oxide is to improve the heat transfer efficiency of common/conventional liquids in terms of fluid flow continuity and thermal conductivity. Choi and colleagues developed the nanofluids theory to increase heat conductivity [63–65]. A study was performed on NFs and their thermal characteristics were compared and analyzed by researchers in association with nanoparticle suspensions. Some of the relevant NPs in NFs are carbon, carbides, and metal oxide nanotubes. Specific properties of NFs, such as enhanced thermal conductivity of these fluids with small particle concentrations, were surprising and could not be explained by existing theories; this property makes them keystones in a wide range of fields, including vehicle thermal management, electronic devices, engine cooling, and heat transfer [45,66–76]. Many researchers studied the heat transfer properties of NFs experimentally and theoretically [77–79]. Many studies have been conducted on NFs' critical functions in heat transfer. According to previous findings, the use of NFs increases heat transfer by up to 30% [80].

Many research have found that ZnO NFs have better thermal properties than water. Haghshenasfar et al. investigated heat transmission of water/zinc oxide NFs in concentric plate and tube heat exchangers and found 14% and 20% increases in heat transfer, respectively, as compared to the base fluid [81]. Lee et al. examined the thermal conductivity of ZnO/EG NFs, proposing a One-Step physical procedure that results in thermal conductivity augmentation of NFs above the analytical level. According to the literature, NFs do not always improve heat transfer. In fact, despite increasing NP concentration, some NFs exhibit decreased amounts of heat transmission. Despite various study efforts on the topic of oxide NFs, little data on zinc oxide NFs exists/is available [82]. Rahmati determined that packing density influences the thermal efficiency of a mechanical draft wet cooling tower employing NFs such as zinc oxide and water [83]. Multiple concentrations, up to 0.7%, have been generated to investigate the impact of nanoparticles on the thermal efficiency of the wet cooling tower. The experiments found that the number of packing layers mostly predicted thermal efficiency, regardless of whether NFs or water was utilized. In any case, the concentration of NPs had only a tiny effect on thermal efficiency. Furthermore, zinc oxide nanoparticles are being explored as diesel oil additives. Pumping power augmentation and friction factor enhancement were discovered by Mousavi et al. by integrating NPs in the base fluid and a 0.4 wt.% concentration of NPs, respectively [84]. Another study found that incorporating NPs into the base fluid boosted pumping power. Furthermore, with a 0.4 wt.% concentration, NPs increased the friction factor. Dinesh et al. investigated the tribological and thermophysical capabilities of zinc oxide-containing engine oil and discovered that NPs addition improved the engine oil's practical qualities [85]. Este et al. conducted engine oil studies with ZnO and NWCNT. They observed that adding nanoparticles to engine oil reduced damage to several components of the engine caused by a cold vehicle short [86].

Wen and colleagues investigated the flow and thermal properties of zinc oxide dispersed water with two different concentrations (0.75% and 1.5%) synthesized in two distinct multiport micro channels. The results showed that the rate of heat transmission increased by up to 48% for higher Re numbers (i.e., Re = 1430) [87].

Kumar et al. examined the thermal performance of ZnO NFs in a plate heat exchanger using symmetric and mixed chevron angles. The authors demonstrate that at 1% concentrations of ZnO nanoparticles, the maximum improvement in thermal performance occurred at the expense of greater concentrations [88].

Ali et al. investigated the cooling performance of a car radiator using zinc oxide NFs, concluding that the best performance occurred at 46% for a 0.2% ZnO NFs volume concentration [89]. Mc cants et al. observed an increase in cooling activities in a hot plate at a zinc oxide volume concentration of 1.6% due to the zinc oxide NF's low viscosity and density [90]. Islam et al. performed a PEM fuel cell cooling procedure with suspensions, finding another zinc oxide NF characteristic that resulted in a 29% cooling capability progression in 0.5 vol.% NFs [91]. Many researchers, including Hozien et al., Li et al., and Sharifipour et al., found that combining EG, water, and EG-water NFs with zinc oxide NPs resulted in increased levels of heat transfer activity [92–94].

Pourfattah and Asadi studied the thermophysical properties of NFs, such as heat transfer capacities, based on ZnO-engine oil, indicating their favorable attributes to a limited thermal extent. Heating and cooling are critical activities in a variety of businesses [95]. Furthermore, various elements, including temperature and fluid flow rate, determine their associated thermal efficiency. As a result, to achieve an efficient heating and cooling process, their various features were thoroughly examined. NFs are innovative fluids developed to meet the needs for NPs involvement in standard heat transmission fluids. Alina Adrina Minea et al. studied the hydrothermal properties of a newly found NFs for HVAC applications based on zinc oxide and polyethylene glycol (PEG). PEG 400-zinc oxide thermophysical characteristics have previously been studied. The temperature changes associated with various NP loadings were then measured. They are now found in computational tools. The computational findings are achieved by employing Raynolds numbers (200–2000) and NPs concentrations (0.5%–5%). The aforementioned conclusions are investigated using dimensionless pressure drop ratio, friction factor, and Nusselt number. The evaluation performance criteria (EC) are also measured and studied. The recently developed NFs eventually boosted heat transfer by up to 16% while decreasing pressure by 13%. Furthermore, many associations between both the friction factor and the Nusselt number are formed as a result of critical operational situations [96]. Zinc oxide NFs were synthesized in polymer matrixes by Bhagat UK et al. They also studied the rheological properties of NFs and used them in heat transmission applications. The heat transmission applications of aqueous-based zinc oxide NFs were investigated, and it was also discovered that temperature propagation in a sonochemically heated system was greatly reduced by zinc oxide NFs. The capacity of ZnO-containing NFs to absorb heat was increased by 30%–40%. ZnO NPs were widely used in the manufacture of NFs after being distinguished/identified using a variety of advanced instruments such as thermogravimetric analysis (TGA), particle dimensions analysis, scanning electron microscope (SEM), X-ray diffraction (XRD), Raman spectroscopy, and ultraviolet (UV)-visible studies. XRD examination revealed that the NPs with diameters of 19–30 nm had a hexagonal crystal structure [75].

4.5 Biomedical applications of zinc oxide nanofluids

Zinc oxide NFs are important in biological and environmental cleanup applications due to their excellent antibacterial and antioxidant properties. This section provides an overview of the most recent studies to learn more about zinc oxide NFs characteristics, green manufacture, and a variety of biomedical applications.

4.5.1 Anticancer effects of zinc oxide nanofluids

Cancer is recognized worldwide as a major cause of mortality in humans [97,98]. One of the best therapeutic techniques for different malignancies is NPs injection into the nearest blood vessel to the neoplasm site, then planting a magnet near the tumor. Along with the planted magnet field with changing features, the NPs behave as heat resources. When the temperature rises to the range of 42°C–45°C, the tumor tissues are devastated [99]. The drug-NP conjugation can also be injected in the vicinity of the tumor. Next, the cancerous tissue absorbs the drug through the high-gradient magnetic field around the tumoral site [100]. High concentrations of NPs result in better drug absorption. Moreover, uniform NPs dispersion in blood circulation is prevented by the magnetic force [101], leading to very low levels of side effects, despite using antitumor drugs at high doses. Cancerous stem cells (CSCs) play a central role in cancer therapeutic resistance, relapse, metastasis, and growth. This paper reviews three zinc oxide NFs, affecting the apoptosis, proliferation, and stemness markers of stem-like cells of breast neoplasm. Flow cytometry and MTS assay, respectively, evaluated the apoptotic and antiproliferative characteristics of zinc oxide NPs on mammospheres enriched by stem-like cells of breast cancer. Real-time polymerase chain reaction (PCR) determined the stemness markers expression, such as ALDH3A1, SOX2, CXCR4, β-catenin, NOTCH1, and WNT1. Signal Transducer and Activator of Transcription 3 and Janus kinase 2 phosphorylation was analyzed by Western blot. Zinc oxide NFs considerably reduce the stemness markers, specifically sample (c) with zinc oxide-148 code in which PEG solution is added with a varied arrangement at the formulation termination, resulting in a significant reduction of the whole markers in comparison with the controls. The reports demonstrated that all researched zinc oxide NFs; especially ZnO-148 could induce apoptosis and decrease the viability of parental and spheroidal cells. Knowing that ZB4 acts as an extrinsic apoptotic pathway blocker and CD95L is a death ligand. The recent data showed that there was no apoptosis induction via the extrinsic pathway. It was also indicated that expression downregulation of the Bcl-XL and Mcl-1 by zinc oxide NPs led to obvious blockage of the JAK/STAT pathway. As discussed earlier, breast CSCs can be destroyed by zinc oxide NFs via JAK/STAT pathway blockage, apoptosis induction, and stemness markers [102]. This article is intended to investigate low doses of zinc oxide NFs and the ultraviolet A (UVA) radiation effects on the doxorubicin (DOX) cellular absorption in the DU145 cell line of the human prostate. The solvothermal technique was used to develop zinc oxide NPs. Then 10% bovine serum albumin was used as a scattering agent. 3-(4,5-dimethylthiazol-2-yl)-2,5-diphenyltetrazolium bromide test evaluated DOX cytotoxic impact alone or combined with various concentrations (0.95–15.6 μg/mL) of zinc oxide NFs, first accompanied by UVA radiation and then in the absence of it on the DU145 cell line. UV-Vis uptake spectroscopy and fluorescence microscopy assessed DOX remaining in and out of the DU145 cells, respectively. Flow cytometry, comet assay, and DAPI (4',6-diamidino-2-phenylindole) staining were used to evaluate UVA radiation and zinc oxide NFs impact on cell cycle arrest and apoptosis induced by DOX. The intracellular delivery and cytotoxicity of DOX were promoted by zinc oxide NFs at low doses (0.95 μg/mL) along with radiation in the DU145 cell line. Despite an increase in the DNA tail and chromatin condensation/fragmentation of DU145 cells with simultaneous zinc oxide NFs and DOX utilization, there were no changes in cell cycle promotion, following UVA radiation. The studies reported that DOX efficiency was increased by zinc oxide NFs when accompanied by radiation in the DU145 cell line, recommending that the mentioned modality combination could be encouraging in the treatment of neoplasms [103]. A new method for breast tumor therapy is unique nanostructures produced from zinc oxide NPs. In this study, surface modification techniques, sol-gel hydrothermal process, and coprecipitation containing fast quenching resulted in the fabrication of two kinds of zinc oxide nanomaterial. 3-(4, 5-dimethylthiazol-2-yl)-2, 5-diphenyl tetrazolium bromide (MTT) test assessed the cytotoxicity impacts on the MCF-7 cells proliferation. By zinc oxide NFs concentration augmentation at 48 and 72 hours of therapy, there was a reduction in the viability of the MCF-7 cell line of the breast tumor. The study, with a P value under 0.05, reported that after 72 hours of therapy using zinc oxide (a) and zinc oxide (c), the level of MCF-7 cell line IC50, respectively, equaled 51.02 and 48.63 μg/mL [104]. Nanoformulations originated from fine penetrable zinc oxide quantum dot nanoparticles suggest noticeable medical applications, specifically in cancer therapeutics. By employing methods, such as rapid cold quenching, sol-gel hydrothermal procedure, and surface functionalization to achieve super small sizes of about 1–4 nm, zinc oxide quantum dot NPs were fabricated. The zinc oxide nanopolymer was used in the NFs fabrication as a PEG cosolvent, water/oil emulsion stabilizer, and wetting agent. The fabricated NFs was distinguished by UV-Vis spectroscopy, zeta potential, band gap energy, photoluminescence, FTIR, and SEM. MTT assessment evaluated the cytonoxious impacts on four cancer cell-lines proliferation. IC50 levels were measured to be 35, 40, 41, and 30 (μg/mL) for HeLa, HT29, MCF-7, and KB44 cell-lines, respectively, proposing cytotoxicity of nanoformulation theory on four cell-lines after 48 hours in a pattern dependent on concentration with ($P < .05$). ZnO NFs destroyed cancerous cells more efficiently than the normal human foreskin fibroblast cells (HFF-2)(IC_{50} = 105 μg/ml). Treatment with zinc oxide NFs led to increased apoptosis by cell viability reduction. The antimicrobial function and skin sensitization experiment were done. Seventy white rat models were tested in vivo for side effects evaluation, which showed only 3–4 abnormal results in the rat liver tissue owing to unusual drug response [105]. Combination therapy in cancer treatment reduces dose-related adverse effects while increasing the therapeutic efficiency of chemotherapy drugs such as DOX. The greatest challenge to reach the highest efficiency in cancer treatment is to conquer DOX side effects. Utilization of zinc oxide NPs is known as an efficient method for a large number of biomedical practices. Achieving a combined anticancer therapy method in order to reach the uttermost DOX efficiency and uptake by cells is the main goal of this study. Zinc oxide nanoparticles were fabricated by solvothermal technique and distinguished by TEM, dynamic light scattering, and XRD.

Scattered in the bovine 10% serum albumin, the NPs of zinc oxide and their toxicity were revealed with and without DOX utilization on breast tissue cancerous DU145 cells. Semiquantitative reverse transcription-PCR and fluorescence microscopy evaluated ZnO NPs' effect on the DOX-mediated catalase mRNA expression and DOX uptake by cells, respectively. The results of MTT showed that in a dose and time-dependent manner, zinc oxide NFs led to a reduction in the viability of DU145 cells. Zinc oxide NFs combined with DOX concurrent therapy indicated a remarkable augmentation in DOX cellular absorption and anticancer function in comparison with DOX usage individually. In addition, DOX and ZnO NFs usage alone or combined led to a decrease in the expression of catalase mRNA in a time-dependent manner. Altogether, the results show the pivotal role of zinc oxide NFs in cell proliferation and delivery of drugs in the DU145 cell line. As a result, NFs of ZnO utilization could be a promising anticancer therapeutic method ready to be combined with chemotherapy [106].

4.5.2 Zinc oxide nanofluid bioimaging

Zinc oxide nanoparticles are promising candidates for cancer diagnosis and imaging because of their various physical, chemical, and magnetic characteristics. Having unique optimal characteristics, such as remedial constituents, numerous sensing transportation capacities, and adjustable surface modifications, NPs have lately been used in bioimaging. Anyhow, NPs must have the capacity to describe brisk alterations of the biologic processes, satisfactory photo-stability, and biocompatibility with the tested biological mechanisms and distinguished optical potentials in the bioimaging field. Speaking about bioimaging, there are two classifications of NPs: NPs belonging to the first class are those that project definite optical signals such as carbon dots and quantum. NPs belonging to the second class are those that need to be labeled by fluorophore to become visible [49]. Sung et al. have revealed a unique in situ bioimaging technique using fluorescence to investigate NFs' real-time micro-flow arrangements inside an expanding and contracting soft micro-tube, which allows researchers to monitor the physiological behavior of the micro-fluids spatiotemporally, as well as channel fluctuation through the peristaltic transportation of suspensions of florescent-labeled nanoparticles via lymphatic vessels of the hind-limb tissues. Using high optical contrast of the fluorescent NFs made the lymphatic micro-tube visible, and then the NFs viscosity and concentration could be measured. The behavioral pattern of the micro-tube NFs mechanics of the mesenteric peristaltic base fluid flow resembles that in the hind-limb lymphatic vessels. Moreover, the expansion and contraction of the micro-tube resulted in forward flow promotion and the development of an inert flow at the uttermost expansion and contraction, corresponding to Poiseuille flow. Also, it was found that greater shear wall stress was associated with augmented axial flow velocity [107].

Shelat et al. demonstrated that labeling mesenchymal stem cells took place after surface adjustment and stabilization of superparamagnetic NPs with biocompatible amino acids. After cytotoxicity evaluation, it was observed that amino acid stem cell magnetic NFs possessed minimum toxicity. It was also revealed that the suggested NFs could be used in magnetic resonance imaging (MRI) scans as a contrast [108]. Yogesh et al. fabricated ZnO NPs by using nanosecond pulsed laser ablation and evaluated their potential to be applied in bioimaging. Based on the results, nanoparticles of graphene oxide can be used for bioimaging purposes due to their biocompatibility and photoluminescent properties [109]. Talev et al. reviewed the applications of zinc oxide NPs in atherosclerosis molecular bioimaging and therapy [110]. Properties such as high stability against solar light and air, lack of contamination, cost-effectiveness, and high level of safety make ZnO NPs an encouraging candidate for imaging purposes. The greatest challenge of researchers is to fabricate florescent NPs with high stability in watery environments. Water molecules had the ability to destroy the surface of ZnO from which the fluorescence visibility comes. This problem can be solved by appropriate surface regulation. The best technique to modify zinc oxide quantum dots is sol-gel application because there has been a progression in wet chemical methods, and ZnO surface defects have not been damaged in room temperature fabrication reactions [111]. In the two-step method binary polymer shell with a hydrophobic polyester inner layer and hydrophilic polyether outer layer was used to coat zinc oxide NPs [112]. For the first time, core-shell NPs of zinc oxide polymer were used successfully in cell imaging due to their fluoroscopy stability against water solutions (quantum yield over 50% for weeks). Also, the injection of nanoparticles into the rats could be possible. There was no observable toxicity in the histopathologic assessments as well as clinical urine and blood test results [113]. Later, other researchers used saline groups and silica shells to coat ZnO quantum dots, which showed more stability in animal blood and cell culture. The synergism of magnetic elements and zinc oxide NPs, in order to achieve altered magnetic and phosphorescent binary probes, led to the detection of labeled deep animal body tissues identification by MRI method. Additionally, it was reported that zinc nanoparticles combined with fundamental and rare earth elements surprisingly showed noticeable fluorescence enhancement. For instance, ZnO NPs, conjugated with Gd^{3+} via the sol-gel route, were applied to label human HeLa cancerous cells of the cervix, being detectable by both fluorescence microscopy and MRI [111].

4.5.3 Antibacterial activity of zinc oxide nanofluids

Pathogen resistance increases more and more as time passes due to common antibacterial drug overuse. Accordingly, discovering new and practical methods to overcome pathogen resistance is essential. In recent years, different suspensions of nanostructures have been found to have antibacterial characteristics [114]. Antibacterial agents have numerous applications in industries, such as food packaging, medical fields, the textile industry, and water disinfection. Among the metal oxide nanoparticles, zinc oxide has been the most effective antibacterial agent, which inhibits the growth of a wide spectrum of bacteria [44]. Most of the previous studies have investigated ZnO NPs' antibacterial impact on human cells and concluded that these NFs are safe, whereas some ZnO NPs' toxic effects have been revealed by FESEM and SEM through ZnO NPs-treated bacterial wall structural study. Finally, the results demonstrated bacterial wall injury by zinc oxide nanoparticles. Antibacterial ions release, specifically Zn^{2+} and reactive oxygen species (ROS) production were associated with ZnO NPs toxicity [115,116]. The studies reported that the stability of zinc oxide NFs has been promoted by using scattering agents, including sodium carbonate, PEG, and deflocculants such as sodium silicate [44]. Moreover, UV and radiation resulted in ROS creation and accordingly more antibacterial activity. As discussed earlier, zinc oxide metallic NFs have a few defects in solubility, excretion, stability, and biocompatibility. It has been observed that dissolved nanoparticle metal ions release was associated with toxicity levels. The morphology and nature of nanoparticles are strongly related to their biological features. Thus considering size, shape, and structure adjustment as well as cytotoxicity reduction can help to enhance their biocompatibility [44].

The electrospun metal oxide/polylactic acid (PLA) NFs fibrous membrane with excellent antibacterial properties, hydrophilicity, and stiffness is synthesized by utilizing solvent-free metal oxide, which has ionic conductivity and amphiphilicity. The mentioned metal oxide/PLA NFs' fibrous membrane displays outstanding mechanical activity with elongation and tensile performance at break of 97.32 and 3.68 MPa at 5 wt.% loading, which equals 8 and 4 times more than that in PLA used alone, respectively. In addition, it was reported that wicking characteristics, wetness absorption capability, antistatic feature, and hydrophilicity were considerably increased due to the electrospun procedure during which the metal oxide moves to the PLA fibers. Additionally, the membrane exhibited rapid water filtration at a rate of 3500 $L/m^2/h$, propelled by the force of gravity. This filtration rate is 10–12 times greater than that of commercially available ultrafiltration membranes. The fabricated fibers expressed separation effectiveness of more than 90% as well as great antibacterial properties against *Staphylococcus aureus* (99%) and *Escherichia coli* (95%), following an ion-interchange reaction using salt solution. Generally, organic NFs are perfect substitutes for biodegradable hydrophilic fabric, based on PLA. They can be used in absorptive biodegradable materials, antibacterial clothes, and water refinement [117]. New antimicrobial drug development is a great concern due to increasing antibiotic resistance amongst nosocomial pathogens. Antibiotics that are conjugated with zinc oxide nanoparticles decreased the pathogen resistance. In this study, the effects of zinc oxide nanoparticles on the antibacterial properties of ceftazidime (CAZ) and ciprofloxacin (CP) against *Acinetobacter baumannii* and *Enterococcus faecalis* were evaluated in an acidic environment with PH of 5.5, respectively. Zinc oxide nanoparticles were fabricated and differentiated by solvothermal technique. In this study, the highest activity against bacterial growth was noticed at 0.125 mg/mL in the MIC90 value of zinc oxide nanoparticles belonging to *E. coli*. Also, the mentioned value for *A. baumannii* was 0.25 mg/mL. The interaction between ZnO NPs and each of the two antibiotics was shown by the Fourier transform infrared spectroscopy spectra of ZnO NPs treated with antibiotics. It was proved that under-inhibitory concentrations, zinc oxide nanoparticles did not affect CAZ and CP antibacterial activities against *A. baumannii* and *E. faecalis*, respectively. It was evaluated that zinc oxide nanoparticles had synergic antibacterial effects with CAZ against *A. baumannii*. CAZ antibacterial function was promoted against *A. baumannii* via Zn^{2+} release, which led to a bacterial permeable membrane [118]. Another recently published study investigated the effects of a sonophotocatalytic procedure inactivation by ZnO NFs utilization, owning time, frequency, and power density. The results demonstrated that the natural light photocatalytic procedure and ultrasonic radiation combination resulted in a 20% increase in the inactivation effectiveness. Taken together, the interaction effectiveness from highest to lowest belongs to sonophotocatalysis, photocatalysis, sonocatalysis, and ultrasonic irradiation. When they are compared, using *E. coli*, sonophotocatalysis inactivation effectiveness is suggested to be the best to apply in practice. Studies have found that ROS play a more pivotal role than Zn^{2+} in bacterial inactivity. It was also revealed that the sonophotocatalytic process remarkably augmented the permeability of the inner membrane and outer membrane of the *E. coli* bacteria cells. The sonoporation phenomenon, zinc oxide sonocatalysis, and acoustic cavitation properties were assisted by the effects of enhanced bacterial inactivation on the sonophotocatalytic process [119]. As the results show, NFs of ZnO possess bacteriostatic characteristics. Furthermore, at high zinc oxide concentrations, there was an obvious association between bacterial cell membranes and zinc oxide NFs [46]. Jalal et al. studied the zinc oxide NPs suspension antibacterial effects on the *E. coli* bacterial cells via a green method, suggesting that the bacterial survival ratio expressed a reduction and had a direct relation with zinc oxide concentration and time [120]. Another study undoubtedly revealed that zinc oxide NPs possess a variety of antibacterial impacts on several bacteria. It is proposed that normal visible light spectrum and size may be associated with zinc oxide antibacterial performances. Additionally, this study highlighted the direct relation between zinc oxide NPs concentration and inhibitory properties. Extraordinary antibacterial activities of zinc NFs against one of the most common food pathogens called *E. coli* variant O157 H7 were also revealed. In detail, the proteins and lipids originating in the bacterial cell membrane are altered by zinc oxide NPs. Also, it was found that bacterial cell membrane deformity, intracellular constituents loss, and eventually cell death are impacts of zinc oxide NPs, which can be a promising efficient antibacterial agent/factor to be used in food and agricultural industries [46].

4.6 Application of zinc oxide nanofluids as photocatalysts

Zinc oxide NFs have been used in many fields, such as cosmetics, fertilizers, varistors, ceramics, paints, rubber, and photocatalysis. Owning exceptional properties, including great free-exciton binding energy, excellent photocatalytic features, potent oxidation capability, and direct and wide band gap in the near-UV spectrum, ZnO NFs seems to be a promising candidate for green environmental system management. These properties also lead to persistence in excitonic emission binding energy at room or higher temperatures. Because the band gap energy of zinc oxide and titanium dioxide (TiO_2) (3.2 eV) are approximately the same, their photocatalytic abilities are also alike. Additionally, using zinc oxide is much more economical for water filtration than titanium dioxide, plus zinc oxide can absorb a broad spectrum of solar rays as well as more light quanta with semiconductive metal oxide [121].

4.6.1 Mechanisms of zinc oxide photocatalysis

Photocatalysis happens when a zinc oxide photocatalyst is emitted by light with energy more than its band gap. Electron excitation from the valence to the conduction band and charge separation is proved by light energy, leaving holes in the valence band. Initially, the photo-generated hþ/e carriers move to the zinc oxide photocatalysis surface. At the same time, hþ and e are combined again and lead to quantum yield reduction. Surface modifications and photocatalyst structures both affect the recombination rate. The arrival of reactive hþ and e to the zinc oxide surface results in reduction reactions and oxidation simplification that produce too much ROS, such as radical (OH^-) and superoxide anion (O^{-2}). The zinc oxide conduction band bottom (0.5 V vs normal hydrogen electrode (NHE)) is much more negative than the redox capability of O_2/O_2 (0.33 V vs NHE). The excited electrons have the ability to generate O_2. Simultaneously, the top part of the zinc oxide valence band (þ2.7 V vs NHE) is much more positive than the redox capability of OH/H_2O (þ2.53 V vs NHE). Therefore these holes oxidize the water molecules and produce highly reactive radicals of hydroxyl. The reactive radicals of OH^- and O^{-2} straightly lead to contaminant biological molecules oxidization into solutions [122]. The large band gap, only active in the presence of UV light radiation, is the most important disadvantage of zinc oxide photocatalyst. Consequently, solar light utilization becomes restricted. Different techniques, including coupling with carbon materials or the semiconductors, depositing noble metals and metal or nonmetal atoms combination have been discovered to conquer the earlier mentioned disadvantage [122]. Recently, NPs production using herbal essence, an interesting and challenging method with many beneficial properties, has impressed the nanoscientists and nanobiotechnologists, surpassing the previous conventional physiochemical methods. In this study, the zinc oxide nanoparticles are made from different copper-dopant concentrations of *Synadenium grantii* leaf essence. Copper nitrate trihydrate and zinc nitrate hexahydrate were as precursors in herb leaves essence to fabricate pure copper-conjugated zinc oxide nanoparticles. In order to evaluate the optical and physical characteristics of the fabricated samples, the measurements of PL, XPS, FTIR, TEM, and XRD were done. Organic pollutants, such as rhodamine B (RhB), indigo carmine (IC), and methylene blue (MB) were used to study the photocatalytic properties of the samples. TEM and XRD analyses confirmed the wurtzite crystal structure of the fabricated samples. Additionally, FTIR analysis confirmed the functional groups' existence in the synthetized samples. Also, the binding energies of dopant ions and host materials were confirmed by XPS analysis. Maximum amounts of the emission were detected at 573, 446, and 424 nm, which are related to oxygen vacancy, zinc interstitial to the zinc vacancy, and electron movement from the deep donor level, respectively. It was found that 3% and 5% of copper-conjugated fabricated samples showed higher photocatalytic behaviors for RhB, IC, and MB organic pollutants. Great photocatalytic activity was performed by the green fabricated zinc oxide nanoparticles, indicating that bio-synthesis is the best way to achieve functional and biocompatible samples [123].

4.7 Conclusion and outlook

Zinc oxide NFs have a wide range of applications. One of the most prominent ones is in the medical and biomedical fields. This study discussed zinc oxide nanosuspensions' different utilizations, including antibacterial activities, illness diagnosis, medical therapy, and medication delivery systems. According to these items, the previous related papers are reviewed, discussed, and analyzed in the current article. The main goal of this review is to organize and summarize the previous research as well as suggest new methods for further investigations about the topic of NFs usage in biomedical diagnosis and therapy, as well as their antibacterial functions. Prolonged stability and agglomeration prevention were the main goals in the medication delivery area. Lots of articles have been published and many researchers have focused on the nanocolloidal drug's bioviability reduction. Having unique characteristics, ZnO nanocrystals have recently been used frequently. Magnetic NP-based hyperthermia is used to make tumoral cells more sensitive against anticancer drugs and therapies, killing them or heating the tumoral tissues. Thermal conductivity enhancement by zinc oxide NFs led to tumoral tissue destruction. Although some anticancer effects of zinc oxide NFs have been revealed, more detailed investigations about each disease and malignancy are needed. Recently, NFs utilization in targeted MRI, the most common imaging method for diagnosis, has interestingly increased. Additionally, ZnO NFs are considered as contrast agents in MRI. To reach the optimal contrast, trying other NFs may be beneficial. According to the literature review and the importance of the previously done studies, finding new utilizations of the zinc oxide NFs in medical and biomedical fields is essential. Also, further research and experiments on ZnO NFs in vivo and in vitro must be done. Besides zinc oxide NFs, different types have to be investigated in the fields of disease diagnosis and treatment, antibacterial therapy, and medication delivery.

References

[1] M. Rostami, M. Rahimi-Nasrabadi, A. Ghaderi, A. Hajiabdollah, H.R. Banafshe, A. S. Nasab, $ZnFe_2O_4$@ L-cysteine-N/RGO as efficient nano-sonosensitizers, pH-responsive drug carriers and surface charge switchable drug delivery system for targeted chemo-sonodynamic therapy of cancer, Diam. Relat. Mater. 133 (2023) 109701.

[2] A. Abedini, M. Rostami, H.R. Banafshe, M. Rahimi-Nasrabadi, A. SobhaniNasab, M.R. Ganjali, Utility of biogenic iron and its bimetallic nanocomposites for biomedical applications: a review, Front. Chem. 10 (2022) 893793.

[3] M. Akbari, M. Nejati, A. Davoodabadi, E.A. Kashi, A.N. Alam, A.S. Nasab, The antibacterial effects of terbium vanadate-silver peroxide nanostructures against surgical wounds infected by *Staphylococcus aureus* and *Pseudomonas aeruginosa* infections in a rat model, Inorg. Chem. Commun. 150 (2023) 110390.

[4] A.M. Sorouri, A. Sobhani-Nasab, M.R. Ganjali, S. Manani, H. Ehrlich, Y. Joseph, et al., Metal tungstates nanostructures for supercapacitors: a review, Appl. Mater. Today 32 (2023)101819.

[5] A.S. Nasab, K. Adib, H. Afshari, M.R. Ganjali, M. Rahimi-Nasrabadi, F. Ahmadi, Synthesis of praseodymium titanate nanoparticles supported on core–shell silica coated magnetite via mild condition and their photocatalytic capability evaluation, J. Mater. Sci.: Mater. Electron. 32 (2021) 13527–13538.

[6] A. Abbasi, H. Khojasteh, A.H. Keihan, K. Adib, A. Sobhani-Nasab, M. Rahimi-Nasrabadi, Co-precipitation synthesis of Ag-doped $NiCr_2O_4$ nanoparticles: investigation of structural, optical, magnetic, and photocatalytic properties, J. Mater. Sci.: Mater. Electron. 32 (2021) 1413–1426.

[7] H. Halakoei, M. Ghalkhani, A. Sobhani-Nasab, M. Rahimi-Nasrabadi, An efficient electrochemical sensor based on $CeVO_4$-$CuWO_4$ nanocomposite for methyldopa, Mater. Res. Express 8 (2021) 085001.

[8] A. Sobhani-Nasab, S.M. Hoseinpour, M. Rahimi-Nasrabadi, S. Pourmasoud, M. Eghbali-Arani, F. Ahmadi, Synthesis of Fe_3O_4/$CdWO_4$/carbon dots heterostructure with excellent visible light photocatalytic stability and activity for degradation of 4-nitrophenol and organic pollutant, J. Mater. Sci.: Mater. Electron. 32 (2021) 26998–27013.

[9] M. Akbari, M.S. Mohammadnia, M. Ghalkhani, M. Aghaei, E. Sohouli, M. Rahimi-Nasrabadi, et al., Development of an electrochemical fentanyl nanosensor based on MWCNT-HA/Cu-H_3BTC nanocomposite, J. Ind. Eng. Chem. 114 (2022) 418–426.

[10] E. Sheibani, A. Hosseini, A. Sobhani Nasab, K. Adib, M.R. Ganjali, S.M. Pourmortazavi, et al., Application of polysaccharide biopolymers as natural adsorbent in sample preparation, Crit. Rev. Food Sci. Nutr. 63 (2023) 2626–2653.

[11] M. Ghalkhani, E.M. Khosrowshahi, E. Sohouli, K. Eskandari, M. Aghaei, M. Rahimi-Nasrabadi, et al., Electrochemical monitoring of carbamazepine in biological fluids by a glassy carbon electrode modified with CuO/$ZnFe_2O_4$/rGO nanocomposite, Surf. Interfaces 30 (2022) 101943.

[12] F. Gandomi, A. Sobhani-Nasab, S. Pourmasoud, M. Eghbali-Arani, M. Rahimi-Nasrabadi, Synthesis of novel Fe_3O_4@SiO_2@Er_2TiO_5 superparamagnetic core–shell and evaluation of their photocatalytic capacity, J. Mater. Sci.: Mater. Electron. 31 (2020) 10553–10563.

[13] M. Eghbali-Arani, S. Pourmasoud, F. Ahmadi, M. Rahimi-Nasrabadi, V. Ameri, A. Sobhani-Nasab, Optimization and detailed stability study on coupling of $CdMoO_4$ into $BaWO_4$ for enhanced photodegradation and removal of organic contaminant, Arab. J. Chem. 13 (2020) 2425–2438.

[14] A. Moradi Hasan-Abad, M.A. Esmaili, A. Ghotaslou, A. Atapour, A. Khoshroo, E. Naghian, A review: electrochemical biosensors for testosterone detection, Anal. Bioanal. Electrochem. 14 (2022) 1060–1077.

[15] A. Moradi Hasan-Abad, M.A. Esmaili, M. Akbari, A.M. Sorouri, L. Hosseinzadeh, A. Sobhaninasab, Voltammetric electrochemical sensor for rapid and convenient morphine detection: a review, Anal. Bioanal. Electrochem. 14 (2022) 1152–1169.

[16] E. Naghian, E. Marzi Khosrowshahi, E. Sohouli, H.R. Pazoki-Toroudi, A. Sobhani-Nasab, M. Rahimi-Nasrabadi, et al., Electrochemical oxidation and determination of antiviral drug acyclovir by modified carbon paste electrode with magnetic CdO nanoparticles, Front. Chem. 8 (2020) 689.

[17] M. Rostami, A.S. Nasab, M. Fasihi-Ramandi, A. Badiei, M.R. Ganjali, M. Rahimi-Nasrabadi, et al., Cur-loaded magnetic $ZnFe_2O_4$@ mZnO-Ox-pg-C_3N_4 composites as dual pH-and ultrasound responsive nano-carriers for controlled and targeted cancer chemotherapy, Mater. Chem. Phys. 271 (2021) 124863.

[18] S. Behvandi, A. Sobhani-Nasab, M.A. Karimi, E. Sohouli, M. Karimi, M.R. Ganjali, et al., Synthesis and characterization of $Sm_2(MoO_4)_3$, $Sm_2(MoO_4)_3$/GO and $Sm_2(MoO_4)_3/C_3N_4$ nanostructures for improved photocatalytic performance and their anti-cancer the MCF-7 cells, Polyhedron 180 (2020) 114424.

[19] M. Nejati, M. Rostami, H. Mirzaei, M. Rahimi-Nasrabadi, M. Vosoughifar, A.S. Nasab, et al., Green methods for the preparation of MgO nanomaterials and their drug delivery, anti-cancer and anti-bacterial potentials: a review, Inorg. Chem. Commun. 136 (2022)109107.

[20] M. Amin Marsooli, M. Rahimi Nasrabadi, M. Fasihi-Ramandi, K. Adib, S. Pourmasoud, F. Ahmadi, et al., Synthesis of magnetic $Fe_3O_4/ZnWO_4$ and $Fe_3O_4/ZnWO_4/CeVO_4$ nanoparticles: the photocatalytic effects on organic pollutants upon irradiation with UV-Vis light, Catalysts 10 (2020) 494.

[21] M. Akbari, M. Rahimi-Nasrabadi, M. Eghbali-Arani, H.R. Banafshe, F. Ahmadi, M. R. Ganjali, CdTe quantum dots prepared using herbal species and microorganisms and their anti-cancer, drug delivery and antibacterial applications; a review, Ceram. Int. 46 (2020) 9979–9989.

[22] M. Rahimi-Nasrabadi, A. Ghaderi, H.R. Banafshe, M. Eghbali-Arani, M. Akbari, F. Ahmadi, et al., Preparation of $Co_2TiO_4/CoTiO_3$/polyaniline ternary nano-hybrids for enhanced destruction of agriculture poison and organic dyes under visible-light irradiation, J. Mater. Sci.: Mater. Electron. 30 (2019) 15854–15868.

[23] A. Sobhaninasab, A. Ghaderi, H.R. Banafshe, M. Bakhshi, E. Sohouli, R. Eshraghi, et al., New synthesis and characterization of $NiMoO_4/Mn(VO_3)_2$ heterostructures for electrochemical detection of dopamine, Anal. Bioanal. Electrochem. 15 (2023) 506–515.

[24] V. Trisaksri, S. Wongwises, Critical review of heat transfer characteristics of nanofluids, Renew. Sustain. Energy Rev. 11 (2007) 512–523.

[25] S. Özerinç, S. Kakaç, A.G. Yazıcıoğlu, Enhanced thermal conductivity of nanofluids: a state-of-the-art review, Microfluidics Nanofluidics 8 (2010) 145–170.

[26] S. Kakaç, A. Pramuanjaroenkij, Review of convective heat transfer enhancement with nanofluids, Int. J. Heat. Mass. Transf. 52 (2009) 3187–3196.

[27] A. Khoobi, F. Shahdost-fard, M. Arbabi, M. Akbari, H. Mirzaei, M. Nejati, et al., Sonochemical synthesis of $ErVO_4/MnWO_4$ heterostructures: application as a novel nanostructured surface for electrochemical determination of tyrosine in biological samples, Polyhedron 177 (2020) 114302.

[28] Y. Li, S. Tung, E. Schneider, S. Xi, A review on development of nanofluid preparation and characterization, Powder Technol. 196 (2009) 89–101.

[29] X.-Q. Wang, A.S. Mujumdar, A review on nanofluids-part I: theoretical and numerical investigations, Braz. J. Chem. Eng. 25 (2008) 613–630.

[30] X.-Q. Wang, A.S. Mujumdar, Heat transfer characteristics of nanofluids: a review, Int. J. Therm. Sci. 46 (2007) 1–19.

[31] Z. Han, F. Cao, B. Yang, Synthesis and thermal characterization of phase-changeable indium/polyalphaolefin nanofluids, Appl. Phys. Lett. 92 (2008) 243104–243106.

[32] W. Yu, H. Xie, W. Chen, Experimental investigation on thermal conductivity of nanofluids containing graphene oxide nanosheets, J. Appl. Phys. 107 (2010) 94317–94320.

[33] W. Yu, D.M. France, D. Singh, E.V. Timofeeva, D.S. Smith, J.L. Routbort, Mechanisms and models of effective thermal conductivities of nanofluids, J. Nanosci. Nanotechnol. 10 (2010) 4824–4849.

[34] A. Mishra, P. Tripathy, S. Ram, H.-J. Fecht, Optical properties in nanofluids of gold nanoparticles in poly (vinylpyrrolidone), J. Nanosci. Nanotechnol. 9 (2009) 4342–4347.

[35] T. Sharma, A.L.M. Reddy, T. Chandra, S. Ramaprabhu, Development of carbon nanotubes and nanofluids based microbial fuel cell, Int. J. Hydrog. Energy 33 (2008) 6749–6754.

[36] X. Fan, H. Chen, Y. Ding, P.K. Plucinski, A.A. Lapkin, Potential of 'nanofluids' to further intensify microreactors, Green. Chem. 10 (2008) 670–677.

[37] L. Zhang, J. Xia, Q. Zhao, L. Liu, Z. Zhang, Functional graphene oxide as a nanocarrier for controlled loading and targeted delivery of mixed anticancer drugs, Small 6 (2010) 537–544.

[38] Y.S. Kim, Y.H. Kim, Application of ferro-cobalt magnetic fluid for oil sealing, J. Magnetism Magnetic Mater. 267 (2003) 105–110.

[39] K.V. Wong, Nanotechnology and Energy, CRC Press, 2017.

[40] E.-K. Lim, E. Jang, K. Lee, S. Haam, Y.-M. Huh, Delivery of cancer therapeutics using nanotechnology, Pharmaceutics 5 (2013) 294–317.

[41] E.-K. Lim, T. Kim, S. Paik, S. Haam, Y.-M. Huh, K. Lee, Nanomaterials for theranostics: recent advances and future challenges, Chem. Rev. 115 (2015) 327–394.

[42] S. Stankic, S. Suman, F. Haque, J. Vidic, Pure and multi metal oxide nanoparticles: synthesis, antibacterial and cytotoxic properties, J. Nanobiotechnol. 14 (2016) 1–20.

[43] F. Yu, Y. Chen, X. Liang, J. Xu, C. Lee, Q. Liang, et al., Dispersion stability of thermal nanofluids, Prog. Nat. Sci.: Mater. Int. 27 (2017) 531–542.

[44] M. Sheikhpour, M. Arabi, A. Kasaeian, A. Rokn Rabei, Z. Taherian, Role of nanofluids in drug delivery and biomedical technology: methods and applications, Nanotechnol. Sci. Appl. 13 (2020) 47–59.

[45] K.V. Wong, O. De Leon, Applications of nanofluids: current and future, Adv. Mech. Eng. 2 (2010) 519659.

[46] W. Yu, H. Xie, A review on nanofluids: preparation, stability mechanisms, and applications, J. Nanomater. 2012 (2012) 1–17.

[47] A. Arshad, M. Jabbal, Y. Yan, D. Reay, A review on graphene based nanofluids: preparation, characterization and applications, J. Mol. Liq. 279 (2019) 444–484.

[48] B. Sanjeevi, K. Loganathan, Synthesis of multi wall carbon nanotubes nanofluid by using two step method, Therm. Sci. 24 (2020) 519–524.

[49] E.M. Hamad, A. Khaffaf, O. Yasin, Z. Abu El-Rub, S. Al-Gharabli, W. Al-Kouz, et al., Review of nanofluids and their biomedical applications, J. Nanofluids 10 (2021) 463–477.

[50] I. Mahbubul, E. Elcioglu, M. Amalina, R. Saidur, Stability, thermophysical properties and performance assessment of alumina–water nanofluid with emphasis on ultrasonication and storage period, Powder Technol. 345 (2019) 668–675.

[51] A. Waris, M. Din, A. Ali, S. Afridi, A. Baset, A.U. Khan, et al., Green fabrication of Co and Co_3O_4 nanoparticles and their biomedical applications: a review, Open. Life Sci. 16 (2021) 14–30.

[52] K. Kombaiah, J.J. Vijaya, L.J. Kennedy, K. Kaviyarasu, R.J. Ramalingam, H.A. Al-Lohedan, Green synthesis of Co_3O_4 nanorods for highly efficient catalytic, photocatalytic, and antibacterial activities, J. Nanosci. Nanotechnol. 19 (2019) 2590–2598.

[53] L. Chen, H. Xie, Y. Li, W. Yu, Nanofluids containing carbon nanotubes treated by mechanochemical reaction, Thermochim. Acta 477 (2008) 21–24.

[54] Y.-j. Hwang, J. Lee, C. Lee, Y. Jung, S. Cheong, C. Lee, et al., Stability and thermal conductivity characteristics of nanofluids, Thermochim. Acta 455 (2007) 70–74.

[55] L. Chen, H. Xie, Surfactant-free nanofluids containing double-and single-walled carbon nanotubes functionalized by a wet-mechanochemical reaction, Thermochim. Acta 497 (2010) 67–71.

[56] K.A. Wepasnick, B.A. Smith, J.L. Bitter, D. Howard Fairbrother, Chemical and structural characterization of carbon nanotube surfaces, Anal. Bioanal. Chem. 396 (2010) 1003–1014.

[57] E. Tang, G. Cheng, X. Ma, X. Pang, Q. Zhao, Surface modification of zinc oxide nanoparticle by PMAA and its dispersion in aqueous system, Appl. Surf. Sci. 252 (2006) 5227–5232.

[58] T. Missana, A. Adell, On the applicability of DLVO theory to the prediction of clay colloids stability, J. Colloid Interface Sci. 230 (2000) 150–156.

[59] I. Popa, G. Gillies, G. Papastavrou, M. Borkovec, Attractive and repulsive electrostatic forces between positively charged latex particles in the presence of anionic linear polyelectrolytes, J. Phys. Chem. B 114 (2010) 3170–3177.

[60] K.S. Hwang, S.P. Jang, S.U. Choi, Flow and convective heat transfer characteristics of water-based Al_2O_3 nanofluids in fully developed laminar flow regime, Int. J. Heat. Mass. Transf. 52 (2009) 193–199.

[61] S. Lee, S.-S. Choi, S. Li, J. Eastman, Measuring thermal conductivity of fluids containing oxide nanoparticles, J. Heat Transfer. 121 (1999) 280–289.

[62] J.A. Eastman, S. Choi, S. Li, W. Yu, L. Thompson, Anomalously increased effective thermal conductivities of ethylene glycol-based nanofluids containing copper nanoparticles, Appl. Phys. Lett. 78 (2001) 718–720.

[63] K. Leong, C. Yang, S. Murshed, A model for the thermal conductivity of nanofluids—the effect of interfacial layer, J. Nanopart. Res. 8 (2006) 245–254.

[64] S.U.-S. Choi, Nanofluid Technology: Current Status and Future Research, Argonne National Lab.(ANL), Argonne, IL, 1998.

[65] S. Choi, Z.G. Zhang, W. Yu, F. Lockwood, E. Grulke, Anomalous thermal conductivity enhancement in nanotube suspensions, Appl. Phys. Lett. 79 (2001) 2252–2254.
[66] W. Minkowycz, E.M. Sparrow, J.P. Abraham, Nanoparticle Heat Transfer and Fluid Flow, CRC Press, 2012.
[67] V. Sharma, R.N. Kumar, K. Thamilarasan, G.V. Bhaskar, B. Devra, Heat reduction fromic engine by using Al_2O_3 nanofluid in engine cooling system, Am. J. Eng. Res. 3 (2014) 173–177.
[68] H. Ma, C. Wilson, B. Borgmeyer, K. Park, Q. Yu, S. Choi, et al., Effect of nanofluid on the heat transport capability in an oscillating heat pipe, Appl. Phys. Lett. 88 (2006) 143116–143120.
[69] S.J. Kim, I.C. Bang, J. Buongiorno, L. Hu, Surface wettability change during pool boiling of nanofluids and its effect on critical heat flux, Int. J. Heat. Mass. Transf. 50 (2007) 4105–4116.
[70] S. You, J. Kim, K. Kim, Effect of nanoparticles on critical heat flux of water in pool boiling heat transfer, Appl. Phys. Lett. 83 (2003) 3374–3376.
[71] J. Ekanayake, M. Jenkins, A three-level advanced static VAr compensator, IEEE Trans. Power Deliv. 11 (1996) 540–545.
[72] M. Kao, C. Lo, T. Tsung, Y. Wu, C. Jwo, H. Lin, Copper-oxide brake nanofluid manufactured using arc-submerged nanoparticle synthesis system, J. Alloy. Compd. 434 (2007) 672–674.
[73] M.-J. Kao, H. Chang, Y.-Y. Wu, T.-T. Tsung, H.-M. Lin, Producing aluminum-oxide brake nanofluids using plasma charging system, J. Chin. Soc. Mech. Eng. 28 (2007) 123–131.
[74] R.A. Taylor, P.E. Phelan, T.P. Otanicar, R. Adrian, R. Prasher, Nanofluid optical property characterization: towards efficient direct absorption solar collectors, Nanoscale Res. Lett. 6 (2011) 1–11.
[75] U. Bhagat, P. More, P. Khanna, Study of zinc oxide nanofluids for heat transfer application, SAJ Nanosci. Nanotechnol. 1 (2015) 1–9.
[76] S.K. Das, S.U. Choi, H.E. Patel, Heat transfer in nanofluids—a review, Heat. Transf. Eng. 27 (2006) 3–19.
[77] J.-H. Lee, K.S. Hwang, S.P. Jang, B.H. Lee, J.H. Kim, S.U. Choi, et al., Effective viscosities and thermal conductivities of aqueous nanofluids containing low volume concentrations of Al_2O_3 nanoparticles, Int. J. Heat. Mass. Transf. 51 (2008) 2651–2656.
[78] P. Ternik, R. Rudolf, Z. Žunič, Numerical study of heat-transfer enhancement of homogeneous water-Au nanofluid under natural convection, Materiali tehnologije 46 (2012) 257–261.
[79] E. Sourtiji, S.F. Hosseinizadeh, Heat transfer augmentation of magnetohydrodynamics natural convection in L-shaped cavities utilizing nanofluids, Therm. Sci. 16 (2012) 489–501.
[80] M. Heyhat, F. Kowsary, A. Rashidi, M. Momenpour, A. Amrollahi, Experimental investigation of laminar convective heat transfer and pressure drop of water-based Al_2O_3 nanofluids in fully developed flow regime, Exp. Therm. Fluid Sci. 44 (2013) 483–489.
[81] F.M. Haghshenas, M.R. Talaie, S. Nasr, Numerical and experimental investigation of heat transfer of ZnO/water nanofluid in the concentric tube and plate heat exchangers, Therm. Sci. 15 (2011) 183–194.

[82] G.-J. Lee, C.K. Kim, M.K. Lee, C.K. Rhee, S. Kim, C. Kim, Thermal conductivity enhancement of ZnO nanofluid using a one-step physical method, Thermochim. Acta 542 (2012) 24–27.
[83] M. Rahmati, Effects of ZnO/water nanofluid on the thermal performance of wet cooling towers, Int. J. Refrig. 131 (2021) 526–534.
[84] S.B. Mousavi, S.Z. Heris, P. Estellé, Viscosity, tribological and physicochemical features of ZnO and MoS_2 diesel oil-based nanofluids: an experimental study, Fuel 293 (2021) 120481.
[85] R. Dinesh, M.G. Prasad, R.R. Kumar, N.J. Santharaj, J. Santhip, A.A. Raaj, Investigation of tribological and thermophysical properties of engine oil containing nano additives, Mater. Today: Proc. 3 (2016) 45–53.
[86] M.H. Esfe, A.A.A. Arani, S. Esfandeh, Improving engine oil lubrication in light-duty vehicles by using of dispersing MWCNT and ZnO nanoparticles in 5W50 as viscosity index improvers (VII), Appl. Therm. Eng. 143 (2018) 493–506.
[87] T. Wen, G. Zhu, K. Jiao, L. Lu, Experimental study on the thermal and flow characteristics of ZnO/water nanofluid in mini-channels integrated with GA-optimized ANN prediction and CFD simulation, Int. J. Heat. Mass. Transf. 178 (2021) 121617.
[88] V. Kumar, A.K. Tiwari, S.K. Ghosh, Effect of chevron angle on heat transfer performance in plate heat exchanger using ZnO/water nanofluid, Energy Convers. Manag. 118 (2016) 142–154.
[89] H.M. Ali, H. Ali, H. Liaquat, H.T.B. Maqsood, M.A. Nadir, Experimental investigation of convective heat transfer augmentation for car radiator using ZnO–water nanofluids, Energy 84 (2015) 317–324.
[90] D.A. McCants, A.M. Hayes, T.C. Paul, J.A. Khan, A.H. Shaaban, Experimental investigation of heat transfer enhancement of CuO–water and ZnO–water nanofluids flowing over a heated plate, J. Therm. Sci. Eng. Appl. 11 (2019) 041015.
[91] R. Islam, B. Shabani, J. Andrews, G. Rosengarten, Experimental investigation of using ZnO nanofluids as coolants in a PEM fuel cell, Int. J. Hydrog. Energy 42 (2017) 19272–19286.
[92] M. Sharifpur, S.O. Giwa, K.-Y. Lee, H. Ghodsinezhad, J.P. Meyer, Experimental investigation into natural convection of zinc oxide/water nanofluids in a square cavity, Heat. Transf. Eng. 42 (2021) 1675–1687.
[93] Y. Li, J. Fernández-Seara, K. Du, Á.Á. Pardiñas, L.L. Latas, W. Jiang, Experimental investigation on heat transfer and pressure drop of ZnO/ethylene glycol-water nanofluids in transition flow, Appl. Therm. Eng. 93 (2016) 537–548.
[94] O. Hozien, W.M. El-Maghlany, M.M. Sorour, Y.S. Mohamed, Experimental study on thermophysical properties of TiO_2, ZnO and Ag water base nanofluids, J. Mol. Liq. 334 (2021) 116128.
[95] A. Asadi, F. Pourfattah, Heat transfer performance of two oil-based nanofluids containing ZnO and MgO nanoparticles; a comparative experimental investigation, Powder Technol. 343 (2019) 296–308.
[96] A.A. Minea, W.M. El-Maghlany, E.Z. Massoud, Heat transfer analysis of nanocolloids based on zinc oxide nanoparticles dispersed in PEG 400, Nanomaterials 12 (2022) 2344.
[97] V. Keyvani, E. Riahi, M. Yousefi, S.-A. Esmaeili, R. Shafabakhsh, A. Moradi Hasan-Abad, et al., Gynecologic cancer, cancer stem cells, and possible targeted therapies, Front. Pharmacol. 13 (2022) 823572.

[98] S. Jafari, M. Ravan, I. Karimi-Sani, H. Aria, A.M. Hasan-Abad, B. Banasaz, et al., Screening and identification of potential biomarkers for pancreatic cancer: an integrated bioinformatics analysis, Pathol. Res. Pract. (2023) 154726.

[99] K.S. Mekheimer, The influence of heat transfer and magnetic field on peristaltic transport of a Newtonian fluid in a vertical annulus: application of an endoscope, Phys. Lett. A 372 (2008) 1657–1665.

[100] Q.A. Pankhurst, J. Connolly, S.K. Jones, J. Dobson, Applications of magnetic nanoparticles in biomedicine, J. Phys. D: Appl. Phys. 36 (2003) R167.

[101] M.R. Habibi, M. Ghassemi, M.H. Hamedi, Analysis of high gradient magnetic field effects on distribution of nanoparticles injected into pulsatile blood stream, J. Magnetism Magnetic Mater. 324 (2012) 1473–1482.

[102] Z. Fakhroueian, S. Rajabi, N. Salehi, M.R. Tavirani, S. Noori, M. Nourbakhsh, Anticancer properties of novel zinc oxide quantum dot nanoparticles against breast cancer stem-like cells, Anti-Cancer Drugs 33 (2022) e311–e326.

[103] A. Azhdari, R. Jalal, Zinc oxide nanofluids: the influence of modality combinations on prostate cancer DU145 cells, J. Cancer Res. Ther. 17 (2021) 393–400.

[104] M.H. Abdolmohammadi, F. Fallahian, Z. Fakhroueian, M. Kamalian, P. Keyhanvar, F.M. Harsini, et al., Application of new ZnO nanoformulation and Ag/Fe/ZnO nanocomposites as water-based nanofluids to consider in vitro cytotoxic effects against MCF-7 breast cancer cells, Artif. Cells Nanomed. Biotechnol. 45 (2017) 1769–1777.

[105] Z. Fakhroueian, R. Vahabpour, M. Assmar, A. Massiha, A. Zahedi, P. Esmaeilzadeh, et al., ZnO Q-dots as a potent therapeutic nanomedicine for in vitro cytotoxicity evaluation of mouth KB44, breast MCF7, colon HT29 and HeLa cancer cell lines, mouse ear swelling tests in vivo and its side effects using the animal model, Artif. Cells Nanomed. Biotechnol. 46 (2018) 96–111.

[106] S. Soleymani, R. Jalal, ZnO nanofluids for the improved cytotoxicity and cellular uptake of doxorubicin, Nanomed. J. 5 (2018) 27–35.

[107] B. Sung, S.H. Kim, S. Lee, J. Lim, J.-K. Lee, K.-S. Soh, Nanofluid transport in a living soft microtube, J. Phys. D: Appl. Phys. 48 (2015) 345402.

[108] R. Shelat, L.K. Bhatt, A. Khanna, S. Chandra, A comprehensive toxicity evaluation of novel amino acid-modified magnetic ferrofluids for magnetic resonance imaging, Amino Acids 51 (2019) 929–943.

[109] G.K. Yogesh, E. Shuaib, P. Roopmani, M.B. Gumpu, U.M. Krishnan, D. Sastikumar, Synthesis, characterization and bioimaging application of laser-ablated graphene-oxide nanoparticles (nGOs), Diam. Relat. Mater. 104 (2020) 107733.

[110] J. Talev, J.R. Kanwar, Iron oxide nanoparticles as imaging and therapeutic agents for atherosclerosis, Semin. Thromb. Hemost. 46 (2020) 553–562. Thieme Medical Publishers.

[111] H.M. Xiong, ZnO nanoparticles applied to bioimaging and drug delivery, Adv. Mater. 25 (2013) 5329–5335.

[112] H.-M. Xiong, Y. Xu, Q.-G. Ren, Y.-Y. Xia, Stable aqueous ZnO@ polymer core – - shell nanoparticles with tunable photoluminescence and their application in cell imaging, J. Am. Chem. Soc. 130 (2008) 7522–7523.

[113] Z.Y. Pan, J. Liang, Z.Z. Zheng, H.H. Wang, H.M. Xiong, The application of ZnO luminescent nanoparticles in labeling mice, Contrast Media Mol. Imaging 6 (2011) 328–330.

[114] N. Beyth, Y. Houri-Haddad, A. Domb, W. Khan, R. Hazan, Alternative antimicrobial approach: nano-antimicrobial materials, Evid. Based Complement. Alternat. Med. 2015 (2015) 246012–246022.

[115] A. Sirelkhatim, S. Mahmud, A. Seeni, N.H.M. Kaus, L.C. Ann, S.K.M. Bakhori, et al., Review on zinc oxide nanoparticles: antibacterial activity and toxicity mechanism, Nano-Micro Lett. 7 (2015) 219–242.

[116] A.H. Mohammadi, A. Sobhani-Nasab, M. Nejati, S. Hadi, M. Behjati, I. Mirzaii-Dizgah, et al., Preparation and characterization of CuO, Ag_2O and ZnO nanoparticles and investigation of their antibacterial and anticancer properties on HCT-116 and C26 cells, Inorg. Chem. Commun. 149 (2023) 110404.

[117] Q. Yu, Y. Qin, M. Han, F. Pan, L. Han, X. Yin, et al., Preparation and characterization of solvent-free fluids reinforced and plasticized polylactic acid fibrous membrane, Int. J. Biol. Macromol. 161 (2020) 122–131.

[118] J. Javedani Bafekr, R. Jalal, In vitro antibacterial activity of ceftazidime, unlike ciprofloxacin, improves in the presence of ZnO nanofluids under acidic conditions, IET Nanobiotechnol. 12 (2018) 640–646.

[119] L. Zhang, H. Qi, Z. Yan, Y. Gu, W. Sun, A.A. Zewde, Sonophotocatalytic inactivation of *E. coli* using ZnO nanofluids and its mechanism, Ultrason. Sonochem. 34 (2017) 232–238.

[120] R. Jalal, E.K. Goharshadi, M. Abareshi, M. Moosavi, A. Yousefi, P. Nancarrow, ZnO nanofluids: green synthesis, characterization, and antibacterial activity, Mater. Chem. Phys. 121 (2010) 198–201.

[121] K.M. Lee, C.W. Lai, K.S. Ngai, J.C. Juan, Recent developments of zinc oxide based photocatalyst in water treatment technology: a review, Water Res. 88 (2016) 428–448.

[122] K. Qi, B. Cheng, J. Yu, W. Ho, Review on the improvement of the photocatalytic and antibacterial activities of ZnO, J. Alloy. Compd. 727 (2017) 792–820.

[123] K. Karthik, A. Raghu, K.R. Reddy, R. Ravishankar, M. Sangeeta, N.P. Shetti, et al., Green synthesis of Cu-doped ZnO nanoparticles and its application for the photocatalytic degradation of hazardous organic pollutants, Chemosphere 287 (2022) 132081.

CHAPTER

Fe_3O_4-based nanofluids

5

Muhammad Hossein Ashoub[1], Mahnaz Amiri[2] and Ghazal Khajouci[2]

[1]*Stem Cells and Regenerative Medicine Innovation Center, Kerman University of Medical Sciences, Kerman, Iran*

[2]*Neuroscience Research Center, Institute of Neuropharmacology, Kerman University of Medical Science, Kerman, Iran*

5.1 General introduction

Nanofluids are two-phase fluids that consist of solid-liquid mixtures with suspended nanoparticles. Due to their greater thermal conductivity than base liquids, they are potential thermal fluids for heat transfer applications. Nanofluids have gotten attention in engineering and industry because of their potential for heat transmission in heat exchangers. The term "nanofluid" was used to describe a colloidal solution composed of a base fluid and solid nanoparticles [1,2]. Previously, nanofluids were employed to boost fluid thermal conductivity by adding micrometer-sized particles to base fluids, but nanosized particles create a stable suspension [3]. The presence of nanoparticles in a conventional fluid may increase heat transfer qualities and boost the fluid's thermal conductivity [4].

In comparison to single-phase pure fluids and suspensions of microparticles, nanofluids have a number of benefits, including a decrease in particle sedimentation, clogging of microchannel passageways, and tube material degradation. Furthermore, nanofluids create stable suspensions with homogeneous nanoparticle dispersion in the host fluid [1,5]. Colloidal liquids called ferrofluids have high magnetic properties that can be manipulated by a magnetic field. They may be used for a variety of things, including heat transmission and dissipation, magneto-optic sensors, dynamic loudspeakers, and MRI [6–11]. Surfactants, carrier liquids, and nanoparticle features all have an impact on these qualities [12]. Micro- or nanoscale magnetic particles floating in a base fluid can be generically categorized as magnetic nanofluids [5,13]. These particles can be divided into magnetic nanofluids and magnetorheological (Mr) fluids, which are physiochemically and magnetically adjustable [14,15]. Magnetorheological fluids (Mr fluids) and ferrofluids are the two different categories of magnetic fluids (MFs). Ferrofluids have particles in the lower nanometer range, while Mr fluids have particles in the micron range. Ferrofluids are useful for a variety of applications and bring up possibilities that are not available with Mr fluids

Nanofluids. DOI: https://doi.org/10.1016/B978-0-443-13625-2.00013-9

because of their increased suspension stability and prolonged suspension period. When exposed to a magnetic field, MFs can be identified by visual inspection. Ferrofluids display the peaks and troughs of the Rosensweig instability, whereas ferromagnetic particles in Mr fluids form chains along flux routes. A ferrofluid in a magnetic field maintained its liquid form for more than 35 years without seizing up, displaying its distinctive spike pattern and durability [16–19]. Numerous biological uses for magnetic particles are found in gel-forming matrices, known as magnetic gels, including biomimetic actuators and medication delivery. Magnetic nanofluids are being used more and more in a variety of fields [20,21]. However, there are still difficulties in making them usable as packaged or field-ready goods. This chapter offers an overview of MFs, including relevant information, important ideas, preparation, applications, and future research possibilities.

5.2 General properties of magnetic fluid

A stabilizer, dispersion medium, and distributed solid magnetic phase make up the complicated system known as MF. The features of its individual parts, which might differ substantially, define its properties. Small magnetic nanoparticles (MNPs) (5–15 nm) are in strong Brownian motion, ensuring uniform distribution and stability of sedimentation in the liquid volume [22]. Repulsive forces between particles are necessary for aggregate and spatial stability. Incorporating a stabilizer (surfactant) into the MF or utilizing an ionic liquid as a carrier are two ways to do this [23]. Polar chemical compounds with small functional groups and long tail chains often make up surfactants. Oleic acid is a well-known MF stabilizer [24,25]. Particle sizes in magnetorheological fluids are more than 30–40 nm, while those in stable MFs range from 5 to 15 nm. Under the influence of magnetic fields, these fluids display a dramatic rise in viscosity and can totally solidify in strong fields. These fluids are thought to be made of "smart" components that respond nonlinearly to outside stimuli. The size of the magnetic particles in MF and magnetorheological fluids determines their classification [22,26]. MFs were first created, studied, and used in the 1960s, decades before the word "nanotechnology" was coined [27,28]. Due to the absence of any remaining magnetization when the magnetic field outside is removed, MFs do not have ferromagnetic characteristics. Because of their tremendous magnetic sensitivity, they are frequently referred to as superparamagnetic. Temperature and spatial parameters affect magnetism, and when a ferromagnetic particle's size falls below a certain value, it transitions from a polydomain state to a single-domain one [13,29]. A single-domain particle's form, temperature, and crystal magnetic anisotropy all affect how big it is. Thermal energy may quickly randomize magnetic dipoles as particle size falls. These tiny superparamagnetic particles, which resemble paramagnetic atoms but have a large intrinsic magnetic moment, are utilized to create

MFs [26]. By accomplishing key physical properties such as colloidal stability, saturation magnetization, viscosity, and operating temperature range, these technologies' efficacy is determined. Positive MFs exhibit high fluidity and stability for a period of 2–5 years [30,31]. Fig. 5.1 indicates the schematic of the ferrofluid and photograph of stable hexane-based ferrofluid, where the bottom clear phase is water (left) and aqueous ferrofluid in which the clear top is hexane (right). Metal oxides and ferrimagnetic metals are frequently employed as the distributed magnetic phase. Metals such as Fe, Co, Ni, and Gd are used to create MFs. Additionally utilized are nitrides, metal oxides, and bimetallic particles. Iron oxide Fe_2O_3 chemical compounds called ferrites and other metal oxides with unique magnetic characteristics are employed as well. Ferrites spinels have the chemical formula MFe_2O_4 whereas ferrites-garnets have MFe_5O_{12}, making them distinct [32,33]. The maximum saturation magnetization is seen in metals, but they are also quite oxidation-sensitive, particularly when they are widely spread. With magnetite being the most prevalent ferrite, nanoparticles of ferrites, such as magnetite $Fe_3O_4 = FeFe_2O_4$, are frequently utilized as the dispersion phase [34,35]. The weight percentage of solids in MFs, which may be up to 25%, determines their magnetic characteristics. Although there is a wide variety of magnetic materials, making colloidal suspensions necessitates that the nanoparticles be well-compatible with surfactants and carrier liquids, which narrows the selection. A polar or nonpolar carrier liquid is a crucial feature of MF because it creates an environment in which the particles may be suspended. Depending on the MF application, different liquids may be used [36,37]. Based on variables including boiling temperature, vapor pressure, and freezing point, a carrier liquid for MNPs (MF) is chosen. The magnetic phase and other materials employed in the device must not chemically react with the carrier liquid. The last component is a liquid

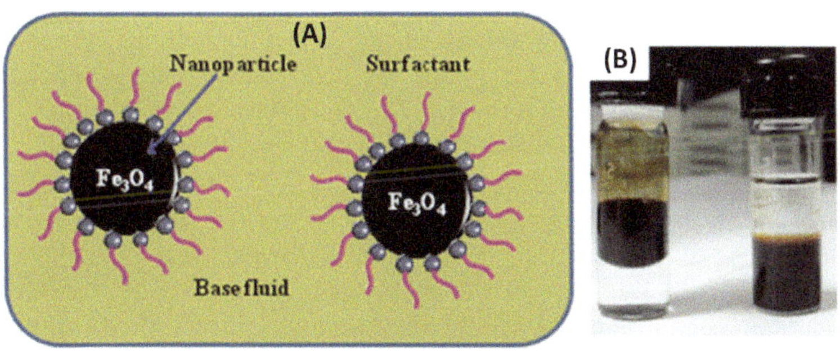

FIGURE 5.1

(A) Schematic of the ferrofluid (B) Photograph of stable hexane-based ferrofluid, where the bottom clear phase is water (left) and aqueous ferrofluid in which the clear top is hexane (right).

Reprinted with permissions from J. Philip, Magnetic nanofluids (ferrofluids): recent advances, applications, challenges, and future directions, Adv. Colloid Interface Sci. 311 (2023) 102810.

carrier stabilizer for MNPs. Stabilizers stop the surface of the nanoparticles from coagulating and oxidizing. Surfactants and polymers are typical stabilizers that are chemically or physically deposited onto the surface of nanoparticles. The most used surfactant for stabilizing MNPs is oleic acid, which is made using traditional chemical processes. Wine, citric acids, fatty acids, sodium oleate, dodecyl amine, and sodium carboxymethylcellulose are other surfactants. The functional characteristics of MFs influence the usage of certain surfactants [24,38–40].

5.3 Preparation and stability of nanofluids

5.3.1 Preparation of nanofluids

Ferrofluids are made up of a surfactant to avoid agglomeration, a lower nanoscale ferromagnetic component, and a carrier liquid to suspend the complex. A steady colloidal solution of ferroMNPs is produced by these elements. Usually, a mixture of magnetite and maghemite makes up the magnetic phase, with magnetite having superior characteristics. Iron and metals, including cobalt, zinc, nickel, and manganese, are examples of other ferromagnetic compounds. Due to their strong magnetic and long-term stability, ferrofluids are used in a variety of applications. Additional investigation on surfactants, carrier liquids, and NP composition is required to improve their stability and magnetism. Iron oxide nanoparticles (IONPs), which have an impact on the ferrofluid's quality, are first created during the synthesis of ferrofluids. Sol-gel pyrolysis, mechanical alloying, hydrothermal processes, mechanochemical synthesis, flame synthesis, thermal decomposition, coprecipitation, mechanical milling, sonochemical approach, and one-pot synthesis are some of the techniques utilized to make IONPs. The most often reported techniques are thermal decomposition and coprecipitation because they are simple and dependable, involve less equipment, produce more consistent findings, and have extensive documentation [40–42]. The creation of ferroMNPs has recently been refined, employing a variety of variables and chemical agents. In this method, metal carbonyls thermally decompose in liquid organic media, gases, or vacuums. Metal carbonyls break down at higher temperatures, producing micron-sized particles. However, in the presence of polymers, these carbonyls break down into nanoparticles of varying sizes, which may be altered by choosing the molecular mass and makeup of the polymer [22]. By breaking down dicobalt octa carbonyl in toluene using methylmethacrylate-ethyl acrylate-vinyl pyrrolidone terpolymer and methyl isobutyl ketone, Thomas created 20 nm cobalt nanoparticles [43]. The ratio of the reaction mixture, the temperature at which breakdown occurs, and the reagent concentration all have a significant impact on nanoparticle production. According to Verdaguer et al., who created magnetic IONPs via laser-induced pyrolysis of Fe $(CO)_5$ vapors in ethylene, the procedure has also been adapted to create IONPs [44]. For the synthesis of monodispersed Fe_3O_4,

$CoFe_2O_4$, $MnFe_2O_4$, and -Fe_2O_3 nanoparticles, Hyeons et al. created a novel method. By adjusting M(CO), oleic ratio, and reflux temperature, nucleated particle size was modified [45,46]. Cheon et al. used dodecyl amine and aerobic conditions to create single-crystalline IONPs in a variety of forms [47]. By utilizing a laser or electric discharge to vaporize compounds, vaporization-condensation-based synthesis enables the condensation of vapor in a liquid media [22]. Fig. 5.2 indicates the schematic of the process of ferrofluid emulsion preparation. In a carrier fluid, Berkowitz and Walter created a ferrofluid containing magnetic metallic particles embedded in polymers. Magnetic metallic particles with a preexisting polymer coating are produced in an organic dielectric liquid by pulsed electric potential between electrodes. Using amorphous metallic particles such as $Fe_{75}Si_{15}B_{10}$, magnetic colloids may be created using this method [48]. In order to vaporize and condense the metal oxide nanoparticles in a diffusion cloud chamber, El-Shall et al. used a pulsed Nd: YAG laser [49]. Magnetite nano pyramid arrays were created utilizing the 2450 MHz microwave plasma chemical vapor deposition method, as proven by Liu et al. [50]. A method called electro-deposition-based synthesis employs electrolysis to prepare metals for deposition [51]. Ferromagnetic metals and alloy-based ferrofluids are produced via the

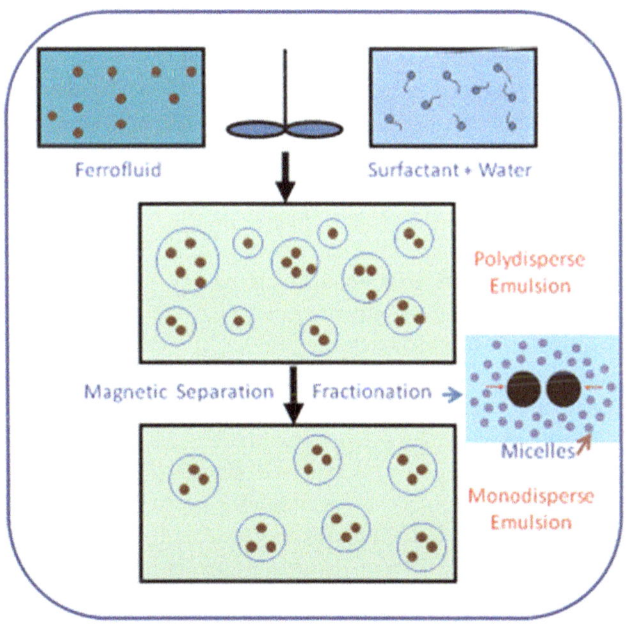

FIGURE 5.2

Schematic of the process of ferrofluid emulsion preparation.

Reprinted with permissions from J. Philip, Magnetic nanofluids (ferrofluids): recent advances, applications, challenges, and future directions, Adv. Colloid Interface Sci. 311 (2023) 102810.

deposition of metals on the cathode in dense, continuous layers [52]. The cathode surface area, current density, electrolyte composition, presence of surfactants, temperature, and agitation are some of the variables affecting this method. The procedure is carried out in a two-layer tank, with an organic medium and surfactant in the top layer and an electrolyte solution in the bottom layer. An electrolytic bath containing metal salt solutions is used for electrolysis, and 1 M KOH (potassium hydroxide) is used for electrochemical oxidation to create $ZnFe_2O_4$ nanoparticles [53]. The sol-gel synthesis includes the aging, drying, dehydration, and breakdown of stable solutions of alkoxide or solvated metal precursors at high temperatures. Stable solutions, gelation, aging, drying, dehydration, and disintegration of the gels are all components of the process [54–56]. Utilizing the solubility of inorganic elements in water at high temperatures and pressures, hydrothermal synthesis enables the crystallization of dissolved materials. Temperature, pressure, reaction time, and salt content all play a role in the process [57,58]. Wang et al. created single-crystalline magnetite nanowires by aging the material while using an external magnetic field [59]. Clusters of spherical magnetic colloidal nanocrystals have been produced hydrothermally in recent years [60]. Rosenberger et al. presented a nonhydrolytic single precursor strategy for dispersable nanocrystals of transition metal oxides via solvothermal synthesis, which uses solvents other than water [61]. Monodisperse Fe_3O_4 nanoparticles were created using an organic phase technique, according to Sun et al. [62]. Fe $(acac)_3$, 1,2-hexadecane-diol, oleic acid, and oleylamine were combined with a new metal acetylacetonate precursor to create MFe_2O_4 nanoparticles with adjustable diameters ranging from 3 to 20 nm [63]. The formation of larger monodispersed nanoparticles was mediated by seeds. In the solvothermal approach, the impact of organic solvents on IONPs has been investigated [64]. A straightforward and affordable way to create nanocrystals with improved yields is microwave-assisted synthesis [65]. With this approach, the reaction vessel's pressure, stirring speed, and temperature may all be precisely controlled. Reaction durations are sped up, product yields are enhanced, and purity or material qualities are improved as a result of the heating effect in microwaves, which is caused by electric field interaction with charged particles in the material. Through the direct coupling of microwave energy to molecules, microwave irradiation effectively raises the temperature of a reaction mixture. This leads to even heating, which promotes homogeneous nucleation and growth. This heating is facilitated by ionic conduction and dipolar polarization. Recent research has looked at the function of microwave-assisted synthesis in producing different inorganic nanocrystals as well as the production of magnetite nanoparticles by the precipitation of iron precursors [66–68].

5.3.2 Ferrofluid stability

Due to their small size, stability is a significant barrier in the practical use of nanofluids. Brownian motion, which ultrafine nanoparticles in nanofluids are subject to, can counteract gravitational force-induced sedimentation. Thermal energy (kT) must be more extensive than the gravitational force exerted on the particle in order to keep it suspended. More stable nanofluids perform better thermally than less stable ones. To create more stable nanofluids, researchers employ stability evaluation and stability augmentation techniques [69,70]. To make nanofluids in heat pipes more stable, researchers employ methods such as PH value changes, surfactant addition, and ultrasonication. A few methods for assessing stability include spectral absorbance measurement, zeta potential, centrifugation, and sedimentation [71,72]. Nanoparticle morphology is determined using characterization techniques such as transmission electron microscopy (TEM), SEM, and X-ray diffraction (XRD) [73]. Nanofluids' thermal conductivity and viscosity are influenced by a number of variables, including small-scale ones such as clustering, particle charge, and micromotion. To ensure the safety and effectiveness of nanofluids in a variety of applications, it is essential to comprehend and manage these concerns because they have both differences and uncertainties [74,75]. The thermal conductivity and dispersion, fluidity, and application of nanofluids are improved by cutting-edge materials, including graphene, carbon nanotubes (CNTs), and gold. Stability is undesirable, though [76,77].

Further research on hybrid nanoparticles is necessary because mixing various nanoparticles can provide distinct qualities from mono nanoparticles [78]. Magnetic dipole-dipole attraction and van der Waals forces cause the aggregation of magnetic particles. These forces are produced through interactions between permanent dipoles and other permanent dipoles, as well as between induced dipoles and other induced dipoles. In colloidal systems, the main elements are London dispersion forces or flickering dipole forces contributing to long-range attractions. Electron cloud polarization changes spontaneously, and there is little attraction between two atoms. The total of the forces produced by the atoms in colloidal particles, however, results in a strong, attractive force when the quantity of atoms is taken into account [22,26]. Utilizing methods such as Dynamic Light Scattering (DLS), zeta potential, and UV-VIS spectrophotometry, one may evaluate the stability of nanofluids. These techniques have low sample volume needs, are straightforward, and are easy to quantify. Fig. 5.3 indicates schematic electrostatic stabilization of Fe_3O_4 nanoparticles by coating.

Aggregate hydrodynamic diameters and nano-to-micron size distributions are both provided by DLS. For practical applications, it is essential to achieve long-term stability in MF, especially for pure metal particles vulnerable to oxidation and deterioration [79–81]. With the help of poisonous substances such as cobalt and acidic media, magnetite particles can disintegrate. By lowering coating thickness and removing magnetic attraction forces, inert coatings can resolve these problems. The use of sonication during MF production may mitigate van der Waals forces. Coated MNPs provide benefits in particular applications, such as chemical inertness, powder storage, and transport, despite physical or chemical synthesis processes costing more and producing less. By changing the characteristics of the base fluid's magnetism, the magnetic powder can drastically change the fluid characteristics in laboratory

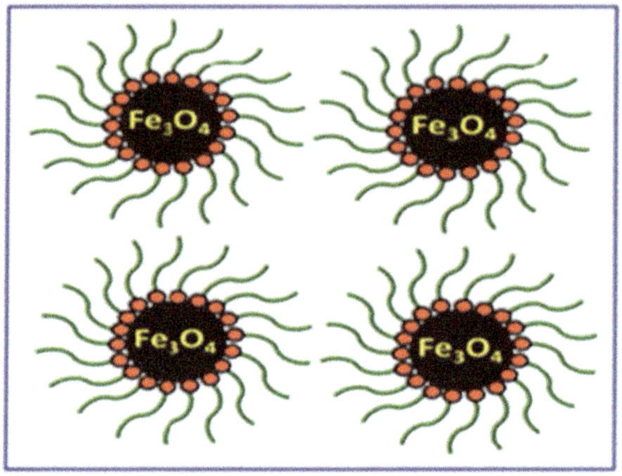

FIGURE 5.3

Schematic electrostatic stabilization of Fe_3O_4 nanoparticles by coating.

Reprinted with permissions from J. Philip, Magnetic nanofluids (ferrofluids): recent advances, applications, challenges, and future directions, Adv. Colloid Interface Sci. 311 (2023) 102810.

experiments. For their manufacture and use, MNPs in silica dioxide, polymers, and carbon coatings are the subject of much research. Biocompatibility and chemical inertness are important requirements for medical and biological applications, which are increasing quickly. MNPs with various coatings can be used to solve specific issues such as targeted medication delivery, staining for magnetic resonance imaging (MRI), or hyperthermia [82,83].

5.3.3 Nanofluids' thermal characteristics

In order to determine qualities such as thermal conductivity, thermal diffusivity, and flow's spatial temperature, it is essential to evaluate a material's specific heat capacity, which assesses its capability to store and exchange heat energy. Researchers quantify the specific heat in nanofluids using twin hot wires and deferential scanning calorimeters. Shape, size, and electrostatic behavior are less important than nanoparticle volume concentration, type, and base fluid, which have a bigger impact. Due to their lower heat capacity as compared to their base fluid, nanoparticles' specific heat drops as their volume fraction increases [84]. The specific heat of five distinct nanofluids, including Al_2O_3, ZnO, TiO_2, CuO, and SiO_2, was investigated in a study using a 60:40 propylene glycol/water ratio. The study discovered that whereas volume concentration greatly affected heat capacity behavior, particle size had no effect on specific heat. Due to the higher thermal conductivity and efficiency of heat transmission at low concentrations, the decrease in specific heat was bearable. However, the heat capacity dropped as the volume percentage of nanoparticles rose [85]. It was also discovered that the specific heat of MgO, ZnO, and ZrO_2 nanofluids based on water and ethylene glycol mixtures increased by 30% in comparison to their base fluids but declined with increasing nanoparticle volume fraction [86]. Due to the high specific heat capacity of CNTs, specific heat can rise with increasing loading. However, some research has shown that when temperatures rise, specific heat capacity falls. The majority of the literature claims that specific heat rises with temperature [85,87,88]. However, other studies show the reverse trend [89,90]. In contrast to single-walled CNT nanofluids, multiwalled CNTs exhibit an increase in specific heat with temperature. Since not all CNT nanofluids exhibit the prior behavior of specific heat, the temperature influence on specific heat is not uniform across all CNT nanofluids [84,91].

5.3.4 Thermal conductivity of nanofluids

Increasing the effectiveness of heat transmission via nanofluids requires a material's thermal conductivity, or *k*, which measures how quickly heat is transferred through a material. Because they transport heat through solids more quickly than their base fluids, nanofluids have greater *k* values than their basis fluids. The most popular technique for determining thermal conductivity is the transient hot wire technique. Fig. 5.4 indicates factors affecting the thermophysical properties of hybrid nanofluids. Researchers have created equations and tested them using particle-fluid mixes and continuum equations. *Maxwell's* model is a pioneering

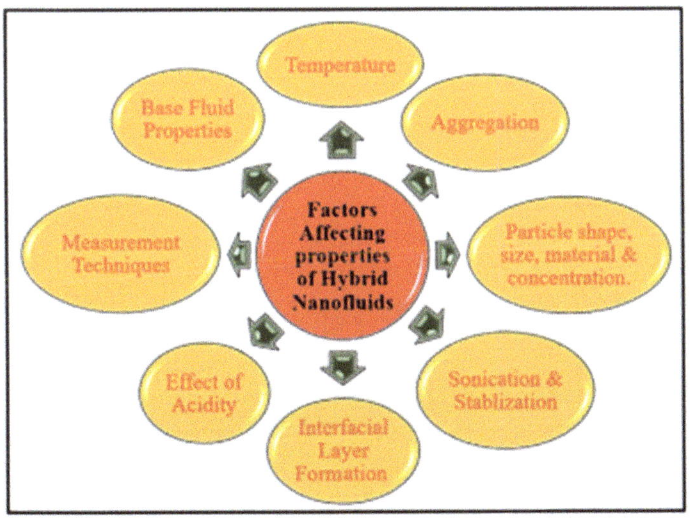

FIGURE 5.4

Factors affecting the thermophysical properties of hybrid nanofluids.

Reprinted with permissions from A.K. Tiwari, V. Kumar, Z. Said, H.K. Paliwal, J. Clean. Prod. 292 (2021) 126031.

approach that estimates the effective thermal conductivity of a fluid mixture of spherical particles at a millimeter to micrometer scale. This equation, which is only relevant for low particle concentrations, incorporates the thermal conductivity of solid particles in the base fluid and their volume fraction relative to the entire fluid [84]. Hamilton and Crosser [92] created a model for every particle shape, whereas Bruggeman [93] presented models for spherical particles. The thermal conductivity of nanofluids in high-volume concentrations and temperatures, however, cannot be predicted by any model. The thermal conductivity of Al_2O_3 nanofluid improved by 32.4% when the volume concentration was raised to 4.3% as a result of the volume fraction effect.

Similarly to single-walled CNT, Al_2O_3 and CuO nanofluids show an increase in thermal conductivity with increasing volume concentration. The thermal conductivity of a nanoparticle is substantially influenced by its size. The greater surface area on smaller particles increases heat conductivity. For instance, compared to Al_2O_3 nanofluids, copper oxide nanofluids with smaller particles demonstrated improved thermal conductivity. However, with other nanofluids, such as SiC nanofluids, this effect is erratic. According to a study, thermal conductivity increased by 15.8% and 22.9% in SiC nanofluids with diameters of 26 and 600 nm, perhaps as a result of nanoparticle clustering. A factor in determining the thermal conductivity of nanofluids is temperature. According to studies, the conductivity of water-based Al_2O_3, CuO, ZnO, and CNT nanofluids, as well as those based on ethylene glycol, increases as the temperature rises [84]. According to Philip et al. [94], tunable characteristics and chainlike structures in magnetite-based nanofluids have increased heat conductivity. This was caused by the effective heat transfer through aggregates that were percolating linear chains. Significant scientific interest was sparked by the revelation of magnetic nanofluids' adjustable thermal characteristics. In magnetically controlled nanofluids, which are helpful for microfluidic devices, Shima et al. [95] demonstrated the capability to modify thermal conductivity to viscosity. This improvement has been noted in suspensions of magnetic graphite nanoflakes, magnetite nanofluids, and water-based ferrofluids. A MEMS-specific, miniature magnetically-driven cooling device was created by Yamaguchi and Iwamoto [96]. MFs and composites have been used in novel ways for heat transfer applications. The efficiency of heat transfer in nanofluids is strongly influenced by nanoparticle dimensions. Nanofluids' thermal conductivity is observed to improve as particle size decreases, according to research. Nanoparticles that are spherical and cylindrical exhibit growing gains in thermal conductivity. Additionally, temperature has a positive impact on the thermal conductivity of nanofluids. To improve thermal conductivity, more materials or surfactants must be added when the temperature rises [22,97].

5.3.5 Viscosity of nanofluids

Although viscosity is a key factor in determining the convective heat transfer coefficient, it is still difficult to predict how viscosity would behave in nanofluids since there is no universal mathematical model or knowledge of the mechanics underlying viscosity [98]. The effective viscosity for suspended rigid spherical solids in liquids is predicted by *Einstein's* model [99] as a function of volume. However, it only forecasts viscosity behavior for sparse particle populations and stiff, spherical particles. Viscosity correlations have undergone several improvements, including the addition of *Batchelor's* Brownian motion [100] and *Brinkman's* model [101] for larger particle concentrations. Experiments have revealed inconsistent outcomes from these models despite these enhancements. When compared to Einstein–Batchelor correlations, greater viscosity levels were found in studies on nanofluids made of alumina and titania [102]. These models do not take temperature impacts into account because they are based on the volume fraction of nanoparticles. Individual correlations created by researchers are only based on volume percent. Nanofluid viscosity has recently been determined using viscometers. The heat transfer coefficient and viscosity of nanofluids are increased by nanoparticle volume concentration. Variations in the concentration of Al_2O_3 in water, Al_2O_3 based on ethylene glycol, and water-based SiC nanofluids increase viscosity, which raises the friction factor. Additionally, nonmetallic nanofluids show the same pattern, where increased carbon nanotube loading causes a rise in viscosity [103–105]. The research on viscosity behavior does, however, contain contradictions. In comparison to pure water, heat transport was shown to be significantly reduced at a volume concentration of 3% in Pak and Cho's investigation [106] on water-based Al_2O_3 and TiO_2 nanofluids. Viscosity is also influenced by variables such as nanoparticle size, shape, and surface chemistry, although these variables have received little attention in the literature. To fully comprehend the effect of nanoparticle concentration on viscosity, more research is required. Since temperature is not taken into account in theoretical models for viscosity, they are only valid for low concentrations and ambient temperatures. According to research, viscosity often reduces with temperature, especially in Al_2O_3 nanofluids. According to studies, the viscosity of CuO, Al_2O_3, and SiO_2 dispersed in water and ethylene glycol reduces exponentially as the temperature rises. The increase in friction makes nanofluids unusable if the rise in viscosity is four times greater than their heat conductivity [84,102,107]. The resistance of fluids to deforming under shear or longitudinal stress is known as viscosity. The two categories of fluids are Newtonian and non-Newtonian. At constant pressure and temperature, Newtonian fluids have shear stress proportional to deformation rate, but non-Newtonian fluids have viscosity tracking shear rate changes. Viscosity affects flow resistance, availability, pressure drop, and pumping power, specifically in nanofluid suspension systems [98,108]. Volume loading, temperature, particle size, and other variables can all have an impact. Effective viscosity ratios, however, differ across a range of studies. Another influence is particle size, with variations brought on by experimental protocols and equipment calibration. Rod-like nanoparticles' viscosity is substantially greater than that of spherical ones. The viscosity of nanofluids can also be influenced by base fluid, surfactant, and PH value. The stability and dispersion of nanofluids can change depending on the PH value, changing the viscosity [107,109]. The impacts of surfactants on the viscosity of nanofluids are unclear; several studies have even noted a modest reduction in drag [110]. In the presence of adjustable magnetic fields, Wang et al. [111] conducted a laboratory evaluation of a water-based Fe_3O_4 nanofluid's viscosity. With the use of an SV-10 viscometer and the coprecipitation process, they created Fe_3O_4 and introduced magnetic fields. Solid volume concentration, magnetic induction, and a drop in temperature all led to an increase in viscosity. To simulate water-Fe_3O_4 nanofluid viscosity and thermal conductivity, Bahiraei and Hangi [112] employed neural networks. Their findings demonstrated that viscosity and thermal conductivity rise with volume concentration and that temperature increases improved thermal conductivity while lowering viscosity. The connection between the conductivity and viscosity and the concentration of nanoparticles is nonlinear [70,113].

5.4 Characterization techniques

Researchers have been enhancing the physical and thermal characteristics of nanofluids to improve heat transmission. Experimental research has revealed connections with different nanoparticles, but the results have been variable and occasionally conflicting. Different methods are used to characterize MNPs and MFs, including XRD,

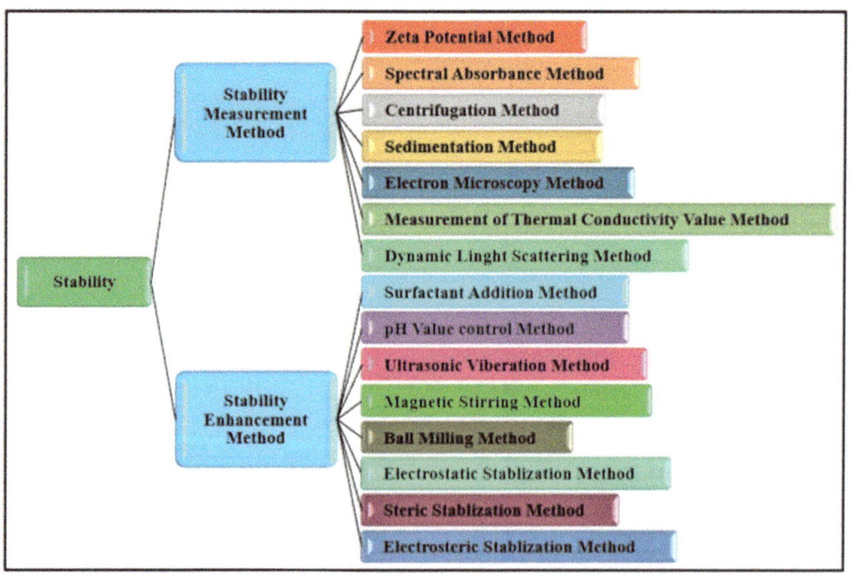

FIGURE 5.5

Stability measurement and enhancement methods.

Reprinted with permissions from A.K. Tiwari, V. Kumar, Z. Said, H.K. Paliwal, J. Clean. Prod. 292 (2021) 126031.

small-angle X-ray scattering (SAXS), TEM, DLS, Thermogravimetric Analysis (TGA), differential scanning calorimetry (DSC), Fourier Transform Infrared Spectroscopy (FTIR), Vibrating Sample Magnetometer (VSM), and atomic force microscopy (AFM). The crystal structure is determined by XRD; the size, shape, and periodicity of the structure are assessed by SAXS; and the hydrodynamic size and dispersion of dispersed particles or emulsion droplets in a liquid media are assessed by DLS. Zeta potential measurements for suspended particles are possible using DLS equipment. Zeta potential provides information on the colloidal dispersion stability, while particle size and shape are assessed by TEM. MNPs' surfactant, moisture, and volatile chemical contents are all measured by TGA. While FTIR assesses functional groups and bonds, weight loss measures the development of surfactant monolayers. Magnetic moment, saturation magnetization, susceptibility, blocking temperature, and anisotropy are among the magnetic parameters that are evaluated by VSM. Magnetic particle shapes and sizes are now often assessed using AFM [22]. Fig. 5.5 indicates stability measurement and enhancement methods.

5.5 Applications of magnetic nanofluids

The usage of nanofluids in many sectors, particularly the automotive and energy industries, is the main topic of this section. An in-depth discussion is provided on the function of nanofluids in various applications. Fig. 5.6 provides a schematic of nanoparticles and their application in the parabolic collector.

5.5.1 Utilization for heat transfer

Although MFs have been utilized in thermal management for many years, it was not until the middle of 2000 that their precise mechanics and adjustable features were completely known. Tunable characteristics and chainlike structures in magnetite-based nanofluids have increased heat conductivity. This was caused by the effective heat transfer through aggregates that were percolating linear chains. Significant scientific interest was sparked by the revelation of magnetic nanofluids' adjustable thermal characteristics. Magnetically adjustable nanofluids offer the singular option of changing thermal conductivity to viscosity ratio, making them valuable for microfluidic device applications [22,30,114,115].

5.5.2 Temperature and pH sensors

Philip and Zaibudeen [116] have created an optical probe that measures pH and temperature using magnetic nanoemulsions containing temperature- or pH-responsive macromolecules. The emulsion is excellent for a variety of applications because it is affordable, demonstrates long-term stability, responds quickly, and is easy to use. The sensor is superior to traditional electrochemical approaches in that

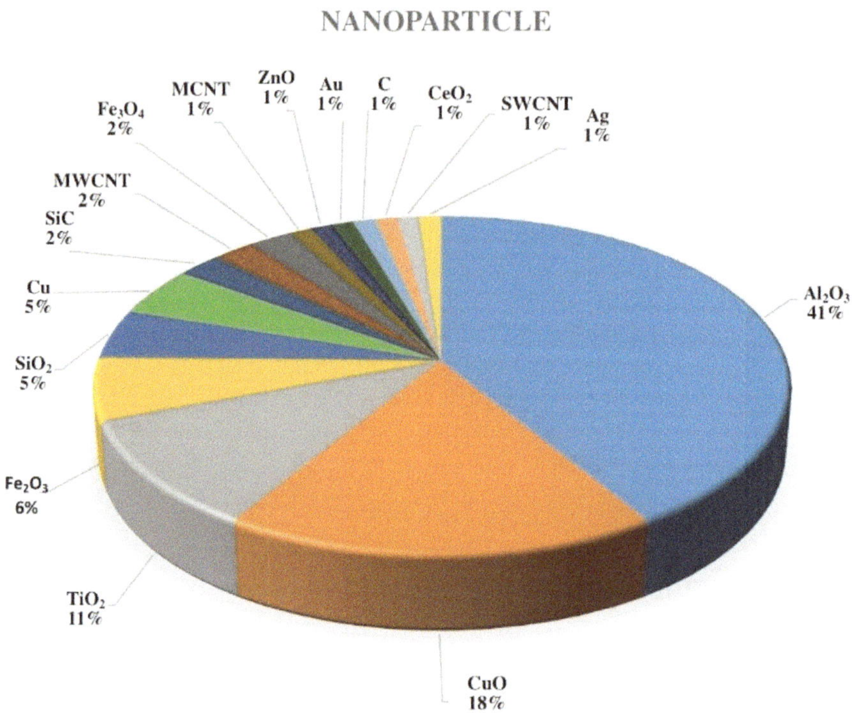

FIGURE 5.6

Share of different nanoparticles on the studies concerning PTC enhancement using nanofluids.

Reprinted with permissions from H. Olia, M. Torabi, M. Bahiraei, M. H. Ahmadi, M. Goodarzi, M.R. Safaei, Appl. Sci. 9 (2019) 463.

it is immune to electrical interferences, small in size, and capable of remote sensing. It includes features that are either noninvasive or minimally invasive. Faster reaction times, wide temperature, and pH sensing ranges are its distinguishing characteristics. The sensor employs anionic polyelectrolytes, such as polyacrylic acid, or water-soluble nonionic macromolecule such as poly (vinyl alcohol-vinyl acetate) as pH and temperature-sensitive macromolecules. Additionally, it utilizes an oil-in-water emulsion with droplets of about 200 nm, stabilized with sodium dodecyl sulfate, and functionalized. To track the nanofluid's reaction to temperature fluctuations, a 130 G constant magnetic field was used.

5.5.3 Tunable optical filter

A ferrofluid emulsion was used by Philip et al. [117] to create a tunable optical filter that makes use of the adjustable optical characteristics of magnetic emulsion droplets. By adjusting the magnetic field intensity, the pass band of the wavelength may be modified. In the experiment, the ferrofluid was stabilized with sodium dodecyl sulfate at a concentration of 0.8 mM under various magnetic fields ranging from 280 to 110 G. The polydispersity of the emulsion was around 10%. By varying the current to the solenoid or magnetic field, the pass band of reflected light may be tuned over a large variety of wavelengths. High polydispersity values result in larger FWHM, and the intensity of the reflected band is dependent on the ferrofluid emulsion's concentration. With the help of a ferrofluid stabilized with polyvinyl alcohol at a concentration of 0.66%, as a function of the applied field strength, the Bragg reflected light's center wavelengths are illustrated. Ferrofluid-based emulsions offer advantages for picking different wavelength bands, according to the investigation [22].

5.5.4 Magnetic hyperthermia-based cancer therapy

In order to treat malignant tissues, a radiofrequency alternating magnetic field is used in conjunction with the injection of MFs, a process known as MF hyperthermia (MFH). Due to loss mechanisms, including hysteresis loss, viscous loss, and Neel-Brown relaxations, the temperature rises as a result [118]. MFH has benefits, including in vitro control, target localization controlled by an external magnetic field, and high vascular escape probability. Due to its therapeutic advantages, magnetic field therapy has quickly acquired popularity as a complementary cancer treatment [119,120]. Without an external magnetic field, MNPs randomly align along basic magnetic axes. When an external magnetic field is added, however, they align with the direction of the field and store magnetic energy. When the field is removed, this energy is released as heat, which enables the moments to realign themselves in accordance with their simple axes. Neel relaxation entails the rotation of a magnetic particle on a large scale, whereas Brownian relaxation includes heating by shear action while retaining magnetic moment orientation. Both actions take place simultaneously. For experimental determination, precise measurements of temperature rise are essential for calculating the specific absorption rate, which assesses the field-induced heating of magnetic nanofluids [22,121,122]. Lahari and colleagues investigated the material parameters affecting the heating effectiveness of magnetic nanofluids [123].

5.5.5 Technology for separation based on magnetic nanomaterials

A magnetic field can easily separate and regenerate magnetite-based nanoparticle adsorbents, making them interesting for water cleanup [124]. These adsorbents may absorb a broad range of contaminants, such as heavy metals, radioactive materials, pharmaceutical residues, antibiotics, dyes, pesticides, and pharmaceutical residues [125]. The functional group determines the specificity, maximum quantity of adsorption, and adsorption rate, whereas magnetic characteristics prevent efficient recovery or separation. This method is frequently used to get rid of different contaminants [126,127]. Under magnetic field gradients, MNPs, and colloids in solution move by a process called magnetophoresis [128]. MNP composites filter organic, inorganic, microbial, and microplastic contaminants from water by combining nanoparticles, polyoxometalate ionic liquid, and magnetic microporous particles. Technology that uses magnetic separation has promise for overcoming current constraints [129]. Using carboxymethyl cellulose and MNPs, Anushree et al. [130] developed cellulose-capped magnetite nanoparticles for cationic dye removal, adding carboxymethyl cellulose to adjust the size and reduce steric hindrance. The physisorption of methylene blue made the nanoparticles efficient in removing dye.

5.5.6 Magnetic sealing

Magnetic sealing offers hazard-proof security for diverse rotating equipment at a lower cost and less impact on the environment than conventional approaches. Low frictional resistance, extended life, and great dependability are characteristics of magnetic nanofluids, which are stable emulsions containing magnetic particles such as magnetite (Fe_3O_4). The created seal can bear 25 times greater pressure than conventional magnetite sealing and can endure continual flow for 286 days [131–133].

5.5.7 Drug delivery

MNPs (IONPs) in ferrofluids offer efficient drug delivery due to their larger reactive area and ability to cross biological barriers. Common coating materials include polymers, chitosan, oleic acid, trisodium citrate, lipids, proteins, and peptides. IONPs can be bound to various drugs through adsorption, dispersion, encapsulation, electrostatic interactions, and covalent attachment. IONPs have been used as carriers for anticancer, immunosuppressive, anticonvulsant, anti-inflammatory, antibiotic, and antifungal agents. Magnetic drug targeting enhances the accumulation and efficacy of IONPs loaded with chemotherapeutic drugs. Advanced agents such as deoxyribonucleic acid (DNA), siRNA, and radionuclides have been combined with IONPs and magnetic drug targeting [134,135]. Due to their chemical stability, low toxicity, biodegradability, and superparamagnetism, IONPs are being extensively researched for magnetically guided drug delivery in cancer therapies. Cancer chemotherapy has shown promise with doxorubicin, hydroxyapatite, PEG, citric acid, and extremely water-dispersible surfactant-stabilized IONPs containing doxorubicin and curcumin. Docetaxel, epirubicin, 5-fluorouracil, gemcitabine, methotrexate, mitoxantrone, and paclitaxel are other anticancer medications that can be connected to IONPs together with carmustine, cetuximab, cytarabine, daunomycin, and daunomycin-docetaxel [40,136–138].

5.5.8 Antifungal and antibacterial properties

Drug resistance has accelerated the development of novel antibacterial and antifungal drugs, such as IONPs, although ferrofluids have already demonstrated their efficacy in these areas. These substances work well against *Proteus vulgaris*, *Staphylococcus aureus*, *Staphylococcus epidermidis*, and *Escherichia coli* [139]. Dimethyl sulfoxide (DMSO) and oleic acid-coated Fe_3O_4/Ag nanohybrid ferrofluids have antibacterial action against *Candida albicans*, *S. aureus*, *E. coli*, and *Bacillus subtilis*. Additionally, the inflammation brought on by liver fibrosis and the fibrogenic activity on hepatic stellate cells are reduced by these nanohybrid ferrofluids. Metal oxides prevent the synthesis of amino acids, lipid peroxidation, and DNA replication by causing oxidative stress and reactive oxygen species (ROS). This harm alters the permeability of the cell membrane and results in permanent harm. Electrostatic interaction between microorganisms and Fe_3O_4/Ag nanohybrid ferrofluids, which have a negative surface charge, attracts them. ROS enter bacterial pores and destroy the harmful DNA and nucleus of the bacteria. In conclusion, Fe_3O_4/Ag nanohybrid ferrofluids kill cells by accessing the DNA and nucleus of microbes while dissolving the cell's outer membrane [40,140,141].

5.6 Conclusion and outlook

This chapter gives a summary of current developments in magnetic nanofluids and discusses their creation, properties, and uses. Ferrofluid seals are perfect for high-rpm feed-through and semiconductor process industries because they have low friction, minor wear, smooth rotary motion, and maximum torque transfer. Potential uses for MFs include cancer therapies based on MFH. Nanofluids have potential in a number of areas, but further study is needed to determine their stability, practicality, and operating circumstances. For commercial applications, data from experiments and lab-size testing must be scaled up to prototype levels. Consequences for the environment must also be looked at and evaluated. The size, morphology, monodispersity, and characteristics of magnetic colloids provide obstacles to synthesis techniques. The use of nanocolloidal particles in field-worthy goods depends on the creation of novel methods for their functionalization and dispersion. However, problems with long-term stability must be addressed. The development of MNPs with strong magnetic characteristics and thin cappings should be prioritized, along with devising novel microfluidic channel technologies and enhancing MF stability in a field. For applications in biomedical domains, including cancer therapy, cell separation, analyte sensors, medication administration, and micro-robots, it is critical to develop more affordable and environmentally friendly synthesis processes for stable, nontoxic, and biocompatible MFs. It is difficult to achieve selectivity for high-sensitivity analyte detection. Hence correct functionalization is required. Additional research is needed on the toxicology, bioavailability, and biodegradability of nanoparticles following utilization. More in-depth research is required before MNPs are widely used in biomedical applications. New-generation nanorobots are best created using MNPs and clusters, and more experimental and theoretical work is required to power and carry out various activities. Understanding complicated processes, including contact forces, scattering, and macromolecule conformational changes, may be possible with the help of magnetic nanofluids. To do this, straightforward production techniques for polydisperse MFs with regulated sizes are required [22,40,75,142,143]. Although ferrofluids' biological uses have evolved greatly in recent years, toxicity and size control remain major obstacles. The synthesis technique must be created or modified according to the requirements of the application. Enhancements are required to address toxicity problems brought on by NPs' oxidative stress, which can harm cell activities, as well as to maintain stability and biocompatibility. Studying the possible toxicity of biological systems exposed to various forms of NPs is crucial to solving these problems [144–146]. It is critical to accurately assess the temperature of ferrofluids during hyperthermia, and numerical models that link the parameters of magnetic NP hyperthermia to thermal response are key. By analyzing a ferrofluid solution and computing temperature distribution in space and time, ideal therapeutic conditions may be identified. IONP contrast agents are less hazardous than gadolinium-based materials, but before they can be used in clinical settings, biocompatibility, biodistribution, and pharmacokinetics must be investigated [147]. IONPs can combine MRI with other imaging modalities, such as positron emission tomography, computed tomography, and tomography [148]. In the biomedical sectors of illness diagnostics, early detection, imaging, drug and gene delivery, and multifunctional therapies, new IONP production techniques can enhance their use. The existing pharmaceutical industry model may shift as a result of the appealing properties of multifunctional IONPs [149]. Since the majority of investigations are carried out in vitro and in vivo, studies are essential for biomedical research. For in vivo applications, new nanofabrication methods and a better comprehension of ferrofluid-biological interactions are required. Biodegradability, excretion capability, and avoiding unfavorable immunological reactions are crucial [38,146].

References

[1] H. Younes, M. Mao, S.S. Murshed, D. Lou, H. Hong, G. Peterson, Nanofluids: key parameters to enhance thermal conductivity and its applications, Appl. Therm. Eng. 207 (2022) 118202–118209.

[2] U. Fayaz, S. Manzoor, A.H. Dar, K.K. Dash, I. Bashir, V.K. Pandey, et al., Advances of nanofluid in food processing: preparation, thermophysical properties, and applications, Food Res. Int. 170 (2023) 112954–112959.

[3] Z. Said, L.S. Sundar, A.K. Tiwari, H.M. Ali, M. Sheikholeslami, E. Bellos, et al., Recent advances on the fundamental physical phenomena behind stability, dynamic motion, thermophysical properties, heat transport, applications, and challenges of nanofluids, Phys. Rep. 946 (2022) 1–94.

[4] E.C. Okonkwo, I. Wole-Osho, I.W. Almanassra, Y.M. Abdullatif, T. Al-Ansari, An updated review of nanofluids in various heat transfer devices, J. Therm. Anal. Calorim. 145 (2021) 2817–2872.
[5] G.F. Smaisim, D.B. Mohammed, A.M. Abdulhadi, K.F. Uktamov, F.H. Alsultany, S. E. Izzat, et al., Nanofluids: properties and applications, J. Sol-Gel Sci. Technol. 104 (2022) 1–35.
[6] S. Schrittwieser, B. Pelaz, W.J. Parak, S. Lentijo-Mozo, K. Soulantica, J. Dieckhoff, et al., Homogeneous biosensing based on magnetic particle labels, Sensors 16 (2016) 828–912.
[7] Y.-T. Chen, A.G. Kolhatkar, O. Zenasni, S. Xu, T.R. Lee, Biosensing using magnetic particle detection techniques, Sensors 17 (2017) 2300–2320.
[8] J. King, The Use of Magnetic Fluids in Loudspeaker Design, Audio Engineering Society Convention. 57 (1977).
[9] L. Martinez, F. Cecelja, R. Rakowski, A novel magneto-optic ferrofluid material for sensor applications, Sens. Actuators A: Phys. 123 (2005) 438–443.
[10] K. Raj, B. Moskowitz, S. Tsuda, New commercial trends of nanostructured ferrofluids, Indian J. Engin. Mater. Sci. 11 (4) (2004) 241–252.
[11] S. Khizar, N.M. Ahmad, H. Saleem, M.A. Hamayun, S. Manzoor, N. Lebaz, et al., Magnetic colloidal particles in combinatorial thin-film gradients for magnetic resonance imaging and hyperthermia, Adv. Polym. Technol. 2020 (2020) 1–18.
[12] M. Imran, A.H. Shaik, A.R. Ansari, A. Aziz, S. Hussain, A.F.F. Abouatiaa, et al., Synthesis of highly stable γ-Fe_2O_3 ferrofluid dispersed in liquid paraffin, motor oil and sunflower oil for heat transfer applications, RSC Adv. 8 (2018) 13970–13975.
[13] M.M. Selim, S. El-Safty, A. Tounsi, M. Shenashen, Review of the impact of the external magnetic field on the characteristics of magnetic nanofluids, Alex. Eng. J. 76 (2023) 75–89.
[14] A.H. Pordanjani, S. Aghakhani, M. Afrand, M. Sharifpur, J.P. Meyer, H. Xu, et al., Nanofluids: physical phenomena, applications in thermal systems and the environment effects—a critical review, J. Clean. Prod. 320 (2021) 128573.
[15] S.A. Suslov, Peculiarities of convective heat and mass transfer in magnetic nanofluids, ASME J. Heat. Mass. Transf. 145 (2023) 030801.
[16] H. Wang, C. Bi, Y. Zhang, L. Zhang, F. Zhou, Compressions of magnetorheological fluids under instantaneous magnetic field and constant area, Sci. Rep. 11 (2021) 8887.
[17] X. Meng, X. Qiu, J. Zhao, Y. Lin, X. Liu, D. Li, et al., Synthesis of ferrofluids using a chemically induced transition method and their characterization, Colloid Polym. Sci. 297 (2019) 297–305.
[18] M. Arana, P.G. Bercoff, S.E. Jacobo, P.M. Zélis, G.A. Pasquevich, Mechanochemical synthesis of MnZn ferrite nanoparticles suitable for biocompatible ferrofluids, Ceram. Int. 42 (2016) 1545–1551.
[19] C. Sarkar, H. Hirani, Synthesis and characterization of antifriction magnetorheological fluids for brake, Def. Sci. J. 63 (2013) 408.
[20] L. Yang, W. Ji, M. Mao, J.-N. Huang, An updated review on the properties, fabrication and application of hybrid-nanofluids along with their environmental effects, J. Clean. Prod. 257 (2020) 120408.
[21] Z. Shi, S. Qing, Z. Luo, Y. Liu, J. Li, Thermal physical and magnetic properties of water-based yolk-shell Fe_3O_4@C nanofluids, Inorg. Chem. Commun. 151 (2023) 110562.

[22] J. Philip, Magnetic nanofluids: recent advances, applications, challenges, and future directions, Adv. Colloid Interface Sci. 311 (2022) 102810.

[23] M. Hemmat Esfe, M. Afrand, S. Esfandeh, Investigation of the effects of various parameters on the natural convection of nanofluids in various cavities exposed to magnetic fields: a comprehensive review, J. Therm. Anal. Calorim. 140 (2020) 2055–2075.

[24] M. Sajid, K. Kalinowska, J. Płotka-Wasylka, Ferrofluids based analytical extractions and evaluation of their greenness, J. Mol. Liq. 339 (2021) 116901.

[25] R. Nayebi, F. Shemirani, Ferrofluids-based microextraction systems to process organic and inorganic targets: the state-of-the-art advances and applications, TrAC. Trends Anal. Chem. 138 (2021) 116232.

[26] S. Vinod, J. Philip, Thermal and rheological properties of magnetic nanofluids: recent advances and future directions, Adv. Colloid Interface Sci. 307 (2022) 102729.

[27] X. Xuan, Recent advances in continuous-flow particle manipulations using magnetic fluids, Micromachines 10 (2019) 744.

[28] S. Bayda, M. Adeel, T. Tuccinardi, M. Cordani, F. Rizzolio, The history of nanoscience and nanotechnology: from chemical–physical applications to nanomedicine, Molecules 25 (2019) 112.

[29] I. Nkurikiyimfura, Y. Wang, B. Safari, E. Nshingabigwi, Electrical and thermal performances of photovoltaic/thermal systems with magnetic nanofluids: a review, Particuology 54 (2021) 181–200.

[30] S. Doganay, R. Alsangur, A. Turgut, Effect of external magnetic field on thermal conductivity and viscosity of magnetic nanofluids: a review, Mater. Res. Express 6 (2019) 112003.

[31] F. Mebarek-Oudina, I. Chabani, Review on nano-fluids applications and heat transfer enhancement techniques in different enclosures, J. Nanofluids 11 (2022) 155–168.

[32] A. Joseph, S. Mathew, Ferrofluids: synthetic strategies, stabilization, physicochemical features, characterization, and applications, ChemPlusChem 79 (2014) 1382–1420.

[33] S. Genc, B. Derin, Synthesis and rheology of ferrofluids: a review, Curr. Opin. Chem. Eng. 3 (2014) 118–124.

[34] R. Kaur, A. Hasan, N. Iqbal, S. Alam, M.K. Saini, S.K. Raza, Synthesis and surface engineering of magnetic nanoparticles for environmental cleanup and pesticide residue analysis: a review, J. Sep. Sci. 37 (2014) 1805–1825.

[35] X. Zhang, L. Sun, Y. Yu, Y. Zhao, Flexible ferrofluids: design and applications, Adv. Mater. 31 (2019) 1903497.

[36] V. Socoliuc, M. Avdeev, V. Kuncser, R. Turcu, E. Tombácz, L. Vékás, Ferrofluids and bio-ferrofluids: looking back and stepping forward, Nanoscale 14 (2022) 4786–4886.

[37] M. Kole, S. Khandekar, Engineering applications of ferrofluids: a review, J. Magnetism Magnetic Mater. 537 (2021) 168222.

[38] M. Imran, A.M. Affandi, M.M. Alam, A. Khan, A.I. Khan, Advanced biomedical applications of iron oxide nanostructures based ferrofluids, Nanotechnology 32 (2021) 422001.

[39] A.O. Ivanov, A. Zubarev, Chain formation and phase separation in ferrofluids: the influence on viscous properties, Materials 13 (2020) 3956.

[40] O. Oehlsen, S.I. Cervantes-Ramírez, P. Cervantes-Avilés, I.A. Medina-Velo, Approaches on ferrofluid synthesis and applications: current status and future perspectives, ACS Omega 7 (2022) 3134–3150.
[41] T.K.O. Vuong, T.L. Le, D.V. Pham, H.N. Pham, T.H. Le Ngo, H.M. Do, et al., Synthesis of high-magnetization and monodisperse Fe_3O_4 nanoparticles via thermal decomposition, Mater. Chem. Phys. 163 (2015) 537–544.
[42] S. Khoramian, M. Saeidifar, A. Zamanian, A. Saboury, Synthesis and characterization of biocompatible ferrofluid based on magnetite nanoparticles and its effect on immunoglobulin G as an immune protein, J. Mol. Liq. 273 (2019) 326–338.
[43] J. Thomas, Preparation and magnetic properties of colloidal cobalt particles, J. Appl. Phys. 37 (1966) 2914–2915.
[44] S. Veintemillas-Verdaguer, M.P. Morales, C.J. Serna, Continuous production of γ-Fe_2O_3 ultrafine powders by laser pyrolysis, Mater. Lett. 35 (1998) 227–231.
[45] T. Hyeon, S.S. Lee, J. Park, Y. Chung, H.B. Na, Synthesis of highly crystalline and monodisperse maghemite nanocrystallites without a size-selection process, J. Am. Chem. Soc. 123 (2001) 12798–12801.
[46] T. Hyeon, Y. Chung, J. Park, S.S. Lee, Y.-W. Kim, B.H. Park, Synthesis of highly crystalline and monodisperse cobalt ferrite nanocrystals, J. Phys. Chem. B 106 (2002) 6831–6833.
[47] J. Cheon, N.-J. Kang, S.-M. Lee, J.-H. Lee, J.-H. Yoon, S.J. Oh, Shape evolution of single-crystalline iron oxide nanocrystals, J. Am. Chem. Soc. 126 (2004) 1950–1951.
[48] A.E. Berkowitz, J.L. Walter, Process for Producing a Ferrofluid, Google Patents, 1983.
[49] M.S. El-Shall, W. Slack, W. Vann, D. Kane, D. Hanley, Synthesis of nanoscale metal oxide particles using laser vaporization/condensation in a diffusion cloud chamber, J. Phys. Chem. 98 (1994) 3067–3070.
[50] F. Liu, P.J. Cao, H.R. Zhang, J.F. Tian, C.W. Xiao, C.M. Shen, et al., Novel nanopyramid arrays of magnetite, Adv. Mater. 17 (2005) 1893–1897.
[51] X. Liu, L. Xie, Y. Liu, P. Zhao, Y. Han, S. Cheng, et al., Rapid preparation of highly stable ZnO-CeO_2/CF cathode by one-step electro-deposition for efficient degradation of ciprofloxacin in electro-Fenton system, Catal. Today 355 (2020) 458–465.
[52] S. Patil, V. Patil, A. Tawade, K. Sharma, A. Bhosale, S. Vhanalkar, Synthesis of MnO_2 electrode material by electro-deposition method for the enhanced capacitive properties, (2023). Available from: http://doi.org/10.21203/rs.3.rs-3056383/v1.
[53] M. Roy, H. Verma, Magnetization anomalies of nanosize zinc ferrite particles prepared using electrodeposition, J. Magnetism Magnetic Mater. 306 (2006) 98–102.
[54] M. Parashar, V.K. Shukla, R. Singh, Metal oxides nanoparticles via sol–gel method: a review on synthesis, characterization and applications, J. Mater. Sci.: Mater. Electron. 31 (2020) 3729–3749.
[55] J.G. Mahy, L. Lejeune, T. Haynes, S.D. Lambert, R.H.M. Marcilli, C.-A. Fustin, et al., Eco-friendly colloidal aqueous sol-gel process for TiO_2 synthesis: the peptization method to obtain crystalline and photoactive materials at low temperature, Catalysts 11 (2021) 768.
[56] A.I. Borhan, A.R. Iordan, D. Ghercă, M.N. Palamaru, 7—Ferrite nanoparticles by sol–gel method, in: J. Pal Singh, K.H. Chae, R.C. Srivastava, O.F. Caltun (Eds.), Ferrite Nanostructured Magnetic Materials, Woodhead Publishing, 2023, pp. 103–119.
[57] G. Yang, S.-J. Park, Conventional and microwave hydrothermal synthesis and application of functional materials: a review, Materials 12 (2019) 1177.

[58] A. Nandagudi, S.H. Nagarajarao, M.S. Santosh, B.M. Basavaraja, S.J. Malode, R.J. Mascarenhas, et al., Hydrothermal synthesis of transition metal oxides, transition metal oxide/carbonaceous material nanocomposites for supercapacitor applications, Mater. Today Sustain. 19 (2022) 100214.

[59] J. Wang, Q. Chen, C. Zeng, B. Hou, Magnetic-field-induced growth of single-crystalline Fe_3O_4 nanowires, Adv. Mater. 16 (2004) 137–140.

[60] C. Anushree, D.N.G. Krishna, J. Philip, Synthesis of Ni doped iron oxide colloidal nanocrystal clusters using poly(N-isopropylacrylamide) templates for efficient recovery of cefixime and methylene blue, Colloids Surf. A: Physicochem. Eng. Asp. 650 (2022) 129616.

[61] J. Rockenberger, E.C. Scher, A.P. Alivisatos, A new nonhydrolytic single-precursor approach to surfactant-capped nanocrystals of transition metal oxides, J. Am. Chem. Soc. 121 (1999) 1–9.

[62] S. Sun, H. Zeng, Size-controlled synthesis of magnetite nanoparticles, J. Am. Chem. Soc. 124 (2002) 8204–8205.

[63] S. Sun, H. Zeng, D.B. Robinson, S. Raoux, P.M. Rice, S.X. Wang, et al., Monodisperse MFe_2O_4 ($M = Fe$, Co, Mn) nanoparticles, J. Am. Chem. Soc. 126 (2004) 273–279.

[64] Z. Li, H. Chen, H. Bao, M. Gao, One-pot reaction to synthesize water-soluble magnetite nanocrystals, Chem. Mater. 16 (2004) 1391–1393.

[65] A. Kumar, Y. Kuang, Z. Liang, X. Sun, Microwave chemistry, recent advancements, and eco-friendly microwave-assisted synthesis of nanoarchitectures and their applications: a review, Mater. Today Nano 11 (2020) 100076.

[66] P.T. Phan, J. Hong, N. Tran, T.H. Le, The properties of microwave-assisted synthesis of metal–organic frameworks and their applications, Nanomaterials 13 (2023) 352.

[67] I.V. Kubrakova, D.V. Pryazhnikov, Microwave-assisted synthesis of nanosized magnetic sorbents, J. Anal. Chem. 76 (2021) 15–25.

[68] S.K. Vijay, R.K. Prabhu, D. Annie, V. Chandramouli, S. Anthonysamy, A. Jain, Microwave-assisted preparation of precursor for the synthesis of nanocrystalline boron carbide powder, Trans. Indian Ceram. Soc. 79 (2020) 244–250.

[69] S. Chakraborty, P.K. Panigrahi, Stability of nanofluid: a review, Appl. Therm. Eng. 174 (2020) 115259.

[70] A.R.I. Ali, B. Salam, A review on nanofluid: preparation, stability, thermophysical properties, heat transfer characteristics and application, SN Appl. Sci. 2 (2020) 1636.

[71] B. Mehta, D. Subhedar, H. Panchal, Z. Said, Synthesis, stability, thermophysical properties and heat transfer applications of nanofluid—a review, J. Mol. Liq. 364 (2022) 120034.

[72] U. Singh, N.K. Gupta, Stability issues and operating limitations of nanofluid filled heat pipe: a critical review, Mater. Today: Proc. (2023). Available from: http://doi.org/10.1016/j.matpr.2023.01.309.

[73] I. Ijaz, E. Gilani, A. Nazir, A. Bukhari, Detail review on chemical, physical and green synthesis, classification, characterizations and applications of nanoparticles, Green. Chem. Lett. Rev. 13 (2020) 223–245.

[74] A. Asadi, F. Pourfattah, I.M. Szilágyi, M. Afrand, G. Żyła, H.S. Ahn, et al., Effect of sonication characteristics on stability, thermophysical properties, and heat transfer of nanofluids: a comprehensive review, Ultrason. Sonochem. 58 (2019) 104701.

[75] B. Bakthavatchalam, K. Habib, R. Saidur, B.B. Saha, K. Irshad, Comprehensive study on nanofluid and ionanofluid for heat transfer enhancement: a review on current and future perspective, J. Mol. Liq. 305 (2020) 112787.

[76] T. Le Ba, O. Mahian, S. Wongwises, I.M. Szilágyi, Review on the recent progress in the preparation and stability of graphene-based nanofluids, J. Therm. Anal. Calorim. 142 (2020) 1145–1172.

[77] A. Sofiah, M. Samykano, A. Pandey, K. Kadirgama, K. Sharma, R. Saidur, Immense impact from small particles: review on stability and thermophysical properties of nanofluids, Sustain. Energy Technol. Assess. 48 (2021) 101635.

[78] H. Adun, D. Kavaz, M. Dagbasi, Review of ternary hybrid nanofluid: aynthesis, stability, thermophysical properties, heat transfer applications, and environmental effects, J. Clean. Prod. 328 (2021) 129525.

[79] M. Sandhya, D. Ramasamy, K. Sudhakar, K. Kadirgama, W. Harun, Ultrasonication an intensifying tool for preparation of stable nanofluids and study the time influence on distinct properties of graphene nanofluids—a systematic overview, Ultrason. Sonochem. 73 (2021) 105479.

[80] P. Yadav, S.M. Gupta, S. Sharma, A review on stabilization of carbon nanotube nanofluid, J. Therm. Anal. Calorim. 12 (2022) 1–25.

[81] A. Abdelrazik, M.A. Sayed, H. Hashim, A.M. Omar, M.H. Helmy, A. Oulguidoum, et al., A guide to the preparation techniques of six classes of metal-, metal oxide-, and carbon-based nanofluids and the implications for their stability, J. Therm. Anal. Calorim. 148 (2023) 1–47.

[82] S. Novopashin, M. Serebryakova, S.Y. Khmel, Methods of magnetic fluid synthesis, Thermophys. Aeromech. 22 (2015) 397–412.

[83] R. Mehta, A concise account of optics of magnetic fluids: from "baravais iron" to ferrofluids, Material Aspects of Ferrofluids, EBIN publication, pp. 37–53.

[84] M. Jama, T. Singh, S.M. Gamaleldin, M. Koc, A. Samara, R.J. Isaifan, et al., Critical review on nanofluids: preparation, characterization, and applications, J. Nanomater. 2016 (2016) 6717624.

[85] J.R. Satti, D.K. Das, D. Ray, Specific heat measurements of five different propylene glycol based nanofluids and development of a new correlation, Int. J. Heat. Mass. Transf. 94 (2016) 343–353.

[86] D. Cabaleiro, C. Gracia-Fernández, J. Legido, L. Lugo, Specific heat of metal oxide nanofluids at high concentrations for heat transfer, Int. J. Heat. Mass. Transf. 88 (2015) 872–879.

[87] R.S. Vajjha, D.K. Das, Measurements of specific heat and density of Al_2O_3 nanofluid, in: AIP Conference Proceedings, American Institute of Physics, 2008, pp. 361–370.

[88] E. De Robertis, E. Cosme, R. Neves, A.Y. Kuznetsov, A. Campos, S. Landi, et al., Application of the modulated temperature differential scanning calorimetry technique for the determination of the specific heat of copper nanofluids, Appl. Therm. Eng. 41 (2012) 10–17.

[89] M. Saeedinia, M. Akhavan-Behabadi, M. Nasr, Experimental study on heat transfer and pressure drop of nanofluid flow in a horizontal coiled wire inserted tube under constant heat flux, Exp. Therm. Fluid Sci. 36 (2012) 158–168.

[90] M.X. Ho, C. Pan, Optimal concentration of alumina nanoparticles in molten Hitec salt to maximize its specific heat capacity, Int. J. Heat. Mass. Transf. 70 (2014) 174–184.

[91] V. Kumaresan, R. Velraj, Experimental investigation of the thermo-physical properties of water–ethylene glycol mixture based CNT nanofluids, Thermochim. Acta 545 (2012) 180–186.
[92] R.L. Hamilton, O. Crosser, Thermal conductivity of heterogeneous two-component systems, Ind. Eng. Chem. Fundamentals 1 (1962) 187–191.
[93] V.D. Bruggeman, Berechnung verschiedener physikalischer Konstanten von heterogenen Substanzen. I. Dielektrizitätskonstanten und Leitfähigkeiten der Mischkörper aus isotropen Substanzen, Annalen der Phys. 416 (1935) 636–664.
[94] J. Philip, P. Shima, B. Raj, Enhancement of thermal conductivity in magnetite based nanofluid due to chainlike structures, Appl. Phys. Lett. 91 (2007) 203108–203111.
[95] P. Shima, J. Philip, B. Raj, Magnetically controllable nanofluid with tunable thermal conductivity and viscosity, Appl. Phys. Lett. 95 (2009) 133112–133119.
[96] H. Yamaguchi, Y. Iwamoto, Energy transport in cooling device by magnetic fluid, J. Magnetism Magnetic Mater. 431 (2017) 229–236.
[97] T. Ambreen, M.-H. Kim, Influence of particle size on the effective thermal conductivity of nanofluids: a critical review, Appl. Energy 264 (2020) 114684.
[98] S. Porgar, H.F. Oztop, S. Salehfekr, A comprehensive review on thermal conductivity and viscosity of nanofluids and their application in heat exchangers, J. Mol. Liq. 386 (2023) 122213.
[99] A. Einstein, Eine neue bestimmung der moleküldimensionen, ETH Zurich, 1905.
[100] G. Batchelor, The effect of Brownian motion on the bulk stress in a suspension of spherical particles, J. Fluid Mech. 83 (1977) 97–117.
[101] H.C. Brinkman, The viscosity of concentrated suspensions and solutions, J. Chem. Phys. 20 (1952) 571.
[102] A.T. Utomo, H. Poth, P.T. Robbins, A.W. Pacek, Experimental and theoretical studies of thermal conductivity, viscosity and heat transfer coefficient of titania and alumina nanofluids, Int. J. Heat. Mass. Transf. 55 (2012) 7772–7781.
[103] B. Aladag, S. Halelfadl, N. Doner, T. Maré, S. Duret, P. Estellé, Experimental investigations of the viscosity of nanofluids at low temperatures, Appl. Energy 97 (2012) 876–880.
[104] S.W. Lee, S.D. Park, S. Kang, I.C. Bang, J.H. Kim, Investigation of viscosity and thermal conductivity of SiC nanofluids for heat transfer applications, Int. J. Heat. Mass. Transf. 54 (2011) 433–438.
[105] J. Albadr, S. Tayal, M. Alasadi, Heat transfer through heat exchanger using Al_2O_3 nanofluid at different concentrations, Case Stud. Therm. Eng. 1 (2013) 38–44.
[106] B.C. Pak, Y.I. Cho, Hydrodynamic and heat transfer study of dispersed fluids with submicron metallic oxide particles, Exp. Heat. Transf. Int. J. 11 (1998) 151–170.
[107] H.P. Hirudayanathan, S. Debnath, M. Anwar, M.B. Johar, N.K. Elumalai, U. Mohammed Iqbal, A review on influence of nanoparticle parameters on viscosity of nanofluids and machining performance in minimum quantity lubrication, Proc. Inst. Mech. Eng. Part E: J. Process. Mech. Eng. 153 (2023) 09544089231189668.
[108] C. Ezekwem, A recent review of viscosity models for nanofluids, Energy Sources Part A: Recovery Util. Environ. Eff. 44 (2022) 1250–1315.
[109] A. Patra, M. Nayak, A. Misra, Viscosity of nanofluids—a review, Int. J. Thermofluid Sci. Technol. 7 (2020) 070202.

[110] W. Safiei, M.M. Rahman, R. Kulkarni, M.N. Ariffin, Z.A. Abd Malek, Thermal conductivity and dynamic viscosity of nanofluids: a review, J. Adv. Res. Fluid Mech. Therm. Sci. 74 (2020) 66–84.

[111] W. Lijun, Y. Wang, X. Yan, X. Wang, B. Feng, Investigation on viscosity of Fe_3O_4 nanofluid under magnetic field, Int. Commun. Heat. Mass. Transf. 72 (2016) 23–28.

[112] M. Bahiraei, M. Hangi, An empirical study to develop temperature-dependent models for thermal conductivity and viscosity of water-Fe_3O_4 magnetic nanofluid, Mater. Chem. Phys. 181 (2016) 333–343.

[113] J. Gbadeyan, E. Titiloye, A. Adeosun, Effect of variable thermal conductivity and viscosity on Casson nanofluid flow with convective heating and velocity slip, Heliyon 6 (2020) 03076–03086.

[114] X. Zhang, Y. Zhang, Experimental study on enhanced heat transfer and flow performance of magnetic nanofluids under alternating magnetic field, Int. J. Therm. Sci. 164 (2021) 106897.

[115] B. Sun, Y. Guo, D. Yang, H. Li, The effect of constant magnetic field on convective heat transfer of Fe_3O_4/water magnetic nanofluid in horizontal circular tubes, Appl. Therm. Eng. 171 (2020) 114920.

[116] A. Zaibudeen, J. Philip, Temperature and pH sensor based on functionalized magnetic nanofluid, Sens. Actuators B: Chem. 268 (2018) 338–349.

[117] J. Philip, T. Jaykumar, P. Kalyanasundaram, B. Raj, A tunable optical filter, Meas. Sci. Technol. 14 (2003) 1289.

[118] A.E. Deatsch, B.A. Evans, Heating efficiency in magnetic nanoparticle hyperthermia, J. Magnetism Magnetic Mater. 354 (2014) 163–172.

[119] M.R. Phadatare, J. Meshram, K. Gurav, J.H. Kim, S. Pawar, Enhancement of specific absorption rate by exchange coupling of the core–shell structure of magnetic nanoparticles for magnetic hyperthermia, J. Phys. D: Appl. Phys. 49 (2016) 095004.

[120] P. Das, M. Colombo, D. Prosperi, Recent advances in magnetic fluid hyperthermia for cancer therapy, Colloids Surf. B: Biointerfaces 174 (2019) 42–55.

[121] M. Suleman, S. Riaz, R. Jalil, A mathematical modeling approach toward magnetic fluid hyperthermia of cancer and unfolding heating mechanism, J. Therm. Anal. Calorim. 146 (2021) 1193–1219.

[122] Z. Shaterabadi, G. Nabiyouni, M. Soleymani, Physics responsible for heating efficiency and self-controlled temperature rise of magnetic nanoparticles in magnetic hyperthermia therapy, Prog. Biophys. Mol. Biol. 133 (2018) 9–19.

[123] S. Ranoo, B. Lahiri, T. Muthukumaran, J. Philip, Enhancement in hyperthermia efficiency under in situ orientation of superparamagnetic iron oxide nanoparticles in dispersions, Appl. Phys. Lett. 115 (2019) 2312–2318.

[124] S. Moosavi, C.W. Lai, S. Gan, G. Zamiri, O. Akbarzadeh Pivehzhani, M.R. Johan, Application of efficient magnetic particles and activated carbon for dye removal from wastewater, ACS Omega 5 (2020) 20684–20697.

[125] T. Sasaki, S.C.W. Sakti, N. Nuryono, P.A. Krisbiantoro, Magnetic separation of pollutants for environmental remediation, Design of Materials and Technologies for Environmental Remediation, Springer, 2022, pp. 571–615.

[126] M. Soylak, O. Ozalp, F. Uzcan, Magnetic nanomaterials for the removal, separation and preconcentration of organic and inorganic pollutants at trace levels and their practical applications: a review, Trends Environ. Anal. Chem. 29 (2021) e00109.

[127] P.C. Pinheiro, A.L. Daniel-da-Silva, H.I. Nogueira, T. Trindade, Functionalized inorganic nanoparticles for magnetic separation and SERS detection of water pollutants, Eur. J. Inorg. Chem. 2018 (2018) 3443–3461.
[128] S.S. Leong, Z. Ahmad, S.C. Low, J. Camacho, J. Faraudo, J. Lim, Unified view of magnetic nanoparticle separation under magnetophoresis, Langmuir 36 (2020) 8033–8055.
[129] A. Misra, C. Zambrzycki, G. Kloker, A. Kotyrba, M.H. Anjass, I. Franco Castillo, et al., Water purification and microplastics removal using magnetic polyoxometalate-supported ionic liquid phases (magPOM-SILPs), Angew. Chem. Int. Ed. 59 (2020) 1601–1605.
[130] C. Anushree, J. Philip, Efficient removal of methylene blue dye using cellulose capped Fe_3O_4 nanofluids prepared using oxidation-precipitation method, Colloids Surf. A: Physicochem. Eng. Asp. 567 (2019) 193–204.
[131] Z. Li, D. Li, A comparative study of magnetic seals by ferrofluids, magnetorheological fluids and magnetic powders, Front. Mater. 9 (2022) 984761.
[132] S. Liu, D. Li, X. He, Z. Zhang, Structure design study of vacuum magnetic fluid seal, Front. Mater. 9 (2022) 932697–932707.
[133] D.K. Devendiran, V.A. Amirtham, A review on preparation, characterization, properties and applications of nanofluids, Renew. Sustain. Energy Rev. 60 (2016) 21–40.
[134] L.S. Arias, J.P. Pessan, A.P.M. Vieira, T.M.T.D. Lima, A.C.B. Delbem, D.R. Monteiro, Iron oxide nanoparticles for biomedical applications: a perspective on synthesis, drugs, antimicrobial activity, and toxicity, Antibiotics 7 (2018) 46.
[135] S.M. Dadfar, K. Roemhild, N.I. Drude, S. von Stillfried, R. Knüchel, F. Kiessling, et al., Iron oxide nanoparticles: diagnostic, therapeutic and theranostic applications, Adv. Drug Deliv. Rev. 138 (2019) 302–325.
[136] B. Dutta, N.G. Shetake, B. Barick, K. Barick, B. Pandey, K. Priyadarsini, et al., pH sensitive surfactant-stabilized Fe_3O_4 magnetic nanocarriers for dual drug delivery, Colloids Surf. B: Biointerfaces 162 (2018) 163–171.
[137] F. Ji, K. Zhang, J. Li, Y. Gu, J. Zhao, J. Zhang, A dual pH/magnetic responsive nanocarrier based on PEGylated Fe_3O_4 nanoparticles for doxorubicin delivery, J. Nanosci. Nanotechnol. 18 (2018) 4464–4470.
[138] P.-C. Liang, Y.-C. Chen, C.-F. Chiang, L.-R. Mo, S.-Y. Wei, W.-Y. Hsieh, et al., Doxorubicin-modified magnetic nanoparticles as a drug delivery system for magnetic resonance imaging-monitoring magnet-enhancing tumor chemotherapy, Int. J. Nanomed. 11 (2016) 2021–2037.
[139] G.R. Rudramurthy, M.K. Swamy, U.R. Sinniah, A. Ghasemzadeh, Nanoparticles: alternatives against drug-resistant pathogenic microbes, Molecules 21 (2016) 836.
[140] L.K. Ruddaraju, S.V.N. Pammi, G. Sankar Guntuku, V.S. Padavala, V.R.M. Kolapalli, A review on anti-bacterials to combat resistance: from ancient era of plants and metals to present and future perspectives of green nano technological combinations, Asian J. Pharm. Sci. 15 (2020) 42–59.
[141] A. Taufiq, R.E. Saputro, H. Susanto, N. Hidayat, S. Sunaryono, T. Amrillah, et al., Synthesis of Fe_3O_4/Ag nanohybrid ferrofluids and their applications as antimicrobial and antifibrotic agents, Heliyon 6 (2020) 05813–05819.
[142] H. Aslam, S. Shukrullah, M.Y. Naz, H. Fatima, H. Hussain, S. Ullah, et al., Current and future perspectives of multifunctional magnetic nanoparticles based controlled drug delivery systems, J. Drug Deliv. Sci. Technol. 67 (2022) 102946.

[143] H. Yasmin, S.O. Giwa, S. Noor, H.Ş. Aybar, Influence of preparation characteristics on stability, properties, and performance of mono-and hybrid nanofluids: current and future perspective, Machines 11 (2023) 112.

[144] S. Liu, B. Yu, S. Wang, Y. Shen, H. Cong, Preparation, surface functionalization and application of Fe_3O_4 magnetic nanoparticles, Adv. Colloid Interface Sci. 281 (2020) 102165.

[145] M. Ramazanov, A. Karimova, H. Shirinova, Magnetism for drug delivery, MRI and hyperthermia applications: a review, Biointerface Res. Appl. Chem. 11 (2021) 8654–8668.

[146] N. Senthilkumar, P.K. Sharma, N. Sood, N. Bhalla, Designing magnetic nanoparticles for in vivo applications and understanding their fate inside human body, Coord. Chem. Rev. 445 (2021) 214082.

[147] I. Raouf, P. Gas, H.S. Kim, Numerical investigation of ferrofluid preparation during in-vitro culture of cancer therapy for magnetic nanoparticle hyperthermia, Sensors 21 (2021) 5545.

[148] M. Salehipour, S. Rezaei, J. Mosafer, Z. Pakdin-Parizi, A. Motaharian, M. Mogharabi-Manzari, Recent advances in polymer-coated iron oxide nanoparticles as magnetic resonance imaging contrast agents, J. Nanopart. Res. 23 (2021) 1–35.

[149] N.S. Vallabani, S. Singh, Recent advances and future prospects of iron oxide nanoparticles in biomedicine and diagnostics, 3 Biotech. 8 (2018) 279.

CHAPTER

SiO_2-based nanofluids

6

Razieh Razavi[1], Mahnaz Amiri[2] and Ghazal Khajouei[2]

[1]*Department of Chemistry, Faculty of Science, University of Jiroft, Jiroft, Iran*

[2]*Neuroscience Research Center, Institute of Neuropharmacology, Kerman University of Medical Science, Kerman, Iran*

6.1 General introduction

Mineral oils and other conventional fluids are excellent lubricants, but their weak thermal chattels limit their usage as industrial coolants. Various methods exist today for increasing the transfer heat rate for conventional fluids, including adding small solid particles (millimeters and micrometers) to conventional fluids to enhance their thermal properties. Despite this, these fluids have been associated with severe problems such as clogging, excessive erosion, pipeline pressure drops, and weak suspension stability. In the past decade, nanosized particles have replaced larger size particles in suspension, resulting in a novel class named nanofluids. Comparing these base- and nanofluids, nanofluids indicated well stability, more thermal conductivity, and minor pressure drop. Additionally, these nanofluids have enhanced the performance parameters of numerous alloys and metals in machining, including milling [1–8], grinding [9–18], drilling [19], and turning [20–23]. Sharma et al. [24] investigated using nanofluid as cutting fluid in numerous processes of machining and found that nanofluid significantly enhanced machining performance. Nanofluids are colloid-like solutions that include nanometer-size particles (1–100 nm) in a base fluid. There are many types of nanoparticles, such as nonmetallic, oxides, metallic, carbides, carbonic, nanoparticle mixtures (hybrid nanoparticles), as well as nanoscale liquid droplets. Base fluids are typically liquids with low viscosity, such as H_2O, or an extremely viscous one like mineral oil, ethylene glycol (EG), or a combination of different liquids. In 1995, Dr. Stephen Choi introduced nanofluids for the first time [25]. Nevertheless, Masuda et al. [26] presented an independent and earlier report on a similar subject. The escalating demand for thermal power is primarily driven by the proliferation of industries that utilize heating and cooling processes. Enhancing the heat transfer rate is crucial for shortening operational times, conserving energy, improving thermal efficiency, and extending the system's lifespan [1]. Historically, H_2O, EG, and oil, which are conventional fluids that are used in forced heat convection, exhibited significantly lower thermal conductivity than solid particles. The incorporation of micrometer-scale particles into the base fluid improved heat transfer.

Nanofluids. DOI: https://doi.org/10.1016/B978-0-443-13625-2.00015-2

However, using these larger particles brought about challenges, including the propensity for clogging, rapid sedimentation, damage to heat transfer equipment, and an unexpected increase in pressure loss [3]. As a result, scientists have shown considerable interest in nanometer-scale particles that can be dispersed within the base fluids. The concept of "micro-mixtures" was first introduced by Choi and his team [2]. These are essentially nanosized particles, with dimensions less than 100 nm, spread in standard solvents such as H_2O, oil, or EG [2,4–7]. Their recent popularity in industries can be attributed to a multitude of benefits, including:

1. enhanced heat transfer due to improved thermal conductivity and an increased surface area,
2. reduced resistance because of their tiny particle size, and
3. remarkable efficiency in rapid temperature regulation systems [8].

The ongoing development of nanoemulsion research has positioned hybrid nanoemulsions at the forefront of recent studies [9–13]. Hybrid nanoemulsions, which can be created by suspending two or more distinct nanoparticles either mixed or combined in the base fluids [9], are seen as an advancement in nanoemulsion research. The primary goal of developing hybrid nanoemulsions is to enhance the characteristics of individual components, potentially achieving significant improvements in thermal or rheological properties. However, experimental and theoretical investigations into convective heat transfer involving hybrid nanoemulsions have been limited to date. Silicon, second only to oxygen in terms of earthly abundance [11], boasts a range of outstanding properties that have driven its use in various industries. From transformative changes in the electronics and communication industries to breakthroughs in silicon chip technology, silicon has revolutionized sectors such as mass communication, space exploration, and aeronautical engineering. Silicon also plays a key role in photovoltaics, with high expectations for silicon-based photovoltaic modules potentially dominating the energy sector, given they currently represent the most efficient method for renewable power generation. Silicon dioxide, or silica, is extensively used in heat and energy transfer applications, gaining increasing recognition as a nanosupplement. Rahman and Padavettan [12] comprehensively examined potential applications for silica nanosupplements, highlighting inherent qualities that make silica nanoparticles (SNPs) a preferred choice across numerous systems. These properties include a low dispersion index, excellent biocompatibility, cost efficiency, hydrophilic tendencies, and a large surface area [13]. Their potential applications span fields such as biomedical science as carriers for enzymes or Deoxyribonucleic Acid (DNA), polymers as fillers or reinforcements, aerospace, and optical imaging, among others. Silica, particularly in the form of silicon dioxide, has played a pivotal role in advancing the application of micro-mixtures. These have been explored across various technical sectors, including the fields of chemistry [14], automobile manufacturing [15], solar heat management [16], and device temperature control [17]. The use of SiO_2 particle-based micro-mixtures has been shown to boost the effectiveness of these applications significantly. A summary of the research related to SiO_2-enriched micro-mixtures illustrates their numerous benefits. Among them are reduced resistance compared to other micro-mixtures, increased heat transfer rates, excellent stability, efficient conversion of light to heat, and superior heat and flow properties. In-depth academic investigations reveal that micro-mixtures founded on quartz display decreased resistance and maintain excellent stability, tackling a major challenge associated with micro-mixtures. However, there remains uncertainty regarding their heat retention capabilities across different applications. Some studies have reported inconsistent results regarding the heat performance of quartz-based micro-mixtures in certain thermal regulation setups, such as vehicle temperature systems. Given quartz micro-mixtures' promising nature and robustness, a thorough analysis of their capabilities is crucial to derive definitive scientific understandings. The current study aims to assess the efficiency of a silica-water nanoemulsion in an experimental model that simulates an automotive cooling system. It also intends to evaluate the reproducibility of repeated experiments. This research effort will channel various nanoemulsion samples through the radiator at different temperatures and flow rates. Adding a negligible volume fraction of nanoparticles dramatically improves heat transfer. Nanofuid's advantages and reasons for improving heat transfer are still under investigation by researchers. Several studies reported the following results [23–27].

- By dispersing nanoparticles, the fluid can conduct heat more effectively. The nanoparticles' volume fraction determines effective thermal conductivity. Accordingly, as the nanoparticles' volume fraction rises, effective thermal conductivity also increases.
- Due to increased particle surface area, nanoparticles interact with base fluid more effectively.
- By dispersing nanoparticles, Brownian motions are generated, resulting in more significant interaction and collisions between the fluids and particles.
- Turbulence and mixing fluctuations are intensified by dispersed nanoparticles.
- For equivalent heat transfer, the pumping power is relatively lower than the base fluid.
- In order to increase solar energy absorption, nanofluids can be used.
- It has better stability than other colloid suspensions.

6.2 Preparation of SiO_2 nanofluid

SNPs are highly beneficial for a vast array of applications, ranging from chemical industries to biomedical applications, biotechnologies, agriculture, environmental

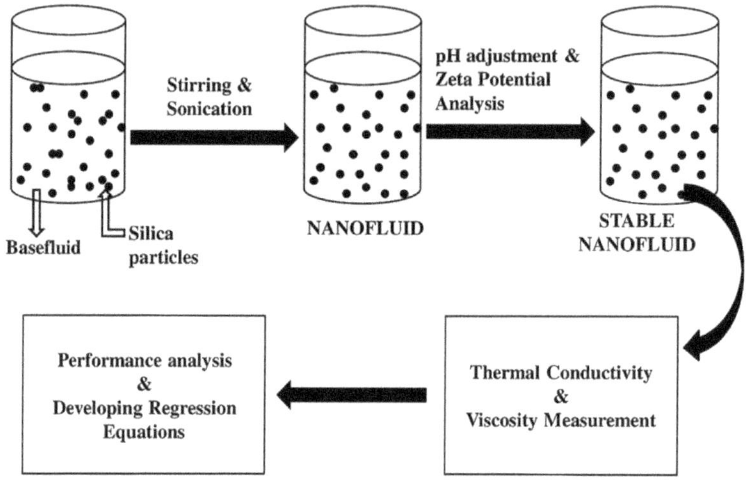

FIGURE 6.1

Photograph of preparation of stable SiO_2 nanofluid.

Reprinted with permissions from T. Rajendra Prasad, Thermal performance of stable SiO_2 nanofluids and regression correlations to estimate their thermophysical properties, J. Indian. Chem. Soc. 99 (2022) 100461.

restoration, and wastewater treatment. Their unique attributes, such as structure with mesoporous, great surface area, perfect pore size, biocompatibility, modifiability, as well as the ability to interlink with polymers, significantly broaden their applicability. Being nontoxic, they are safely used in biomedical research. Furthermore, their ability to bind molecules on both their internal and external surfaces positions them as exceptional carriers for biological and nonbiological materials. Fig. 6.1 indicated the preparation of stable SiO_2 nanofluid.

6.2.1 SNP production (chemical techniques)

Historically, SNPs have been produced using methods such as the sol-gel process, Stober's method, flame-based preparation, and microemulsion methods. While these chemical routes are versatile and adaptable, they can be expensive and complicated. Take the inverse microemulsion method, for instance, which employs surfactants that solubilize in water to create micelle structures. Though effective, extracting surfactants from the final products is both complex and costly [16]. Nevertheless, nanoparticles synthesized this way have proven useful, especially as coatings to attach functional entities [25,26]. Another commonly employed method for SNP production is chemical vapor condensation (CVC) [27]. In the CVC process, $SiCl_4$ reacts with hydrogen and oxygen. This method boasts the flexibility to tweak the characteristics of the nanoparticles, such as shape and size, based on specific needs. It is especially favored for producing powdered nanoparticles [28–30]. The sol-gel technique, renowned for producing silica gel and silica [31,32], primarily relies on TEOS or inorganic salts such as sodium silicate. These undergo hydrolysis and condensation, typically catalyzed by an acid or base [33,34]. While using metal alkoxides (such as TEOS or TMOS), the resulting hydrolysis leads to the formation of Si–O–H that subsequently polymerizes to form the structure of silica. Stober is credited for pioneering what is now referred to as Stober's technique. Utilizing a silica alkoxide alcoholic solution, he proficiently fashioned spherical SNPs, achieving sizes ranging from 5 to 2000 nm, with ammonia as a catalyst [35]. Over the years, this method has been refined, becoming even more flexible in tailoring parameters and achieving desired features [36,37]. Notably, while acid-catalyzed processes tend to produce gels, Stober's method typically yields uniformly-sized SNPs [35]. Although these traditional methods have their merits, they often come with high costs, employ hazardous chemicals, and require significant energy input. As a result, recent trends lean toward exploring greener, biogenic methods [22] (Fig. 6.1).

6.2.2 Synthesis of SNPs (biogenic methods)

SNPs can be fabricated using natural processes involving microorganisms and organic substances, including bacteria, fungi, phytoplankton, and botanical metabolites [38]. The biological synthesis of silica often leverages proteins such as silicatein and silaffin, forming engineered cationic polypeptides. In particular, the fungus Fusarium oxysporum has been spotlighted in research for its capability to produce silica under ambient conditions in a water-based anionic complex [39,40]. In recent years, the potential of utilizing various biological waste materials, such as residues from rice, sugarcane, and rice straw, for SNP synthesis has been explored [14,41–44]. The process commences with silica extraction from the waste, subsequently transforming it into a sodium silicate solution [45]. This technique demands the optimization of several parameters, including temperature, pH, and duration, to ensure efficiency [46]. This innovative method transforms waste biomass into valuable byproducts such as SNPs, signifying a vital advancement in sustainable materials science [47]. Ensuring the purity of the resulting silica involves multiple stages. Initially, the biomass is thoroughly rinsed with purified water to expunge residual contaminants. This is followed by a leaching process using various agents to eradicate metallic pollutants within the biomass [48,49]. The presence of these contaminants, if unchecked, could negatively influence the physicochemical properties of the synthesized silica particles [41]. Leaching agents typically include HCl, HNO_3, and H_2SO_4, with HCL proving especially effective, capable of eliminating up to 99% of metal contaminants [49–53]. In addition to acidic treatments, alkaline and salt treatments, such as sodium hydroxide and $KMnO_4$, have also been explored for their ability to remove impurities [54].

Following leaching, the biomass undergoes calcination. The quality of the produced silica is contingent upon factors such as temperature, duration, and airflow during this process [55]. For example, when rice husks are calcined at 600 degrees, the resulting silica exhibits amorphous characteristics. At 700 degrees,

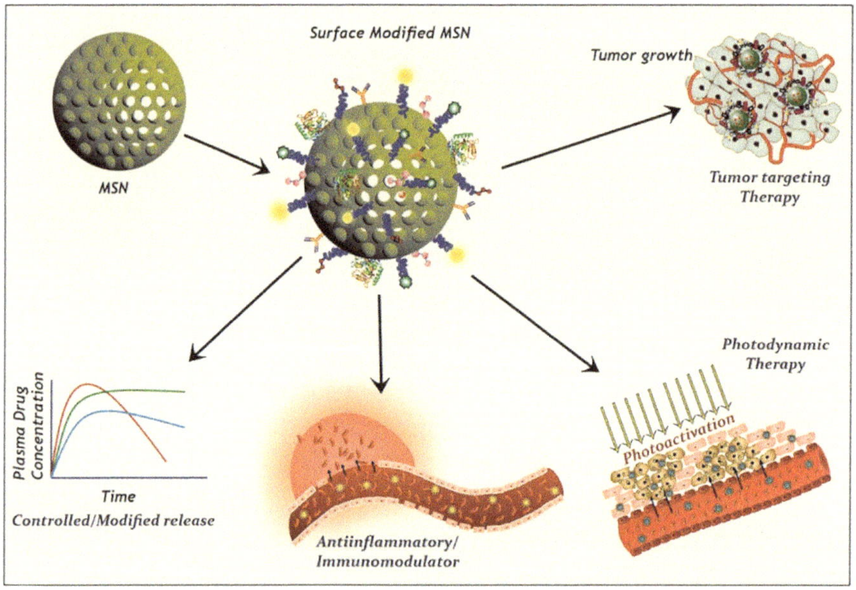

FIGURE 6.2

Photograph of biomedical application of nanofluid.

Reprinted with permissions from H. Ahmed, et al., Biomedical applications of mesoporous silica nanoparticles as a drug delivery carrier, J. Drug Deliv. Sci. Technol. 76 (2022) 103729.

a mixture of amorphous and crystalline features is observed, while temperatures exceeding 700 degrees generally yield a predominantly crystalline structure [56–59]. Numerous eco-friendly synthesis pathways have been examined to fabricate amorphous SNPs from biological sources such as rice husk and sugar beet bagasse [22]. Research involving rice husk has yielded semicrystalline porous and regular SNPs [29,60]. Studies with sugar beet bagasse have similarly resulted in the successful production of SNPs [61] (Fig. 6.2).

6.3 Chemical applications of SNPs

There are specific sites on SNPs that facilitate their functionalization. For instance, mesoporous silica nanoparticles (MSNs) have three main functionalization sites, namely the walls of the pore, the entrance of the pore, and the particle interior/exterior [16,62]. MSNs are formed according to the mechanism presented in Fig. 6.3. In most cases, particles of silica are reacted by cocondensation or else postpreparation joining with organo-substituted trialkoxysilanes [17]. Mesopores are not blocked by nonsiliceous groups when silica particles are functionalized on the pore walls. Alkoxysilanes bind to silanol groups on the surface during

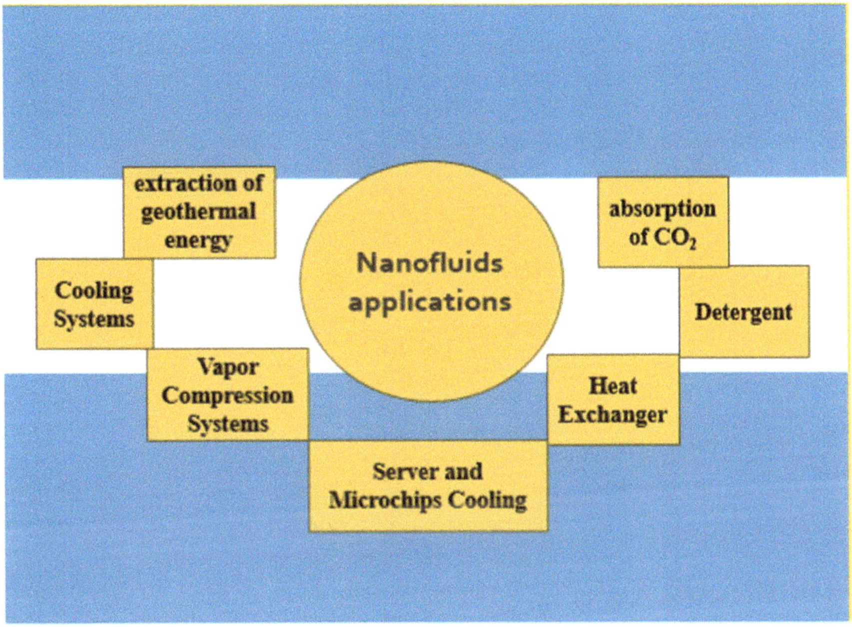

FIGURE 6.3

The various applications of nanofluid.

Reprinted with permissions from A.H. Pordanjani et al., Nanofluids: physical phenomena, applications in thermal systems and the environment effects—a critical review, J. Clean. Prod. 320 (2021) 128573.

functionalization. Similarly, many metals have been reacted on silica particles [63,64]. For instance, aluminum is commonly functionalized on MSNs used in catalytic-related applications. Zhai et al. [52] synthesized aluminosilicate nanoparticles in this regard. Polyethylene glycol was used to limit the particle size to twenty nm and the particles were mesoporous. Zhao et al. [65] also used hematite nanocores to synthesize magnetic MSNs. In this study, ibuprofen was delivered via these particles. MSNs were synthesized using quantum dots (QDs) (CdSe/ZnS), as well as magnetic nanomaterials, by researchers [66] prepared MSNs entrenched with TiO_2 as well as Fe_3O_4. Catalytic reactions were conducted using the prepared material. As an antimycobacterial agent against Mycobacterium tuberculosis, In 2020, Ref. [67] produced silica with silver nanoparticles in their base recently. Additionally, the researchers [68] synthesized respective metallic SNPs using Ag, Cu, and other salt, demonstrating their potential as antibacterial agents. MSNs can be immobilized on both their inner and outer surfaces, which is one of their advantages. Accordingly, a study by Lin et al. [69] has effectively immobilized fluorophores onto the inner surface of MSNs and grown dense polymers on the external surface for neurotransmitter detection. In a similar way, Chen et al. [70] reported more recent developments. They advanced hybrid silylated fluorophores in the core of MSNs and detected their effectiveness in contrast to tumor cells. It has been demonstrated that numerous MSNs are used in a single-pot catalytic reaction, each immobilized with a different compound. Chemically incompatible MSNs are often found to participate without neutralizing one another, as reported by Huang et al., in which 4-nitrobenzaldehyde methyl acetal is changed to aldehyde with two different MSN species immobilized with a remarkable yield of 97.7%. In a similar way, Reis et al. [71] confirmed controlling release and loading of molecules at the pore entrance of MSNs, potentially contributing to drug delivery [72].

Regarding the biomedical and biotechnological potential of SNPs, the study of drug delivery is regarded as one of the most essential aspects. Researchers recognize the importance of using engineered nanostructures for targeted drug delivery to patients [72]. In the body, these nanostructures are able to operate as carriers and target specific tissues or organs. As a result of their large surface area, as well as structure with porous, MSNs gained significant recognition in this regard. In recent years, these elements employed as nanocarriers for the delivery of drugs [18,20,22,73–75]. A study was conducted by Aughenbaugh et al. [76] employing silica xerogels as a potential sample of carriers for drugs to study drug release. Additionally, hydrophobicity reduces the absorption of drugs when taken orally, which can be enhanced when SNPs are used as carriers [72]. In this case, it may be because SNPs have a negative charge on their surface, whereas hydrophobic drugs have a positive charge [28]. Moreover, biocompatibility, the loading capacity, and possible modification of the functional groups of these nanoparticles make them particularly efficient [16]. Accordingly, Zhang et al. [77] employed SNPs to carry the hydrophobic drug telmisartan. The SNP is found to improve the permeability of the drug. Similarly, Samira et al. [78] and Nihal et al. [79] demonstrated that SNP-conjugated curcumin performed significantly better than curcumin (an anticancer drug) alone. As a result of effectively attaching thermosensitive as well as magnetic nanoparticles on the silica particles surface coated with Fe_3O_4, Lein et al. [80] prepared nanoparticles for controlled drug delivery. Also, SNPs with surface modifications have been studied to capture boron neutrons [81–85]. Due to their extensive surface area and their minor size, SNPs can be used as biomarkers for detecting specific biomolecules. Accordingly, QDs are widely used, but heavy metal toxicity and their insolubility in water have prompted researchers to seek alternative solutions. Consequently, some researchers have covalently attached compounds, including fluorophores and the dye tetramethyl rhodamine isothiocyanate, to Si nanoparticles. The invented materials demonstrate H_2O soluble and fewer or nontoxic conduct, which are positive outcomes, making them an acceptable alternative for QDs [86–90]. Researchers have also used SNPs' adhesive properties to synthesize glue-like materials, which are less invasive than standard tissue adhesives such as cyanoacrylate [22]. SNPs also have the remarkable property of cell differentiation, which could be used to treat obesity [19,91]. Due to its sensitivity and presence in biological tissues, the fluorine nucleus (19 F) is used to identify even the most minor details, leading to its grafting onto the SNPs, which allows the fluorine to immobilize both the inner and external particles layers [92–95]. Specific detection is required for medical diagnoses and research that interact and sense minute targets such as proteins, enzymes, DNA, and Messenger RNA (mRNA). As a result, nanotechnology developments are supporting and contributing to medical research [96]. Several applications of nanotechnology have already been demonstrated, including diagnosis, drug delivery, medical imaging, treatment of cancer, and treatment of diabetes [19,60,75,97]. A drug delivery system based on SNPs was developed by Maleki et al. [98]. The researchers also demonstrated that the same system could entrap colorless water-soluble drugs such as isoniazid. As a result, researchers have recently studied more H_2O-soluble drugs that can be effectively immobilized on Si nanoparticles and applied for delivery of drugs [99–103]. Fig. 6.2 represented the biomedical application of nanofluid.

6.4 Mono-phase and dual-phase methods of nanofluid preparation

Nanofluid fabrication primarily relies on two distinct approaches: the mono-phase and dual-phase methods [1]. The mono-phase method emphasizes the simultaneous generation and nanoparticle dispersion inside the fluid base. This is typically achieved using methods such as vapor deposition or liquid-phase chemical reactions. By forgoing steps such as drying, preservation, distribution, and transport, this method reduces particle agglomeration and ensures a uniform dispersion within the base fluid [15]. However, this approach comes with its own set of challenges. The risk of residual chemicals remaining in the nanofluid due to incomplete reactions raises concerns about its purity [88]. Moreover, its utility is generally limited to fluids with a low evaporation rate. In contrast, the dual-phase method, often preferred for nanofluid synthesis, begins with the production of nanoparticles, tubes, fibers, or rods. This is usually done through vapor deposition or gas-phase condensation, resulting in a dry nanoparticulate substance. Subsequently, this dry substance is introduced into the base fluid using methods such as magnetic stirring, ultrasonication, high-intensity mixing, stabilization, and bead milling. By bypassing steps such as drying, preservation, transport, and nanoparticle distribution, the vapor deposition method within the dual-phase approach promotes the creation of uniform nanofluids. This method encourages the direct evaporation and deposition of nanoparticles into the main fluid, resulting in a clean and evenly dispersed product while minimizing the chances of nanoparticle agglomeration. However, the dual-phase method is not without its challenges, including the potential for leftover substances in the nanofluids and increased operational costs. Compared to the 1-step method, the 2-step approach provides a more cost-effective solution for the commercial production of nanofluids. However, this approach has a significant downside: nanoparticles have a tendency to cluster together owing to their high surface area and attractive forces, leading to agglomeration within the fluid. This clustering reduces thermal conductivity and increases the likelihood of microchannel sedimentation and blockage. To mitigate these issues, surfactants are often used to steady nanoparticles inside the fluids. Despite these challenges, the method's flexibility—accommodating a wide variety of particles, including oxide particles and carbon nanotubes (CNT)—combined with its straightforwardness in the nanofluid formulation [89] makes it attractive for industrial use. This approach stands out as the most economical technique for large-scale nanofluid manufacture. In the two-step technique, nanoparticles are created using various methods and then dispersed into the base fluid to formulate the desired nanofluid. This manufacturing process proves to be both cost-effective and scalable. The primary challenge in the two-step method is managing the inherent tendency of nanoparticles to aggregate. To combat this issue, surfactants are employed to enhance nanoparticle stability within the fluid. Given these advantages, this method is established as the standard commercial procedure for nanofluid production and is widely endorsed by researchers in the field of nanofluid development.

6.5 Nanofluids applications

Nanofluids have gained prominence as potent tools for enhancing the properties of base liquids, particularly in improving thermal conductivity performance. Their versatility has sparked interest across various industries, from petroleum extraction and renewable energy to medical innovations and thermal performance optimization. Consequently, fields such as energy, physics, chemical sciences, and healthcare are fervently exploring their potential. In engineering, SiO_2-based water nanofluids are prized for their cost-effectiveness, simplicity in production, adjustable properties, and environmental sustainability.

Li's research group [5] pioneered an application by developing a detergent-free nanofluid infused with water to enhance oil recovery using reactive SNPs. These particles exhibited impressive traits, reducing oil-water boundary tensions, increasing interface barriers, preventing oil adherence on rocky surfaces, and altering core wetness. Such findings illuminate nanofluid's potential in refining oil recovery processes. Concurrently, Minakov's team [6] studied the effects of SNPs on drilling fluid properties, such as rheology, pressure drop, and debris transport in vertical drilling pathways with stable flows. Their work spotlighted a 2.5fold increase in pressure drops but a noteworthy 17% rise in debris transportation. Further, Dayo Afekare and colleagues [7] employed atomic force microscopy to examine silica-based nanofluids' impact on the wetness shift in conventional sandstone samples at microscopic scales. The results alluded to potential enhancements in oil extraction, steered by fluid-rock and fluid-fluid interactions (Fig. 6.4). Fig. 6.4 indicated the oil recovery application of SiO_2 nanofluid.

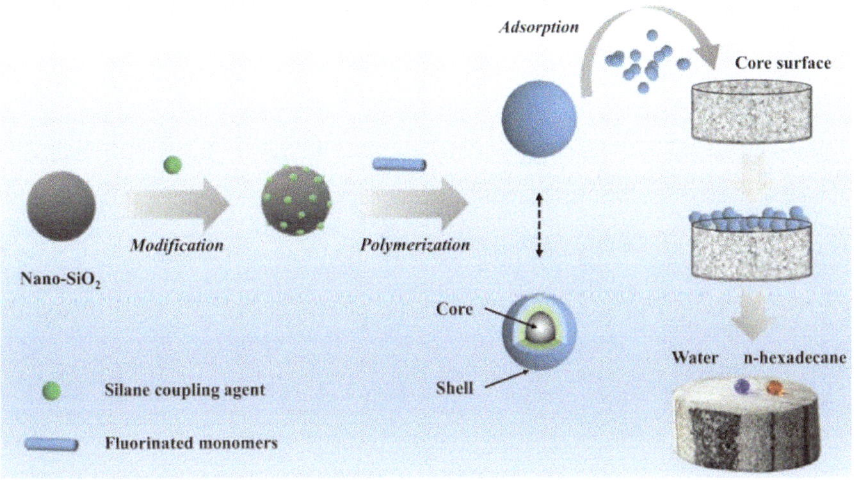

FIGURE 6.4

Oil recovery application of SiO_2 nanofluid.

Reprinted with permissions from L. Liang, Fluoropolymer-coated SiO_2 nanoparticle-based nanofluids for oil recovery, Colloids Surf. A: Physicochem. Eng. Asp. 637 (2022) 128242.

However, nanofluids face practical challenges, including limited stability, particle agglomeration, and settling during application, hindering their widespread adoption. Nanofluid stability is often gauged by monitoring sedimentation over a set period [23]. Recent studies are actively unraveling factors affecting nanofluid stability and their underlying mechanisms. Kumar's group [8] found that SiO_2 aggregates exhibited prolonged suspension in the presence of TiO_2 compared to pure SiO_2 clusters, as validated by increased absorption and electro-conductivity readings. Similarly, Wang and team [9] delved into the stability of CuO-infused water nanofluids, examining stabilization factors, nanoparticle forces, equilibrium dynamics, and detergent roles. They proposed enhancing nanofluid stability by reducing nanoparticle surface energy, narrowing the density difference between nanoparticles and the base liquid or decreasing the liquid's viscosity. Amin Asadi's team [10] explored the effects of ultrasound exposure duration on the stability and heat transfer of MWCNT-infused water nanofluids. Their research identified the optimal ultrasound exposure duration for such nanofluids using zeta potential measurements and visual evaluations. Similarly, Azadeh Ghadimi and her researchers [11] investigated the stability impact of detergents and ultrasound treatments on titanium oxide suspensions. Their comprehensive study, employing various tools such as Ultraviolet (UV)-visible spectroscopy and electron microscopy (EM), concluded that a combination of a 3-hour ultrasonic treatment with 0.1% detergent weight yielded the most stable mixture, maintaining its efficacy for up to a month. Fig. 6.5 indicated the stability of nanofluids.

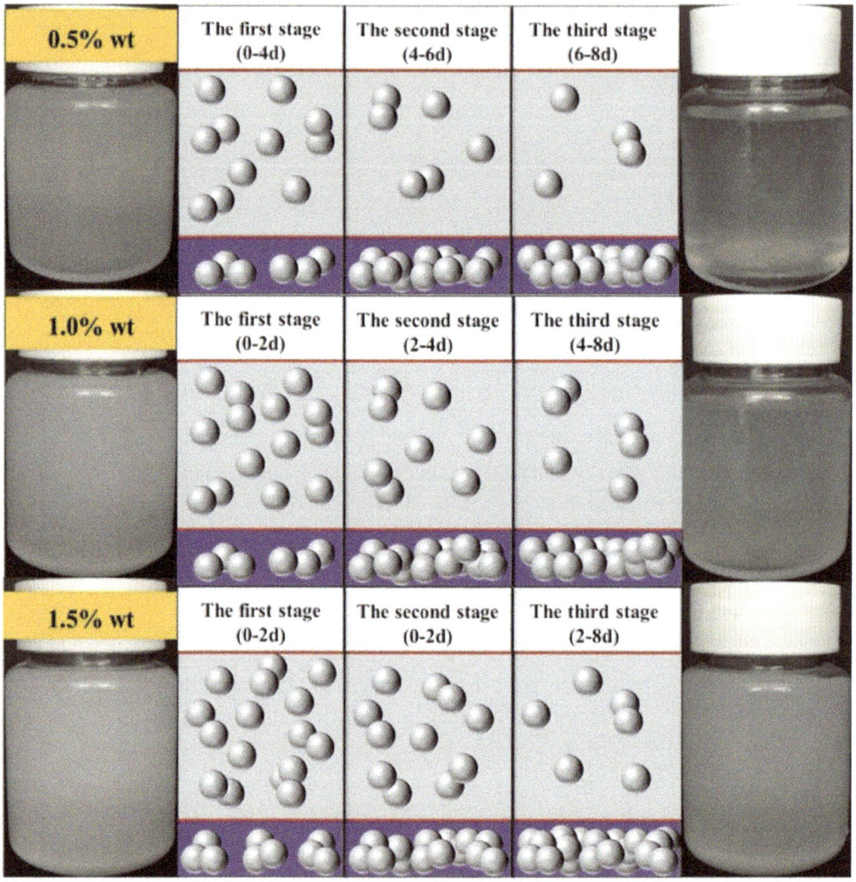

FIGURE 6.5

Photograph of stability of nanofluid.

Reprinted with permissions from T. Zhang, Effect of particle concentration on the stability of water-based SiO_2 nanofluid, Powder Technol. 379 (2021) 457–465.

6.6 Parameters that effected on nanofluids applications

The rheological properties of nanofluids, or their flow characteristics, have become a popular focus of study in recent years. Researchers have been particularly interested in exploring how factors such as the variety, density, size, shape of particles, operating temperature, and base fluids can influence these properties. Generally, it has been observed that the nanofluid's viscosity tends to increase with the nanoparticles' concentration. For example, Pak and Cho [3] found a significant 300-fold rise in viscosity for an alumina-H_2O nanofluid in comparison to the base fluid. However, other studies have reported a more modest increase in viscosity—around 85%—for similar particle concentrations and sizes [4,5]. Interestingly, researchers have found that the increase in viscosity tends to be relative to the average diameter of the particle [9,11]. For instance, Nguyen et al. [12] found that nanofluids with 36 nm particles had viscosity measurements that were 5% lower than those of nanofluids with 47 nm particles. The morphology and structure of the microscopic particles within nanofluids play a pivotal role in determining their flow resistance. Timofeeva and her associates [13] observed that nanofluids containing elongated or rod-like particles (such as plates) exhibited higher viscosity at equivalent volume percentages. Like many fluids, nanofluids generally become less viscous when heated [14,15]. Namburu et al. [11] documented a marked decline in viscosity with increasing temperature for SiO_2-ethylene glycol/water composite nanofluids within the range of 35°C–50°C. These nanofluids exhibited anomalous flow characteristics at cooler temperatures, but they transitioned to standard flow properties upon heating. Conversely, Namburu's subsequent research [16] found that copper oxide nanoparticles in an EG/H_2O mixture retained typical flow characteristics, even at lower temperatures. Nevertheless, the flow behaviors of nanofluids are not uniform, and they arise from various causes. For instance, Chen and colleagues [17] observed a marginal decrease in flow resistance for carbon tube-water nanofluids up to a volume concentration of 0.4%, beyond which the viscosity increased. This initial reduction in resistance was ascribed to the lubricating effect of the nanoparticles. Furthermore, the base fluid's composition and attributes significantly influence nanofluids' flow dynamics [15]. Research by Timofeeva's team [18] suggested that the characteristics of the base fluid more influence flow resistance in nanofluids than the properties of the nanoparticles. Most nanofluids with water as their primary component demonstrated regular flow characteristics at low concentrations. However, deviations in flow were evident at higher concentrations and with nonaqueous base fluids. Chen's team [19] emphasized that TiO_2 nanofluids with an EG foundation maintained regular flow, with the viscosity increase being independent of temperature changes. They theorized that particle agglomeration contributed to the detected rise in viscosity. Conflicting to many previous studies, He and his associates [20] reported pseudoplastic properties in TiO_2-H_2O nanofluid, observing a quick decline in shear viscosity as the shear rate increased. Yonder a shear rate of 100 s^{-1}, the viscosity stabilized. Phuoc and Massoudi [21], while investigating Fe_2O_3-water nanofluids with a polyvinylpyrrolidone dispersant, observed interesting rheological attributes. At diluted concentrations, the nanofluids showed Newtonian behavior, but a transition to pseudoplasticity was apparent beyond a certain loading. In several industrial activities, fluids are subjected to higher pressures and temperatures than those typically studied in nanofluid research. For instance, high-pressure water jets, used in processes such as the preparation of surface, existing hydro-demolition, oil extraction, and industrial processes, operate under harsh conditions [22]. High-pressure cooling, with pressures reaching 200 MPa, has proven effective in reducing heat in cutting processes, thereby boosting productivity and machining efficiency [22,23]. The potential benefits of using nanofluids in these areas are intriguing due to their superior thermal and rheological characteristics and the possible enhancements that nanoparticles could bring to machining efficiency. Drilling fluids is pivotal in sectors such as the gas, oil, and geothermal manufacturing. These fluids' operating pressure and temperature can vary significantly, influenced by factors such as borehole depth and surrounding thermal conditions. In some instances, these could soar as high as 272 MPa and 300°C [24]. For instance, in the Persian Gulf, typical values hover around 48 MPa (equivalent to about 7000 psi) and 110°C [25,26]. Incorporating nanofluids in such scenarios appears promising, given their potential to enhance drilling fluids' viscosity, flow dynamics, mechanical strength, and thermal attributes [27]. A specific investigation by Sedaghatzadeh et al. [28] delved into the impact of integrating CNT on a drilling fluid's thermal and flow characteristics. Their findings revealed that the addition of CNTs improved the fluid's thermal conductivity, with an increase of up to 31% observed at 53°C. Furthermore, there was a noticeable increment in flow resistance corresponding to borehole depth, peaking at 12.2% at a depth of 2365 m. However, the study did not elucidate the influence of pressure on flow resistance [28].

6.7 Characterization techniques

Fluid dynamics play a pivotal role in various industrial sectors, daily life, and scientific fields. One significant global challenge is enhancing fluid quality, leading to the development of nanofluids that contain dispersed nanoparticles. Nanofluids demonstrate considerable improvements in multiple fluid properties, such as heat transfer capability, durability, and consistency, compared to traditional engineered fluids [1]. As a result, nanofluids are employed in diverse engineering situations, including porous materials [2,3], among others. However, producing these fluids can be complex, with concerns such as clumping posing major challenges in nanofluid production [4]. Evaluating the quality of the resulting nanofluid can also prove difficult. The primary methodologies for nanofluid synthesis include the 1-step and 2-step processes. The 1-step technique merges the manufacture of nanoparticles with the nanofluids synthesis. Nanoparticles are created through physical vapor deposition or nanoparticle concentration from the vapor phase in a small vapor pressure fluid (VEROS) [5]. Ongoing research [6] is centered on the proposed direct evaporation condensation technique, offering superior control over nanoparticle size and producing a stable nanofluid without the necessity for additives. On the other hand, the 2-step technique, which is the most regularly employed approach for nanofluid synthesis, involves obtaining nanoparticles through various procedures and then dispersing the nanopowder into the liquid using various techniques [7–10]. The nanofluids' stability is paramount, especially considering the significant economic and societal benefits they offer as innovative fluid additives across various domains. These fluids are indispensable in functions such as regulating engine temperatures, managing vehicle heat, reducing boiler exhaust gas temperatures, optimizing heat exchange systems, and enhancing complex electronic devices [11]. In heat transfer applications, particle diffusion rates correlate directly with the fluid's temperature. An increased diffusion rate augments particle interactions, leading to a higher likelihood of particle agglomeration. High-temperature submissions may necessitate additional measures to counteract nanofluid inconsistencies. In medical transfer mechanisms, unsought particle sedimentation or accumulation might obstruct the flow of nanofluids through channels. The uniformity of nanofluids predominantly hinges on the attributes of the nanoparticles and the base liquid, encompassing their shape, chemical composition, and intrinsic properties [12,13]. One crucial strategy for maintaining stable nanofluids is pH regulation [14,15]. Methods, such as EM, dynamic light scattering, centrifugal liquid sedimentations, field flow fractionation, and inductively coupled plasma–mass spectroscopy system (ICP-Ms), have been championed for nanoparticle analysis [14,15]. Evaluating nanoparticle quality can be accomplished by separating nanostructured particles' super-aggregates through sonication and a nanofluid droplet's evaporation process. Adjusting the suspension's pH levels using a surface activator has also been proposed by researchers [16]. Such methodologies, among others, depict potential procedural blueprints for creating an end product with a specific scale of particle sizes uniformly dispersed in the liquid. In addition, postevaporation visual inspection of samples (examining the remnants from nanopowder) could provide insights about the liquid medium itself (the flow patterns within a static droplet) and the attributes of the powder (particle movements and the varied patterns generated). The clumping process changes in the solution or liquid postevaporation due to complex rearrangement activities among nanoparticles, such as atomic irregularities on the surface of the NPs, electron-deflecting interactions, and environmental evaporation factors. Efforts to understand the effects of liquid and particle movement dynamics, which allow for different nanoparticle pattern formations postliquid evaporation, have been documented in Ref. [18]. Three competing convective methods have been evaluated. Three significant stages observed during the evaporation sequence include (1) evaporation dominated by the liquid, (2) the drying phase, and (3) nanoparticle stain development.

A burgeoning research avenue is the nonlinear analysis of nanofluids, delving into issues like particle aggregation, convective heat transfer, critical heat flashpoints, and chilled fluid thermal transmission, using fractal analysis as a foundation. One study [20] presented three novel fractal models, manifesting a strong correlation between the fractal predictions and empirical data. Research conducted by the group in Ref. [21] delved into the fractal dimensions of aggregates and the heat conduction dynamics in nanofluids. Their fractal-based exploration revealed that flow resistance accelerates faster than thermal conductivity as aggregations grow, rendering highly aggregated nanofluids less effective. In contrast, the research in Ref. [22] delineated complex flow patterns based on their fractal dimensions. This paper presents insights from the analysis of dried nanofluid droplet characteristics. SiO_2 nanoparticles spread in H_2O and C_3H_8O through sonication. Following this, these nanofluids are characterized using standard techniques and evaluating parameters such as contact angle, zeta potential, pH, and particle size. Individual droplets were then allowed to dry for microscopic examination. Acquired scanning electron microscope (SEM) images underwent a nonlinear multifractal analysis, validating the quality of the nanofluid based on nanoparticle sizes and distribution. A notable highlight of this study is its effort to establish the efficacy of using nonlinear multifractal analysis to determine the quality of synthesized nanofluids by inspecting their dried droplet samples.

6.8 Applications of nanofluid

The preceding sections delved into nanofluids' genesis, characteristics, and features. We now shift our focus to their practical applications. These nanoenhanced liquids find use in a variety of sectors, from automotive to energy. Some standout applications include cooling in electrical, mechanical, and electronic devices; improving thermal exchange in power generation and industrial operations; energy extraction from exhaust gases; and building temperature control, thermal storage, solar energy systems, water treatment, cooling solutions, and defense and space applications. They also act as lubricants in machinery and medical devices. The following sections provide a deeper understanding of nanofluids' role in these areas.

6.8.1 Nanofluids in cooling systems

Traditional water-based cooling faces multiple challenges, necessitating fluids that offer superior heat transfer capabilities. While solids generally have better thermal conductivity than liquids, they are expected to provide improved thermal properties. Nanofluids emerge as top contenders to replace both water and standard coolants, given their advantageous properties for improved heat exchange. The presence and concentration of nanoparticles in a fluid can affect its thermal behavior, transfer efficiency, and flow resistance. As such, understanding these effects is crucial before using nanofluids. The subsequent sections delve into the specifics of nanofluids in cooling.

6.8.2 Nanofluids in vapor compression systems

Water-based nanofluids can enhance heat transfer while retaining water's advantages. Numerous studies have investigated the application of nanofluids in household cooling systems. One such study used R134a as a coolant, combined with mineral polyolester oil (POE oil) infused with TiO_2 nanoparticles for lubrication. The results highlighted a 26% drop in energy usage in comparison to setups using

R134a with standard POE oil. Additionally, there was a marked decrease in power consumption and a significant increase in cooling capacity. The difference in the cooling cycle's efficiency was attributed to improved lubricant properties and the incorporation of nanoparticles with R134a. Another study examined a home cooling system using TiO_2-R600a nano coolant, observing that the combination of TiO_2 with R600a enabled more efficient and consistent operation, reducing energy consumption by up to 10%. Similarly, a study focused on a refrigerator using the Al_2O_3-R134a nanocoolant found that its performance exceeded that of a solely R134a-based system. With just a 0.2% volume mixture, energy consumption was reduced by 10.30%, and the addition of nanosized Al_2O_3 led to better heat exchange.

6.8.3 Nanofluids in server and microchips cooling

In order to remain effective, electronics need efficient cooling. The results of experiments have demonstrated that alumina nanofluid can improve heat transfer in microchannels. Additionally, it has also proven to be efficient in microchip cooling applications. Nanofluid was used to test a new design of a cooler containing a microchannel heat sink, which showed decreased heat resistance and gradients of temperature between the warmed microchannel walls and the coolant. A cooler designed with a microchannel heat sink was used to investigate how water-Al_2O_3 nanofluid enhances heat transfer. An analysis of Cu nanoparticle performance was also conducted with a silicon microchannel heat sink.

6.8.4 Nanofluids in the application of heat exchanger

- Heat exchangers using nanofluids instead of conventional fluids are promising. In the design of heat exchangers, developing new, highly efficient heat exchanger fluids is essential. A nanofluid can improve heat transfer more than twice with a volume fraction of less than 0.3%. An analysis concentrated on thoroughly characterizing the nanofluid's properties so that its suitability for use in a specific heat exchanger could be determined. Recent experimental research on nanofluid utilization in heat exchangers has concluded that the flow type within the heat exchanger is a critical factor in nanofluid suitability [104]. When the heat exchanger works under turbulent conditions, nanofluids can be helpful if a minimum increase in fluid viscosity is achieved, which seems extremely difficult. However, Nusselt numbers were improved by employing alumina-water and Titania-water nanofluids under turbulent flow conditions. As a result of a study, multi-walled carbon nanotubes (MWCNTs) scattered in water demonstrated improved critical convective heat transfer. In this study, volume concentration and flow conditions were factors affecting the improvement. An investigation was conducted on conductive heat transfer in turbulent flows by utilizing copper/water nanofluids. Study results showed at least a 39% improvement with a volume concentration of 1.5%.

6.8.5 Nanofluid detergent

Reflected light and digital video microscopy were employed to determine the technique of dispersion dynamics in polystyrene nanofluids, and it was shown that nanofluids might function as a detergent; nevertheless, further research is needed in this area. The same nanofluid type can also remove oil spills and commercial oil extraction.

6.8.6 Application of nanofluids in the extraction of geothermal energy

Geothermal systems are able to transfer heat more effectively when zinc oxide nanofluids are used. Using nanofluids enhances the entire power generation cycle efficiency associated with geothermal energy. The use of geothermal energy is uncommon among renewable energy sources. Currently, only 1.5% of geothermal energy available worldwide can be extracted. Despite this, nanofluids can enhance geothermal energy extraction to increase Rankine cycle power generation efficiently. The pipes used to transport geothermal fluid at temperatures between 500°C and 1000°C can be cooled using nanofluids. Additionally, nanofluids may act as "fluid superconductors," allowing them to serve as a working fluid that can convert energy into proper forms.

6.8.7 Nanofluids for the absorption of CO_2

The most significant greenhouse gas is CO_2, which accounts for about 77% of all global emissions. Global warming is intensified by excessive CO_2 emission. Burning fossil fuels and producing cement are the two primary sources of CO_2 emissions [2]. It is possible to reduce CO_2 emissions by separating and collecting it, utilizing it, or storing it to be used later. CO_2 can be captured from a variety of gas streams using a variety of separation methods. CO_2 capture technologies include adsorption, absorption, cryogenics, and membrane separation [3,4]. CO_2 absorption by gas-liquid has made significant progress, some of which are proposed to decrease energy consumption. In addition to gas-liquid absorption of carbon dioxide, the approach can also be used to absorb carbon dioxide from other renewable energy sources. Solvents-based novel absorption methods tend to improve absorption by taking advantage of physical and/or chemical phenomena. In a methanol-based solution, Pineda et al. investigated Al_2O_3 and SiO_2's effects on CO_2 absorption rates using a tray column. Their experiments showed that nanofluids containing SiO_2 and Al_2O_3 had an increased absorption capacity by 9.4% and 9.7%, respectively, at the optimum concentration of 0.05% of the nanoparticles. The absorption performance of methanol-based nanofluids with different concentrations of alumina and silica particles was also examined by Lee et al. Research has shown that the highest CO_2 absorption compared to pure methanol is obtained by dispersing 0.01 vol.% Al_2O_3 or 0.01 vol.% SiO_2 in pure methanol.

Accordingly, the absorption process will be improved by 4.5% and 5.6% more, respectively [100]. According to Peng et al., another study used the transient hot-wire technique to examine the nanofluids' thermal conductivity in which Al_2O_3 and SiO_2 are dispersed in methanol. At 0.005−0.5 vol.% nanoparticle concentration, the maximum change in thermal conductivity was 14.29%. An analysis of nanofluids containing Al_2O_3 dispersed in methanol-based liquids was conducted by Jung et al. using a bubble column. The results showed that the nanofluid absorbed 8.3% more than the pure base fluid [101].

6.8.8 Biomedical applications of nanofluids

Medical technology is ever-advancing, and the burgeoning field of nanofluids is gaining traction in the biomedical sphere. These fluids offer the potential to mitigate the side effects traditionally seen in conventional radiation treatments for cancer. Specifically, iron-based nanoparticles have the potential to guide medicines or radiation precisely to target cells. Steering these particles through blood vessels with external magnetic fields helps minimize inadvertent damage to healthy cells. Additionally, surgeries can be distressing for the body. Utilizing nanofluids to cool surgical areas has demonstrated an improvement in patient recovery times and a reduction in potential damage to organs. Remarkably, these fluids can also be warmed, offering a technique to target and eliminate tumors or abnormal cells while sparing surrounding healthy tissue. Further studies accentuate the antimicrobial capabilities of nanofluids. For instance, ZnO nanofluids have been pinpointed as effective agents against the *Escherichia coli* bacteria. There is a discernible link between heightened antimicrobial activity, the increased concentration of nanoparticles, and their reduced size. In the evolving field of nanodrug delivery, nanofluids show promise in ensuring drug-sustained and controlled release.

6.8.9 Optical applications

Optical filters are essential for isolating specific light wavelengths. With ferrofluids, it is possible to select across diverse portions of the wavelength spectrum, be it infrared, UV, or even the visible range. Custom ferrofluid mixtures can intricately adjust the requisite wavelength intervals, bandwidths, and reflectivity levels.

6.9 Preserving nanofluid stability

The clustering of nanoparticles can lead to various challenges, ranging from blocking tiny channels to reducing the nanofluids' heat transfer efficiency. Numerous strategies have been formulated to monitor nanofluid stability. One foundational approach is the sedimentation technique, where a nanofluid is deemed stable if its concentration remains consistent over extended periods. A straightforward visual inspection can also serve to evaluate the uniformity of nanofluids. The following sections delve deeper into the various methods to ensure and evaluate nanofluid stability (Fig. 6.5).

6.9.1 Zeta potential analysis

The voltage difference between the dispersing agent and the stationary fluid layer is referred to as the zeta charge. This charge plays a critical role in predicting the stability of colloidal mixtures. A higher zeta charge indicates a more stable colloidal mixture and vice versa.

6.9.2 Light absorption analysis

light absorption analysis offers a powerful method, especially when used alongside zeta charge measurement, to evaluate the stability of nanofluids. Generally, there is a direct relationship between the density of nanoparticles in the fluid and their absorption capacity. If nanoparticles dispersed in primary liquids have specific absorption bands within the 190–1100 nm range, the stability of the nanofluid can be reliably determined using UV-vis spectral analysis.

6.9.3 Strategies to enhance nanofluid stability

6.9.3.1 Using detergents in nanofluids

Detergents, also known as dispersants, are used to enhance the stability of nanofluids. They are particularly important for stabilizing nanofluid mixtures produced by the bi-phase method. By reducing the surface tension of liquids, detergents increase nanofluid stability and prevent the agglomeration of nanoparticles in primary liquids. The bi-phase method is commonly used in the industry due to its simplicity and cost-effectiveness for large-scale nanofluid production. Detergents feature a hydrophobic tail (long hydrocarbon chain) and a hydrophilic head, facilitating optimal dispersion or interaction between particles. If the base liquid is polar, water-soluble detergents are chosen; otherwise, oil-soluble ones are preferred.

However, detergents have certain drawbacks. For example, dispersants can contaminate thermal transfer agents, producing bubbles when heat. Additionally, introducing detergents might cause nanoparticle swelling, thereby reducing the inherent thermal conductivity of the nanofluid. Therefore, the process of incorporating detergents into nanofluids requires careful optimization.

6.9.3.2 Particle modification techniques: the detergent-free method

An alternative strategy involves the use of tailored nanoparticles, presenting a promising avenue for achieving long-term nanofluid stability called the detergent-free method. Some researchers have experimented with modified silica (SiO_2) nanoparticles by attaching silanes directly to the nanoparticle surface in various solutions. This resulted in unique nanofluid properties, with no sediment layer observed on the heated surface after boiling. Other researchers attached hydrophilic groups to the nanoparticle surface through mechanochemical modification, producing nanofluids with properties such as noncontamination, extreme smoothness, reduced tackiness, improved thermal conductivity, and strong stability. These advanced nanofluids could be used as coolants in modern heating systems.

When particles disperse within a medium, they tend to cluster, forming larger aggregates. These clusters can eventually settle due to gravity. The sustained stability of nanofluids means these particles do not aggregate and settle quickly. The aggregation rate depends on how often particles collide and the likelihood of them sticking together upon collision. Derjaguin and colleagues theorize that the balance between van der Waals attractive forces and the repulsion from the electrical double layer determines a particle's stability in a mixture. These forces act when particles approach each other because of random movements. Suppose attraction exceeds repulsion, and particles collide, destabilizing the mixture. However, if repulsion prevails, the mixture remains stable.

6.9.4 Flow characteristics of nanofluids

The flow behavior of any fluid is governed by the relationship between shear stress (τ) and shear rate (γ). Shear stress is the tangential force exerted per unit area, while the shear rate describes the change in deformation over time. Viscosity (η), defined as the ratio of shear stress to shear rate, represents the friction between adjacent fluid layers during flow. Fluids can be categorized as Newtonian or non-Newtonian, including pseudoplastic, Bingham, and dilatant behaviors. Newtonian fluids have a constant viscosity irrespective of shear rate, indicating a linear relationship between stress and shear rate. On the other hand, non-Newtonian fluids may exhibit viscosity changes with varying shear rates, following models like the Bingham plastic.

6.9.4.1 Non-Newtonian and Newtonian behavior of nanofluids

Any fluid's rheological flow behavior can be explained by the relationship between shear stress (τ) and shear rate (γ). Shear stress is the tangential force applied per unit area, while shear rate is the change in shear strain per unit time. Viscosity (η) is the ratio of shear stress to shear rate, also known as a measure of resistance offered by the adjacent layers to one another during the liquid suspension flow. Fluid behavior can be classified into Newtonian and non-Newtonian (Bingham plastic, pseudoplastic, Bingham, and Dilatant). In Newtonian behavior, the viscosity stays constant with the shear rate, and the stress demonstrates a linear relationship with the shear rate, whereas in non-Newtonian behavior, the viscosity may change with the shear rate, and the correlation between stress and shear rate follows Bingham plastic behavior. Additionally, it provides insight into nanoparticle structure, potentially helpful for predicting nanofluid thermal conductivity. Rheometers are used to measure rheological behavior [52,63–65,105]. Viscometers have been used by some researchers [67] to measure viscosity. Currently, viscometers are regarded as insufficient because they cannot determine the shear dependence characteristic, particularly for low-viscosity liquid-based nanofluids that contain nonspherical particles. It has been shown by Richmond et al. [61] that mixing TiO_2 with SiO_2/water nanofluid alters its flow behavior from Newtonian to non-Newtonian. The TiO_2/water nanofluid generally shows shear thinning behavior [62], except for the behavior reported by Penkavova et al. [67]. Even at high shear rates, the TiO_2/EG nanofluid displays Newtonian behavior. Depending on the base fluid, MWCNT nanofluid displayed Newtonian and non-Newtonian behavior. MWCNT mixed in oil, water, resin, and EG showed explicit shear thinning behavior [78–80] except for MWCNT/EG [77]. All MWCNT temperatures and volume fractions studied show Newtonian behavior for glycerol-based and silicone oil-based fluids [106]. MWCNT nanofluids with a high volumetric concentration exhibit non-Newtonian behavior, whereas nanofluids with a lower concentration display Newtonian behavior [70,71]. It was found that all the suspensions that contained nanoparticles of SiO_2 exhibited Newtonian behavior [81–86]. Al_2O_3/water nanofluid exhibits non-Newtonian behavior [87,88], whereas Al_2O_3/EG and Al_2O_3/PG behave as Newtonian fluids [89–91]. Shear thinning behavior is observed in water-based nanofluids containing microsized Al_2O_3 particles. The rheological behavior of different nanofluids enriched with nanoparticles, including $BaTiO_3$, CuO, Ni, Al, Ag, graphite, graphene, $CaCO_3$, gold, Yttrium oxide, and carbon black powder. Stable nanofluids are often prepared with surfactants. In order to achieve a uniform particle packing structure within a suspension, stabilizing the nanofluids is typically regarded as essential. According to Phuoc et al. [76] and Wang et al. [79], adding a surfactant to MWCNT/water nanofluid changed its flow behavior from Newtonian to non-Newtonian.

For a higher shear rate, mixing NH4PA as a surfactant in $BaTiO_3$/distilled water nanofluid [100] changed its flow behavior from Newtonian to dilatant. Even though surfactant did not impact the flow behavior of Ni/terpineol nanofluid [102], it changed the rheological behavior of $BaTiO_3$/ethanol isopropanol from pseudoplastic to dilatant [101]. The flow behavior of nanofluids impacts their pressure profiles and provides insights into the distribution of nanoparticles, aiding in predicting their heat transfer characteristics. This behavior is typically measured using instruments called rheometers. While some researchers have used viscometers to determine viscosity, these are now considered less accurate, especially for determining shear dependence in nanofluids based on low-viscosity fluids with nonspherical particles. Research by Richmond and associates revealed that introducing TiO_2 into a SiO_2/water nanofluid changes its flow behavior from Newtonian to non-Newtonian. TiO_2/water nanofluids primarily display a shear-thinning behavior. However, a TiO_2/EG nanofluid retains a Newtonian behavior even at higher shear rates.

6.9.4.2 MWCNT nanofluid flow dynamics

MWCNT nanofluids exhibited Newtonian (regular) and non-Newtonian (irregular) flow behaviors, depending on the base fluid used. Notable shear-thinning behaviors were observed when MWCNT was combined with water, oil, resin, and EG, with the exception of its pairing with EG. Nanofluids formulated with glycerol and silicone oil consistently displayed Newtonian behaviors across various MWCNT volume fractions and temperature conditions. Nanofluids with higher concentrations of MWCNT demonstrated nonNewtonian behaviors, while those with lower concentrations exhibited Newtonian flow. SiO_2 nanoparticle mixtures consistently displayed Newtonian behavior. The Al_2O_3/water combination was non-Newtonian, whereas mixtures such as Al_2O_3/EG and Al_2O_3/PG showed Newtonian behaviors. A water-based nanofluid containing larger Al_2O_3 particles demonstrated shear-thinning characteristics.

The introduction of surfactants into nanofluids can dramatically modify their flow characteristics. Both Phuoc et al. and Wang et al. observed a transition from Newtonian to non-Newtonian flow in MWCNT/water nanofluids upon surfactant addition. Similarly, the inclusion of the surfactant (NH4PA) into a $BaTiO_3$/distilled water nanofluid caused a shift in its flow behavior from Newtonian to shear-thickening at elevated shear rates.

6.10 The effect of various factors on thermal conductivity

A majority of the initial nanofluid research was conducted at Argonne National Laboratory in the USA. Research at this stage focused primarily on thermal conductivity under macroscopically static conditions. Some researchers have found in their studies that nanoparticles added to conventional fluids significantly increased their thermal conductivity compared to the base fluids [27−39,107]]. According to Saidur et al. [40], nanofluids' thermal conductivity increases with increasing the volumetric concentration of particles in the base fluid. Nanoparticles possess a higher thermal conductivity than base fluids, so mixing them with base fluids may alter fluid thermophysical properties [25,41]. Nevertheless, it has been found in different experiments that an increase in thermal conductivity may be offset by an increase in viscosity, leading to a slight drop in pressure [42−45]. A nanofluid's thermal conductivity is determined mainly by its nanoparticles, base fluid, particle size, temperature, and particle shape (Fig. 6.6). Fig. 6.6 represented the thermal conductivity of nanofluids.

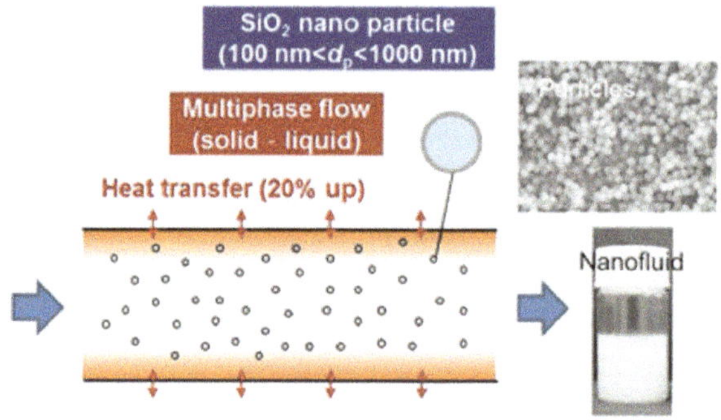

FIGURE 6.6

Photograph of thermal conductivity of nanofluid.

Reprinted with permissions from S. Hashimoto, Anomalous enhancement of convective heat transfer with dispersed SiO_2 particles in ethylene glycol/water nanofluid, Int. J. Heat and Mass Transf. 150 (2020) 119302.

6.10.1 Base fluid

Depending on the base fluid's thermal conductivity, nanofluids have different thermal conductivities. Accordingly, if the base fluid has a high thermal conductivity, the nanofluid will have a higher thermal conductivity.

6.10.2 Nanoparticles

Due to their chemical stability, metal oxides are usually used as dispersed particles in base fluids. Aside from that, pure metal is costly. It is the nanoparticle type in the base fluid that determines nanofluid thermal conductivity. Accordingly, the most preferred nanoparticles are those with high thermal conductivity.

6.10.3 Concentration of nanoparticles

Increasing nanoparticle concentration increases nanofluid thermal conductivity [76–78]. In most cases, researchers provided a thermal conductivity model according to the concentration of the nanoparticle. Researchers have established both nonlinear and linear relationships between nanoparticular concentration and thermal conductivity. In the case of Fe_2O_3/water nanofluids, Zhu et al. [73] found a nonlinear relationship. A study by Barbés et al. [74] on CuO nanofluid found a linear relationship. Ding et al. [76] studied MWCNT nanofluids and found an increase in thermal conductivity that was nonlinear. In the case of glycol-based CNT nanofluids, Xie et al. [77] also observed a nonlinear relationship.

6.10.4 Particle size

Nanofluid is made from nanoparticles with diameters ranging from 1 to 100 nm. There is also an inverse relationship between nanofluid thermal conductivity and nanoparticle size [78]. With decreasing nanoparticle size, the nanofluid's thermal conductivity improves. Brownian motion is more noticeable when particles are small. This results in an increase in chaotic movement in the base fluid, increasing nanofluid thermal conductivity. If the particle size decreases, the surface-to-volume ratio increases [79]. A study by Teng et al. [80] examined how Al_2O_3 nanoparticle size affects the thermal conductivity of Al_2O_3/water nanofluids, resulting in a rise in thermal conductivity.

6.10.5 Particle shape

According to the Hamilton–Crosser equation for thermal conductivity, nanoparticle shape also affects thermal conductivity. There are various particle shapes, such as cylindrical or rod, spherical, blade, and brick. Timofeeva et al. [81] studied the impact of particle shape on Al_2O_3/EG-water, revealing that nanoparticles with a cylindrical shape had the most significant thermal conductivity. Brick-shaped nanoparticles' thermal conductivity is higher than those with platelet and blade shapes [82].

6.10.6 Temperature

Nanofluid thermal conductivity depends on temperature [83]. As the temperature increases, the nanofluid's thermal conductivity also increases. As temperature increases, Brownian motion becomes more vigorous, leading to a rise in thermal conductivity. The thermal conductivity of Al_2O_3/water and CuO/water nanofluids increased with increasing temperature, as found by Das et al. [84]. The same result was found by Li and Peterson [85] applying Al_2O_3/water nanofluids. Using TiO_2/water nanofluid at temperatures between 15°C and 30°C, Duangthongsuk and Wongwises [86] found increased thermal conductivity. Graphene/DO and CNT/DO nanofluids exhibit increased thermal conductivity as temperature increases, according to Naddaf and Heris.

6.10.7 Surfactant

Surfactants enhance nanofluid stability by reducing nanoparticle agglomeration. By reducing thermal conductivity, surfactants increase the stability of nanofluids. CuO/water nanofluid with Sodium Dodecylbenzene Sulfonate (SDBS) surfactant was investigated by Khairul et al. Their results showed that thermal conductivity decreased after surfactant was added. Nanofluid thermal conductivity decreases with increasing concentrations of surfactant [88]. The thermal conductivity of the surfactant is lower than that of the base fluid. This results in a decrease in effective thermal conductivity. As a result, surfactants should be applied at optimum concentrations.

6.11 Conclusion and outlook

SNPs have significantly contributed to the field of nanotechnology due to their unique properties, including mesoporous structure, high surface area, customizable particle and pore sizes, diverse morphology, and biocompatibility. These features have allowed SNPs to find extensive applications in various sectors, such as agriculture, food preservation, biomedical applications, and catalytic reactions. The efficacy of SNPs as encapsulation agents for numerous bioactive molecules has enhanced the prospects of targeted drug delivery. Moreover, the compatibility of SNPs with a variety of materials, both polymeric and nonpolymeric, has broadened their functional applications. Specifically, MSNs have showcased potential as carriers for targeted chemotherapy in cancer treatment. While these applications have seen success, multiple avenues are ripe for deeper exploration:

1. Future research could focus on discovering more stabilizing compounds that synergize well with MSNs. Such agents can be used for single-step catalytic reactions without disturbing the overall reactivity or altering the system's environment.
2. The potential of SNPs to encapsulate colorless and low water-soluble drugs presents an exciting area of study, which could revolutionize targeted drug delivery.
3. An intriguing prospect is identifying seeds, such as maize, that might gain antifungal characteristics when integrated with SNPs.
4. The observed phenomenon where SNPs bolster resilience against salt stress for certain plants, but decrease chlorophyll levels in others, requires more in-depth analysis.
5. Considering the success of the chitosan/silica blend film in prolonging the freshness of Loquat fruit, its efficacy with other fruits should be evaluated.
6. A more comprehensive array of enzymes should be examined for potential integration into MSN structures.
7. Considering the success of specific aminosilanes with SNPs for dye extraction from wastewater, the efficiency of other aminosilanes should also be probed.

With the ongoing growth and application of nanotechnology, these research directions can help maximize the potential benefits of SNPs and MSNPs across various fields.

References

[1] N. Suresh Kumar Reddy, P. Venkateswara Rao, Experimental investigation to study the effect of solid lubricants on cutting forces and surface quality in end milling, Int. J. Mach. Tools Manuf. 46 (2006) 189–198. Available from: https://doi.org/10.1016/j.ijmachtools.2005.04.008.

[2] K.-H. Park, B. Ewald, P.Y. Kwon, Effect of nano-enhanced lubricant in minimum quantity lubrication balling milling, J. Tribol. 133 (2011). Available from: https://doi.org/10.1115/1.4004339.

[3] A.A.D. Sarhan, M. Sayuti, M. Hamdi, Reduction of power and lubricant oil consumption in milling process using a new SiO_2 nanolubrication system, Int. J. Adv. Manuf. Technol. 63 (2012) 505–512. Available from: https://doi.org/10.1007/s00170-012-3940-7.

[4] M. Sayuti, A.A.D. Sarhan, T. Tanaka, M. Hamdi, Y. Saito, Cutting force reduction and surface quality improvement in machining of aerospace duralumin AL-2017-T4 using carbon onion nanolubrication system, Int. J. Adv. Manuf. Technol. 65 (2013) 1493–1500. Available from: https://doi.org/10.1007/s00170-012-4273-2.

[5] B. Rahmati, A.A.D. Sarhan, M. Sayuti, Morphology of surface generated by end milling AL6061-T6 using molybdenum disulfide (MoS_2) nanolubrication in end milling machining, J. Clean. Prod. 66 (2014) 685–691. Available from: https://doi.org/10.1016/j.jclepro.2013.10.048.

[6] M. Sayuti, O.M. Erh, A.A.D. Sarhan, M. Hamdi, Investigation on the morphology of the machined surface in end milling of aerospace AL6061-T6 for novel uses of SiO_2 nanolubrication system, J. Clean. Prod. 66 (2014) 655–663. Available from: https://doi.org/10.1016/j.jclepro.2013.11.058.

[7] M. Sayuti, A.A.D. Sarhan, M. Hamdi, An investigation of optimum SiO_2 nanolubrication parameters in end milling of aerospace Al6061-T6 alloy, Int. J. Adv. Manuf. Technol. 67 (2013) 833–849. Available from: https://doi.org/10.1007/s00170-012-4527-z.

[8] B. Rahmati, A.A.D. Sarhan, M. Sayuti, Investigating the optimum molybdenum disulfide (MoS2) nanolubrication parameters in CNC milling of AL6061-T6 alloy, Int. J. Adv. Manuf. Technol. 70 (2014) 1143–1155. Available from: https://doi.org/10.1007/s00170-013-5334-x.

[9] R. Razavi, M. Amiri, H.A. Alshamsi, T. Eslaminejad, M. Salavati-Niasari, Green synthesis of Ag nanoparticles in oil-in-water nano-emulsion and evaluation of their antibacterial and cytotoxic properties as well as molecular docking, Arab. J. Chem. 14 (2021) 103323–103329. Available from: https://doi.org/10.1016/j.arabjc.2021.103323.

[10] B. Shen, A.J. Shih, S.C. Tung, Application of nanofluids in minimum quantity lubrication grinding, Tribol. Trans. 51 (2008) 730–737. Available from: https://doi.org/10.1080/10402000802071277.

[11] M. Alberts, K. Kalaitzidou, S. Melkote, An investigation of graphite nanoplatelets as lubricant in grinding, Int. J. Mach. Tools Manuf. 49 (2009) 966–970. Available from: https://doi.org/10.1016/j.ijmachtools.2009.06.005.

[12] S. Prabhu, B.K. Vinayagam, Nano surface generation of grinding process using carbon nano tubes, Sadhana 35 (2010) 747–760. Available from: https://doi.org/10.1007/s12046-010-0048-3.

[13] P.-H. Lee, J.S. Nam, C. Li, S.W. Lee, An experimental study on micro-grinding process with nanofluid minimum quantity lubrication (MQL), Int. J. Precis. Eng. Manuf. 13 (2012) 331–338. Available from: https://doi.org/10.1007/s12541-012-0042-2.

[14] P. Kalita, A.P. Malshe, S. Arun Kumar, V.G. Yoganath, T. Gurumurthy, Study of specific energy and friction coefficient in minimum quantity lubrication grinding using oil-based nanolubricants, J. Manuf. Process. 14 (2012) 160–166. Available from: https://doi.org/10.1016/j.jmapro.2012.01.001.

[15] P. Kalita, A.P. Malshe, K.P. Rajurkar, Study of tribo-chemical lubricant film formation during application of nanolubricants in minimum quantity lubrication (MQL)

grinding, CIRP Ann. 61 (2012) 327–330. Available from: https://doi.org/10.1016/j.cirp.2012.03.031.

[16] C. Mao, X. Tang, H. Zou, X. Huang, Z. Zhou, Investigation of grinding characteristic using nanofluid minimum quantity lubrication, Int. J. Precis. Eng. Manuf. 13 (2012) 1745–1752. Available from: https://doi.org/10.1007/s12541-012-0229-6.

[17] C. Mao, H. Zou, X. Huang, J. Zhang, Z. Zhou, The influence of spraying parameters on grinding performance for nanofluid minimum quantity lubrication, Int. J. Adv. Manuf. Technol. 64 (2013) 1791–1799. Available from: https://doi.org/10.1007/s00170-012-4143-y.

[18] C. Mao, J. Zhang, Y. Huang, H. Zou, X. Huang, Z. Zhou, Investigation on the effect of nanofluid parameters on MQL grinding, Mater. Manuf. Process. 28 (2013) 436–442. Available from: https://doi.org/10.1080/10426914.2013.763970.

[19] J.S. Nam, P.-H. Lee, S.W. Lee, Experimental characterization of micro-drilling process using nanofluid minimum quantity lubrication, Int. J. Mach. Tools Manuf. 51 (2011) 649–652. Available from: https://doi.org/10.1016/j.ijmachtools.2011.04.005.

[20] P. Vamsi Krishna, R.R. Srikant, D. Nageswara Rao, Experimental investigation on the performance of nanoboric acid suspensions in SAE-40 and coconut oil during turning of AISI 1040 steel, Int. J. Mach. Tools Manuf. 50 (2010) 911–916. Available from: https://doi.org/10.1016/j.ijmachtools.2010.06.001.

[21] J. Yan, Z. Zhang, T. Kuriyagawa, Effect of nanoparticle lubrication in diamond turning of reaction-bonded SiC, Int. J. Autom. Technol. 5 (2011) 307–312. Available from: https://doi.org/10.20965/ijat.2011.p0307.

[22] S. Khandekar, M.R. Sankar, V. Agnihotri, J. Ramkumar, Nano-cutting fluid for enhancement of metal cutting performance, Mater. Manuf. Process. 27 (2012) 963–967. Available from: https://doi.org/10.1080/10426914.2011.610078.

[23] M. Sayuti, A.A.D. Sarhan, F. Salem, Novel uses of SiO_2 nano-lubrication system in hard turning process of hardened steel AISI4140 for less tool wear, surface roughness and oil consumption, J. Clean. Prod. 67 (2014) 265–276. Available from: https://doi.org/10.1016/j.jclepro.2013.12.052.

[24] A.K. Sharma, A.K. Tiwari, A.R. Dixit, Progress of nanofluid application in machining: a review, Mater. Manuf. Process. 30 (2015) 813–828. Available from: https://doi.org/10.1080/10426914.2014.973583.

[25] S. Gholamrezae, M. Amiri, O. Amiri, M. Salavati-Niasari, H. Moayedi, Ultrasound-accelerated synthesis of uniform $SrMnO_3$ nanoparticles as water-oxidizing catalysts for water splitting systems, Ultrason. Sonochem. 62 (2020) 104899. Available from: https://doi.org/10.1016/j.ultsonch.2019.104899.

[26] H. Masuda, A. Ebata, K. Teramae, N. Hishinuma, Alteration of thermal conductivity and viscosity of liquid by dispersing ultra-fine particles. Dispersion of Al_2O_3, $SiO_2$and TiO_2 ultra-fine particles, Netsu Bussei. 7 (1993) 227–233. Available from: https://doi.org/10.2963/jjtp.7.227.

[27] T. Rajendra Prasad, K. Rama Krishna, K.V. Sharma, C. Naga Bhaskar, Thermal performance of stable SiO_2 nanofluids and regression correlations to estimate their thermophysical properties, J. Indian. Chem. Soc. 99 (2022) 100461. Available from: https://doi.org/10.1016/j.jics.2022.100461.

[28] S.M.S. Murshed, K.C. Leong, C. Yang, Investigations of thermal conductivity and viscosity of nanofluids, Int. J. Therm. Sci. 47 (2008) 560–568. Available from: https://doi.org/10.1016/j.ijthermalsci.2007.05.004.

[29] E. Ettefaghi, A. Rashidi, H. Ahmadi, S.S. Mohtasebi, M. Pourkhalil, Thermal and rheological properties of oil-based nanofluids from different carbon nanostructures, Int. Commun. Heat. Mass. Transf. 48 (2013) 178–182. Available from: https://doi.org/10.1016/j.icheatmasstransfer.2013.08.004.

[30] X.-Q. Wang, A.S. Mujumdar, A review on nanofluids – part II: experiments and applications, Braz. J. Chem. Eng. 25 (2008) 631–648. Available from: https://doi.org/10.1590/S0104-66322008000400002.

[31] L. Qiang, X. Yimin, Convective heat transfer and flow characteristics of Cu-water nanofluid, Sci. China Ser. E-Technol. Sci. 45 (2002) 408–412.

[32] S.U.S. Choi, Z.G. Zhang, W. Yu, F.E. Lockwood, E.A. Grulke, Anomalous thermal conductivity enhancement in nanotube suspensions, Appl. Phys. Lett. 79 (2001) 2252–2254. Available from: https://doi.org/10.1063/1.1408272.

[33] S. Kakaç, A. Pramuanjaroenkij, Review of convective heat transfer enhancement with nanofluids, Int. J. Heat. Mass. Transf. 52 (2009) 3187–3196. Available from: https://doi.org/10.1016/j.ijheatmasstransfer.2009.02.006.

[34] V. Trisaksri, S. Wongwises, Critical review of heat transfer characteristics of nanofluids, Renew. Sustain. Energy Rev. 11 (2007) 512–523. Available from: https://doi.org/10.1016/j.rser.2005.01.010.

[35] A. Nasiri, M. Shariaty-Niasar, A.M. Rashidi, R. Khodafarin, Effect of CNT structures on thermal conductivity and stability of nanofluid, Int. J. Heat. Mass. Transf. 55 (2012) 1529–1535. Available from: https://doi.org/10.1016/j.ijheatmasstransfer.2011.11.004.

[36] J.-H. Lee, K.S. Hwang, S.P. Jang, B.H. Lee, J.H. Kim, S.U.S. Choi, et al., Effective viscosities and thermal conductivities of aqueous nanofluids containing low volume concentrations of Al_2O_3 nanoparticles, Int. J. Heat. Mass. Transf. 51 (2008) 2651–2656. Available from: https://doi.org/10.1016/j.ijheatmasstransfer.2007.10.026.

[37] D. Zheng, J. Wang, Z. Chen, J. Baleta, B. Sundén, Performance analysis of a plate heat exchanger using various nanofluids, Int. J. Heat. Mass. Transf. 158 (2020), 119993. Available from: https://doi.org/10.1016/j.ijheatmasstransfer.2020.119993.

[38] S.M.S. Murshed, K.C. Leong, C. Yang, Enhanced thermal conductivity of TiO_2—water based nanofluids, Int. J. Therm. Sci. 44 (2005) 367–373. Available from: https://doi.org/10.1016/j.ijthermalsci.2004.12.005.

[39] J.A. Eastman, S.U.S. Choi, S. Li, W. Yu, L.J. Thompson, Anomalously increased effective thermal conductivities of ethylene glycol-based nanofluids containing copper nanoparticles, Appl. Phys. Lett. 78 (2001) 718–720. Available from: https://doi.org/10.1063/1.1341218.

[40] A.K. Tiwari, P. Ghosh, J. Sarkar, H. Dahiya, J. Parekh, Numerical investigation of heat transfer and fluid flow in plate heat exchanger using nanofluids, Int. J. Therm. Sci. 85 (2014) 93–103. Available from: https://doi.org/10.1016/j.ijthermalsci.2014.06.015.

[41] R. Saidur, K.Y. Leong, H.A. Mohammed, A review on applications and challenges of nanofluids, Renew. Sustain. Energy Rev. 15 (2011) 1646–1668. Available from: https://doi.org/10.1016/j.rser.2010.11.035.

[42] X.-Q. Wang, A.S. Mujumdar, Heat transfer characteristics of nanofluids: a review, Int. J. Therm. Sci. 46 (2007) 1–19. Available from: https://doi.org/10.1016/j.ijthermalsci.2006.06.010.

[43] L. Godson, B. Raja, D. Mohan Lal, S. Wongwises, Enhancement of heat transfer using nanofluids—an overview, Renew. Sustain. Energy Rev. 14 (2010) 629–641. Available from: https://doi.org/10.1016/j.rser.2009.10.004.

[44] J. Sarkar, A critical review on convective heat transfer correlations of nanofluids, Renew. Sustain. Energy Rev. 15 (2011) 3271–3277. Available from: https://doi.org/10.1016/j.rser.2011.04.025.

[45] W. Daungthongsuk, S. Wongwises, A critical review of convective heat transfer of nanofluids, Renew. Sustain. Energy Rev. 11 (2007) 797–817. Available from: https://doi.org/10.1016/j.rser.2005.06.005.

[46] D. Wen, G. Lin, S. Vafaei, K. Zhang, Review of nanofluids for heat transfer applications, Particuology 7 (2009) 141–150. Available from: https://doi.org/10.1016/j.partic.2009.01.007.

[47] L. Cheng, E.P. Bandarra Filho, J.R. Thome, Nanofluid two-phase flow and thermal physics: a new research frontier of nanotechnology and its challenges, J. Nanosci. Nanotechnol. 8 (2008) 3315–3332. Available from: https://doi.org/10.1166/jnn.2008.413.

[48] M. Shanbedi, S.Z. Heris, A. Amiri, E. Hosseinipour, H. Eshghi, S.N. Kazi, Synthesis of aspartic acid-treated multi-walled carbon nanotubes based water coolant and experimental investigation of thermal and hydrodynamic properties in circular tube, Energy Convers. Manag. 105 (2015) 1366–1376. Available from: https://doi.org/10.1016/j.enconman.2015.09.002.

[49] A. Naddaf, S. Zeinali Heris, Experimental study on thermal conductivity and electrical conductivity of diesel oil-based nanofluids of graphene nanoplatelets and carbon nanotubes, Int. Commun. Heat. Mass. Transf. 95 (2018) 116–122. Available from: https://doi.org/10.1016/j.icheatmasstransfer.2018.05.004.

[50] S.Z. Heris, F. Mohammadpur, O. Mahian, A.Z. Sahin, Experimental study of two phase closed thermosyphon using Cuo/water nanofluid in the presence of electric field, Exp. Heat. Transf. 28 (2015) 328–343. Available from: https://doi.org/10.1080/08916152.2014.883448.

[51] G. Huminic, A. Huminic, Numerical study on heat transfer characteristics of thermosyphon heat pipes using nanofluids, Energy Convers. Manag. 76 (2013) 393–399. Available from: https://doi.org/10.1016/j.enconman.2013.07.026.

[52] S.R. Zhai, C.S. He, D. Wu, Y.H. Sun, Hydrothermal synthesis of mesostructured aluminosilicate nanoparticles assisted by binary surfactants and finely controlled assembly process, J. Non Cryst. Solids 353 (2007) 1606–1611. Available from: https://doi.org/10.1016/j.jnoncrysol.2007.01.032.

[53] W. Zhao, J. Gu, L. Zhang, H. Chen, J. Shi, Fabrication of uniform magnetic nanocomposite spheres with a magnetic core/mesoporous silica shell structure, J. Am. Chem. Soc. 127 (2005) 8916–8917. Available from: https://doi.org/10.1021/ja051113r.

[54] Y. Huang, B.G. Trewyn, H.-T. Chen, V.S.-Y. Lin, One-pot reaction cascades catalyzed by base- and acid-functionalized mesoporous silica nanoparticles, N. J. Chem. 32 (2008) 1311. Available from: https://doi.org/10.1039/b806664g.

[55] D. Videira-Quintela, F. Guillén, G. Montalvo, O. Martin, Silver, copper, and copper hydroxy salt decorated fumed silica hybrid composites as antibacterial agents, Colloids Surf. B Biointerfaces 195 (2020) 111216. Available from: https://doi.org/10.1016/j.colsurfb.2020.111216.

[56] D.R. Radu, C.-Y. Lai, J.W. Wiench, M. Pruski, V.S.-Y. Lin, Gatekeeping layer effect: a poly (lactic acid)-coated mesoporous silica nanosphere-based fluorescence probe for detection of amino-containing neurotransmitters, J. Am. Chem. Soc. 126 (2004) 1640–1641. Available from: https://doi.org/10.1021/ja038222v.

[57] M. Maurel, T. Montheil, J. Martin, L. Chaar, V. Guzman-Gonzalez, M. Couvet, et al., Design of PEGylated three ligands silica nanoparticles for multi-receptor targeting, Nanomaterials 11 (2021) 177. Available from: https://doi.org/10.3390/nano11010177.
[58] N.S. Zaharudin, E.D. Mohamed Isa, H. Ahmad, M.B. Abdul Rahman, K. Jumbri, Functionalized mesoporous silica nanoparticles templated by pyridinium ionic liquid for hydrophilic and hydrophobic drug release application, J. Saudi Chem. Soc. 24 (2020) 289–302. Available from: https://doi.org/10.1016/j.jscs.2020.01.003.
[59] I.I. Slowing, B.G. Trewyn, S. Giri, V.S.-Y. Lin, Mesoporous silica nanoparticles for drug delivery and biosensing applications, Adv. Funct. Mater. 17 (2007) 1225–1236. Available from: https://doi.org/10.1002/adfm.200601191.
[60] S. Montalvo-Quirós, S. Gómez-Graña, M. Vallet-Regí, R.C. Prados-Rosales, B. González, J.L. Luque-Garcia, Mesoporous silica nanoparticles containing silver as novel antimycobacterial agents against *Mycobacterium tuberculosis*, Colloids Surf. B Biointerfaces 197 (2021) 111405. Available from: https://doi.org/10.1016/j.colsurfb.2020.111405.
[61] H.A. Ab Wab, K. Abdul Razak, N.D. Zakaria, Properties of amorphous silica nanoparticles colloid drug delivery system synthesized using the micelle formation method, J. Nanopart. Res. 16 (2014) 2256. Available from: https://doi.org/10.1007/s11051-014-2256-y.
[62] D. Dhinasekaran, R. Raj, A.R. Rajendran, B. Purushothaman, B. Subramanian, A. Prakasarao, et al., Chitosan mediated 5-Fluorouracil functionalized silica nanoparticle from rice husk for anticancer activity, Int. J. Biol. Macromol. 156 (2020) 969–980. Available from: https://doi.org/10.1016/j.ijbiomac.2020.04.098.
[63] Y. Zhang, J. Wang, X. Bai, T. Jiang, Q. Zhang, S. Wang, Mesoporous silica nanoparticles for increasing the oral bioavailability and permeation of poorly water soluble drugs, Mol. Pharm. 9 (2012) 505–513. Available from: https://doi.org/10.1021/mp200287c.
[64] S. Mashayekhi, S. Rasoulpoor, S. Shabani, N. Esmaeilizadeh, H. Serati-Nouri, R. Sheervalilou, et al., Curcumin-loaded mesoporous silica nanoparticles/nanofiber composites for supporting long-term proliferation and stemness preservation of adipose-derived stem cells, Int. J. Pharm. 587 (2020) 119656. Available from: https://doi.org/10.1016/j.ijpharm.2020.119656.
[65] Y. Wang, Y. Xu, J. Yang, X. Qiu, N. Li, Y. Zhu, et al., Carborane based mesoporous nanoparticles as a potential agent for BNCT, Mater. Chem. Front. 5 (2021) 2771–2776. Available from: https://doi.org/10.1039/D0QM00867B.
[66] F. Tamanoi, S. Chinnathambi, M. Laird, A. Komatsu, A. Birault, T. Takata, et al., Construction of boronophenylalanine-loaded biodegradable periodic mesoporous organosilica nanoparticles for BNCT cancer therapy, Int. J. Mol. Sci. 22 (2021) 2251. Available from: https://doi.org/10.3390/ijms22052251.
[67] G. Vares, V. Jallet, Y. Matsumoto, C. Rentier, K. Takayama, T. Sasaki, et al., Functionalized mesoporous silica nanoparticles for innovative boron-neutron capture therapy of resistant cancers, Nanomedicine 27 (2020) 102195. Available from: https://doi.org/10.1016/j.nano.2020.102195.
[68] E. Fanizza, H. Zhao, S. De Zio, N. Depalo, F. Rosei, A. Vomiero, et al., Encapsulation of dual emitting giant quantum dots in silica nanoparticles for optical ratiometric temperature nanosensors, Appl. Sci. 10 (2020) 2767. Available from: https://doi.org/10.3390/app10082767.

[69] X. Lin, M. Kong, N. Wu, et al., Measurement of temperature distribution at the nanoscale with luminescent probes based on lanthanide nanoparticles and quantum dots, ACS Appl. Mater. Interfaces 12 (2020) 52393–52401. http://www.scientific.net/AMR.47-50.528.
[70] H. Xie, E. Chen, Y. Ye, S. Xu, T. Guo, Highly stabilized gradient alloy quantum dots and silica hybrid nanospheres by core double shells for photoluminescence devices, J. Phys. Chem. Lett. 11 (2020) 1428–1434. Available from: https://doi.org/10.1021/acs.jpclett.9b03578.
[71] S.R.R. dos Reis, S.R. Pinto, F.D. de Menezes, R. Martinez-Manez, E. Ricci-Junior, L.M.R. Alencar, et al., Senescence and the impact on biodistribution of different nanosystems: the discrepancy on tissue deposition of graphene quantum dots, polycaprolactone nanoparticle and magnetic mesoporous silica nanoparticles in young and elder animals, Pharm. Res. 37 (2020) 40. Available from: https://doi.org/10.1007/s11095-019-2754-9.
[72] C.V. Ha, C. Anh Tuan, N. Thi Bich Ngoc, T.H. Nhung, N.Q. Liem, V.T. Kim Lien, Synthesis and optical characterizations of the fluorescence silica nanoparticles containing quantum dots, VNU J. Sci. Mathematics – Phys. 36 (2020). Available from: https://doi.org/10.25073/2588-1124/vnumap.4476.
[73] X. Lin, M. Kong, N. Wu, Y. Gu, X. Qiu, X. Chen, et al., Measurement of temperature distribution at the nanoscale with luminescent probes based on lanthanide nanoparticles and quantum dots, ACS Appl. Mater. Interfaces 12 (2020) 52393–52401. Available from: https://doi.org/10.1021/acsami.0c15697.
[74] R. Rani, P. Malik, S. Dhania, T.K. Mukherjee, Recent advances in mesoporous silica nanoparticle-mediated drug delivery for breast cancer treatment, Pharmaceutics 15 (2023) 227. Available from: https://doi.org/10.3390/pharmaceutics15010227.
[75] M. Wen, M. Chen, K. Chen, P.-L. Li, C. Lv, X. Zhang, et al., Superhydrophobic composite graphene oxide membrane coated with fluorinated silica nanoparticles for hydrogen isotopic water separation in membrane distillation, J. Memb. Sci. 626 (2021) 119136. Available from: https://doi.org/10.1016/j.memsci.2021.119136.
[76] S. Su, L. Chen, L. Hao, H. Chen, X. Zhou, H. Zhou, Fluorinated sodium carboxymethyl cellulose nanoparticles as carrier for improving adhesion and sustaining release of AVM, J. Macromol. Sci. Part A 58 (2021) 219–231. Available from: https://doi.org/10.1080/10601325.2020.1840922.
[77] C. Guo, H. Ding, M. Xie, H. Zhang, X. Hong, L. Sun, et al., Multifunctional superamphiphobic fluorinated silica with a core-shell structure for anti-fouling and anticorrosion applications, Colloids Surf. A Physicochem. Eng. Asp. 615 (2021) 126155. Available from: https://doi.org/10.1016/j.colsurfa.2021.126155.
[78] N.S. Elbialy, S.F. Aboushoushah, B.F. Sofi, A. Noorwali, Multifunctional curcumin-loaded mesoporous silica nanoparticles for cancer chemoprevention and therapy, Microporous Mesoporous Mater. 291 (2020), 109540. Available from: https://doi.org/10.1016/j.micromeso.2019.06.002.
[79] L. Zhang, F. Gu, J. Chan, A. Wang, R. Langer, O. Farokhzad, Nanoparticles in medicine: therapeutic applications and developments, Clin. Pharmacol. Ther. 83 (2008) 761–769. Available from: https://doi.org/10.1038/sj.clpt.6100400.
[80] A. Maleki, H. Kettiger, A. Schoubben, J.M. Rosenholm, V. Ambrogi, M. Hamidi, Mesoporous silica materials: from physico-chemical properties to enhanced dissolution of poorly water-soluble drugs, J. Controlled Rel. 262 (2017) 329–347. Available from: https://doi.org/10.1016/j.jconrel.2017.07.047.

[81] M. Mozafarinia, S. Karimi, M. Farrokhnia, J. Esfandiari, In vitro breast cancer targeting using Trastuzumab-conjugated mesoporous silica nanoparticles: towards the new strategy for decreasing size and high drug loading capacity for drug delivery purposes in MSN synthesis, Microporous Mesoporous Mater. 316 (2021) 110950. Available from: https://doi.org/10.1016/j.micromeso.2021.110950.

[82] R. Zhang, X. Wang, N. Fan, J. Li, Enhanced anticancer performances of doxorubicin loaded macro-mesoporous silica nanoparticles with host-metal-guest structure, Microporous Mesoporous Mater. 310 (2021) 110589. Available from: https://doi.org/10.1016/j.micromeso.2020.110589.

[83] C.M. Day, M.J. Sweetman, S.M. Hickey, Y. Song, Y. Liu, N. Zhang, et al., Concept design, development and preliminary physical and chemical characterization of tamoxifen-guided-mesoporous silica nanoparticles, Molecules 26 (2021) 219. Available from: https://doi.org/10.3390/molecules26010219.

[84] M. Liu, J. Liu, Y. Wang, W. Wang, H. Zhu, H. Guo, et al., Redox-responsive mesoporous silica nanoparticles based on fluorescence resonance energy transfer for anticancer drug targeting and real-time monitoring, J. Mater. Res. 36 (2021) 1883−1898. Available from: https://doi.org/10.1557/s43578-021-00252-z.

[85] L. Guo, J. Ping, J. Qin, M. Yang, X. Wu, M. You, et al., A comprehensive study of drug loading in hollow mesoporous silica nanoparticles: impacting factors and loading efficiency, Nanomaterials 11 (2021) 1293. Available from: https://doi.org/10.3390/nano11051293.

[86] H. Zhu, C. Zhang, S. Liu, Y. Tang, Y. Yin, Effects of nanoparticle clustering and alignment on thermal conductivities of Fe_3O_4 aqueous nanofluids, Appl. Phys. Lett. 89 (2006). Available from: https://doi.org/10.1063/1.2221905.

[87] B. Barbés, R. Páramo, E. Blanco, C. Casanova, Thermal conductivity and specific heat capacity measurements of CuO nanofluids, J. Therm. Anal. Calorim. 115 (2014) 1883−1891. Available from: https://doi.org/10.1007/s10973-013-3518-0.

[88] L.S. Sundar, M.J. Hortiguela, M.K. Singh, A.C.M. Sousa, Thermal conductivity and viscosity of water based nanodiamond (ND) nanofluids: an experimental study, Int. Commun. Heat. Mass. Transf. 76 (2016) 245−255. Available from: https://doi.org/10.1016/j.icheatmasstransfer.2016.05.025.

[89] Y. Ding, H. Alias, D. Wen, R.A. Williams, Heat transfer of aqueous suspensions of carbon nanotubes (CNT nanofluids), Int. J. Heat. Mass. Transf. 49 (2006) 240−250. Available from: https://doi.org/10.1016/j.ijheatmasstransfer.2005.07.009.

[90] H. Xie, L. Chen, Adjustable thermal conductivity in carbon nanotube nanofluids, Phys. Lett. A 373 (2009) 1861−1864. Available from: https://doi.org/10.1016/j.physleta.2009.03.037.

[91] S. Akilu, K.V. Sharma, A.T. Baheta, R. Mamat, A review of thermophysical properties of water based composite nanofluids, Renew. Sustain. Energy Rev. 66 (2016) 654−678. Available from: https://doi.org/10.1016/j.rser.2016.08.036.

[92] A. Kazemi-Beydokhti, S.Z. Heris, N. Moghadam, M. Shariati-Niasar, A.A. Hamidi, Experimental investigation of parameters affecting nanofluid effective thermal conductivity, Chem. Eng. Commun. 201 (2014) 593−611. Available from: https://doi.org/10.1080/00986445.2013.782291.

[93] T.-P. Teng, Y.-H. Hung, T.-C. Teng, H.-E. Mo, H.-G. Hsu, The effect of alumina/water nanofluid particle size on thermal conductivity, Appl. Therm. Eng. 30 (2010) 2213−2218. Available from: https://doi.org/10.1016/j.applthermaleng.2010.05.036.

[94] H.J. Kim, S.-H. Lee, J.-H. Lee, S.P. Jang, Effect of particle shape on suspension stability and thermal conductivities of water-based bohemite alumina nanofluids, Energy 90 (2015) 1290–1297. Available from: https://doi.org/10.1016/j.energy.2015.06.084.

[95] W. Duangthongsuk, S. Wongwises, Measurement of temperature-dependent thermal conductivity and viscosity of TiO_2-water nanofluids, Exp. Therm. Fluid Sci. 33 (2009) 706–714. Available from: https://doi.org/10.1016/j.expthermflusci.2009.01.005.

[96] S.K. Das, N. Putra, P. Thiesen, W. Roetzel, Temperature dependence of thermal conductivity enhancement for nanofluids, J. Heat. Transf. 125 (2003) 567–574. Available from: https://doi.org/10.1115/1.1571080.

[97] C.H. Li, G.P. Peterson, Experimental investigation of temperature and volume fraction variations on the effective thermal conductivity of nanoparticle suspensions (nanofluids), J. Appl. Phys. 99 (2006). Available from: https://doi.org/10.1063/1.2191571.

[98] W. Duangthongsuk, S. Wongwises, An experimental study on the heat transfer performance and pressure drop of TiO_2-water nanofluids flowing under a turbulent flow regime, Int. J. Heat. Mass. Transf. 53 (2010) 334–344. Available from: https://doi.org/10.1016/j.ijheatmasstransfer.2009.09.024.

[99] M.A. Khairul, K. Shah, E. Doroodchi, R. Azizian, B. Moghtaderi, Effects of surfactant on stability and thermo-physical properties of metal oxide nanofluids, Int. J. Heat. Mass. Transf. 98 (2016) 778–787. Available from: https://doi.org/10.1016/j.ijheatmasstransfer.2016.03.079.

[100] S. Wen, P. Wang, L. Wang, Preparation and antifouling performance evaluation of fluorine-containing amphiphilic silica nanoparticles, Colloids Surf. A Physicochem. Eng. Asp. 611 (2021) 125823. Available from: https://doi.org/10.1016/j.colsurfa.2020.125823.

[101] S. Gupta, F. Martoïa, L. Orgéas, P. Dumont, Ice-templated porous nanocellulose-based materials: current progress and opportunities for materials engineering, Appl. Sci. 8 (2018) 2463. Available from: https://doi.org/10.3390/app8122463.

[102] V. Penkavova, J. Tihon, O. Wein, Stability and rheology of dilute TiO_2-water nanofluids, Nanoscale Res. Lett. 6 (2011) 273. Available from: https://doi.org/10.1186/1556-276X-6-273.

[103] B.C. Pak, Y.I. Cho, Hydrodynamic and heat transfer study of dispersed fluids with submicron metallic oxide particles, Exp. Heat. Transf. 11 (1998) 151–170. Available from: https://doi.org/10.1080/08916159808946559.

[104] Y. Xuan, W. Roetzel, Conceptions for heat transfer correlation of nanofluids, Int. J. Heat. Mass. Transf. 43 (2000) 3701–3707. Available from: https://doi.org/10.1016/S0017-9310(99)00369-5.

[105] W. Aughenbaugh, S. Radin, P. Ducheyne, Silica sol-gel for the controlled release of antibiotics. II. The effect of synthesis parameters on the in vitro release kinetics of vancomycin, J. Biomed. Mater. Res. 57 (2001) 321–326. Available from: https://doi.org/10.1002/1097-4636(20011205)57:3 < 321: AID-JBM1174 > 3.0.CO; 2-S.

[106] Q.-Z. Xue, Model for effective thermal conductivity of nanofluids, Phys. Lett. A 307 (2003) 313–317. Available from: https://doi.org/10.1016/S0375-9601(02)01728-0.

[107] J. Wang, J. Zhu, X. Zhang, Y. Chen, Heat transfer and pressure drop of nanofluids containing carbon nanotubes in laminar flows, Exp. Therm. Fluid Sci. 44 (2013) 716–721. Available from: https://doi.org/10.1016/j.expthermflusci.2012.09.013.

CHAPTER

Carbon nanotubes/graphene-based nanofluids

7

Zahra Fakharshakeri[1], Raed Muslim Mhaibes[2] and Hossein Safardoust-Hojaghan[1]

[1]*Institute of Nano Science and Nano Technology, University of Kashan, Kashan, Iran*

[2]*Department of Biochemistry, College of Medicine, Misan University, Misan, Iraq*

7.1 General introduction

The incorporation of nanoparticles with a low volume fraction and good dispersion in liquids, known as nanofluids (NFs) [1–3], holds immense promise for enhancing efficiency [4,5], conserving energy [6–8], and reducing costs of thermal management equipment [9–11]. NFs are a recent advancement in the field of improving the thermal conductivity (TC) of liquids, which has been an ongoing effort for over a century [12–16]. The limited TC of conventional heat transfer fluids (HTFs) poses a significant obstacle to improving the performance of engineering equipment [16–18]. What initially started with particles ranging from millimeters to micrometers in size during the late 19th century has now evolved into exploring nanoparticles, delving into the realm of minuscule dimensions [19,20]. Specifically, NFs represent a novel category of HTFs based on nanotechnology. They are created by stably suspending particles, fibers, or tubes measuring nanoscale in length within traditional HTFs at certain concentrations [21,22]. This concept and term were introduced by Choi in the early 1990s [23]. Unlike HTFs containing larger particles, NFs possess both stability and dilution properties. Their success as engineered HTFs lies in their ability to synergize solid and liquid materials through mechanisms that are still not fully understood. As a result, these fascinating fluids exhibit remarkable thermal properties that have captivated students, scientists, and engineers worldwide—prompting them to explore the potential applications offered by NFs [24–26]. The literature reports various types of particles used in the preparation of NFs. These include nonmetallic particles such as titanium dioxide [27,28], silica [29,30], aluminum oxide [31,32], SiC [33], CuO [34–36], ZnO [37,38], Fe_3O_4 [39,40], and AlN [41]; metallic particles such as Cu [42], Ag [43], and Au [44,45]; and different particle shapes such as carbon nanotubes (CNTs) [46,47], nanodroplets [48], nanofibers [49], and nanorods [50,51]. The commonly used base fluids are water [52], oil [53,54], ethylene glycol [55], and mineral oil [56]. In addition to their high surface/volume ratio and adsorption capability, carbon-based nanostructures also

Nanofluids. DOI: https://doi.org/10.1016/B978-0-443-13625-2.00011-5

exhibit excellent TC, making them ideal for use in NF applications [57–60]. The incorporation of CNTs and graphene into NFs has been proven to considerably increase their TC, making them attractive for use in HTFs and cooling systems [61–63]. Moreover, the unique properties of CNTs and graphene also make them suitable for use in various applications, including energy storage, catalysis, and sensors [64]. Overall, the use of carbon-based materials in NFs and other applications plays a key role in increasing energy efficiency, decreasing costs, and addressing environmental challenges. Ongoing research in this field promises even more exciting advancements in the future. In the following, the principles related to the synthesis and application of CNTs and graphene-based nanomaterials are investigated comprehensively.

7.2 Carbon nanotubes

CNTs are cylindrical carbon-based structures that have gained more attention recently for their interesting features and feasible applications [65]. These structures are formed from a single layer of graphene rolled into a tube, with diameters ranging from a few nanometers to several micrometers [66]. CNTs show attractive mechanical, thermal, and electrical characteristics, making them excellent candidates for various applications [67]. One of the most promising applications of CNTs is in the field of electronics. Due to their high electrical conductivity, CNTs have been used as components in transistors, sensors, and other electronic devices [68–70]. CNTs also exhibit sufficient TC, which enables them to be used in heat sinks and other thermal management applications [62,71]. In addition to their electrical, optical, and thermal features, CNTs also have high strength and stiffness [72]. These properties make them ideal for use in the NF field. CNT composites have been shown to exhibit improved mechanical properties, such as increased stiffness and strength, as well as enhanced electrical and TC [73–76].

Despite their many promising properties and potential applications in NFs, there are still challenges associated with the synthesis and functionalization of CNTs. Current routes for synthesizing CNTs are high-cost and time-consuming, and the resulting materials may contain defects that affect their properties [77,78]. Additionally, functionalizing CNTs for application in NF can be challenging due to their complex structure and surface chemistry [79].

The application of CNTs in NFs has found more attention in recent decades. The literature in this field demonstrates a growing interest in the subject, with a notable increase in the quantity of publications. Numerous studies have investigated various aspects of using CNTs in NFs, including their synthesis, characterization, dispersion, stability, and heat transfer properties [80]. The breadth of available published literature is extensive, covering different applications such as heat transfer enhancement in engineering systems, biomedical applications, and energy storage [81–83]. Additionally, researchers have explored the influence of factors such as nanotube concentration, NF preparation methods, and different NF base fluids on the overall performance. Moreover, the depth of the available literature is commendable, with a multitude of experimental and theoretical studies providing insights into the behavior and potential applications of CNTs in NFs [61]. The research attempts have provided a better understanding of the underlying mechanisms, such as the role of nanotube aspect ratio, functionalization, and surface modification for improved dispersion and stability [84,85]. Additionally, investigations into the interactions between CNTs and other nanomaterials have highlighted potential synergistic effects that could further enhance the performance of NFs. These advancements indicate a maturing field with researchers focusing on specific aspects to deepen their understanding. One of the key challenges in utilizing CNTs in NFs is achieving stable suspensions and proper dispersion of CNTs within the base fluid. Several techniques have been employed to address this issue, such as functionalization of CNTs, surfactant stabilization, and sonication. The stability and dispersion of CNT-based NFs directly affect their thermal and fluid flow properties, making it a critical aspect to consider in practical applications. Also, the TC of CNTs is exceptionally high, ranging from 2000 to 6000 W/m K, surpassing that of common metals such as copper and silver [86–89]. As the TC of solid particles greatly influences the overall TC of NFs, utilizing NFs with CNTs as the working fluid in heat transfer applications is ideal due to their favorable thermophysical properties [90].

Caihang Liang et al. studied the heat transfer performances of different fluids in a pulsating heat pipe (PHP). The fluids tested were ultrapure multiwalled carbon nanotubes (MWCNTs) NF, water, *n*-butanol self-rewetting MWCNTs NF, and *n*-butanol self-rewetting fluid. The goal was to test their thermal activity under various conditions such as inclination angles, heating powers, and ambient temperatures. The work findings showed that all the functional working fluids typically outperformed ultrapure water in terms of thermal transport. The self-rewetting fluid (SRF) showed the most significant improvement, with a maximum enhancement ratio of 22% at higher thermal inputs. The NF only showed an advantage at lower heating powers, with a 16% enhancement at 35 W. The SRNF demonstrated considerable improvement in thermal transfer activity at low heating powers, with an increase ranging from 7% to 17%, peaking at 35 W. The inclination angle also had a notable effect on the thermal efficiency of different working fluids in the PHP. Lower ambient temperatures were found to enhance heat transfer in the PHP when using functional working fluids. Overall, this study provides valuable insights into the performance of different fluids in a PHP system under various conditions [91].

The study by Sarafraz et al. focused on investigating the effects of aqueous MWCNTs NFs on pool boiling heat transfer. The researchers specifically examined the impact of using modified surfaces with diamond-like morphology microfins of different geometrical properties. The findings showed that the heat transfer coefficient on the plain surface decreased, while it significantly improved by 77% and 56% for wt.% = 0.3 and wt.% = 0.1, respectively, on the microfinned surfaces. These modified surfaces also showed a lower fouling rate compared to the plain surface. It was observed that the irregular and nonuniform fouling layer formed on the modified surfaces increased the bubble creation and number of active nucleations, leading to improved heat transfer. In contrast, the continuous fouling on the plain surface resulted in a significant decrease in heat transfer coefficient due to increased thermal resistance. Furthermore, the deposition of CNTs on the surface altered its characteristics, resulting in a decrease in static contact angle and an increase in surface wettability. This change led to a significant enhancement of up to 95% in critical heat flux for wt.% = 0.3. As the mass dosage of NFs increased, there was a corresponding decrease in static contact angle. The study also found that modified surfaces facilitated more intense bubble formation, which discreetly departed from the surface. This phenomenon contributed to an elevation in the heat transfer coefficient for local agitation and an increase in the available heat transfer area [92].

In another work, the enhancement of TCs in synthetic engine oil, water, and ethylene glycol using MWCNTs, copper oxide (CuO), and copper (Cu) NFs was investigated. The physical mixing method and chemical reduction method are used to synthesize the NFs. The findings demonstrate that NFs with low amounts of copper, copper oxide, or CNTs have significantly higher TC than the base solvent. The improvement in TC for CuO and CNT NFs is approximately linear with the volume fraction. However, an intensive dependence of TC on the calculated time is seen for copper–water fluid. The system activity of a water chiller subject to MWCNTs/water NF is also studied, revealing that dynamic effects such as nanostructure distribution can augment the system activity. The coefficient of activity of the water chiller is elevated by 5.15% in comparison with nanoparticle-free fluid [93].

7.2.1 Synthesis of carbon nanotubes and carbon nanotubes nanofluids

Currently, CNTs are primarily used in coating and as additives in nanocomposites. However, there has been a significant increase in their application as heat transfer enhancers in NFs. The synthesis and fabrication of CNTs can be approached through two methods: top-down and bottom-up. The bottom-up method involves starting with atoms or molecules and gradually building up to nanoscale materials. This approach is often referred to as the natural approach because it mimics how nature creates structures through self-assembly. Self-assembly utilizes physical forces at the nanodimension to combine basic units into bulk materials. On the other hand, the top-down method involves scaling down bulk materials or structures to nanodimension structures [94]. The vapor deposition route, specifically chemical vapor deposition (CVD), is commonly employed for scale-up production of CNTs using fluidized bed reactor systems [95–97]. Additionally, in recent years, the laser ablation and arc discharge methods have also generally utilized for CNT synthesis [98].

There are two routes for the synthesis of a CNT-based NF: the one-step route and the two-step route. In the one-step route, nanotubes are produced and NFs are synthesized simultaneously. On the other hand, the two-step method involves first preparing CNT and then dispersing them into a base fluid or solvent in the next step. Common base fluids such as water and ethylene glycol oil are commonly applied as coolants and heat storage mediums; however, their thermal and physical ability have limitations. By adding nanoparticles to these fluids, their thermophysical properties can be further enhanced. These nanoparticle-dispersed fluids have various names depending on their specific role or base fluid used, such as nanolubricants, nanocoolants, nanogreases, nanorefrigerants, heat-storage NFs, and ionanofluids [99–102].

Li et al. prepared CNT-based NF for H_2S absorption. First, the synthesis of deep eutectic solvent (DES) involved combining ChCl (0.1 mol) and urea (0.2 mol) in a beaker and stirring the mixture at 80°C until a transparent and homogeneous solution was achieved. This resulting liquid was referred to as DES. To create alkaline DES systems, ethanolamine was added to DES. The NF system was produced by introducing CNTs and amino-functionalized CNTs (ACNTs) with varying weights to the DES system. The mixture was stirred for 30 minutes and treated with intermittent sonication for 40 minutes at ambient conditions using an ultrasonic cleaner. The resulting uniform NF systems were obtained. The production routes of the ACNTs, DES, as well as the NF systems are illustrated in Fig. 7.1 [103].

The surface functional groups of CNTs were analyzed before and after acid treatment and amino functionalization using FT-IR. The results revealed that the peaks at 1580 and 1630 cm^{-1} in the CNT spectrum corresponded to the carbon-carbon, while the peaks at 2920 and 3440 cm^{-1} represented O—H and C—H absorption peaks, respectively. After treatment with acid, new peaks were revealed at 1190 and 1710 cm^{-1}, indicating the presence of carboxylic acid (COOH) groups through stretching vibrations of C=O and C—O bonds. This confirmed the presence of —COOH groups on the surface of CNTs after treatment with acid. In the ACNT spectrum, an absorption peak at 3300 cm^{-1} indicated the presence of N—H bonds. Additionally, a peak caused by silicon–oxygen appeared at 1120 cm^{-1}, confirming that (3-aminopropyl)triethoxysilane had linked to CNTs. Raman spectra were also collected for both CNTs and ACNTs, showing slight shifts in the D band (1340 cm synthesis and fabrication of CNT can be approached through two methods) and G band (1380 cm^{-1}) of CNTs, indicating chemical functionalization of the CNTs. These results demonstrated the successful grafting of amino functional groups onto the surface of CNTs through amination modification. To further analyze the surface changes of CNTs, EDS analysis was conducted to determine element compositions before and after amination modification. The results showed that nitrogen and silicon elements revealed on the surface of ACNTs, with N content at 1.23 w% and Si content at 5.07 w%, confirming the successful grafting of amino functional groups onto the surface of CNTs. SEM was used to characterize the CNTs and ACNTs. It is evident that

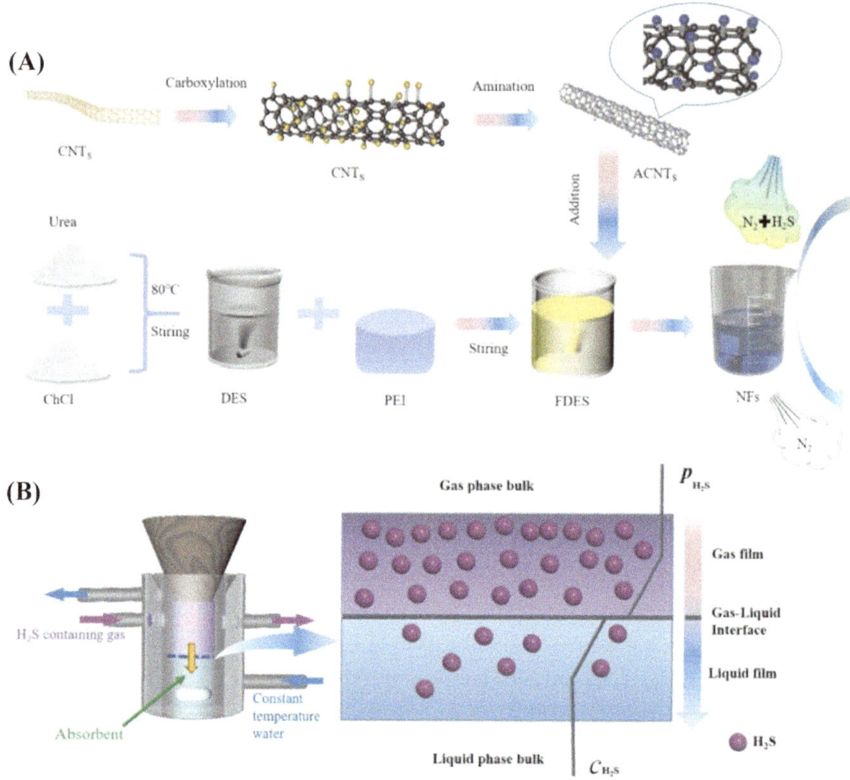

FIGURE 7.1

The synthesis process of the ACNTs, FDES, and NF systems (A); the diagram of the double mixed gradientless gas–liquid reactor (B). *ACNTs*, Amino-functionalized carbon nanotubes; *FDESs*, functionalized deep eutectic solvents; *NF*, nanofluid.

Reproduced with permissions from M. Li, L. Ren, Z. Gu, P. Gao, W. Sun, X. Dong, F. Liu, B. Wang, Z. Zhang, X. Liu, P. Gao, Insight into the enhancement effect of amino functionalized carbon nanotubes on the H_2S removal performance of nanofluid system, J. Hazard. Mater. 458 (2023) 131977.

both CNTs and ACNTs possess consistent tube-like structures with uniform morphology. To gain further insight into the characteristics of the ACNTs prepared in this study, a transmission electron microscopy image was obtained. The image reveals that the ACNTs exhibit excellent dispersion, allowing for even distribution in a liquid medium.

As depicted in Fig. 7.2A and C, the N_{H2S} levels in both NF systems experienced a notable increase after the addition of nanostructures, leading to a considerable reduction in mass transfer resistance. However, for CNTs and ACNTs, the N_{H2S} levels exhibited a pattern of initially increasing and then decreasing, indicating an optimal dosage of nanostructures. According to Fig. 7.2A, the best dosage

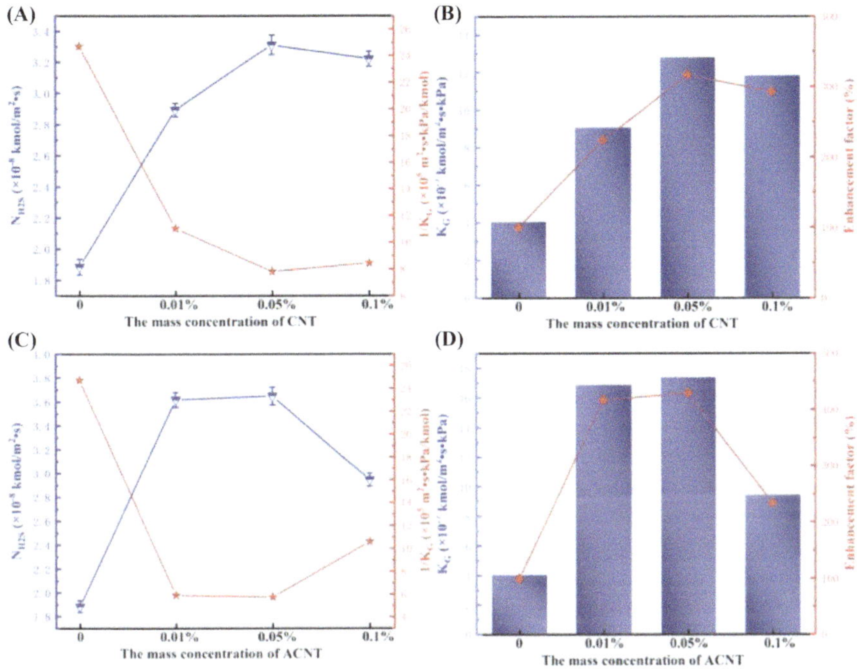

FIGURE 7.2

(A) The effect of the mass concentration of CNT on the absorption rate (NA) and mass transfer resistance ($1/K_G$); (B) the effect of the mass concentration of CNT on the total mass transfer coefficient (K_G) and enhancement factor; (C) the effect of the mass concentration of ACNT on the absorption rate (NA) and mass transfer resistance ($1/K_G$); and (D) the effect of the mass concentration of ACNT on the total mass transfer coefficient (K_G) and enhancement factor. $T = 30°C$, $CH_2S = 1000$ mg/m^3, $r_G = 1400$ r/min, $r_L = 800$ r/min (r_G: gas phase stirring speed, r_L: liquid phase stirring speed).

Reproduced with permissions from M. Li, L. Ren, Z. Gu, P. Gao, W. Sun, X. Dong, F. Liu, B. Wang, Z. Zhang, X. Liu, P. Gao, Insight into the enhancement effect of amino functionalized carbon nanotubes on the H_2S removal performance of nanofluid system, J. Hazard. Mater. 458 (2023) 131977.

of CNTs is 0.05%. In Fig. 7.2C, for the ACNT-based system, the N_{H2S} levels at ACNT concentrations of 0.05% and 0.01% were presented. As the nanoparticle concentration continued to increase, the N_{H2S} levels in CNTs and ACNTs started to decrease but remained considerably higher than those in the bare sample without CNTs and ACNTs within the experimental dosage scale. The changes in K_G and enhancement factors (β) are illustrated in Fig. 7.2B and D. It is observed that in the CNT system, the highest β reached 316%, whereas in the ACNTs system, when ACNTs dosage were at 0.05% and 0.01%, the β was determined 429% and 416%, respectively. These findings are in good agreement with the dynamic absorption findings of H_2S, determining that nanoparticle presence substantially decreases gas–liquid mass transfer resistance and enhances gas–liquid mass transfer rate. It is important to note that as nanoparticle dosage continued to increase, although N_{H2S} began to decrease, this decline was much less pronounced than during the dynamic absorption process [103].

In another work, a two-step process was employed to design the NFs. To achieve MWCNTs–COOH/deionized water NFs with a volume fraction of 0.05 vol. for the tests, the specific dosages of sodium dodecylbenzene sulfonate and Triton X-100 were measured based on the mentioned fraction. The resulting mixture was stirred using an electromagnetic stirrer for 2 hours. MWCNTs–COOH masses were then introduced to the DIW-Triton X-100/SDBS mixture according to the specified fractions. To ensure proper dispersion and homogeneity of the NFs, the mixture underwent ultrasonic treatment in a water bath before adding the MWCNTs–COOH. During this process, MWCNTs–COOH were introduced to the mixture at a rate of 1 g/h in each step, totaling 0.1 per step. After each addition, the suspension underwent ultrasonication with varying amounts of energy, resulting in stable and homogeneous NFs. The temperature of the bath was maintained at approximately 25°C ± 5°C via the ultrasonication process. They examined six different geometry characteristics of CNTs, which are mentioned in Table 7.1.

Fig. 7.3 presents the change of effective TC with temperature for all samples. The temperature range considered for the experiments is 10°C–45°C. This range was chosen to avoid potential bond destruction between the surfactant and CNTs at high temperatures, which could lead to instabilities. As shown in Fig. 7.3, all samples exhibit higher effective TC compared to the CNT-free water. Results provide

Table 7.1 Properties for the selected nanotubes.

Properties	Samples					
	A	B	C	D	E	F
Outer diameter (nm)	<8	10–20	50<	<8	10–20	50<
Inner diameter (nm)	2–5	5–10	5–15	2–5	5–10	5–15
Length (μn)	10–30	10–30	10–30	0.5–2	0.5–2	0.5–2
Aspect ratio	1250–3750	500–3000	200–400	70–250	20–200	10–40
Special surface area (m^2/g)	500<	200<	40<	500<	200<	40<
Thermal conduction coefficient (W/m K) [48]	3000	3000	3000	3000	3000	3000
True density (g/cm^3)	−2.1	−2.1	−2.1	−2.1	−2.1	−2.1
Tap density (g/cm^3)	0.27	0.22	0.18	0.27	0.22	0.18
Purity (%)	95<	<95<	95<	5<	95<	95<
−COOH content (wt.%)	3.86	2.00	0.49	3.86	2.00	0.49

Reproduced with permissions from A.N. Omrani, E. Esmaeilzadeh, M. Jafari, A. Behzadmehr, Effects of multi walled carbon nanotubes shape and size on thermal conductivity and viscosity of nanofluids, Diamond Relat. Mater. 93 (2019) 96–104.

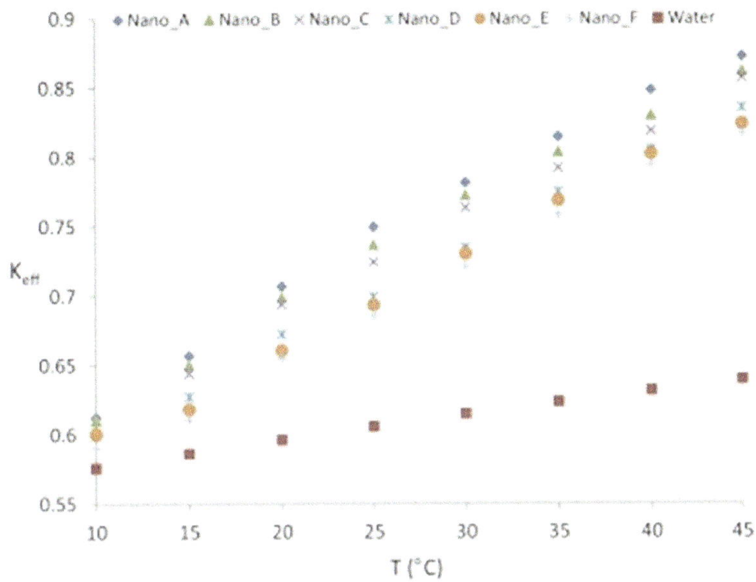

FIGURE 7.3

Variation of relative thermal conductivity with MWCNT aspect ratio-comparison between experimental relative thermal conductivity and some theoretical models at 25°C.

Reproduced with permissions from A.N. Omrani, E. Esmaeilzadeh, M. Jafari, A. Behzadmehr, Effects of multi walled carbon nanotubes shape and size on thermal conductivity and viscosity of nanofluids, Diamond Relat. Mater. 93 (2019) 96–104.

the enhanced relative TC of NFs compared to the base fluid. It can be observed from this diagram that sample A, which has the maximum length/dimension among all samples at 45°C, shows the highest enhancement of 36%. On the other hand, sample F with the minimum length/dimension at 10°C exhibits the least improvement of about 3% among all samples. The experimental results indicate that as the temperature of NFs increases, their TC also elevates in all samples. This increase shows a linear behavior with respect to NFs temperature [104]. The next section is focused on the synthesis and application of graphene-based NFs.

7.3 Graphene

Graphene is a 2D allotrope of carbon with a single layer of carbon atoms arranged in a hexagonal lattice. It is the basic structural element of other carbon allotropes such as graphite, charcoal, and CNTs [105]. Graphene possesses several unique properties that make it highly important and valuable in various fields: (1) Strength and flexibility: Graphene is incredibly strong, around 200 times stronger than steel. It can withstand large amounts of stress without breaking and is also highly flexible, allowing it to be stretched without damage [106]. (2) Electrical conductivity: Graphene is an attractive conductor of electricity for its unique electronic structure. Electrons can move through graphene at extremely high speeds, making it a good candidate for applications in electronics and energy storage [107,108]. (3) TC: Graphene has exceptional TC, allowing it to efficiently transfer heat. This property makes it useful for applications such as thermal management in electronics and energy storage devices [109,110]. (4) Transparency: Despite being a single layer of atoms, graphene is transparent to visible light, as well as other wavelengths across the electromagnetic spectrum. This property makes it useful for applications in different fields such as optoelectronics and energy storage. (5) Chemical stability: Graphene is chemically stable and resistant to many corrosive substances, making it suitable for use in harsh environments [111]. (6) Large surface area: Due to its two-dimensional structure, graphene has an enormous surface area relative to its volume. This property makes it useful for applications such as energy storage devices (batteries and supercapacitors) where a high surface area is desirable [112].

The importance of graphene lies in its potential to revolutionize various industries: electronics, energy storage, composite materials, water filtration, and biomedical applications. The investigations show a significant increase in publications on graphene NFs in the past decade, indicating their growing importance in the field of engineering. These NFs have shown great potential for various applications, including automobiles, coolants, refrigerators, solar devices, cosmetics, and drug delivery [63]. Fig. 7.4 provides a visual representation of some of these applications. While there have been numerous studies on the thermal and rheological properties of graphene NFs, as well as different modes of heat transfer such as boiling, there is currently no comprehensive literature on their limitations and applications [113]. Specifically, this section focuses on presenting the results achieved by the scientific community in relation to graphene NFs. Additionally, an analysis has been conducted to determine how different parameters influence the expected findings and activity, allowing for an evaluation of the advantages and disadvantages associated with using graphene NFs in this specific field. Furthermore, it is anticipated that further improvements can be made to enhance the performance of these identified applications [114].

Yu et al. aimed to increase the functionalities of carbon fiber-reinforced cement mortar by incorporating graphene NF additive. The results demonstrated that the combination of carbon fibers and graphene NF additive had a significant synergistic effect on various characteristics of the cement mortar. The carbon fiber-reinforced cement mortar modified with graphene NF additive exhibited remarkable improvements, with flexural and compressive strengths increasing by 101% and 126%, respectively, compared to the control, which only saw increases of 25% and 45%. Moreover, the carbon fiber-reinforced cement mortar modified with graphene NF additive showed much higher thermal and electrical conductivities, with increases of 36% and 68%, respectively, compared to the control's

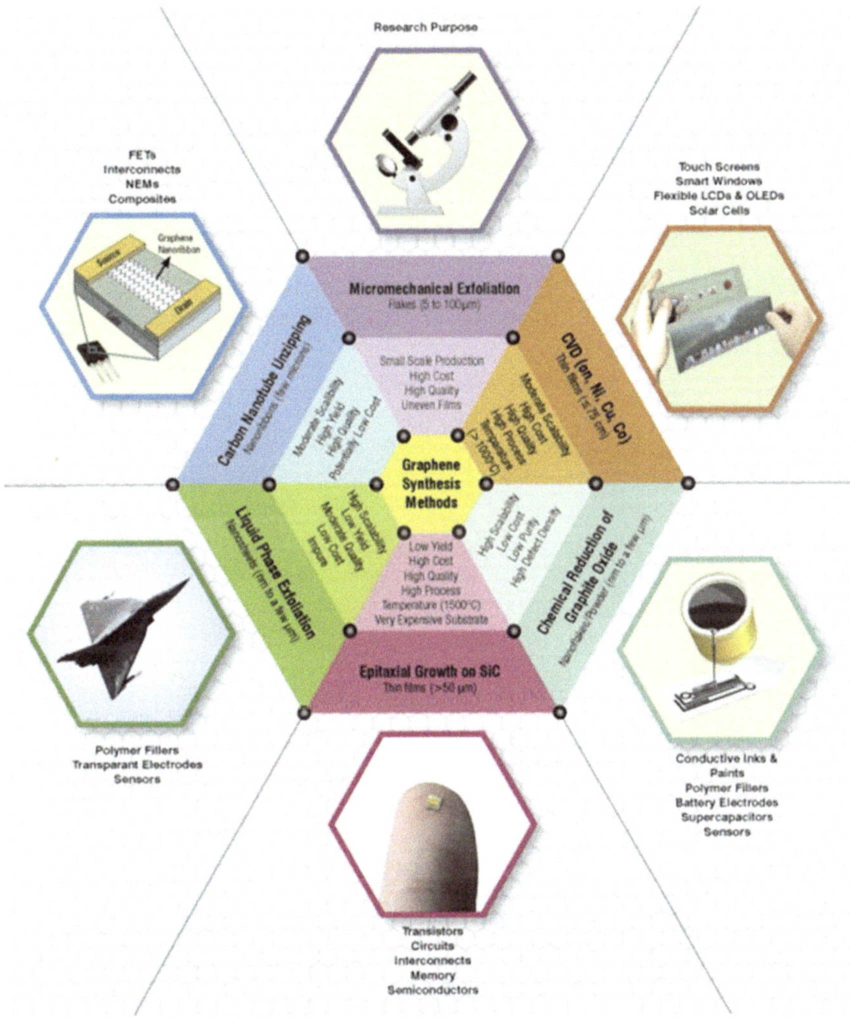

FIGURE 7.4

Scheme depicting various conventional synthesis methods of graphene along with their important features, and their current and prospective applications.

Reproduced with permissions from G. Mittal, V. Dhand, K.Y. Rhee, S.-J. Park, W.R. Lee, A review on carbon nanotubes and graphene as fillers in reinforced polymer nanocomposites, J. Ind. Eng. Chem. 21 (2015) 11–25.

increases of 27% and 19%. Additionally, the carbon fiber-reinforced cement mortar modified with graphene NF additive exhibited an unexpected increase in service life for cathodic protection systems, surpassing that of the control by more than 200%. These findings highlight the suitability of graphene NF material for high-efficiency cathodic protection system applications [115].

7.3.1 Synthesis of graphene and graphene nanofluids

Graphene has been extensively researched in NFs and has proven to be beneficial in this field. The main limitation in graphene-based NFs lies in producing high-quality graphene sheets on a large scale. Several routes have been developed and optimized for commercial-scale preparation of graphene. Numerous review papers have been reported on the preparation and applications of graphene. Research indicates that attempts to produce large-scale graphene date back to 1975. The reported studies focus on three main preparation methods of graphene (CVD, chemical, and mechanical exfoliation) as these have already been extensively reviewed. Fig. 7.4 provides a detailed overview of various synthesis routes and their general applications in modern life [113] (Fig. 7.5).

In the reported work, the synthesis of graphene nanoplatelet (GNP)-based NF involves a two-step method, where nanoparticles are dispersed in base phase change materials (PCMs) and prepared by applying chemical dispersants. To begin, surfactants such as cetyltrimethyl ammonium bromide are diffused into a certain amount of a standard using sonication and magnetic stirrer. The mass concentrations of additives, such as surfactants and GNPs, are measured using a weighing device. Next, the GNPs are distributed in surfactant solutions for 45 minutes by applying the magnetic stirrer. To confirm a colloidal mixture, probe sonication is employed for 60 minutes at short intervals to prevent overheating. The nanoenhanced PCMs

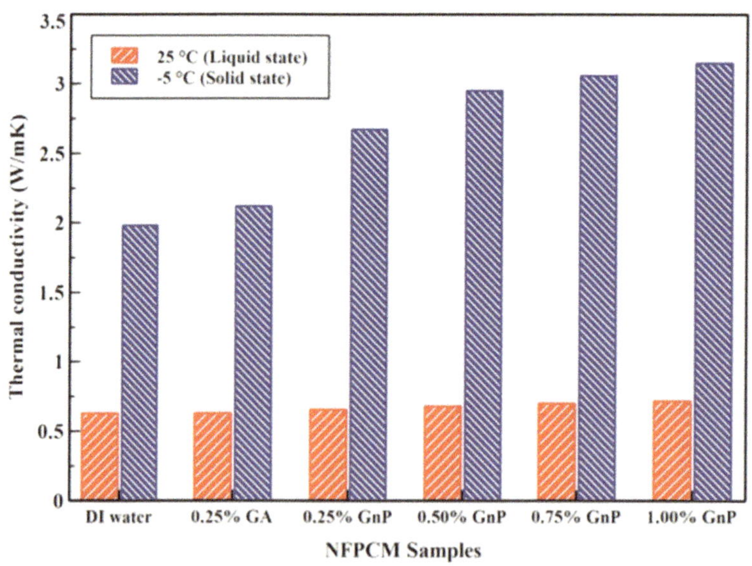

FIGURE 7.5

Thermal conductivity of phase change materials at different mass concentrations of additives.

Reproduced with permissions from G. Mittal, V. Dhand, K.Y. Rhee, S.-J. Park, W.R. Lee, Synthesis, stability, and heat transfer behavior of water and graphene nanoplatelet-based nanofluid for cool thermal storage applications, J. Ind. Eng. Chem. 21 (2015) 11–25.

are prepared by finding sufficient mass concentrations relative to the base fluid. Results showed that the TC of PCMs is shown at temperatures of 25°C and −5°C, with varying concentrations of GNPs. The results indicate that the TC increased by 59.1%, 54.6%, 49%, and 34.6% for GNPs introduction of 1.00, 0.75, 0.50, and 0.25 wt.%, respectively. These refinements can be related to the high heat-conductive surface area of GNP-based NF, which enhances the heat transfer rate as the concentration increases. Additionally, GNP-based NF has minimal interfacial thermal resistance, leading to improved solid-state heat transfer. The same process has been observed for water-based NF PCMs [116].

Legrand et al. prepared graphene nanoflakes through homogeneous nucleation by decomposing a carbon precursor in the thermal plasma reactor. After preparation in the plasma reactor, the graphene nanoflakes undergo functionalization downstream in the plasma recombination zone, where oxygen is added to the main plasma stream. This allows for tunable oxygen functionalization, with surface values reaching up to 14.2 at.%. While nonfunctionalized graphene nanoflakes are hydrophobic, oxygen-functionalized exhibit complete stability when dispersed directly in water or ethanol without the need for a surfactant. The oxygen-functionalized graphene nanoflakes maintain their structural integrity even after the introduction of hydrophilic groups on their surface, and they remain stably dispersed in water and ethanol over an extended period of time (Fig. 7.6).

FIGURE 7.6

Observed stability after 6 weeks for (A) oxygen-functionalized GNFs (O-GNFs) with 14.2 at.% oxygen in ethanol, (B) O-GNFs with 14.2 at% oxygen in water, (C) O-GNFs with 6.4 at% oxygen in water, and (D) nonfunctionalized GNFs in water. (A color version of this figure can be viewed online.)

Reproduced with permissions from U. Legrand, N.Y. Mendoza Gonzalez, P. Pascone, J.L. Meunier, D. Berk, Synthesis and in-situ oxygen functionalization of deposited graphene nanoflakes for nanofluid generation, Carbon 102 (2016) 216–223.

The NFs that are produced can have various applications. Additionally, they can undergo wet chemistry treatment to graft atoms or molecules onto the surface or edges of the graphene nanoflakes [117].

7.4 Carbon nanotube/graphene-based nanofluids

One of the important applications of CNTs/graphene-based NFs is in heat transfer systems. The high TC of these nanomaterials allows for efficient heat dissipation, making them ideal candidates for cooling electronic devices and enhancing the performance of heat exchangers. Additionally, their excellent mechanical strength makes them suitable for use in lightweight structural materials. Furthermore, CNTs/graphene-based NFs have shown promise in energy storage devices such as supercapacitors and batteries. Their large surface area and high electrical conductivity enable faster charge-discharge rates and increased energy density compared to traditional materials. Despite their immense potential, there are still challenges that need to be addressed before widespread commercialization can occur. Issues such as scalability, cost-effectiveness, and environmental impact must be carefully considered. In addition, there is a collaborative impact when utilizing both CNTs and graphene as fillers simultaneously. The lengthy and intertwined morphology of the CNTs allows them to act as a scaffold between the graphene layers when introduced to the NF. This process effectively prevents agglomeration and stacking between the graphene layers, resulting in a three-dimensional carbon network within the NF. This, in turn, increases the contact area between the carbon nanomaterials and NFs, ultimately improving the mechanical, thermal, and electrical properties of the material by enhancing interfacial strength and contact surface [49,118].

7.4.1 Synthesis of graphene/carbon nanotubes nanofluids

The graphene/CNT nanocomposites are commonly prepared and applied as NF using various technologies such as CVD, electrophoretic deposition, and in situ chemical reduction [119–121]. However, the high specific surface energy of this hybrid often leads to significant agglomeration within the fluid system. This agglomeration hinders the enhancement of comprehensive properties in NF. Therefore, the applied synthesis procedure plays a key role in the characteristics of graphene/CNT-based NF.

There are several methods for synthesizing graphene/CNT composites, including CVD, coprecipitation, and chemical functionalization. One common synthesis method is the CVD method, where a gaseous carbon source, such as methane or ethylene, is decomposed on a metal catalyst surface at high temperatures. This process leads to the growth of graphene sheets and CNTs simultaneously. The CNTs then become embedded within the graphene sheets, forming a composite structure. Another synthesis method involves solution mixing, where a dispersion of CNTs and graphene sheets is prepared using a solvent, such as ethanol or water. The CNTs and graphene sheets are dispersed through ultrasonication or other mechanical means to ensure uniform mixing. The resulting solution is then evaporated, leaving behind a solid composite material. Chemical functionalization can also be used to synthesize graphene/CNT composites. In this method, the individual graphene sheets and CNTs are functionalized with certain chemicals, such as acids or organic compounds, to modify their surface properties. The functionalized graphene and CNTs are then mixed together and subjected to heat treatment, leading to the formation of a composite material. The coprecipitation method involves the simultaneous growth of graphene sheets and CNTs within a solution. Graphene oxide (GO) and CNT precursors are mixed together, followed by the addition of reducing agents, leading to the preparation of reduced graphene oxide (rGO) and CNTs. This method provides a simple and scalable synthesis route for graphene/CNT NFs [122,123].

Van Trinh et al. prepared graphene/CNT-based NFs via a chemical route. The process for preparing the NFs is depicted in Fig. 7.7. Carboxyl-functionalized graphene was obtained by treating graphene sheets with a mixture of HNO_3 and H_2SO_4 at 70°C for 5 hours. The resulting solution was filtered and cleaned with distilled water. Gr–COOH was then distributed in ethylene glycol using ultrasonication for 30 minutes to obtain a Gr–COOH solution with a volume fraction of 0.07%. MWCNTs were treated with hydroxyl groups utilizing a common route. Hydroxyl groups treated MWCNTs were prepared applying the same route as

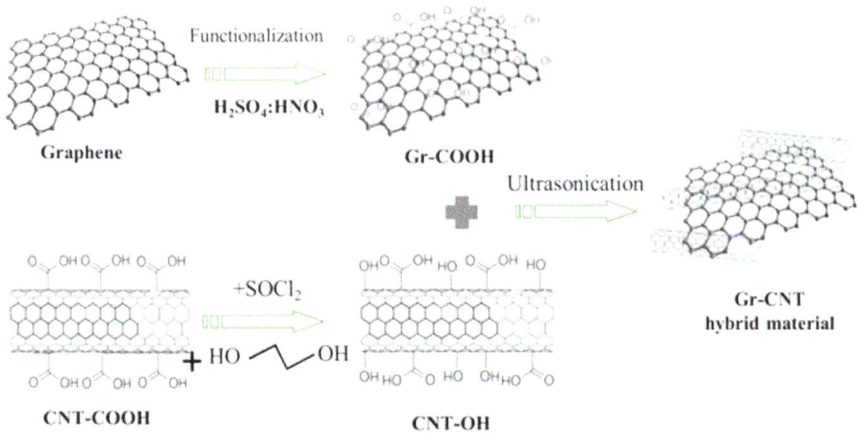

FIGURE 7.7

Schematic view of the preparation process of Gr–CNT hybrid material.

Reproduced with permissions from P. Van Trinh, N.N. Anh, N.T. Hong, P.N. Hong, P.N. Minh, B.H. Thang, Experimental study on the thermal conductivity of ethylene glycol-based nanofluid containing Gr-CNT hybrid material, J. Mol. Liq. 269 (2018) 344–353.

carboxyl-functionalized graphene, followed by suspension in $SOCl_2$ solution and washing with tetrahydrofuran and distilled water to obtain MWCNTs−COCl. Hydroxyl groups treated MWCNTs were obtained by mixing MWCNT−COCl with ethylene glycol (EG) at 120°C for 8 hours, followed by cleaning and dispersion in EG using ultrasonication to obtain hydroxyl groups treated MWCNTs solution with a volume fraction of 0.07%. Finally, NFs, including carboxyl-functionalized graphene−MWCNTs hybrid structures with different dosages, were synthesized by dispersing the carboxyl-functionalized graphene and hydroxyl groups treated MWCNTs solutions with a volume fraction ratio of 1:1 for carboxyl-functionalized graphene and MWCNTs. Higher dosages were not synthesized to restrict the effect of aggregation on TC [124]. The TC and enhancement of ethylene glycol-based NFs, including hydroxyl groups treated MWCNTs, carboxyl-functionalized graphene, and graphene−CNTs with the same dosage of 0.07 vol.% are depicted in Fig. 7.8.

The TC of as-obtained NFs elevates with the temperature, due to the increment of Brown motion and the decline of nanoadditive aggregation in NFs. The TC enhancement (%K) of NFs was determined as $\%K = 100\% \times (K_{nf} - K_{bf})/K_{bf}$, with K_{bf} and K_{nf} being the TC of ethylene glycol solvent and NFs, respectively. At 30°C, the $\%K$ values were 18%, 10%, and 5.2% for NFs containing graphene−CNTs, carboxyl-functionalized graphene, and hydroxyl groups treated MWCNTs, respectively. The $\%K$ of Gr−CNT hybrid NF was nearly 3.5 and 2 times higher than that of CNTs NF and graphene NF, respectively. The findings revealed that adding hybrid materials can improve the TC of NFs better than monophasic structures. As the temperature rose to 50°C, the increase in TC was

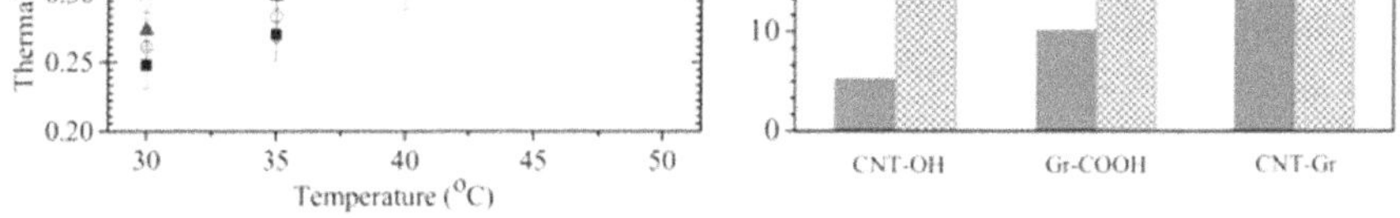

FIGURE 7.8

(A) Thermal conductivity and (B) thermal conductivity enhancement of EG-based nanofluid containing carbon nanotube (CNT)−OH, Gr−COOH, and Gr−CNT.

Reproduced with permissions from P. Van Trinh, N.N. Anh, N.T. Hong, P.N. Hong, P.N. Minh, B.H. Thang, Experimental study on the thermal conductivity of ethylene glycol-based nanofluid containing Gr-CNT hybrid material, J. Mol. Liq. 269 (2018) 344–353.

determined to be 50%, 37%, and 22% for NFs containing graphene–CNTs, graphene, and CNTs, respectively [124].

The solar exfoliation method was utilized to prepare graphene from graphite oxide. The initial step involved obtaining graphite oxide through Hummer's method from Bay carbon graphite. The resulting graphite oxide was then subjected to intense solar radiation using a 90 mm diameter convex lens. To make the graphene suitable for dispersion in polar base fluids, it was treated with concentrated nitric acid through refluxing. MWNTs were also prepared to utilize catalytic CVD technique and then purified and functionalized through air oxidation and refluxing in concentrated nitric acid. The functionalization process for MWNT was similar to that of graphene. A composite of graphite oxide-f-MWNT (GO-f-MWNT) was obtained by refluxing a mixture of equal amounts of graphite oxide and functionalized MWNT in concentrated nitric acid. This composite was then subjected to solar exfoliation, resulting in sG-f-MWNT. These products were used to synthesize NFs using a two-step route, where f-sG and sG-f-MWNT were dispersed separately in de-ionized water and ethylene glycol base fluids through ultrasonicating. The TC of the prepared graphene–MWNT composite was found to be improved. This improvement can be related to the combination of the high TC of both one-dimensional MWCNTs and two-dimensional graphene. Previous studies have explained the significant increase in TC of CNTs using the percolation model. Based on this model, CNTs form interconnected networks that act as conductive paths. The elongated shape and high aspect ratio of nanotubes are ideal for developing such a network. As the percolation path increases, the effective TC also increases. Therefore, CNTs with a higher length/dimension ratio would lead to a longer percolation path and higher enhancement in TC. In a similar vein, a combination of graphene networks and high-aspect-ratio MWCNTs can function as superior conducting pathways for heat conduction, improving thermal transfer. The following explains why graphene-f-MWNT NFs perform better than pure graphene NFs: the cohesive energy (van der Waals interaction) between layers in as-grown graphene is highly hydrophobic, and it grows dramatically with increasing sheet size. The instability of graphene NFs may be adversely affected by this rise, which frequently results in irreversible aggregation without electrostatic or steric protection. In order to solve this problem, concentrated acid is used to functionalize graphene, making it soluble in polar base fluids. Nevertheless, the inherent high heat conductivity of pure graphene might be diminished by the addition of insulating functional groups. On the other hand, adding MWCNTs can increase the spacing between graphene sheets, effectively preventing their agglomeration and improving stability [125]. In other work, graphene-wrapped MWCNTs were synthesized using a CVD technique via S. S. Jyothirmayee Aravind and S. Ramaprabhu. The precursors used were graphene oxide (GO) and a misch metal-based alloy hydride catalyst called $MmNi_3$. The synthesis process involved grinding a mixture of $MmNi_3$ and GO in a 1:1 ratio in a mortar, which was then loaded into the center of a quartz tubular reactor. The sample was heated to 500°C in an argon atmosphere for approximately

30 minutes before introducing hydrogen gas. Afterward, the temperature was increased to 700°C, and acetylene gas (C_2H_2) was introduced as the carbon source. The obtained samples were treated with n concentrated nitric acid at 60°C for 6 hours to eliminate the $MmNi_3$ particles. The nitric acid-treated products were then washed with deionized water to remove impurities and separated. Finally, the purified graphene-wrapped MWCNTs, named graphene–MWCNTs, were dried at 60°C under vacuum conditions. Results showed the relationship between TC of graphene–MWCNTs NFs and temperature and volume fraction in deionized (DI) water and ethylene glycol. The highest temperature used for measuring TC was 50°C, and it is evident that the absolute TC of graphene–MWCNTs NFs increases with temperature. This trend is similar to what has been observed in polar liquids-based NFs, where the increased Brownian motion of graphene–MWCNTs at higher temperatures contributes to the increment in TC. At 25°C and 50°C, a low volume fraction of 0.04% in DI water-based graphene–MWCNTs NFs resulted in an increase in TC by 11.3% and 97.5%, respectively. Similarly, EG-based NFs exhibited enhancements of 13.7% and 24% in TC at a volume amount of 0.04% nanostructures at 25°C and 50°C, respectively. Previous studies on NF TC have consistently shown that the enhancement becomes more significant with increasing nanoparticle loading, and graphene–MWCNTs NFs follow this pattern as well. Notably, substantial improvements in TC are seen for NFs with more volume fractions of graphene–MWCNTs. The nonlinear trend of TC can be interpreted similarly to that observed in CNTs NFs because graphene-wrapped MWCNTs also possess a one-dimensional nature. This nonlinearity can be attributed to the nonlinear increase in heat transfer distance within graphene–MWCNTs as the number of effective contact points or amount of particles increases. This purified sample, without any further acid treatment or functionalization, was used for the synthesis of NFs [126].

7.5 Conclusion and outlook

CNTs and graphene-based NFs have gained considerable attention in the field of nanotechnology for their excellent properties and potential applications in various industries. Both CNTs and graphene are carbon-based materials, but they possess distinct structures and properties. The incorporation of graphene/CNT nanocomposites into NFs opens up new avenues for various applications. These materials offer improved TC, enhanced energy storage capabilities, superior mechanical properties, and unique electrical conductivity and transparency. Optimizing the synthesis, stability, and dispersion of graphene/CNT nanocomposites in NFs to harness their full potential are key factors in graphene/CNT-based NFs. One of the key areas where graphene/CNT nanocomposites in NFs are being explored is in thermal management. The exceptional TC of graphene and CNTs can be leveraged in NFs to improve heat transfer efficiency. This enhanced heat transfer performance makes these materials suitable for applications such as advanced cooling systems, heat exchangers, and thermal energy storage. Also, the graphene/CNT-based NFs can be applied in energy storage devices, sensors and flexible electronics, supercapacitors, and batteries.

References

[1] S. Chakraborty, P.K. Panigrahi, Stability of nanofluid: a review, Appl. Therm. Eng. 174 (2020) 115259.

[2] Y. Li, S. Tung, E. Schneider, S. Xi, A review on development of nanofluid preparation and characterization, Powder Technol. 196 (2009) 89–101.

[3] M.M. Selim, S. El-Safty, A. Tounsi, M. Shenashen, Review of the impact of the external magnetic field on the characteristics of magnetic nanofluids, Alex. Eng. J. 76 (2023) 75–89.

[4] Y. Chen, Y. Zhang, H. Lan, C. Li, X. Liu, D. He, Electric field combined nanofluid to enhance photothermal efficiency of the direct absorption solar collector, Renew. Energy 215 (2023) 118988.

[5] R. Kiruba, A. Kingson Solomon Jeevaraj, Rheological characteristics and thermal studies of EG based Cu:ZnO hybrid nanofluids for enhanced heat transfer efficiency, Chem. Phys. Impact (2023) 100278.

[6] C. Liu, L. Geng, T. Xiao, Q. Liu, S. Zhang, H.M. Ali, et al., Recent advances of plasmonic nanofluids in solar harvesting and energy storage, J. Energy Storage 72 (2023) 108329.

[7] Z. Said, M.A. Sohail, A.K. Pandey, P. Sharma, A. Waqas, W.-H. Chen, et al., Nanotechnology-integrated phase change material and nanofluids for solar applications as a potential approach for clean energy strategies: progress, challenges, and opportunities, J. Clean. Prod. 416 (2023) 137736.

[8] S. Liu, S. Ma, Y. Liu, Y. Wang, Analysis of the energy conversion properties and applications of nanofluids: a review, Energy Rep. 8 (2022) 175–184.

[9] W. Chen, J. Zhou, F. Zhang, X. Li, J. Guo, Stability study of low-cost carbon quantum dots nanofluids with saline water and their application investigation for the performance improvement of solar still, Diamond Relat. Mater. 138 (2023) 110194.

[10] M. Hemmat Esfe, A. Alirezaie, D. Toghraie, Thermal conductivity of ethylene glycol based nanofluids containing hybrid nanoparticles of SWCNT and Fe_3O_4 and its price-performance analysis for energy management, J. Mater. Res. Technol. 14 (2021) 1754–1760.

[11] Q. Xiong, S. Altnji, T. Tayebi, M. Izadi, A. Hajjar, B. Sundén, et al., A comprehensive review on the application of hybrid nanofluids in solar energy collectors, Sustain. Energy Technol. Assess. 47 (2021) 101341.

[12] M.H. Ahmadi, A. Mirlohi, M. Alhuyi Nazari, R. Ghasempour, A review of thermal conductivity of various nanofluids, J. Mol. Liq. 265 (2018) 181–188.

[13] Y. Zhang, X. Xu, Predicting the thermal conductivity enhancement of nanofluids using computational intelligence, Phys. Lett. A 384 (2020) 126500.

[14] T. Ambreen, M.-H. Kim, Influence of particle size on the effective thermal conductivity of nanofluids: a critical review, Appl. Energy 264 (2020) 114684.

[15] F. Esmaeilzadeh, A.S. Teja, A. Bakhtyari, The thermal conductivity, viscosity, and cloud points of bentonite nanofluids with *n*-pentadecane as the base fluid, J. Mol. Liq. 300 (2020) 112307.
[16] P.B. Maheshwary, C.C. Handa, K.R. Nemade, S.R. Chaudhary, Role of nanoparticle shape in enhancing the thermal conductivity of nanofluids, Mater. Today: Proc. 28 (2020) 873–878.
[17] D.-H. Yoo, K. Hong, H.-S. Yang, Study of thermal conductivity of nanofluids for the application of heat transfer fluids, Thermochim. Acta 455 (2007) 66–69.
[18] M. Ghanbarpour, E.B. Haghigi, R. Khodabandeh, Thermal properties and rheological behavior of water based Al_2O_3 nanofluid as a heat transfer fluid, Exp. Therm. Fluid Sci. 53 (2014) 227–235.
[19] J. Hulla, S. Sahu, A. Hayes, Nanotechnology: history and future, Hum. Exp. Toxicol. 34 (2015) 1318–1321.
[20] C.P. Poole, F.J. Owens, Introduction to Nanotechnology, Wiley-Interscience, New York, 2003, pp. 145–150.
[21] L.S. Sundar, Synthesis and characterization of hybrid nanofluids and their usage in different heat exchangers for an improved heat transfer rates: a critical review, Eng. Sci. Technol. 44 (2023) 101468.
[22] M. Sheikholeslami, D.D. Ganji, Chapter 1 – Nanofluid: definition and applications, in: M. Sheikholeslami, D.D. Ganji (Eds.), Applications of Nanofluid for Heat Transfer Enhancement, William Andrew Publishing, 2017, pp. 1–52.
[23] S.U.-S. Choi, Nanofluid Technology: Current Status and Future Research, Argonne National Lab. (ANL), Argonne, IL, 1998.
[24] A. Tarafdar, R. Sirohi, T. Negi, S. Singh, P.C. Badgujar, N. Chandra Shahi, et al., Nanofluid research advances: preparation, characteristics and applications in food processing, Food Res. Int. 150 (2021) 110751.
[25] A.G.N. Sofiah, M. Samykano, A.K. Pandey, K. Kadirgama, K. Sharma, R. Saidur, Immense impact from small particles: review on stability and thermophysical properties of nanofluids, Sustain. Energy Technol. Assess. 48 (2021) 101635.
[26] L. Xing, Y. Ha, R. Wang, Z. Li, Recent advances of solar thermal conversion with wide absorption spectrum based on plasmonic nanofluids, Sol Energy 262 (2023) 111858.
[27] İ. Özcan, A. Ozsoy, A.E. Özgür, M. Karaboyaci, An investigation of serial connected U-pipe evacuated tube solar collector performance using TiO_2/water nanofluid, Appl. Therm. Eng. 233 (2023) 121088.
[28] R. Zhou, M. Wang, Z. Jin, S. Li, Modeling and experimental verification of the enhancement of TiO_2 nanofluid on ammonia falling film absorption process, Int. J. Therm. Sci. 184 (2023) 107917.
[29] M. Hemmat Esfe, S. Alidoust, D. Toghraie, Comparison of thermal conductivity of water-based nanofluids with various combinations of MWCNT, CuO, and SiO_2 nanoparticles for using in heating systems, Case Stud. Therm. Eng. 42 (2023) 102683.
[30] C. Dasa Manikanta, R. Saravanan, Comparison of SiO_2 nanofluid and conventional coolant for flood cooling in CNC turning of INCONEL 800 to improve surface finish and machining rate, Mater. Today: Proc. 69 (2022) 1048–1054.
[31] T.K. Murtadha, Effect of using Al_2O_3/TiO_2 hybrid nanofluids on improving the photovoltaic performance, Case Stud. Therm. Eng. 47 (2023) 103112.

[32] Y.-P. Hu, F.-J. Wang, Y.-C. Zhang, Y.-R. Li, M.-H. Li, Oscillatory natural convection of Al_2O_3-water nanofluid near its density maximum in a narrow horizontal annulus, Int. Commun. Heat. Mass. Transf. 136 (2022) 106207.

[33] G. Zhang, H. Chen, G. Xiao, M. Yi, Z. Chen, J. Zhang, et al., Effect of SiC nanofluid minimum quantity lubrication on the performance of the ceramic tool in cutting hardened steel, J. Manuf. Process. 84 (2022) 539–554.

[34] Q. Liu, Y. Tao, L. Shi, T. Zhou, Y. Huang, Y. Peng, et al., Experimental investigation on the use of CuO/water nanofluid in horizontal spiral-coil ground heat exchanger, Int. J. Refrig. 149 (2023) 204–223.

[35] D. Shah, S. Mukherjee, A. Shrivastava, P. Chaudhuri, R. Patel, S. Rimza, Preparation, stabilization, and thermal characterization of CuO/water nanofluids applicable to domestic solar water heater, Mater. Today: Proc. (2023).

[36] T. Dagdevir, V. Ozceyhan, A comprehensive second law analysis for a heat exchanger tube equipped with the rod inserted straight and twisted tape and using water/CuO nanofluid, Int. J. Therm. Sci. 181 (2022) 107765.

[37] M. Leena, S. Srinivasan, Effects of rare earth doped on thermal conductivity of ZnO-water nanofluid by ultrasonic velocity measurements, Mater. Lett. 219 (2018) 220–224.

[38] B. Yang, Y. Shi, X. Ma, X. Yu, Effects of mixed anionic/cationic surfactants on ZnO nanofluid, J. Mol. Liq. 377 (2023) 121530.

[39] R.B. Dehkordi, D. Toghraie, M. Hashemian, F. Aghadavoudi, M. Akbari, Molecular dynamics simulation of ferro-nanofluid flow in a microchannel in the presence of external electric field: effects of Fe_3O_4 nanoparticles, Int. Commun. Heat. Mass. Transf. 116 (2020) 104653.

[40] Y. Peng, M.B. Ghahnaviyeh, M.N. Ahmad, A. Abdollahi, S.A. Bagherzadeh, H. Azimy, et al., Analysis of the effect of roughness and concentration of Fe_3O_4/water nanofluid on the boiling heat transfer using the artificial neural network: An experimental and numerical study, Int. J. Therm. Sci. 163 (2021) 106863.

[41] G. Żyła, J. Fal, Experimental studies on viscosity, thermal and electrical conductivity of aluminum nitride–ethylene glycol (AlN–EG) nanofluids, Thermochim. Acta 637 (2016) 11–16.

[42] L. Dou, B. Ding, Q. Zhang, G. Kou, M. Mu, Numerical investigation on the thermal performance of parabolic trough solar collector with synthetic oil/Cu nanofluids, Appl. Therm. Eng. 227 (2023) 120376.

[43] H.D. Koca, S. Doganay, A. Turgut, Thermal characteristics and performance of Ag-water nanofluid: application to natural circulation loops, Energy Convers. Manage. 135 (2017) 9–20.

[44] L. Zhang, Y. Jing, P. Qu, W. Wang, X. Yao, Molecular dynamics simulations of the microscopic mechanism of thermal conductivity enhancement of ethylene Glycol–Water–Au nanofluids, Appl. Surf. Sci. 609 (2023) 155389.

[45] H. Loulijat, H. Moustabchir, Numerical study of the effects of Brownian motion and interfacial layer on the viscosity of nanofluid (Au-H_2O), J. Mol. Liq. 350 (2022) 118221.

[46] S. Sonawane, Investigation of turbulent heat transfer performance of aviation turbine fuel multi-wall carbon nanotube nanofluid, Adv. Powder Technol. 34 (2023) 104079.

[47] F. Selimefendigil, Mixed convection in a lid-driven cavity filled with single and multiple-walled carbon nanotubes nanofluid having an inner elliptic obstacle, Propuls. Power Res. 8 (2019) 128–137.

[48] W. Qiang, B. Wang, Q. Li, W. Wang, Molecular dynamics simulation of wetting and evaporation characteristics for sessile nanofluid nanodroplets, Chem. Phys. Lett. 695 (2018) 112–118.

[49] S.U. Ilyas, R. Shamsuddin, T. Kai Xiang, P. Estellé, R. Pendyala, Rheological profile of graphene-based nanofluids in thermal oil with hybrid additives of carbon nanotubes and nanofibers, J. Mol. Liq. 376 (2023) 121443.

[50] S. Farooq, D. Rativa, Z. Said, R.E.d. Araujo, High performance blended nanofluid based on gold nanorods chain for harvesting solar radiation, Appl. Therm. Eng. 218 (2023) 119212.

[51] L. Yang, X. Xu, W. Jiang, K. Du, A new thermal conductivity model for nanorod-based nanofluids, Appl. Therm. Eng. 114 (2017) 287–299.

[52] T. Hayat, S. Qayyum, M. Imtiaz, A. Alsaedi, Comparative study of silver and copper water nanofluids with mixed convection and nonlinear thermal radiation, Int. J. Heat. Mass. Transf. 102 (2016) 723–732.

[53] A. Roushan, U.S. Rao, P. Sahoo, K. Patra, S.S. Rawat, Wear behavior of AlTiN coated WC tools in micromilling of Ti_6Al_4V alloy using vegetable oil-based nanofluids, Tribol. Int. 188 (2023) 108825.

[54] S. Tiwari, M. Amarnath, M.K. Gupta, Synthesis, characterization, and application of Al_2O_3/coconut oil-based nanofluids in sustainable machining of AISI 1040 steel, J. Mol. Liq. 386 (2023) 122465.

[55] J. Fal, J. Sobczak, R. Stagraczyński, P. Estellé, G. Żyła, Electrical conductivity of titanium dioxide ethylene glycol-based nanofluids: impact of nanoparticles phase and concentration, Powder Technol. 404 (2022) 117423.

[56] N.S. Suhaimi, M.F.M. Din, M.T. Ishak, A.R.A. Rahman, J. Wang, M.Z. Hassan, Performance and limitation of mineral oil-based carbon nanotubes nanofluid in transformer application, Alex. Eng. J. 61 (2022) 9623–9635.

[57] Y. Yao, C. Ma, J. Wang, W. Qiao, L. Ling, D. Long, Rational design of high-surface-area carbon nanotube/microporous carbon core–shell nanocomposites for supercapacitor electrodes, ACS Appl. Mater. Interfaces 7 (2015) 4817–4825.

[58] S. Zhang, H. Wang, J. Liu, C. Bao, Measuring the specific surface area of monolayer graphene oxide in water, Mater. Lett. 261 (2020) 127098.

[59] S. Xin, N. Yang, F. Gao, J. Zhao, L. Li, C. Teng, Three-dimensional polypyrrole-derived carbon nanotube framework for dye adsorption and electrochemical supercapacitor, Appl. Surf. Sci. 414 (2017) 218–223.

[60] M. Pavía, K. Alajami, P. Estellé, A. Desforges, B. Vigolo, A critical review on thermal conductivity enhancement of graphene-based nanofluids, Adv. Colloid Interface Sci. 294 (2021) 102452.

[61] M. Ghalandari, A. Maleki, A. Haghighi, M. Safdari Shadloo, M. Alhuyi Nazari, I. Tlili, Applications of nanofluids containing carbon nanotubes in solar energy systems: a review, J. Mol. Liq. 313 (2020) 113476.

[62] S.M.S. Murshed, C.A. Nieto de Castro, Superior thermal features of carbon nanotubes-based nanofluids – a review, Renew. Sustain. Energy Rev. 37 (2014) 155–167.

[63] H. Lin, Q. Jian, X. Bai, D. Li, Z. Huang, W. Huang, et al., Recent advances in thermal conductivity and thermal applications of graphene and its derivatives nanofluids, Appl. Therm. Eng. 218 (2023) 119176.

[64] Z. Li, A. Kan, K. Wang, Y. He, N. Zheng, W. Yu, Optical properties and photothermal conversion performances of graphene based nanofluids, Appl. Therm. Eng. 203 (2022) 117948.

[65] M. Dresselhaus, G. Dresselhaus, R. Saito, Physics of carbon nanotubes, Carbon 33 (1995) 883–891.

[66] M.F. De Volder, S.H. Tawfick, R.H. Baughman, A.J. Hart, Carbon nanotubes: present and future commercial applications, Science 339 (2013) 535–539.

[67] A. Eatemadi, H. Daraee, H. Karimkhanloo, M. Kouhi, N. Zarghami, A. Akbarzadeh, et al., Carbon nanotubes: properties, synthesis, purification, and medical applications, Nanoscale Res. Lett. 9 (2014) 1–13.

[68] N. Gupta, S.M. Gupta, S. Sharma, Carbon nanotubes: synthesis, properties and engineering applications, Carbon Lett. 29 (2019) 419–447.

[69] M.N. Norizan, M.H. Moklis, S.Z.N. Demon, N.A. Halim, A. Samsuri, I.S. Mohamad, et al., Carbon nanotubes: functionalisation and their application in chemical sensors, RSC Adv. 10 (2020) 43704–43732.

[70] H. Wan, Y. Cao, L.-W. Lo, J. Zhao, N. Sepulveda, C. Wang, Flexible carbon nanotube synaptic transistor for neurological electronic skin applications, ACS Nano 14 (2020) 10402–10412.

[71] T. Ji, Y. Feng, M. Qin, W. Feng, Thermal conducting properties of aligned carbon nanotubes and their polymer composites, Compos. A: Appl. Sci. Manuf. 91 (2016) 351–369.

[72] M. Wiśniewska, A.M. Laptev, M. Marczewski, V. Leshchynsky, G. Lota, I. Acznik, et al., Influence of carbon nanotubes on thermal and electrical conductivity of zirconia-based composite, Ceram. Int. 49 (2023) 15442–15450.

[73] V. Kumar, J.K. Madhukesh, A.M. Jyothi, B.C. Prasannakumara, M. Ijaz Khan, Y.-M. Chu, Analysis of single and multi-wall carbon nanotubes (SWCNT/MWCNT) in the flow of Maxwell nanofluid with the impact of magnetic dipole, Computational Theor. Chem. 1200 (2021) 113223.

[74] A. Abidi, Z. Jokar, S. Allahyari, F. Kolahi Sadigh, S. Mohammad Sajadi, P. Firouzi, et al., Improve thermal performance of simulated-body-fluid as a solution with an ion concentration close to human blood plasma, by additive zinc oxide and its composites: ZnO/carbon nanotube and ZnO/hydroxyapatite, J. Mol. Liq. 342 (2021) 117457.

[75] M. Shokrieh, R. Rafiee, A review of the mechanical properties of isolated carbon nanotubes and carbon nanotube composites, Mech. Compos. Mater. 46 (2010) 155–172.

[76] M.T. Hassan, A.M. Esawi, S. Metwalli, Effect of carbon nanotube damage on the mechanical properties of aluminium–carbon nanotube composites, J. Alloys Compd. 607 (2014) 215–222.

[77] R. Dubey, D. Dutta, A. Sarkar, P. Chattopadhyay, Functionalized carbon nanotubes: synthesis, properties and applications in water purification, drug delivery, and material and biomedical sciences, Nanoscale Adv. 3 (2021) 5722–5744.

[78] M.H. Davood Abadi Farahani, V. Vatanpour, Chapter 4 – Polymer/carbon nanotubes mixed matrix membranes for water purification, in: S. Thomas, D. Pasquini, S.-Y. Leu, D.A. Gopakumar (Eds.), Nanoscale Materials in Water Purification, Elsevier, 2019, pp. 87–110.

[79] P. Yadav, S.M. Gupta, S.K. Sharma, A review on stabilization of carbon nanotube nanofluid, J. Therm. Anal. Calorim. 147 (2022) 6537–6561.

[80] V.Y. Rudyak, A.V. Minakov, M.I. Pryazhnikov, Preparation, characterization, and viscosity studding the single-walled carbon nanotube nanofluids, J. Mol. Liq. 329 (2021) 115517.

[81] Z. Lyu, F. Pourfattah, A.A.A. Arani, A. Asadi, L.K. Foong, On the thermal performance of a fractal microchannel subjected to water and kerosene carbon nanotube nanofluid, Sci. Rep. 10 (2020) 7243.
[82] A. Dillon, Carbon nanotubes for photoconversion and electrical energy storage, Chem. Rev. 110 (2010) 6856–6872.
[83] L. Sun, X. Wang, Y. Wang, Q. Zhang, Roles of carbon nanotubes in novel energy storage devices, Carbon 122 (2017) 462–474.
[84] P. Yadav, S.M. Gupta, S. Sharma, A review on stabilization of carbon nanotube nanofluid, J. Therm. Anal. Calorim. (2022) 1–25.
[85] W.U. Rehman, Z.M.A. Merican, A.H. Bhat, B.G. Hoe, A.A. Sulaimon, O. Akbarzadeh, et al., Synthesis, characterization, stability and thermal conductivity of multi-walled carbon nanotubes (MWCNTs) and eco-friendly jatropha seed oil based nanofluid: an experimental investigation and modeling approach, J. Mol. Liq. 293 (2019) 111534.
[86] Z. Han, A. Fina, Thermal conductivity of carbon nanotubes and their polymer nanocomposites: a review, Prog. Polym. Sci. 36 (2011) 914–944.
[87] K. Chu, H. Guo, C. Jia, F. Yin, X. Zhang, X. Liang, et al., Thermal properties of carbon nanotube–copper composites for thermal management applications, Nanoscale Res. Lett. 5 (2010) 868–874.
[88] F. Yan, L. Liu, M. Li, M. Zhang, L. Shang, L. Xiao, et al., One-step electrodeposition of Cu/CNT/CF multiscale reinforcement with substantially improved thermal/electrical conductivity and interfacial properties of epoxy composites, Compos. A: Appl. Sci. Manuf. 125 (2019) 105530.
[89] L.M. Hoyos-Palacio, D.P.C. Castro, I.C. Ortiz-Trujillo, L.E.B. Palacio, B.J.G. Upegui, N.J.E. Mora, et al., Compounds of carbon nanotubes decorated with silver nanoparticles via in-situ by chemical vapor deposition (CVD), J. Mater. Res. Technol. 8 (2019) 5893–5898.
[90] D.K. Lee, J. Yoo, H. Kim, B.H. Kang, S.H. Park, Electrical and thermal properties of carbon nanotube polymer composites with various aspect ratios, Materials (Basel) 15 (2022) 1356.
[91] C. Liang, X. Yao, M. Liu, J. Du, Y. Hong, L. Zhao, Comparisons of thermal performances in a pulsating heat pipe by using a nanofluid and a self-rewetting nanofluid with carbon nanotubes, Case Stud. Therm. Eng. 41 (2023) 102589.
[92] M.M. Sarafraz, F. Hormozi, Experimental investigation on the pool boiling heat transfer to aqueous multi-walled carbon nanotube nanofluids on the micro-finned surfaces, Int. J. Therm. Sci. 100 (2016) 255–266.
[93] M. Liu, M.C. Lin, C. Wang, Enhancements of thermal conductivities with Cu, CuO, and carbon nanotube nanofluids and application of MWNT/water nanofluid on a water chiller system, Nanoscale Res. Lett. 6 (2011) 297.
[94] A. Nag, S.C. Mukhopadhyay, Fabrication and implementation of carbon nanotubes for piezoresistive-sensing applications: a review, J. Sci.: Adv. Mater. Devices 7 (2022) 100416.
[95] B. Lin, X. Chen, J. Chen, S. Chen, R. Lu, S. Liang, et al., Facile synthesis of homogeneously dispersed carbon nanotubes on TC4 alloy powder by in-situ CVD and its growth behavior, J. Mater. Res. Technol. 24 (2023) 9928–9938.
[96] I.V. Novikov, E.M. Khabushev, D.V. Krasnikov, A.V. Bubis, A.E. Goldt, S.D. Shandakov, et al., Residence time effect on single-walled carbon nanotube synthesis in an aerosol CVD reactor, Chem. Eng. J. 420 (2021) 129869.

[97] G. Machado, C. Coelho, Vertically-aligned carbon nanotube at low pressure by cold-wall thermal CVD using a two-phase deposition step, Carbon Trends 5 (2021) 100087.

[98] H. Ribeiro, M.C. Schnitzler, W.M. da Silva, A.P. Santos, Purification of carbon nanotubes produced by the electric arc-discharge method, Surf. Interfaces 26 (2021) 101389.

[99] B. Sanjeevi, K. Loganathan, Synthesis of multi wall carbon nanotubes nanofluid by using two step method, Therm. Sci. 24 (2020) 519–524.

[100] X. Li, W. Chen, C. Zou, The stability, viscosity and thermal conductivity of carbon nanotubes nanofluids with high particle concentration: a surface modification approach, Powder Technol. 361 (2020) 957–967.

[101] W.T. Urmi, A. Shafiqah, M.M. Rahman, K. Kadirgama, M.A. Maleque, Preparation methods and challenges of hybrid nanofluids: a review, J. Adv. Res. Fluid Mech. Therm. Sci. 78 (2020) 56–66.

[102] K. Aglawe, R. Yadav, S. Thool, Preparation, applications and challenges of nanofluids in electronic cooling: a systematic review, Mater. Today: Proc. 43 (2021) 366–372.

[103] M. Li, L. Ren, Z. Gu, P. Gao, W. Sun, X. Dong, et al., Insight into the enhancement effect of amino functionalized carbon nanotubes on the H_2S removal performance of nanofluid system, J. Hazard. Mater. 458 (2023) 131977.

[104] A.N. Omrani, E. Esmaeilzadeh, M. Jafari, A. Behzadmehr, Effects of multi walled carbon nanotubes shape and size on thermal conductivity and viscosity of nanofluids, Diamond Relat. Mater. 93 (2019) 96–104.

[105] R. Ye, J.M. Tour, Graphene at fifteen, ACS Nano 13 (2019) 10872–10878.

[106] A. Razaq, F. Bibi, X. Zheng, R. Papadakis, S.H.M. Jafri, H. Li, Review on graphene-, graphene oxide-, reduced graphene oxide-based flexible composites: from fabrication to applications, Materials 15 (2022) 1012.

[107] M. Cao, D.B. Xiong, L. Yang, S. Li, Y. Xie, Q. Guo, et al., Ultrahigh electrical conductivity of graphene embedded in metals, Adv. Funct. Mater. 29 (2019) 1806792.

[108] M.G. Saborio, S. Cai, J. Tang, M.B. Ghasemian, M. Mayyas, J. Han, et al., Liquid metal droplet and graphene co-fillers for electrically conductive flexible composites, Small 16 (2020) 1903753.

[109] X. Huang, C. Zhi, Y. Lin, H. Bao, G. Wu, P. Jiang, et al., Thermal conductivity of graphene-based polymer nanocomposites, Mater. Sci. Eng.: R: Rep. 142 (2020) 100577.

[110] H. Lin, Q. Jian, X. Bai, D. Li, Z. Huang, W. Huang, et al., Recent advances in thermal conductivity and thermal applications of graphene and its derivatives nanofluids, Appl. Therm. Eng. (2022) 119176.

[111] W. Zheng, X. Zhao, W. Fu, Review of vertical graphene and its applications, ACS Appl. Mater. Interfaces 13 (2021) 9561–9579.

[112] P. Xu, Q. Gao, L. Ma, Z. Li, H. Zhang, H. Xiao, et al., A high surface area N-doped holey graphene aerogel with low charge transfer resistance as high performance electrode of non-flammable thermostable supercapacitors, Carbon 149 (2019) 452–461.

[113] G. Mittal, V. Dhand, K.Y. Rhee, S.-J. Park, W.R. Lee, A review on carbon nanotubes and graphene as fillers in reinforced polymer nanocomposites, J. Ind. Eng. Chem. 21 (2015) 11–25.

[114] E. Sadeghinezhad, M. Mehrali, R. Saidur, M. Mehrali, S. Tahan Latibari, A.R. Akhiani, et al., A comprehensive review on graphene nanofluids: recent research, development and applications, Energy Convers. Manage. 111 (2016) 466–487.
[115] H. Yu, Y. Lei, C. Pei, L. Wei, J.-H. Zhu, F. Xing, Enhancing the mechanical and functional performance of carbon fiber reinforced cement mortar by the inclusion of a cost-effective graphene nanofluid additive, Cem. Concr. Compos. 134 (2022) 104777.
[116] P. Sundaram, A. Kalaisselvane, A. Sathishkumar, P. GaneshKumar, S.C. Kim, R. Prabakaran, Synthesis, stability, and heat transfer behavior of water and graphene nanoplatelet-based nanofluid for cool thermal storage applications, J. Energy Storage 64 (2023) 107219.
[117] U. Legrand, N.Y. Mendoza Gonzalez, P. Pascone, J.L. Meunier, D. Berk, Synthesis and in-situ oxygen functionalization of deposited graphene nanoflakes for nanofluid generation, Carbon 102 (2016) 216–223.
[118] N. Sezer, S.A. Khan, M. Koç, Amelioration of the pool boiling heat transfer performance via self-assembling of 3D porous graphene/carbon nanotube hybrid film over the heating surface, Int. J. Heat Mass Transf. 145 (2019) 118732.
[119] J. Jyoti, T.K. Gupta, B.P. Singh, M. Sandhu, S.K. Tripathi, Recent advancement in three dimensional graphene-carbon nanotubes hybrid materials for energy storage and conversion applications, J. Energy Storage 50 (2022) 104235.
[120] X. Zang, Y. Jiang, M. Sanghadasa, L. Lin, Chemical vapor deposition of 3D graphene/carbon nanotubes networks for hybrid supercapacitors, Sens. Actuators, A: Phys. 304 (2020) 111886.
[121] K. Mohan, D. Narsimhaswamy, V. Ravi, Graphene/carbon-based materials for advanced energy conversion applications, Adv. Nat. Sci.: Nanosci. Nanotechnol. 13 (2022) 033005.
[122] K. Kaur, K. Jeet, Electrical conductivity of water-based nanofluids prepared with graphene–carbon nanotube hybrid, Fullerenes Nanotubes Carbon Nanostruct. 25 (2017) 726–734.
[123] Y.H. Chai, S. Yusup, V.S. Chok, S. Irawan, J. Singh, B. Chin, Comparison of rheological properties of graphene/carbon nanotube hydrogenated oil based biodegradable drilling fluid, in: IOP Conference Series: Materials Science and Engineering, IOP Publishing, 2017, pp. 012042.
[124] P. Van Trinh, N.N. Anh, N.T. Hong, P.N. Hong, P.N. Minh, B.H. Thang, Experimental study on the thermal conductivity of ethylene glycol-based nanofluid containing Gr-CNT hybrid material, J. Mol. Liq. 269 (2018) 344–353.
[125] S.S.J. Aravind, S. Ramaprabhu, Graphene–multiwalled carbon nanotube-based nanofluids for improved heat dissipation, RSC Adv. 3 (2013) 4199–4206.
[126] S. Jyothirmayee Aravind, S. Ramaprabhu, Graphene wrapped multiwalled carbon nanotubes dispersed nanofluids for heat transfer applications, J. Appl. Phys. 112 (2012).

CHAPTER

Fluid flow and heat transfer simulations of nanofluids in heat sinks

8

Aida Salimi[1], Mohammad Olfati[2] and Morteza Khoshvaght-Aliabadi[3]

[1]*Department of Mechanical Engineering, West Tehran Branch, Islamic Azad University, Tehran, Iran*

[2]*Mechanical Engineering Department, Faculty of Engineering, Razi University, Kermanshah, Iran*

[3]*Department of Chemical Engineering, Shahrood Branch, Islamic Azad University, Shahrud, Iran*

8.1 General introduction

Consumer electronics, aerospace, and automotive industries all rely on thermal management to ensure optimal operation, safety, and longevity of electronic devices, machinery, and systems. Its primary objective is to maximize performance, efficiency, and reliability while minimizing risks and environmental impacts. As a result of technological advancements, consumer demands, and the need for more portable and efficient devices, microelectronics has become increasingly miniaturized, resulting in higher component density, greater power density, faster switching speeds, and smaller form factors. As a result, heat flux has increased, making thermal management even more crucial to ensure reliable and optimal performance.

Thermal management involves both active and passive methods. A typical active technique requires external energy sources and often involves moving parts, while passive techniques operate without external energy sources, making them more reliable and cost-effective.

Heat sinks are a significant part of passive cooling devices. They consist of metal fins attached to hot components of microelectronic devices. The increased surface area of heat sinks allows heat to dissipate into the surrounding air via convection. In order to design effective heat sinks, factors such as heat dissipation requirements, device design, and specific applications must be considered. As a cooling fluid, air or liquid is commonly used in heat sinks.

Increasing the thermal properties of the operating fluids used in thermal management systems is one of the most efficient ways to enhance heat dissipation from these devices. The use of nanofluids offers a promising solution. Nanoparticle suspensions are made of metallic or nonmetallic nanoparticles suspended in a base

Nanofluids. DOI: https://doi.org/10.1016/B978-0-443-13625-2.00008-5

fluid, such as water or oil. Compared with traditional cooling fluids, nanofluids can offer several benefits and enhance cooling performance. Nevertheless, their use may also present some challenges, including nanoparticle aggregation, increased viscosity, and cost considerations. Nanofluids have tremendous potential for improving heat sink performance, but careful selection and formulation are required.

Thermal management is essential for optimizing the performance and longevity of electronic devices. Passive cooling techniques, especially heat sinks with nanofluids, are promising approaches for addressing the growing heat dissipation challenges in modern microelectronics. With continued research and development, nanofluids can revolutionize thermal management and pave the way for even more powerful and efficient electronic devices in the future.

8.2 Thermal management of electronic devices

In order to ensure the proper functioning, reliability, and longevity of electronic components and systems, thermal management is crucial. During operation, electronic devices generate heat due to the flow of electrical current through their components. A failure to manage this heat can result in component failure, malfunctions, or even permanent damage if it is not adequately controlled [1].

However, many electronic components, such as central processing units (CPUs) and graphics processing units (GPUs) in computers, have optimal operating temperatures. Too much heat can negatively affect the performance of these components. Keeping these components within their ideal temperature range ensures that they operate at peak performance. Furthermore, excessive heat can accelerate the aging and degradation of electronic components, causing them to fail more frequently. Device reliability is enhanced through proper thermal management, which reduces stress on components and prolongs their lifespan [2].

Moreover, in some devices, such as smartphones and laptops, the system may automatically reduce its performance when it reaches a certain temperature threshold. This process is called thermal throttling. With effective thermal management, thermal throttling can be minimized or eliminated, allowing the device to run optimally for extended periods. In high-speed electronic circuits, overheating can cause data errors or communication problems. Maintaining signal quality and enhancing system performance can be achieved by managing heat effectively. An increase in energy efficiency can be achieved by operating at a cooler temperature. Electronic components perform better and consume less power when kept within their ideal temperature range. However, fire or electrical hazards can arise from overheating electronic devices. Thermal management minimizes these risks and ensures the safety of users and the environment [3,4].

Due to effective thermal management, electronic devices can be more compact and densely packed, as they can handle heat better. A more space-efficient and innovative design is possible as a result.

As technology advances and innovation continues, electronic devices have become smaller. Integrated circuits, or microchips, allow multiple electronic components to be integrated into a single chip. It allowed electronic circuits to be drastically reduced in size and fit more functionality into smaller spaces. In addition, semiconductor manufacturing processes have improved, leading to smaller, more powerful transistors, which enable faster and more efficient electronic components. Additionally, nanoscale engineering has opened up new possibilities for creating smaller, more capable electronic components [5,6].

As electronic components, such as resistors, capacitors, and inductors, have become smaller over time, the size of devices has decreased overall. As electronic devices became smaller, their energy efficiency also improved, resulting in longer battery life and increased portability.

In microelectronic devices, heat flux is higher due to several factors, including miniaturization and increased complexity. Microelectronic devices generate more heat because they contain a large number of components, such as transistors, resistors, capacitors, and diodes, in a small space. High component density produces concentrated heat within a small space, resulting in a higher heat flux. Microelectronic devices often require higher power inputs as they become more powerful and efficient. Heat flux increases proportionally with higher power density, measured in watts per unit area. Microelectronic devices also operate at extremely high speeds, with transistors switching on and off very quickly. Due to faster switching, more energy is dissipated dynamically, causing more heat to be generated. Thermal mass is reduced as devices are smaller and substrates are thinner. Therefore, any heat generated by the device has less volume to dissipate, leading to a higher heat flux [7–9]. Therefore, microelectronics requires cooling methods to dissipate heat generated by electronic devices and to ensure the devices' longevity and proper functioning. The failure rate of electronic devices increases nearly exponentially with increasing operating temperatures, as shown in Fig. 8.1. Each 1°C increase in the temperature range of 70°C–80°C reduces the reliability of these devices by 5% [10–12].

It has become increasingly important to use effective cooling techniques as microelectronic devices have become more powerful and compact.

8.3 Cooling methods

Microelectronics commonly uses the following cooling methods [7,12,14–17]:

1. *Heat sinks*: Metal fins are attached to the hot components of the microelectronic device to provide passive cooling. Through increased surface area and convection, they promote heat dissipation into the surrounding environment [7,12].
2. *Fans*: Fans are often used to increase airflow over heat sinks or other heat-dissipating components when active cooling is being used. Using fans to

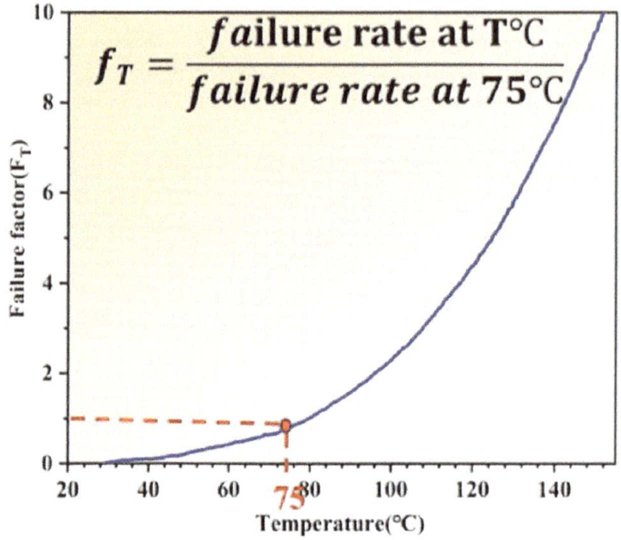

FIGURE 8.1

Relationship between failure rate of electronic devices and temperature [13].

Reprinted with permission from L. Du, W. Hu, An overview of heat transfer enhancement methods in microchannel heat sinks, Chem. Eng. Sci. 280 (2023) 119081.

force air over hot surfaces enhances the efficiency of convective heat transfer [7,14].

3. *Liquid cooling*: This involves circulating a liquid, such as water or a coolant, through pipes or channels in contact with the heat-generating components. As the heat is absorbed by the liquid, it is transferred away from the device. Especially for high-power microelectronic devices, liquid cooling can be more efficient than air cooling [14,15].
4. *Vapor chamber*: A vapor chamber is a two-dimensional flat heat pipe that efficiently spreads heat. It usually consists of a sealed container with a small amount of liquid inside. During heating, the liquid evaporates, carrying heat to cooler regions where it condenses and releases it [14,15].
5. *Phase change cooling*: A phase change cooling system absorbs heat from a cooling medium, such as refrigerants or other volatile compounds. Heat is transferred and cooled efficiently using this method [16].
6. *Thermoelectric cooling (Peltier effect)*: An electric current is used to transfer heat between two sides of a thermoelectric cooler. In high-power applications, these devices are less efficient because they are compact and can provide localized cooling [15].
7. *Heat pipes*: A heat pipe is a passive heat transfer device that transfers heat efficiently by utilizing a phase change in a working fluid. Heat is

commonly removed from high-power microelectronic components with these devices [14].

8. *Microfluidic cooling*: A microfluidic cooling system circulates a coolant through microchannels and micropumps. Individual components or small areas with high heat flux can be cooled with this approach [17].
9. *Synthetic jet cooling*: In synthetic jet cooling, periodic air jets are generated using electromechanical devices, improving heat transfer in restricted spaces. In particular, this method is useful for cooling microelectronic components that are tightly packed [12,15].
10. *Heat spreaders*: Heat spreaders are materials with high thermal conductivity that distribute heat evenly across a large surface area, preventing temperature gradients and promoting efficient heat dissipation [16].

Cooling methods are chosen based on a number of factors, including the power density, the form factor, and the intended use of the microelectronic device. Achieving reliable operation and preventing thermal-related issues in microelectronics requires effective thermal management. Among different methods of cooling in specific applications, heat sinks are often considered superior to other cooling methods due to several distinct advantages they offer, including efficient heat dissipation, passive cooling, compact size, cost-effectiveness, noise-free operation, no moving parts low energy consumption, reliability, and ease of installation [12,14,15].

8.3.1 Heat sinks

Heat sinks serve a vital role in electronics by effectively dissipating heat generated by components such as CPUs, GPUs, power transistors, and integrated circuits. Their primary function is to prevent overheating, a condition that can severely compromise the performance and longevity of these sensitive electronic elements. By efficiently redirecting and dispersing heat away from these components, heat sinks not only safeguard their operational integrity but also enhance the overall reliability and functionality of electronic devices.

At the core of heat sinks lies the fundamental principle of facilitating the efficient transfer of heat from a heated component to a substantially larger surface area. This essential process ensures that the heat generated by the component can be effectively dispersed into the surrounding environment. Such heat transfer plays a pivotal role in maintaining the component's temperature within safe operating limits, thereby not only preserving its operational integrity but also extending its overall lifespan. Heat sinks, as a result, emerge as indispensable thermal management solutions, critical for the optimal functioning and longevity of various electronic and mechanical systems [13].

8.3.1.1 Microchannel heat sinks

Microchannel heat sink structures represent a promising avenue for achieving highly efficient heat transfer. This groundbreaking concept originated with the pioneering work of Tuckerman and Pease at Stanford University in 1981. Their innovative idea revolved around the premise that the reduction of the hydraulic diameter of channels could lead to a substantial enhancement in heat transfer efficiency. Through meticulous experimentation, they achieved a remarkable milestone, demonstrating the capacity to attain a cooling rate of 790 W/cm^2 while maintaining an impressively low maximum thermal resistance (R_T) of merely 0.1°C/W. This seminal achievement not only revolutionized the field of thermal management but also laid the cornerstone for the widespread adoption of microchannel heat sinks, offering a potent solution to the formidable challenges posed by heat dissipation in diverse applications [18].

Heat is carried away by coolant flowing through a series of microchannels in a microchannel heat sink. Several advantages make them an ideal choice for cooling electronic products, including a large heat transfer area, compact structure, low coolant requirement, and low operating cost [19–21]. Microchannel heat sinks have been the subject of an increasing number of articles in the last 11 years, as illustrated in Fig. 8.2 [13].

Geometrically, the microchannel heat sink has micron-sized channels. Flowing in noncircular pipes and channels is strongly influenced by the hydraulic diameter. Research is often classified according to Kandlikar and Grande [22] and Mehendale et al. [23], with their specific nomenclature shown in Table 8.1.

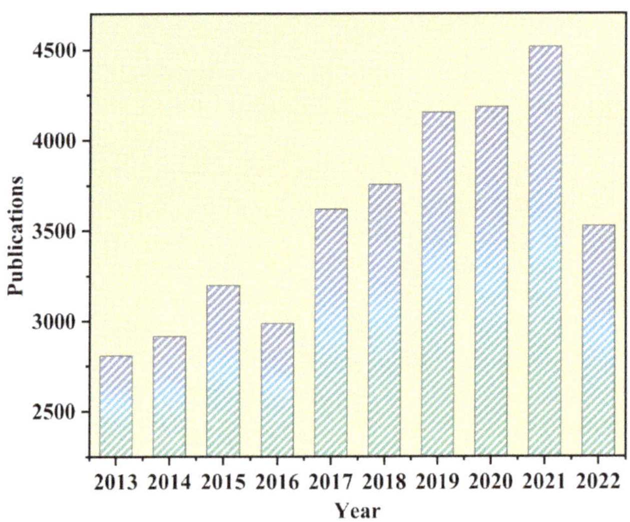

FIGURE 8.2

The research results of the number of papers on microchannels during 2013–23 come from the Web of Science database (as of May 2023) [13].

Reprinted with permission from L. Du, W. Hu, An overview of heat transfer enhancement methods in microchannel heat sinks, Chem. Eng. Sci. 280 (2023) 119081.

Table 8.1 Classification of channels by Mehendale, Kandlikar, and Grande [13].

Mehendale et al. [23]		**Kandlikar and Grande [22]**	
Conventional channels	$D_h > 6$ mm	Conventional channels	$D_h > 3$ mm
Compact passages	1 mm $< D_h \leq 6$ mm	Minichannels	200 μm $< D_h \leq 3$ mm
Meso-channels	100 μm $< D_h \leq 1$ mm	Microchannels	10 μm $< D_h \leq 200$ μm
Microchannels	1 μm $< D_h \leq 100$ μm	Transitional channels	0.1 μm $< D_h \leq 10$ μm
–	–	Molecular nanochannels	$D_h \leq 1$ μm

Reprinted with permission from L. Du, W. Hu, An overview of heat transfer enhancement methods in microchannel heat sinks, Chem. Eng. Sci. 280 (2023) 119081.

Channels are often misclassified due to confusion caused by "microchannel" and "mini-channel."

Solid microchannel heat sinks, such as rectangular, triangular, and trapezoidal, have been extensively studied, yet suffer from poor heat transfer performance, high wall superheats, and unstable two-phase flow. Enhancing microchannel heat sinks offers a solution to improve heat transfer performance to address these issues. There are three types of microchannel heat sinks: active, passive, and composite. In active cooling methods, external energy is used to enhance heat transfer, such as surface vibration, electrohydrodynamic, and magnetohydrodynamics. The passive method, on the other hand, relies on convection and flow disturbances occurring naturally, which can be achieved by altering or adding to the structure of microchannel heat sinks, such as inserts, surface modifications, extensions, or additives. It is easier to implement, more cost-effective, and more reliable for the industry to use passive methods. A composite method maximizes heat transfer by combining passive and active methods simultaneously.

In the past few decades, microchannels have been investigated for improving their heat transfer efficiency. Geometric modifications, jet impingement, nanofluidic, flow boiling, and magnetohydrodynamics have all contributed to this improvement [24–26].

8.3.1.1.1 Microchannel heat sink structure

Microchannel heat sinks' heat transfer performance is greatly influenced by their channel type. Different types of microchannels are used to improve the heat transfer performance of microchannel heat sinks by increasing the effective heat transfer area and reducing the convective thermal resistance.

1. *Straight microchannel heat sinks:* A straight microchannel heat sink dissipates heat from heat-generating components, particularly in electronic devices such

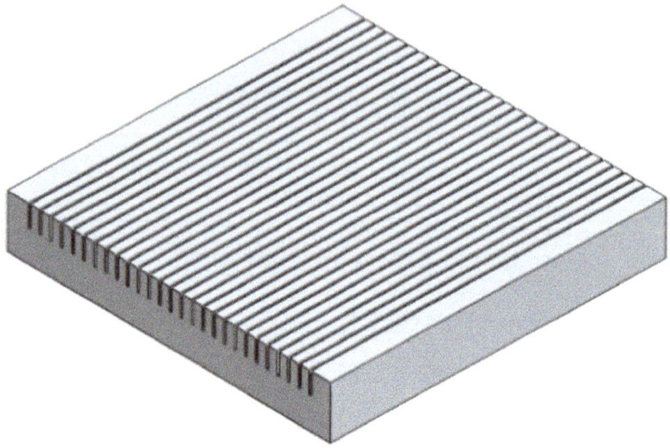

FIGURE 8.3

Schematic of straight microchannel heat sink [41].

Reprinted with permission from N.H. Naqiuddin, L.H. Saw, M.C. Yew, F. Yusof, T.C. Ng, M.K. Yew, Overview of micro-channel design for high heat flux application, Renew. Sustain. Energy Rev. 82 (2018) 901–914.

as CPUs, GPUs, and power electronics (Fig. 8.3). A cooling fluid (usually a liquid) flows through a network of straight, parallel microchannels in these heat sinks. The primary objective of straight microchannel heat sinks is to efficiently transfer heat from the source to the cooling fluid. Compared to more complex designs such as zigzag or serpentine channels, straight microchannel heat sinks are relatively straightforward. Due to this simplicity, manufacturing can be easier and costs can be reduced. Straight microchannels still enhance convective heat transfer by increasing the contact area between the fluid and the heat sink walls, even though they are not as intricate as some other designs. This results in effective cooling and heat dissipation. Convoluted microchannels tend to have higher pressure drops than straight ones. This can be beneficial whenever minimizing pressure drop is important. Fluid flows relatively uniformly across straight channels on heat sinks, helping to ensure even cooling over heat-generating components [27–30].

2. *Curved microchannel heat sinks*: Cooling solutions based on curved microchannels are used to remove heat from electronic components (Fig. 8.4). A cooling fluid (usually liquid) flows through a network of small, interconnected channels. Heat transfer efficiency can be enhanced by the curvature of these channels compared to traditional straight microchannel heat sinks. The curvature of the microchannel walls generates turbulence, increasing the amount of fluid that can contact the channel walls. This improves convective heat transfer compared to straight channels, resulting in better cooling. Due to curvatures, secondary flows such as Dean vortices can be produced, contributing to the mixing of cooling fluids. This mixing can

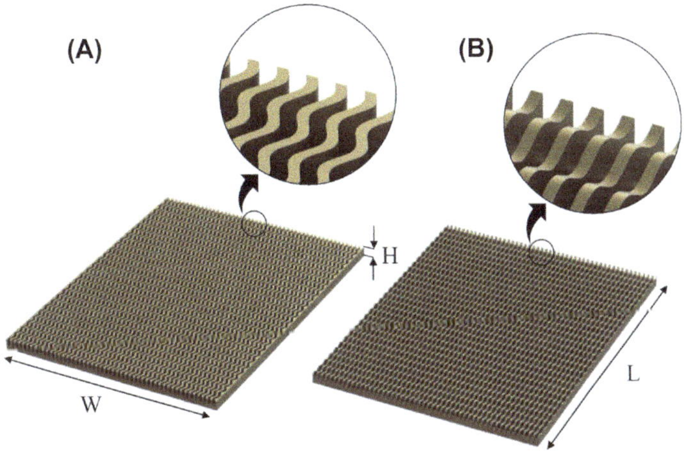

FIGURE 8.4

Schematic of two types of curved microchannel heat sinks (A) left–right wavy microchannel (B) up–down wavy microchannel [42].

Reprinted with permission from M. Jamshidmofid, M. Olfati, A. Arabkoohsar, A. Abbassi, Entropy generation and exergy destruction in two types of wavy microchannels working with various aqueous nanofluids using a multi-phase mixture model, J. Therm. Anal. Calorim. 148 (2023) 3711–3729.

reduce hot spots, distributing heat more evenly across the channel walls. However, due to increased resistance to fluid flow in curved microchannels, they can cause higher pressure drops than straight channels. The design must balance enhanced heat transfer with manageable pressure drops. In applications requiring curved microchannel heat sinks, where space is limited and efficient cooling is crucial, electronics use them extensively, such as CPUs and GPUs in computers, power electronics, and light emitting diode (LED) lighting systems. However, they may pose challenges in terms of fluid flow distribution and manufacturing complexity [31–33].

3. *Double-layer microchannel heat sinks*: Microchannel heat sinks with double layers, also referred to as multilayer microchannel heat sinks, provide advanced cooling solutions for electronics to dissipate heat more effectively (Fig. 8.5). These heat sinks stack multiple layers of microchannels on top of each other to improve heat transfer efficiency. When multiple layers of microchannels are stacked, the effective surface area for heat transfer is significantly increased compared to single-layer designs. A greater contact area between the cooling fluid and the heat sink walls enhances heat dissipation. Double-layer microchannel heat sinks can achieve a more uniform temperature distribution across the heat sink and the device that generates heat. Preventing hot spots that can degrade the performance and lifespan of electronic components is especially important. Increasing the number of layers

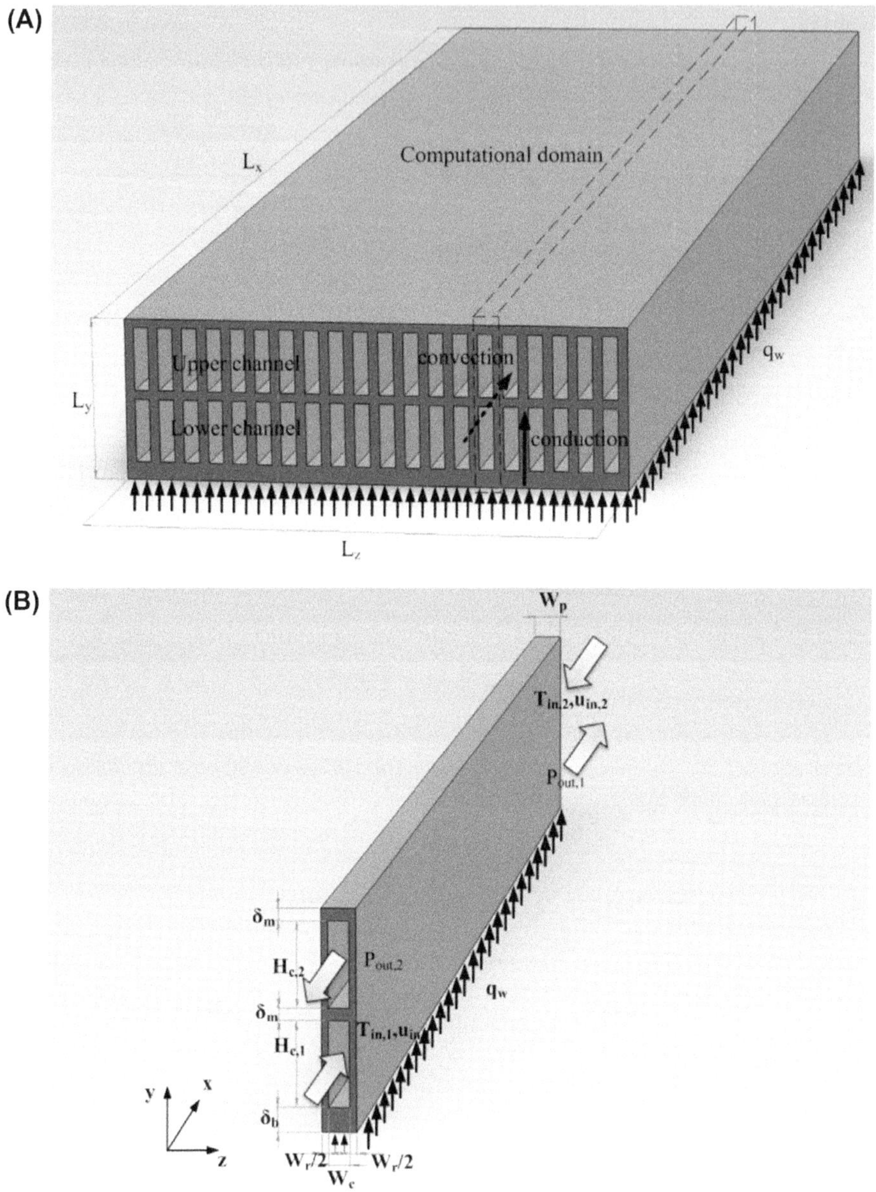

FIGURE 8.5

Schematics of a double-layered microchannel heat sink: (A) geometry and (B) computational domain [43].

Reprinted with permission from T.-C. Hung, W.-M. Yan, W.-P. Li, Analysis of heat transfer characteristics of double-layered microchannel heat sink, Int. J. Heat Mass Transfer 55 (2012) 3090–3099.

reduces the risk of concentrated heat accumulation and thermal runaway by providing additional paths for heat to spread and dissipate. Fabricating microchannel heat sinks with double layers can be more challenging than with single layers. Advanced manufacturing techniques such as microfabrication and micromachining are important for aligning and integrating multiple layers. These heat sinks are particularly useful in high-performance applications such as high-power electronics, data centers, and automotive electronics [34–36].

4. *Manifold microchannel heat sinks*: Manifold microchannel heat sinks, typically used in electronics and high-power devices, efficiently dissipate heat from heat-generating components (Fig. 8.6). These heat sinks incorporate a manifold structure that directs and distributes the cooling fluid to a network of microchannels in a controlled manner, unlike traditional single-layer microchannel heat sinks. This approach allows for optimized fluid distribution and enhanced heat transfer. The manifold structure acts as an inlet/outlet chamber that evenly distributes the cooling fluid to multiple microchannels. Maintaining consistent cooling performance and avoiding uneven heat distribution are particularly important, and the manifold design minimizes pressure variations across microchannels. Microchannel heat sinks contribute to a more uniform temperature distribution across the heat-generating component, preventing localized overheating and extending the device's lifespan. They are ideal for thermal management in complex systems such as high-performance computing, LED lighting, and automotive electronics [36–39].
5. *Zigzag-serpentine microchannel heat sinks*: Microchannel heat sinks with zigzag serpentine patterns dissipate heat from high-power electronic components, such as CPUs, GPUs, power amplifiers, and other heat-generating devices (Fig. 8.7). These heat sinks use a zigzag or serpentine pattern of interconnected microchannels to enhance heat transfer efficiency and provide effective cooling. The zigzag or serpentine pattern increases the heat sink's effective surface area. With this expanded surface area, the cooling fluid can better contact the heat sink walls, enhancing heat transfer. Fluid flow is made turbulent by the tortuous path of the zigzag or serpentine channels, promoting mixing and disrupting the formation of stagnant boundary layers, thereby enhancing convective heat transfer. Convoluted designs ensure consistent cooling of the entire heat-generating component, preventing localized hot spots. However, due to their convoluted path, convoluted microchannels may result in greater pressure drop than straight microchannels [40].

8.3.1.1.2 Internal reinforcement structures in microchannel heat sinks

Internal reinforcement structures within microchannel heat sinks play a pivotal role in elevating their overall performance, bolstering structural integrity, and optimizing fluid flow dynamics. These strategically designed structures are instrumental in addressing critical challenges such as pressure drop mitigation and uniform heat distribution. Commonly encountered types of internal reinforcement in

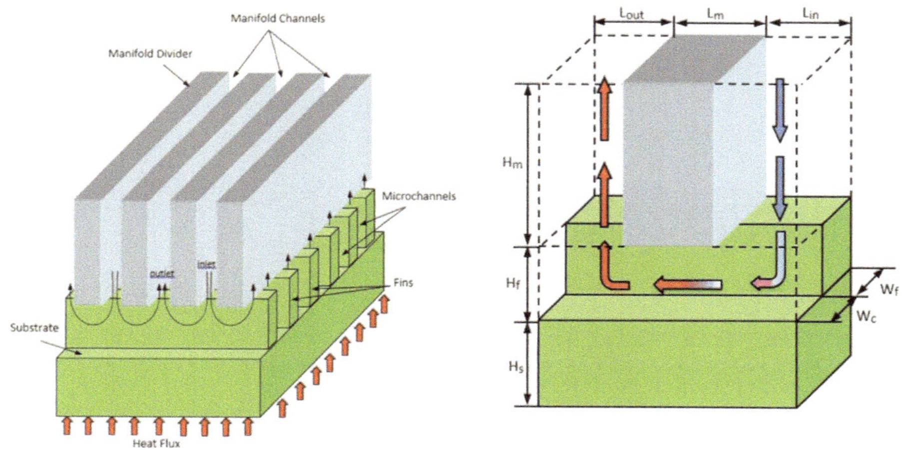

FIGURE 8.6

Schematic views of a manifold microchannel [44].

Reprinted with permission from C. Green, P. Kottke, X. Han, C. Woodrum, T. Sarvey, P. Asrar, X. Zhang, Y. Joshi, A. Fedorov, S. Sitaraman, A review of two-phase forced cooling in three-dimensional stacked electronics: technology integration, J. Electron. Packag. 137 (2015) 040802.

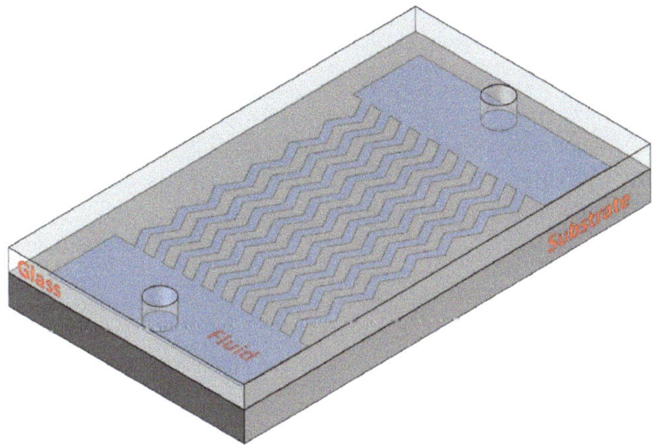

FIGURE 8.7

Schematic views of a zigzag-serpentine microchannel heat sink [32].

Reprinted with permission from F. Alnaimat, D. Varghese, B. Mathew, Investigation of thermal and hydraulic performance of MEMS heat sinks with zig-zag microchannels, Int. J. Thermofluids 16 (2022) 100213.

microchannel heat sinks encompass various configurations and designs, each tailored to specific thermal management objectives and application requirements [13,45]:

1. *Pin fins*: Pin fins are small protrusions or posts positioned within microchannels to increase the area available for heat transfer. Turbulence is enhanced in the fluid flow, reducing the boundary layer effect near the wall of the channel and improving convective heat transfer [46,47].
2. *Vortex generators*: These are small features that generate controlled swirls or vortices in fluid flows. By disrupting the thermal boundary layer, these vortices enhance mixing and improve heat transfer [48].
3. *Baffles and bifurcations*: A baffle is a partition or wall inserted within a microchannel to redirect or divide fluid flow. Flows can be split at bifurcations into multiple streams, improving heat distribution and reducing temperature variations [49].
4. *Micro-obstacles*: An obstacle is a small protrusion or groove added to the walls of a channel. Mixing and heat transfer are enhanced, and pressure drops are minimized as a result of their disturbances in the flow.
5. *Ribs and grooves*: Ribbed or grooved structures on the channel walls increase the heat transfer surface area and promote turbulent flow, leading to improved heat dissipation [50,51].
6. *Cross-connectors*: These structures link adjacent microchannels, allowing flow and temperature to be distributed more evenly [52].

7. *Cavities or reservoirs*: Cavities or reservoirs built into the microchannels can temporarily store and release coolant, helping mitigate sudden fluctuations in flow rate and temperature [53,54].
8. *Nanostructured surfaces*: Integrating nanostructured surfaces with unique patterns at the microchannel walls can enhance heat transfer due to increased surface area and improved fluid behavior at the nanoscale [55].
9. *Phase change enhancements*: Adding internal structures that facilitate phase change processes (such as boiling and evaporation) can significantly enhance heat transfer [56].

It is often necessary to design and refine these internal reinforcement structures based on numerical simulations and optimization techniques to achieve optimal performance. These internal reinforcement structures were designed to address the challenges associated with microchannel heat sinks, such as pressure drop, uneven heat distribution, and limited heat transfer rates. By enhancing fluid mixing, increasing surface area, and improving flow behavior, these structures improve thermal management solutions. Designing and implementing these systems require careful consideration of trade-offs between improved heat transfer and potential increases in pressure drop and manufacturing complexity.

As shown in Fig. 8.8, based on the microchannel heat sink structure, internal reinforcement structure, surface treatment, and material types, the research progress, experimental methods, and development trends are summarized, reviewed, and predicted [13].

8.3.1.1.3 Material used in microchannel heat sinks

A microchannel heat sink can be made from a variety of materials, but some of the most common are:

1. *Copper*: Due to its high thermal conductivity and good mechanical properties, copper is an excellent heat conductor and is widely used in microchannel heat sinks. The heat is transferred efficiently away from the source of heat [57].
2. *Aluminum*: Another popular material for microchannel heat sinks is aluminum. The material is not as thermally conductive as copper, but it is lighter and less expensive, making it a suitable choice for some applications [58].
3. *Stainless steel*: When corrosion resistance is crucial, stainless steel is used. Due to its durability, it is suitable for certain harsh environments, despite not being as thermally conductive as copper or aluminum [57].
4. *Nickel-based alloys*: Nickel-based alloys are used in specific applications where corrosion resistance and higher temperatures are required [57].
5. *Composite materials*: To meet specific performance requirements, some microchannel heat sinks are composed of composite materials such as metals, ceramics, or polymers [59,60].

Material selection is influenced by a number of factors, including thermal performance requirements, weight limitations, cost considerations, and the operating

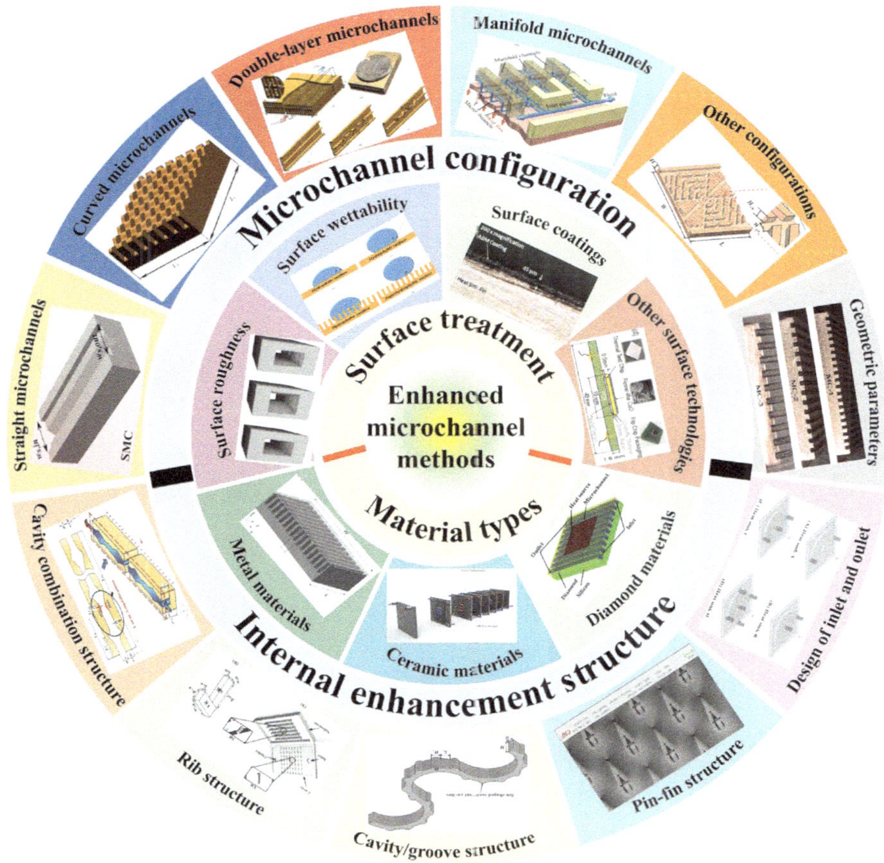

FIGURE 8.8

Summary of structural parameters of enhanced microchannel [13].

Reprinted with permission from L. Du, W. Hu, An overview of heat transfer enhancement methods in microchannel heat sinks, Chem. Eng. Sci. 280 (2023) 119081.

environment of the microchannel heat sink. There are advantages and limitations to each material, and each material is selected according to its specific application requirements.

8.3.1.1.4 Geometrical parameters of microchannel heat sinks

Heat transfer performance and overall effectiveness are highly dependent on the geometrical parameters of microchannel heat sinks. Here are a few key geometrical parameters:

1. *Channel dimensions*: Microchannel size and shape greatly influence heat transfer. A rectangular channel, a triangular channel, a trapezoidal channel, and a circular channel are all common channel shapes. Flow resistance and heat transfer surface area are impacted by channel width, depth, and aspect ratio [61,62].
2. *Channel spacing*: Fluid flow and heat transfer depend on the spacing between adjacent microchannels, known as pitch or interchannel spacing [63–65].
3. *Hydraulic diameter*: In fluid mechanics, hydraulic diameter is used to characterize the flow in noncircular channels. In determining fluid flow rate and heat transfer performance, it is calculated based on channel cross-sectional area and perimeter [31,42].
4. *Aspect ratio*: The aspect ratio describes the relationship between the channel length and hydraulic diameter of microchannel heat sinks. The performance of heat transfer is affected by pressure drop, flow distribution, and overall heat transfer efficiency [61,66].
5. *Fin structure (if applicable)*: Fin structures may be applied to the channel walls of microchannel heat sinks to enhance heat transfer. Several parameters affect the efficiency of heat transfer, including fin height, thickness, and spacing [42,67].
6. *Inlet and outlet geometry*: Flow uniformity and pressure drop can be affected by the design of inlet and outlet geometries [61,68].
7. *Heat sink size*: Microchannel heat sinks should have a size and shape that fits the specific application while dissipating sufficient heat [42].
8. *Material properties*: The material chosen for the microchannel heat sink will influence its thermal conductivity, mechanical strength, and compatibility with the cooling liquid [13].

These geometrical parameters must be optimized based on the specific cooling requirements, available space, and application type. Thermal management of electronic devices requires designing microchannel heat sinks with the right geometrical parameters.

8.3.1.1.5 Coolants used in microchannel heat sinks

The heat dissipated by microchannel heat sinks can be efficiently cooled with various coolants. There are a number of factors that influence the choice of coolant, including the application, temperature requirements, and others. Microchannel heat sinks commonly use the following coolants:

1. *Water*: The high heat capacity and excellent thermal conductivity of water make it a popular coolant in microchannel heat sinks. In addition to being efficient, it is also cost-effective. The electrical conductivity of this material can, however, be a problem for some applications [13].
2. *Air*: In microchannel heat sinks, air can be used as a coolant to dissipate low to moderate amounts of heat. Systems that require only modest cooling or cannot use water-based cooling often employ air cooling [13].
3. *Liquid coolants*: Besides water, microchannel heat sinks can be cooled with ethylene glycol and propylene glycol solutions. When higher cooling performance is needed, these coolants provide enhanced thermal properties [69,70].
4. *Refrigerants*: Microchannel heat sinks use coolants such as R134a or R410a in certain specialized applications. For specific thermal applications, these refrigerants provide efficient cooling [69,70].
5. *Dielectric fluids*: Coolants based on fluorocarbons are used where electrical insulation is crucial. In electronic cooling, it is essential to avoid electrical short circuits by using these nonconductive coolants [69,70].
6. *Phase change materials (PCM)*: PCM are substances that go through phases (from solid to liquid or vice versa) at specific temperatures. Thermal energy can be stored and released during phase change, making it useful for passive cooling in microchannel heat sinks [56].
7. *Nanofluids*: These fluids are composed of a base fluid (e.g., water and oil), as well as nanoscale particles (usually nanoparticles). As a result of adding nanoparticles to traditional heat transfer fluids, they can have improved thermal properties, such as increased thermal conductivity, improved heat transfer rates, and enhanced heat dissipation capabilities. Thus, nanofluids are an excellent choice for cooling electronics, such as heat sinks used in computers and power electronics [71].

Coolants are selected based on factors such as thermal conductivity, specific heat capacity, electrical conductivity, safety considerations, cost, and environmental impact. Choosing the right coolant for microchannel heat sinks is crucial based on their specific requirements and constraints.

8.4 Nanofluids

The invention of nanofluids can be perceived as a breakthrough in the heat transfer realm. In these advanced heat transfer fluids, nanoparticles on the nanoscale are blended with a conventional liquid coolant, such as water, oil, or ethylene glycol. A nanoparticle may be metallic or nonmetallic, such as copper, aluminum oxide, titanium dioxide, or carbon nanotubes.

The thermal and electrical properties of the nanofluid are significantly altered when nanoparticles are infused into the base fluid. In recent years, nanofluids have been attracting considerable attention as they exhibit superior heat transfer capabilities. The high thermal conductivity, specific heat capacity, and convection heat transfer coefficients of these materials make them highly effective in a variety of heat transfer applications [72].

There is increasing interest in nanofluids and they have been found to have a wide range of applications across a variety of industries. It is particularly useful for cooling electronic devices, optimizing heat exchanger performance, enhancing the efficiency of solar collectors, and enhancing the dissipation of heat in automotive engines.

Nanofluids present unique challenges when it comes to their production and stabilization. Before widespread adoption in a variety of industrial settings, researchers continue to examine their long-term stability and potential environmental impacts. Nanofluids must be harnessed to their full potential, but practical considerations and potential limitations must be taken into account [73].

8.4.1 Different types of nanofluids

Here are some common types of nanofluids based on the types of nanoparticles used:

1. *Metal-based nanofluids*: In these nanofluids, metal nanoparticles (e.g., copper, aluminum, silver, gold, nickel, and iron) are incorporated into a base fluid. As a result of their improved thermal and electrical conductivity, they are ideal for cooling electronic devices, heat exchangers, and conductive applications [73].
2. *Oxide-based nanofluids*: These nanofluids contain oxide nanoparticles (e.g., silica, alumina, titania, and zinc oxide) within a base fluid. Heat transfer and UV protection applications benefit from their enhanced thermal conductivity and UV absorption properties [73].
3. *Nitride-based nanofluids*: Nanoparticles made of nitride materials (e.g., silicon nitride, aluminum nitride, and titanium nitride) are dispersed in a base fluid. In addition to providing enhanced thermal conductivity, they also provide electrical insulation [74].
4. *Polymer-based nanofluids*: Using polymer-based nanoparticles or polymer composites in a base fluid is the most common method of making nanofluids. In biomedical applications and surface modification, they provide tunable rheological properties and controlled release [74].
5. *Hybrid nanofluids*: Hybrid nanofluids combine different types of nanoparticles for enhanced performance and properties [72].
6. *Magnetic nanofluids*: Also known as ferrofluids, these fluids contain magnetic nanoparticles dispersed in a nonmagnetic base fluid. Due to their ability to be

magnetized, they are suitable for a wide range of applications, including heat transfer and biomedicine [72].

7. *Graphene nanofluids*: These nanofluids combine graphene nanoparticles with a base fluid to achieve exceptional thermal and electrical properties. Heat transfer, advanced energy storage, electronics, and biomedical applications are among their applications.
8. *Carbon nanotube-based nanofluids*: These nanofluids are based on carbon nanotubes, which are known for their high thermal conductivity and electrical conductivity. Their application ranges from heat transfer to electronics to mechanical components to biomedical advancements.

The production and stabilization of nanofluids present unique challenges, and further research is required to address their long-term stability and potential environmental impacts before they are widely adopted in various industrial environments. Despite this, nanofluids have immense potential for revolutionizing heat transfer and various other applications in the future.

8.4.2 Preparation

Achieving stable nanoparticle dispersion within the base fluid is a critical aspect of nanofluid preparation. The uniform distribution of nanoparticles throughout the base fluid is essential for maintaining stability. To enhance stability, it is crucial to blend the nanoparticles thoroughly with the base fluid. In addition to mechanical mixing, surfactants are often employed to further enhance stability. Fig. 8.9 provides a visual representation of the dispersion of nanoparticles within the base fluid. The nanofluid's volume concentration can be calculated using Eqs. (8.1) and (8.2), serving as essential tools for quantifying the nanoparticle content within the nanofluid mixture [75,76].

For conventional nanofluids:

$$\Phi = \frac{\left(\frac{W_{np}}{\rho_{np}}\right)}{\left(\frac{W_{np}}{\rho_{np}}\right) + \left(\frac{W_{bf}}{\rho_{bf}}\right)} \tag{8.1}$$

For hybrid nanofluids:

$$\Phi = \frac{\left(\frac{W_{np1+np2}}{\rho_{np1+np2}}\right)}{\left(\frac{W_{np1+np2}}{\rho_{np1+np2}}\right) + \left(\frac{W_{bf}}{\rho_{bf}}\right)} \tag{8.2}$$

where ρ indicates the density, ϕ shows the volume concentration, and W is the weight (*bf*: base fluid and *np*: nanoparticles) [77,78].

There are two distinct approaches to nanofluid preparation—the one-step method and the two-step method. For nanofluids to perform efficiently and effectively, uniform and stable dispersion of nanoparticles is crucial. The two methods are as follows:

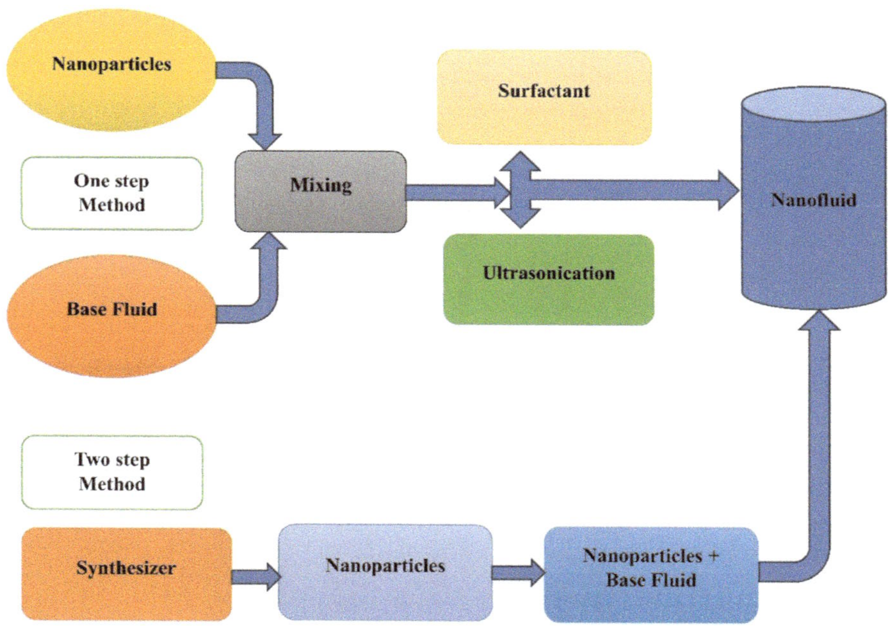

FIGURE 8.9

Methods of dispersion of nanoparticles in the base fluid [72].

Reprinted with permission from G. Sriharan, S. Harikrishnan, H.F. Oztop, A review on thermophysical properties, preparation, and heat transfer enhancement of conventional and hybrid nanofluids utilized in micro and mini channel heat sink, Sustain. Energy Technol. Assess. 58 (2023) 103327.

8.4.2.1 Single-step method

Nanoparticles are dispersed directly into the base fluid via a single mixing step in the one-step method. Nanofluids are typically created using single types of nanoparticles or when the nanoparticles have similar surface properties [79,80].

8.4.2.1.1 Step-by-step process

1. *Selection of base fluid and nanoparticles*: The optimal base fluid and nanoparticles are selected based on the nanofluid's properties and application.
2. *Surface treatment (if required)*: Surface treatment may be required for some nanoparticles to enhance their dispersion and stability.
3. *Mixing*: The nanoparticles are dispersed in the base fluid with appropriate techniques such as high-shear mixing or ultrasonication.
4. *Characterization and quality control*: To ensure proper dispersion and stability of the nanoparticles, the nanofluid is thoroughly characterized following mixing.

8.4.2.2 Two-step method

Alternatively, the two-step method involves preparing a stable nanoparticle suspension, or "nanoparticle fluid," which is then diluted with the base fluid to form the final nanofluid. When dealing with nanoparticles prone to agglomeration, or when preparing hybrid nanofluids, this approach is often preferred [81,82].

8.4.2.2.1 Step-by-step process

1. *Preparation of nanoparticle fluid*: Using high-shear mixing or ultrasonication, nanoparticles are dispersed into a carrier liquid (dispersion medium). In the final nanofluid, the carrier liquid differs from the base fluid.
2. *Dilution*: To achieve the desired nanoparticle concentration in the final nanofluid, the nanoparticle fluid is diluted with the desired base fluid.
3. *Characterization and quality control:* Once the nanofluid is diluted, it is meticulously characterized to ensure that the nanoparticles will disperse and remain stable.

When preparing nanofluids with multiple types of nanoparticles or when dealing with challenging nanoparticles, the two-step method provides superior control over nanoparticle dispersion and stability.

As a result, both the one-step and two-step methods serve specific purposes when preparing nanofluids. A variety of factors influence the selection of a method, including nanoparticle types, surface properties, and desired nanofluid properties.

8.4.3 Thermophysical characterization

1. *Thermal conductivity*: Nanofluids exhibit a significant increase in thermal conductivity over their base fluid counterparts. A nanofluid's thermal

conductivity is markedly boosted by nanoparticles, including copper, aluminum, alumina, titania, carbon nanotubes, and graphene. As part of the characterization process, experiments are typically conducted to determine the effective thermal conductivity, using methods such as the transient hot-wire method and the thermal probe method.

2. *Specific heat capacity*: In comparison to pure base fluids, nanofluids may exhibit different specific heat capacities. A material's specific heat capacity measures the amount of energy needed to raise its temperature by a certain amount. Specific heat capacity can increase or decrease depending on the type and concentration of nanoparticles. To accurately measure specific heat capacity, differential scanning calorimetry or heat pulse methods are used [83].
3. *Viscosity*: The addition of nanoparticles can alter the viscosity of nanofluids. In practical applications, viscosity changes are crucial to understanding flow behavior and pumping power requirements. In order to determine the viscosity of nanofluids, various viscometric methods are used, such as the Brookfield viscometer and rotational rheometry [84].
4. *Density*: Dispersion of nanoparticles in the base fluid can alter the density of the nanofluid. Pycnometers and density meters are used in the process of characterizing materials [85].
5. *Thermal diffusivity*: The ability of nanofluids to conduct heat relative to their ability to store it can also be altered by their thermal diffusivity. Laser flash experiments are used to determine thermal diffusivity [86].
6. *Surface tension*: Nanoparticles can alter nanofluid surface tension. Surface tension changes are measured using techniques such as the pendant drop method [73].
7. *Stability and sedimentation*: Achieving the long-term behavior of nanofluids requires understanding their stability, including the tendency of nanoparticles to agglomerate or settle. Nanoparticle dispersion can be evaluated by stability tests and sedimentation measurements [87].

In order to accurately characterize these thermophysical properties, systematic experiments encompassing different nanoparticle concentrations, sizes, and base fluid conditions must be conducted. To comprehend the underlying physics of nanofluids and predict their behavior under various scenarios, numerical simulations and theoretical models are utilized. Various engineering applications, such as electronics cooling, solar thermal systems, and heat exchangers, require extensive studies to optimize nanofluid properties and design highly efficient heat transfer systems [72,73].

8.4.4 Exploring potential mechanisms of nanofluids

The mechanisms behind nanofluids' unique behavior and enhanced properties require a deep understanding of the physical and chemical processes involved. As colloidal suspensions of nanoparticles in a base fluid, nanofluids affect thermophysical characteristics and heat transfer performance significantly. Understanding the intricacies of nanofluids is paramount to optimizing their applications [72,73].

1. *Brownian motion*: Thermal energy induces random particle movements in nanofluids through Brownian motion. It prevents undesirable agglomeration and sedimentation over time by enhancing nanoparticle dispersion and stability [74].
2. *Increased surface area*: Nanoparticles have a high surface area-to-volume ratio, which makes them more likely to interact with fluid molecules surrounding them. As a result, this enlarged interfacial area contributes to altered properties and unique heat transfer characteristics [72,73].
3. *Formation of interfacial layer*: Due to molecular interactions, a thin "interfacial layer" forms between nanoparticles and the base fluid. It affects properties such as thermal conductivity and viscosity of nanofluids and plays a pivotal role in influencing their behavior [88].
4. *Nanoparticle clustering*: When certain conditions are met, nanoparticles can form clusters or agglomerates. Nanoparticle concentration and temperature determine how much clustering occurs. During flow and heat transfer processes, these clusters can have a significant impact on the overall properties of the nanofluid [89].
5. *Synergistic effects*: It is noteworthy that some nanofluids exhibit synergistic effects, in which the combined properties of nanoparticles and base fluid are enhanced beyond what they could have achieved on their own. A substantial improvement in thermal conductivity and heat transfer can be observed as a result of these effects [90].
6. *Thermophoresis and solutal marangoni effect*: When temperature and concentration gradients are present, nanoparticles undergo thermophoresis (movement in response to temperature differences) and the solutal Marangoni effect (movement due to solute concentration gradients). The distribution and movement of nanoparticles within nanofluids can be significantly influenced by these intriguing phenomena [72].
7. *Nanoparticle shape and size effects*: Nanoparticle shape and size play a significant role in determining the behavior of nanofluids. The heat transfer characteristics of anisotropic nanoparticles (e.g., nanotubes) differ from those of isotropic nanoparticles (e.g., spheres) [73].
8. *Surface modification*: Modifying nanoparticle surface properties through functionalization influences their interaction with the base fluid and other nanoparticles profoundly. As a result, this has implications for dispersion stability and overall nanofluid behavior [74].

In order to optimize nanofluid properties and tailor their performance to meet specific engineering applications, an understanding of these mechanisms is necessary. In order to study these intricacies, researchers use a combination of experimental techniques, theoretical modeling, and numerical simulations, ultimately expanding our understanding of nanofluid behavior and their potential applications in a wide range of engineering fields.

8.4.5 Microchannel heat sinks with nanofluid for electronics cooling

As a result of their compact size, large surface area, and high heat transfer efficiency, microchannel heat sinks have shown great promise as efficient cooling solutions for electronics. Their heat-removal capabilities can be further enhanced when combined with nanofluids, making them even more effective at cooling electronic components. Earlier, it was mentioned that nanofluids can improve the thermal conductivity of the base fluid, which can result in a reduction of thermal resistance and temperature difference between the heated microchannel surface and the operating fluid. The result is an improved cooling performance and more efficient heat transfer. Due to nanofluids' enhanced heat transfer rate, heat is dissipated more efficiently from electronic components, reducing the risk of overheating and potential damage. Microchannel heat sinks can be significantly enhanced with nanofluids, resulting in more efficient cooling of electronic components. Furthermore, nanoparticles in nanofluids reduce thermal resistance, allowing heat to dissipate more quickly. Nanofluids help electronic devices operate at lower temperatures by reducing the temperature difference between heated surfaces and coolants. Moreover, microchannel heat sinks and nanofluids combine to provide compact and efficient cooling solutions, especially for electronics with limited space. In practical applications, it is crucial to carefully consider nanofluid selection and stability. To ensure long-term stability and reliable operation of nanofluid technology, challenges such as nanoparticle agglomeration, sedimentation, and potential material compatibility issues must be addressed [72–74].

A combination of nanofluids and microchannel heat sinks holds promise for advanced cooling solutions in electronics and other high-heat dissipation applications. Future research and development in this field may result in further improvements and widespread adoption of this technology.

8.5 Simulation of nanofluid in microchannel heat sinks

The simulation of nanofluids in microchannel heat sinks involves using computational techniques to model the flow of fluid and heat transfer within a heat sink containing nanoscale particles suspended in a base fluid. In comparison with traditional heat transfer fluids, this type of simulation can provide insights into nanofluids' thermal performance enhancements. An overview of the steps involved in simulating nanofluid behavior in a heat sink is given below [91–103]:

1. *Problem definition*: Define both the geometry of the heat sink and the computational domain, as well as the properties of the nanofluid (e.g., particle size, concentration, and base fluid properties).
2. *Fluid dynamics simulation*: Simulate the fluid flow within the heat sink using computational fluid dynamics (CFD). In order to understand how nanofluid

flows through the heat sink channels, Navier–Stokes equations must be solved. Boundary conditions, such as inlet and outlet velocities and pressures, will need to be specified.

3. *Heat transfer simulation*: Create a simulation of heat transfer within a heat sink. During this process, the heat conduction equation must be solved to determine how heat is transferred from the heat source to the fluid and then carried away. Heat transfer properties can be affected by nanoparticles in fluids.
4. *Nanoparticle behavior modeling*: Create models for the behavior of nanoparticles within fluids. Using methods such as the Brownian motion equation and taking diffusion and interactions between particles into account, it may be possible to track the motion of individual nanoparticles.
5. *Thermal conductivity enhancement*: Implement the enhanced thermal conductivity of the nanofluid due to the presence of nanoparticles. In the heat transfer calculation, this may require modifying the fluid's effective thermal conductivity.
6. *Validation and sensitivity analysis*: Verify simulation results against experimental data if available. Using sensitivity analyses, determine how parameters (particle size, concentration, etc.) affect the heat transfer performance of nanofluids in microchannel heat sinks.
7. *Postprocessing and visualization*: Identify the heat transfer enhancement achieved by nanofluids in the heat sink by analyzing and visualizing the simulation results. Temperature distributions, velocity profiles, and other relevant parameters can be visualized.
8. *Optimization and design*: You can optimize the nanofluid composition and heat sink design according to simulation results.

CFD, heat transfer, and nanotechnology expertise are needed to simulate nanofluids in microchannel heat sinks. Moreover, due to factors such as particle agglomeration and particle interaction, modeling nanofluid behavior can be challenging.

For these types of simulations, software tools such as ANSYS Fluent, COMSOL Multiphysics, and OpenFOAM are commonly used. It is important to remember that the accuracy of your simulation depends on the quality of your input data, the models chosen, and the assumptions made during the simulation setup [104].

8.5.1 Different models of simulating nanofluids in microchannel heat sinks

A variety of models is employed when simulating nanofluids in microchannel heat sinks to capture the complex interactions between nanoparticles and the base fluid, as well as their effect on fluid flow and heat transfer. Simulations of nanofluids in microchannel heat sinks commonly use the following models:

1. *Two-phase approach*: In this approach, solid nanoparticles are one phase of a two-phase fluid, whereas the base fluid is another. Volume fractions capture the interaction between the phases, and equations for each phase's properties, such as density, viscosity, and thermal conductivity, are solved separately. To model this approach, Eulerian–Eulerian or Eulerian–Lagrangian methods can be used [103].
2. *Single-phase approach with effective properties*: By introducing nanoparticles into the nanofluid, it is treated as a single-phase fluid with enhanced properties. Based on empirical or theoretical correlations, effective properties, such as thermal conductivity and viscosity, are modified. Nanoparticle concentration and size are typically considered in these correlations [91].
3. *Brownian motion and particle dispersion models*: Due to thermal fluctuations, nanoparticles in nanofluids move in Brownian motion. In Brownian motion models, nanoparticles are predicted to move randomly within a fluid. For the purpose of capturing the dispersion and distribution of nanoparticles within a heat sink, this is important [105,106].
4. *Thermal conductivity enhancement models*: Several models predict the enhanced thermal conductivity of nanofluids. Most commonly, the Maxwell–Garnett effective medium theory is used to estimate the effective thermal conductivity based on the nanoparticle and fluid properties [106].
5. *Agglomeration and sedimentation models*: There are a variety of forces that can cause nanoparticles in nanofluids to agglomerate (clump together) or sediment (settle). To predict the long-term stability and behavior of nanofluids within microchannel heat sinks, models that consider agglomeration and sedimentation are essential [106].
6. *Non-Newtonian behavior models*: The addition of nanoparticles to the base fluid can alter its rheological characteristics, leading to non-Newtonian behavior. In high nanoparticle concentrations, models that account for non-Newtonian viscosity and shear-thinning effects may be necessary [91,106].
7. *Molecular dynamics (MD) simulations*: MD simulations can be used for highly detailed molecular simulations. MD simulations provide a detailed understanding of nanoparticle–fluid interactions by tracking the motion of individual atoms and molecules. Simulating MD systems, however, requires a lot of computing power and may only be suitable for small-scale systems [106,107].
8. *Monte Carlo simulations*: It is possible to model Brownian motion and interaction between nanoparticles and fluid molecules using Monte Carlo simulations. Nanoparticle dispersion and transport can be studied with these simulations [106].
9. *Hybrid models*: Different modeling approaches are combined in some simulations to capture the behavior of nanofluids. For example, a CFD simulation for fluid flow combined with a separate model for nanoparticle dispersion might provide more accurate results [106].

Choosing a model depends on the specific goals of your simulation, the level of detail required, and the computing resources you have available. The behavior of nanofluids in microchannel heat sinks can be accurately captured by a combination of models. Validating your simulation results against experimental data is crucial to ensuring the accuracy and reliability of your models [91,108].

Among the myriad methods employed to simulate the intricate behavior of nanofluids within microchannel heat sinks, conventional approaches stand prominently. These approaches, widely embraced by researchers, can be broadly categorized into two fundamental subgroups: single-phase and two-phase models, as elegantly depicted in Fig. 8.10. Within the subsequent subsections, each of these models shall be succinctly elucidated, unraveling their essence and significance in the realm of nanofluid simulation.

8.5.1.1 Homogeneous model

The homogeneous model represents a streamlined and simplified approach employed to simulate the behavior of nanofluids within various heat transfer applications, including microchannel heat sinks. Within this model, nanofluids are treated as single-phase fluids, wherein the presence of nanoparticles imparts enhanced properties to the overall fluid. However, it is important to note that this model operates under the assumption of homogeneity, effectively overlooking the discrete nature of nanoparticles suspended within the fluid, thereby providing a useful approximation for practical analysis and computational simplicity while simplifying the complex interactions that occur at the nanoparticle level [91]. The homogeneous model works as follows [91,109]:

8.5.1.1.1 Basic assumptions

1. *Nanoparticles are homogeneously distributed*: This assumption assumes that nanoparticles form no aggregates or clusters in the base fluid. At high nanoparticle concentrations, this simplification may not accurately reflect real-world behavior [91,92].
2. *Local thermal equilibrium*: Microparticles and the base fluid will be in local thermal equilibrium, which means their temperatures will remain constant. Nanoparticles with relatively small sizes can be considered valid under this assumption [92].
3. *No interaction between particles*: Particle–particle interactions, such as agglomeration and sedimentation, can influence the behavior of nanofluids over time, which is ignored by the homogeneous model [91].

8.5.1.1.2 Enhanced properties

In the context of modeling nanofluids, the consideration of effective thermal conductivity and effective viscosity plays a pivotal role in accounting for the presence of nanoparticles within the homogeneous model. Leveraging the

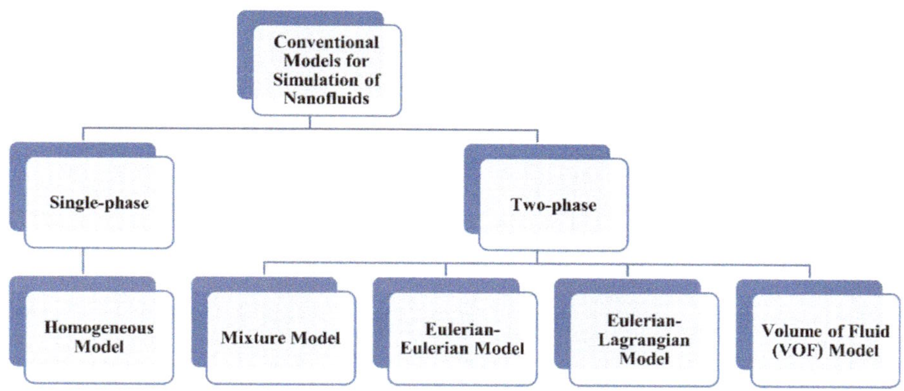

FIGURE 8.10

Conventional models for simulation of nanofluids.

Maxwell–Garnett effective medium theory, the estimation of a nanofluid's effective thermal conductivity takes a particular form, expressed as:

$$k_{eff} = k_{base}(1 + 2.5\varphi) \tag{8.3}$$

wherein k_{eff} symbolizes the effective thermal conductivity of the nanofluid, k_{base} represents the thermal conductivity of the base fluid, and φ signifies the volume fraction of nanoparticles. This framework extends beyond thermal conductivity, as the homogeneous model can similarly employ analogous equations to estimate effective viscosity and various other pertinent properties, affording a comprehensive framework for modeling nanofluid behavior and offering insights into the multifaceted dynamics of these advanced fluids [110].

The most conventional equations applicable for the determination of the properties involved in fluid dynamics and heat transfer are as follows [92]:

$$\rho_{nf} = (1 - \varphi)\rho_{bf} + \varphi\rho_{np} \tag{8.4}$$

$$c_{nf} = (1 - \varphi)c_{bf} + \varphi c_{np} \tag{8.5}$$

$$\frac{\mu_{nf}}{\mu_{bf}} = \frac{1}{(1-\varphi)^{2.5}} \tag{8.6}$$

where subscripts *nf*, *bf*, and *np* refer to the nanofluids, base fluid, and nanoparticle, respectively. Similar to the mentioned properties, there are some correlations for the calculation of thermal conductivity. As an example, for spherical particles at low volume fractions and randomly distributed particles, the Maxwell model can be used as follows [91]:

$$k_{nf} = k_{bf} + 3\varphi\frac{k_{np} - k_{bf}}{k_{np} + 2k_{bf}}k_{bf} \tag{8.7}$$

One of the most important reasons for the attractiveness of this model for the numerical simulation of nanofluid is its simplicity of use and low computational cost. While the homogeneous model is a valuable tool for initial assessments and design insights, it is essential to be aware of its limitations and consider more advanced models for a comprehensive understanding of nanofluid behavior in microchannel heat sinks [92].

8.5.1.1.3 Advantages and limitations

The advantages of homogeneous models include their simplicity and computational efficiency. Nanofluid performance in microchannel heat sinks can be quickly assessed and estimated with this method. In cases where particle interactions and aggregation are significant, the homogeneous model may not capture the full complexity of nanofluid behavior due to its simplifications.

It is imperative to remember that the homogeneous model provides a simplified representation of nanofluid behavior. Other models, such as the two-phase approach or molecular dynamics simulations, might be more suitable for more accurate and detailed simulations, especially at higher nanoparticle concentrations [91,92].

To ensure the reliability of predictions made using the homogeneous model, the results must be validated against experimental data.

8.5.1.2 Mixture model

In microchannel heat sinks and other heat transfer applications, the mixture model is a more advanced approach to simulate nanofluids. Mixture models treat nanofluids as two-phase systems, consisting of a fluid phase and a nanoparticle phase, as opposed to the homogeneous model. In this model, interactions between the two phases are accounted for providing a more accurate representation of the behavior of nanofluids. Using a mixture model, here is how it works [107]:

8.5.1.2.1 Basic concept

In the mixture model, the nanofluid is composed of two phases: continuous fluid (base fluid) and dispersed solid (nanoparticles). Each phase has its own set of conservation equations for mass, momentum, and energy. Drag forces between fluids and particles, heat transfer between them, and particle motion are taken into account in the model.

8.5.1.2.2 Key equations

1. *Continuity equations*: To determine the mass fractions of the fluid and particle phases, separate continuity equations must be solved [107].
2. *Momentum equations*: Considering factors such as particle size and concentration, momentum equations account for drag forces between fluid and particles [94].
3. *Energy equations*: For each phase, separate energy equations are solved to account for heat transfer. The temperature difference between the two phases must also be considered.
4. *Particle transport*: Models typically include equations that account for particle motion caused by Brownian motion and turbulent dispersion. Fluid dispersal and transport are determined by these equations.

The governing equations for this model are represented here according to a study by Garoosi et al. [94]. Considering a two-dimensional domain, equations for continuity, momentum, energy, and volume fraction can be expressed as follows:

$$\nabla.\vec{V}_m = 0 \tag{8.8}$$

$$\nabla\rho_m\vec{V}_m\vec{V}_m = -\nabla P + \nabla.\mu_m\nabla\vec{V}_m + \nabla.\left[\sum_{k=1}^{n}\varphi_k\rho_k\vec{V}_{dr,k}\vec{V}_{dr,k}\right] + (\rho\beta)_m(T - T_c)g \tag{8.9}$$

$$\nabla.\left[\sum_{k=1}^{n}(\rho_k C_{pk})\varphi_k \vec{V}_k T\right] = \nabla.k_m \nabla T \tag{8.10}$$

$$\nabla.\left(\varphi_p \rho_p \vec{V}_m\right) = -\nabla.(\varphi_p \rho_p \vec{V}_{dr,p}) \tag{8.11}$$

In these equations, P and T refer to the dimensional pressure and temperature, respectively. V_m denoted mass average velocity that is determined as follows:

$$\vec{V}_m = \frac{\sum_{k=1}^{n} \varphi_k \rho_k \vec{V}_k}{\rho_m} \tag{8.12}$$

$\vec{V}_{dr,k}$ denotes the secondary phase k drift velocity, here nanoparticles in the base fluid, and is obtained as follows:

$$\vec{V}_{dr,k} = \vec{V}_{np} - \vec{V}_m \tag{8.13}$$

$\vec{V}_{pf}$, relative or slip velocity refers to the second phase velocity relative to the primary phase (*bf*) velocity and is determined as follows:

$$\vec{V}_{pf} = \vec{V}_{np} - \vec{V}_{bf} \tag{8.14}$$

The drift velocity relates to the relative velocity as expressed in Eq. (8.15).

$$\vec{V}_{dr,p} = \vec{V}_{pf} - \sum_{k=1}^{n} \frac{\varphi_k \rho_k}{\rho_m} \vec{V}_{fk} \tag{8.15}$$

Drag function and relative velocity can be determined by using Eqs. (8.13) and (8.14) as follows:

$$f_{drag} = \begin{cases} 1 + 0.15\mathrm{Re}_p^{0.687} & \mathrm{Re}_p \le 1000 \\ 0.0183\mathrm{Re}_p & \mathrm{Re}_p > 1000 \end{cases} \tag{8.16}$$

$$\vec{V}_{pf} = \frac{\rho_{np} d_p^2}{18\mu_f f_{drag}} \frac{\left(\rho_{np} - \rho_m\right)}{\rho_{np}} \left(g - \left(\vec{V}_m.\nabla\right)\vec{V}_m\right) \tag{8.17}$$

8.5.1.2.3 Advantages and limitations

There are several advantages to the mixture model over the homogeneous model [94]:

1. *Realistic nanoparticle behavior*: Based on the mixture model, nanoparticles are modeled as separate phases, taking into account their particle–particle interactions, agglomeration, and sedimentation.
2. *Enhanced accuracy*: When there are high concentrations of nanoparticles and particle interactions play a key role, the mixture model provides a higher degree of accuracy.

3. *Versatility*: It can simulate several aspects of nanofluid behavior, including heat transfer enhancement, pressure drop, and particle dispersion.

The mixture model, though, has a higher computational complexity than the homogeneous model because it involves solving additional equations for each phase. A model's accuracy is also determined by the accuracy of input data, which includes nanoparticle properties and interphase interactions.

8.5.1.2.4 Simulation software

ANSYS Fluent, COMSOL Multiphysics, and OpenFOAM are among the numerical simulation packages that can simulate nanofluids in microchannel heat sinks using the mixture model. They allow for solving governing equations for each phase and handling complex particle–fluid interactions.

When using a mixture model, it is crucial to validate simulation results against experimental data to ensure their validity [94,107].

8.5.1.3 Eulerien–Eulerian model

Multiphase flows, including nanofluids in microchannel heat sinks, are simulated using the Eulerian–Eulerian model, a CFD approach. The dispersed nanoparticle phase as well as the fluid phase are regarded as continuous phases with separate governing equations. Eulerian–Eulerian models are particularly useful for simulating fluid flow and heat transfer where the dispersed phase (nanoparticles) plays a significant role. The Eulerian–Eulerian model simulates nanofluids in microchannel heat sinks as follows [95]:

8.5.1.3.1 Basic concept

Fluids and nanoparticles are treated separately in the Eulerian–Eulerian model, with their own sets of mass, momentum, and energy conservation equations. Within the heat sink, these equations describe how each phase behaves and interacts [97].

8.5.1.3.2 Key equations

1. *Continuity equations*: In each computational cell, the fluid and nanoparticle phases are solved separately using continuity equations.
2. *Momentum equations*: Drag forces between phases are accounted for by the momentum equations for each phase. The effects of particle concentration, particle size, and relative velocities between phases are included here.
3. *Energy equations*: To account for heat transfer between nanoparticles and fluids, separate energy equations are solved for each phase.
4. *Particle transport*: Equations are provided to describe the movement of nanoparticles within a fluid and their dispersion. Particle–particle interactions, Brownian motion, and turbulent dispersion governing equations for Eulerian–Eulerian two-phase model for nanofluids are provided here

according to the study by Rezaei Gorjaei and Rahmani [95]. In these methods, continuity equations are expressed as follows:

$$\nabla.\left(\varphi_{bf}\rho_{bf}v_{bf}\right) = 0 \tag{8.18}$$

$$\nabla.\left(\varphi_{np}\rho_{np}v_{np}\right) = 0 \tag{8.19}$$

Based on the definition of volume concentration, Eq. (8.17) can be written as follows:

$$\varphi_{bf} + \varphi_{np} = 1 \tag{8.20}$$

Momentum equations in this approach are expressed as follows:

$$\nabla.\left(\varphi_{bf}\rho_{bf}v_{bf}v_{bf}\right) = \varphi_{bf}\nabla P + \mu_{bf}\nabla.\left(\varphi_{bf}\nabla v_{bf}\right) + F_d + F_{vm} \tag{8.21}$$

$$\nabla.\left(\varphi_{np}\rho_{np}v_{np}v_{np}\right) = -\varphi_{np}\nabla P + \mu_{np}\nabla.\left(\varphi_{np}\nabla v_{np}\right) - F_d - F_{vm} + F_{col} \tag{8.22}$$

In the above equations, $F_{d,}$ $F_{vm,}$ and F_{col} are drag, added mass (virtual mass) force, and particle–particle forces, respectively. The virtual mass force is determined by using Eq. (8.20) as follows:

$$F_{vm} = 0.5\varphi_{np}\rho_{bf}\frac{D}{Dt}\left(\vec{V}_{bf} - \vec{V}_{np}\right) \tag{8.23}$$

where D is the material derivative. Particle–particle force is determined as follows:

$$F_{col} = G(\varphi_{bf})\nabla\varphi_{bf} \tag{8.24}$$

where G refers to the modulus of particle–particle interaction and is determined as follows:

$$G = \exp\left(-600\left(\varphi_{bf} - 0.376\right)\right) \tag{8.25}$$

In order to determine drag force, Eq. (8.26) is applied as follows:

$$F_d = -\frac{3}{4}C_d\frac{\varphi_{bf}(1-\varphi_{bf})}{d_{np}}\left|\vec{V}_{bf} - \vec{V}_{np}\right|\varphi_{bf}^{-2.65}\left(\vec{V}_{bf} - \vec{V}_{np}\right) \tag{8.26}$$

where

$$C_d = \begin{cases} \dfrac{24 + 3.6\mathrm{Re}_{np}^{0.697}}{\mathrm{Re}_{np}}, & \mathrm{Re}_p = \dfrac{\varphi_{bf}\rho_{bf}d_{np}}{\mu_{bf}}\left|\vec{V}_{bf} - \vec{V}_{np}\right| < 1000 \\ 0.44, & \mathrm{Re}_p = \dfrac{\varphi_{bf}\rho_{bf}d_{np}}{\mu_{bf}}\left|\vec{V}_{bf} - \vec{V}_{np}\right| \geq 1000 \end{cases} \tag{8.27}$$

8.5.1.3.3 Advantages and limitations

For simulating nanofluids in microchannel heat sinks, the Eulerian–Eulerian model offers several advantages [97–99]:

1. *Realistic multiphase behavior*: By capturing the interactions between fluid and nanoparticle phases, this model can be used where particle behavior significantly affects heat transfer and fluid flow.
2. *Detailed insights*: This model sheds light on how nanoparticles disperse, cluster, and influence the fluid behavior within the heat sink.
3. *Comprehensive heat transfer analysis*: With the Eulerian–Eulerian model, heat transfer enhancements due to nanoparticles can be analyzed through a comprehensive analysis of particle concentration and particle–fluid interactions.

Compared to simpler models, such as the homogeneous model, the Eulerian–Eulerian model has a higher computational demand. In addition to increasing simulation time and resource consumption, it requires solving a greater number of equations.

8.5.1.3.4 Simulation software

ANSYS Fluent, COMSOL Multiphysics, and OpenFOAM provide capabilities for simulating nanofluids in microchannel heat sinks using the Eulerian–Eulerian model. A multiphase flow model and the correct solver settings are included in these tools, making it possible to capture the behavior of nanoparticles in fluids accurately [98,99].

As with any modeling approach, validation against experimental data is essential to ensure the accuracy and reliability of the simulation results obtained using the Eulerian–Eulerian model.

8.5.1.4 Eulerian–Lagrangian model

Multiphase flows, including nanofluids in microchannel heat sinks, can also be simulated using the Eulerian–Lagrangian model in CFD. It is especially suitable for situations in which the dispersed phase (nanoparticles) has a minor impact on the overall fluid flow, by tracking nanoparticles as they move through the fluid [111].

Here is how the Eulerian–Lagrangian model simulates nanofluids in microchannel heat sinks:

8.5.1.4.1 Basic concept

Fluid properties are averaged over computational cells in the Eulerian–Lagrangian model. On the other hand, the nanoparticle phase is modeled using the Lagrangian framework, which tracks individual nanoparticles as they move through the fluid. When nanoparticles are sparse or their behavior needs to be closely examined, this model is useful.

8.5.1.4.2 Key equations

Fluid phase equations (Eulerian): These equations describe the behavior of the base fluid based on continuity, momentum, and energy. By averaging properties over the computational domain, the Eulerian approach is used to solve these equations.

Nanoparticle phase equations (Lagrangian): The nanoparticles are considered to be discrete Lagrangian particles that move through the fluid. Each particle's equations of motion are solved by accounting for drag forces, Brownian motion, and other forces. Over time, these equations track the position, velocity, and temperature of each particle. The governing equations of this model are represented here according to a study by Ambreen et al. [111].

For the fluid phase, following equations can be denoted:

$$\nabla.\left(\rho_{bf}V\right) = 0 \tag{8.28}$$

$$\nabla.(\rho VV) = -\nabla P + \nabla.(\mu \nabla V) + S_u \tag{8.29}$$

where S_u refers to the source phase term and can be determined by using Eq. (8.30).

$$S_u = \sum n_{np} \frac{m_{np}}{\delta v} \frac{dV_{np}}{dt} \tag{8.30}$$

where n_{np} refers to the nanoparticles' number in the computational cell. $\frac{dV_{np}}{dt}$ is obtained by applying Eq. (8.31) as follows:

$$\frac{dV_{np}}{dt} = F_{Brownian} + F_{lift} + F_{drag} + F_{gravity} + F_{thermophoresis} + F_{Pressure} + F_{Virtual} \tag{8.31}$$

Drag force of the nanoparticle is estimated through the Stokes–Cunningham drag law, as represented in Eq. (8.32).

$$F_{drag} = \frac{18\mu_{bf}}{d_{np}^2 \rho_{np} C_c} \tag{8.32}$$

In Eq. (8.32), C_c refers to the Cunningham correction factor and is calculated by applying Eq. (8.33).

$$C_c = 1 + \frac{2\lambda_{np}}{d_{np}} \left(1.257 + 0.4e^{-\left(\frac{1.1d_{np}}{2\lambda}\right)}\right) \tag{8.33}$$

where λ_{np} refers to the nanoparticle's molecular average free path. Other forces represented in Eq. (8.32) are determined by employing the following equations.

$$F_{lift} = \frac{5.188 v_{bf}^{0.5} \rho_{bf} d_{ij} (\mu_{bf} - \mu_{np})}{\rho_{np} d_{np} (d_{ik} d_{ki})^{0.25}} \tag{8.34}$$

$$F_{Brownian} = \frac{216 V K_B T}{\pi^2 \rho_{bf}^5 C_c \left(\frac{\rho_{np}}{\rho_{bf}}\right)^2} \tag{8.35}$$

$$F_{thermophoresis} = \frac{7.02\pi d_{np}\mu_f^2(K + 21.18K_n)}{m_{np}\rho_{bf}T(1 + 3.42K_n)(1 + 2K + 4.36K_n)} \times \Delta T \tag{8.36}$$

$$F_{virtual} = 0.5\frac{\rho_{bf}}{\rho_{np}}\frac{d(\mu_{bf} - \mu_{np})}{dt} \tag{8.37}$$

$$F_{gravity} = \frac{g(\rho_{np} - \rho_{bf})}{\rho_{np}} \tag{8.38}$$

$$F_{pressure} = \left(\frac{\rho_{bf}}{\rho_{np}}\right)\mu_{np}\frac{d\mu_f}{dx} \tag{8.39}$$

where $K_n = \frac{2\lambda}{d_{np}}$ and d_{ij} refer to the Knudson number and deformation tensor, respectively.

8.5.1.4.3 Advantages and limitations

When simulating nanofluids in microchannel heat sinks, the Eulerian–Lagrangian Model offers several advantages [111]:

1. *Particle tracking*: By tracking individual nanoparticles, this model can study particle behavior, dispersion, and interactions in detail.
2. *Sparse particle distributions*: When nanoparticle concentrations are low and individual behavior is important, this technique is useful.
3. *Flexible for various particles*: As the model can simulate a variety of particle types, it is not limited to nanoparticles.

When dealing with large numbers of particles, the Eulerian–Lagrangian model is computationally demanding. The computational cost can increase significantly due to the need to solve individual equations of motion for each particle.

8.5.1.4.4 Simulation software

CFD packages such as ANSYS Fluent, COMSOL Multiphysics, and OpenFOAM enable the simulation of nanofluids in microchannel heat sinks with Eulerian–Lagrangian model. For tracking individual nanoparticles within a fluid, these tools provide the necessary solver settings.

The Eulerian–Lagrangian model must be validated against experimental data, as with any modeling approach, to ensure accuracy and reliability.

8.5.1.5 Volume of fluid model

Volume of fluid (VOF) models are widely used in CFD to simulate multiphase flows, including nanofluid behavior in microchannel heat sinks. Using this model, one can gain insights into how fluids flow and how heat is transferred in microchannels containing nanofluids. It is particularly suitable for capturing the interface between different fluid phases [101].

To simulate nanofluids in microchannel heat sinks, the VOF model works as follows [102]:

8.5.1.5.1 Basic concept

In a computational cell, the VOF model tracks the volume fraction of each fluid phase. In nanofluids, it tracks the volume fractions of the base fluid and nanoparticles within each cell. Fluid–nanoparticle interfaces can be accurately simulated using this method [103].

8.5.1.5.2 Key equations

1. *Volume fraction equations*: Fluid and nanoparticle volume fraction equations are solved separately. Using these equations, we can calculate how the volume fractions change over time as a result of fluid flow and particle dispersion.
2. *Momentum equations*: Nanoparticles are taken into account when solving the momentum equations for the fluid phase. The distribution and motion of nanoparticles influence the fluid velocity field.
3. *Energy equations*: To account for heat transfer between fluid and nanoparticle phases, separate energy equations are solved.

The governing equations of this model are represented according to the study by Sidik et al. [101].

In the VOF model, the interface between the nanoparticle and base fluid is followed by a continuity equation solving for the base fluid phase volume fraction and is expressed as follows:

$$\nabla.\left(\varphi_{bf}\rho_{bf}\vec{u}_{bf}\right) = 0 \tag{8.40}$$

By considering the following constraint, the nanoparticle phase volume fraction is calculated:

$$\varphi_{bf} + \varphi_{np} = 1 \tag{8.41}$$

Furthermore, just a momentum equation is solved and the obtained field of velocity is distributed among the considered phases.

$$\nabla.\left(\rho_{nf}u_{nf}u_{nf}\right) = -\nabla P + \nabla.\left(\mu_{nf}\nabla u_{nf}\right) + \rho_{nf}g \tag{8.42}$$

Eq. (8.39) is used as an energy equation in the VOF model.

$$\nabla.\left(\rho_{nf}c_{p,nf}T_{nf}u_{nf}\right) = \nabla.(k_{nf}\nabla T_{nf}) \tag{8.43}$$

The properties of the nanofluid are determined by using the following equations:

$$\rho_{nf} = \varphi_{bf}\rho_{bf} + \varphi_{np}\rho_{np} \tag{8.44}$$

$$\mu_{nf} = \varphi_{bf}\mu_{bf} + \varphi_{np}\mu_{np} \tag{8.45}$$

$$k_{nf} = \varphi_{bf}k_{bf} + \varphi_{np}k_{np} \tag{8.46}$$

$$C_{p,nf} = \frac{\varphi_{bf}\rho_{bf}C_{p,f} + \varphi_{np}\rho_{np}C_{p,np}}{\varphi_{bf}\rho_{bf} + \varphi_{np}\rho_{np}} \tag{8.47}$$

8.5.1.5.3 Advantages and limitations

For simulating nanofluids in microchannel heat sinks, the VOF model offers several advantages [102,103]:

1. *Accurate interface tracking*: The model is well-suited to capture fluid behavior and heat transfer across fluid interfaces because it excels at tracking the interface between different fluid phases.
2. *Realistic multiphase behavior*: Numerous multiphase interactions can be accurately represented by the VOF model, such as the influence of nanoparticle movement on fluid flow.
3. *Applicability to microchannels*: VOF models are especially useful for simulating fluid flow in microchannels, such as heat sinks.

VOF models can, however, be computationally demanding, especially when capturing fine fluid interface details requires a large number of cells.

8.5.1.5.4 Simulation software

CFD software packages such as ANSYS Fluent, COMSOL Multiphysics, and OpenFOAM provide the VOF model as an option for simulating multiphase flows, including nanofluids in microchannel heat sinks. These tools offer the necessary solver settings to accurately track fluid interfaces and particle distribution [103].

As with any modeling approach, validation against experimental data is crucial to ensure the accuracy and reliability of the simulation results obtained using the VOF model.

8.5.2 Examples of simulations conducted using the described methods

Kumar and Sarkar [112] investigated heat transfer within a minichannel heat sink using hybrid nanofluids. They examined flow rate, Reynolds number, and temperature effects through a two-phase model. Their findings showed a modest improvement in heat transfer with hybrid nanofluids, with the Al_2O_3 (10:0) nanofluid exhibiting the most significant enhancement (8.5% numerically and 12.8% experimentally). Interestingly, lower temperatures and higher nanoparticle concentrations led to increased pressure drops, challenging expectations. Additionally, mixing nanoparticles of similar sizes did not boost heat transfer as anticipated, emphasizing the complexity of nanofluid behavior in practical applications [112].

Kumar and Sarkar [113] conducted a numerical simulation study focusing on heat transfer and pressure drop within a minichannel heat sink. Their investigation employed a two-phase mixture model, considering both an Al_2O_3 nanofluid and a hybrid Al_2O_3-multiwalled carbon nanotube (MWCNT) nanofluid. To validate their model, they conducted experiments, which proved instrumental in their analysis. The research encompassed an array of critical parameters, including channel size, composition, and Reynolds number, aiming to uncover the most effective heat transfer configurations. Notably, the study revealed that the two-phase mixture model outperformed the single-phase homogeneous model. The most promising results were achieved with a 0.01% concentration of the (Al_2O_3 + MWCNT) hybrid nanofluid, exhibiting a remarkable 15.6% enhancement in heat transfer efficiency without incurring a significant increase in pressure, showcasing the potential for improved thermal performance in minichannel heat sinks through nanofluid utilization [113].

Bahiraei et al. [114] explored second-law phenomena within elliptical pin-fin heat sinks using a two-phase simulation model. Their study aimed to improve the design of heat sinks for high-performance electronics. Through numerical modeling, they found that higher fin density reduced thermal irreversibility, a critical factor for enhancing heat sink efficiency. However, at higher Reynolds numbers, increased fin density led to more frictional irreversibility, indicating the need for a delicate balance in heat sink design. The introduction of nanoparticles improved temperature uniformity and reduced thermal entropy generation but also introduced frictional irreversibility. To minimize overall irreversibility, they emphasized the importance of adjusting fin density based on Reynolds numbers. Moreover, the Bejan number decreased gradually with increasing fin density, shedding light on the relationship between design parameters and thermodynamic efficiency in heat sinks. These findings provide valuable insights for optimizing heat sink performance in electronic cooling applications [114].

Sarlak et al. [115] undertook a rigorous investigation, harnessing numerical methods to optimize a double-layer microchannel heat sink. Employing the finite volume method and a sophisticated two-phase mixture model, their study involved meticulous modeling of heat and fluid flow within solid walls and nanofluids, specifically focusing on non-Newtonian carbon-based nanofluids. Their objective centered on the reduction of thermal resistance while maintaining a consistent pumping power, a goal pursued through direct search optimization techniques. To validate their model's accuracy, experiments were conducted. Notably, the study highlighted the significant impact of carbon nanoparticles' high conductivity and the shear-thinning behavior of nanofluids, which collectively contributed to a marked reduction in thermal resistance when carbon-based nanofluids were employed in lieu of oxide-metallic counterparts. This insightful research demonstrated that the incorporation of non-Newtonian carbon-based nanofluids could enhance the thermal performance of microchannel heat sinks by a substantial 14%, underscoring the potential for innovative nanofluid applications in thermal management systems [115].

Wang et al. [116] employed a novel hybrid approach to optimize parameters in a nanofluid minichannel heat sink. They developed a numerical model based on the mixture model, which generated a comprehensive dataset for machine learning purposes. Employing support vector regression, Gaussian process regression (GPR), and random forest models, they established correlations between input parameters (Reynolds number, volume fraction, and heat flow density) and desired output values (pressure drop and average temperature of the heating wall). Among these models, GPR emerged as the most suitable, exhibiting high *R*-squared (R^2) values, specifically 0.9939 for the average temperature of the heating wall and 0.9985 for pressure drop. Additionally, to address varying operational conditions, they leveraged the NSGA-II multiobjective optimization algorithm to determine optimal volume fraction values, identifying approximately 3% for low Reynolds numbers and approximately 2% for high Reynolds numbers, showcasing the versatility and potential of their hybrid approach in optimizing nanofluid minichannel heat sinks [116].

Jamshidmofid et al. [98] undertook a comprehensive study to evaluate the advantages of employing a water-based nanofluid containing a hybrid nanocomposite comprised of reduced graphene oxide and cobalt oxide in a microchannel heat sink featuring rectangular ribs and sinusoidal cavities. They developed an intricate two-phase mixture numerical model using a finite volume technique with a coupled approach. Throughout their simulations, they considered a range of Reynolds numbers spanning from 100 to 500, with varying nanofluid concentrations ranging from 0.05% to 0.1%. The impact of temperature on thermophysical properties was also taken into account. Remarkably, despite reduced vortices, nanofluids demonstrated a substantial enhancement in thermal characteristics, primarily owing to their significantly improved thermal conductivity. Notable outcomes included a maximum increase of 14% in the convection heat transfer coefficient, a 2.3% reduction in solid temperature, a 27% improvement in temperature uniformity, and a 25% reduction in thermal resistance, respectively. Importantly, the nanofluid displayed a more pronounced enhancement in heat transfer compared to pressure drop, making it an attractive option for practical applications. Ultimately, they recommended the use of a 0.2% nanofluid at a Reynolds number of 500 due to its superior thermal properties with minimal associated pressure increase, highlighting its potential in enhancing microchannel heat sink performance [98].

Mohammadpour et al. [117] made significant strides in thermal performance improvement by developing a three-dimensional (3D) microchannel heat sink that employed a combination of cooling techniques. Their innovative approach involved utilizing a validated multiphase mixture model, focusing on Al_2O_3 particles suspended in water to investigate nanofluid flow alongside the integration of synthetic jets within the setup. This comprehensive study encompassed an exploration of both convective and conductive heat transfer processes, taking into account variables such as particle concentration, jet configuration, and actuation phase. The results of their investigation revealed noteworthy findings. Jet configurations that operated out of phase exhibited superior performance in terms of convective heat transfer enhancement, whereas in-phase actuation proved more effective for conductive heat transfer improvement. Particularly, when utilizing in-phase actuation in conjunction with nanofluids, significant enhancements in convective heat transfer were achieved. The cumulative outcome of their innovative approach resulted in an impressive 54% overall thermal enhancement, marking a substantial advancement in microchannel heat sink technology and highlighting its potential for enhancing thermal performance in various applications [117].

In their research, Bazkhane and Zahmatkesh [118] conducted a comprehensive simulation and analysis of hydrodynamics and heat transfer within microchannel heat sinks employing alumina–water nanofluids. The investigation encompassed various design factors, including the influence of nanofluids and porous substrates on channel walls (both vertically and horizontally oriented). To model porous media, the researchers employed the Darcy–Brinkman–Forchheimer model, coupled with a 3D solid-fluid conjugate model utilizing a two-phase mixture model for the nanofluid. Their analysis extended beyond surface temperature on the CPU, encompassing critical parameters such as thermal resistance, pumping power, and the figure of merit. The Taguchi method and analysis of variance were employed to assess design sensitivity. Remarkably, the study underscored that microchannel heat sinks and porous substrates exerted the most significant influence on CPU surface temperature, thermal resistance, and the figure of merit, emphasizing their pivotal roles in enhancing thermal performance [118].

Moraveji and Ardehali [119] conducted a CFD study focused on laminar forced convection utilizing Al_2O_3 nanofluid with a particle size of 33 nm at various concentrations (0.5%, 1%, and 6%) within a minichannel heat sink, covering a Reynolds number range of 130–1600. To simulate this complex process, they employed four different models (single phase, VOF, mixture, and Eulerian) and discretized the governing equations using the finite volume method. The study primarily sought to investigate the impact of nanoparticle concentration and Reynolds number on nanofluid heat transfer enhancement. Notably, the research revealed that single-phase models exhibited lower accuracy compared to two-phase models, with minimal discrepancies observed among different two-phase models. Among these, mixture models demonstrated a harmonious blend of precision, computational efficiency, and reduced runtime, leading to their selection for estimating dimensionless Nusselt numbers and friction factors, emphasizing their suitability for predicting heat transfer performance in this minichannel heat sink configuration [119].

Ghale et al. [120] conducted a numerical investigation employing CFD modeling to explore laminar forced convection heat transfer within straight microchannels, using water and alumina nanofluids as the working fluids. The study encompassed two distinct phases. First, both single-phase and two-phase mixture models were applied to predict heat transfer and hydrodynamics parameters within a microchannel heat sink, with the two-phase approach proving superior in terms of predictive accuracy. In the second phase, the impact of rib turbulators on fluid flow and heat transfer was examined. The findings revealed that ribbed microchannels outperformed their simple counterparts, exhibiting higher Nusselt numbers and lower friction coefficients, with the degree of enhancement increasing proportionally with rib width. This comprehensive study not only contributes to the understanding of nanofluid heat transfer but also highlights the benefits of geometric modifications, such as ribbed structures, in enhancing heat transfer efficiency within microchannels [120].

In 2014, under the leadership of Yue-Tzu Yang, a rigorous study was undertaken, meticulously delving into numerical simulations of fluid dynamics within a trapezoidal micro-channel heat sink. This comprehensive investigation spanned both laminar and turbulent cooling scenarios, utilizing a CuO/water nanofluid as the heat transfer medium. Employing the finite-volume method for the discretization of the Navier–Stokes equations in conjunction with conjugate energy equations, the research meticulously compared CFD predictions for CuO/water nanofluids using both single-phase and two-phase (mixture model) approaches. The findings notably highlighted the superior accuracy of the latter approach, aligning closely with experimental measurements. Moreover, the study unveiled that nanofluids, influenced by particle volume fraction and flow rate, demonstrated a notable reduction in thermal resistance when compared to pure water in laminar flow conditions. However, it also observed a marginal increase in pressure drop within nanofluid-cooled micro-channel heat sinks during laminar flow, emphasizing the intricate balance between enhanced heat transfer and potential trade-offs associated with nanofluid-based cooling solutions [121].

In 2021, Mehdi Bahiraei and his research team conducted a comprehensive investigation into the flow of silver–water nanofluid through an elliptical pin-fin heat sink with the aim of enhancing electronic cooling. Their study encompassed the exploration of various nanoparticle volume fractions, fin densities, and Reynolds numbers. Employing two-phase mixture models alongside four-equation transition shear stress transport models tailored for turbulent flow, their research provided compelling evidence that the addition of nanoparticles yielded positive effects on critical parameters including heat transfer coefficients, surface temperatures, thermal resistances, and temperature uniformities. Notably, within specific conditions, a mere 0.4% increase in nanoparticle volume fraction translated into a remarkable 16.2% reduction in thermal resistance and a substantial 25.5% increase in heat transfer coefficient. Furthermore, the study underscored that higher fin density significantly augmented the heat transfer coefficient. However, it also revealed that heightened nanoparticle volume fraction or fin density correlated with increased pumping power requirements and a subsequent rise in the figure of merit, underscoring the intricate balance involved in optimizing the cooling performance of electronic systems through nanofluid-based solutions [122].

Ambreen et al. [123] conducted a meticulous investigation into the influence of nanofluids within an aluminum foam heat sink. Employing a multiphase Eulerian model and the Darcy–Forchheimer–Brinkman model, the researchers systematically examined the interplay of substrate porosity and nanoparticle content on nanofluid performance. Their study encompassed a comprehensive analysis of heat transfer and entropy generation, shedding light on the remarkable enhancements achieved through increased substrate porosity, all while incurring only a minimal pressure drop. Notably, the investigation revealed that higher Reynolds numbers had a substantial impact on media permeability. Furthermore, the utilization of nanofluids proved to be a pivotal factor, leading to significant increases in Nusselt numbers, with the most pronounced improvements observed at Reynolds numbers exceeding 1800, underscoring the potential of nanofluids to enhance heat transfer efficiency within aluminum foam heat sinks [123].

Radwan and Ahmed [124] embarked on a comprehensive endeavor, devising an intricate 3D model to elucidate nanofluid flow dynamics within a broad microchannel heat sink. This model ingeniously amalgamated a two-phase multiphase (Eulerian–Eulerian) model with a thermal framework tailored for concentrator photovoltaic systems. Through a rigorous process of numerical simulations and validation, the study systematically explored the multifaceted influence of nanoparticle types, volume fractions, and flow Reynolds numbers on solar cell performance. Notably, Silicon Carbide–Water nanofluids exhibited the advantage of lowering solar cell temperatures in comparison to Al_2O_3–water nanofluids. The introduction of nanoparticle volume fractions notably contributed to improved electrical efficiency, temperature reduction within the cells, and enhanced uniformity of performance. The study uncovered a critical threshold for the flow Reynolds number, beyond which excessive values resulted in a reduction in performance. The outcome of their meticulous analysis revealed that, contingent on the specific combination of flow Reynolds number and solar concentration ratio, the integration of a 4% Silicon Carbide–Water nanofluid led to a noteworthy 8°C–3°C reduction in local solar cell temperatures compared to the use of pure water, showcasing the potential for nanofluid-based solutions to significantly enhance solar cell efficiency and thermal management [124].

Fani et al. [125] undertook a comprehensive investigation, delving into the influence of spherical nanoparticles on the thermal performance and pressure drop of nanofluids within trapezoidal microchannel heat sinks. Their study involved the simulation of copper-oxide nanoparticles dispersed in water, employing an Eulerian–Eulerian two-phase numerical approach. Within their investigation, a spectrum of nanoparticle diameters ranging from 100 to 200 nm and volume concentrations spanning from 1% to 4% were considered. Notably, the research unearthed intriguing findings, showcasing that an increase in nanoparticle diameter from 100 to 200 nm corresponded to a 15% rise in pressure drop, particularly evident at higher volume concentrations and a Reynolds number of 500. Conversely, the study observed that larger nanoparticles led to reduced heat transfer, emphasizing that the thermal performance was primarily influenced by the base fluid properties, providing valuable insights into the intricacies of nanofluid behavior in microchannel heat sinks [125].

Kalteh et al. [126] conducted a meticulous investigation into laminar convective heat transfer within a rectangular microchannel heat sink, with a focus on the utilization of an alumina–water nanofluid as the heat transfer medium. Notably, the research found that the two-phase Eulerian–Eulerian method outperformed the single-phase model, exhibiting superior agreement with both numerical predictions and experimental data. This underlined the effectiveness of employing the two-phase approach for simulating nanofluid heat transfer. The study further revealed that Nusselt numbers exhibited a noticeable increase with elevated Reynolds numbers, greater volume concentrations, and the utilization of smaller nanoparticle sizes. Importantly, the investigation also highlighted that differences in velocity and temperature between the two phases were negligible, shedding light on critical factors influencing nanofluid-based convective heat transfer within microchannel heat sinks [126].

Naphon et al. [127] embarked on a comprehensive investigation into nanofluid jet impingement heat transfer and associated pressure drop within a microchannel heat sink, employing a combination of CFD and artificial neural networks (ANN). The Levenberg–Marquardt backpropagation algorithm played a pivotal role in fine-tuning and optimizing the model. Notably, the outcomes of this study were meticulously benchmarked against empirical data using the Eulerian two-phase approach model. The results demonstrated that the optimal ANN model exhibited a striking degree of agreement with measured data, with deviations well within 1.5% for both Nusselt number and pressure drop, and a maximum error of merely 1.25%. This innovative approach not only showcased the remarkable accuracy of ANN in predicting heat transfer and pressure drop but also demonstrated its versatility by applying it to analyze microchannel heat sinks of various configurations, underscoring its potential as a powerful tool for optimizing thermal management systems [127].

Ambreen et al. [111] conducted a detailed investigation utilizing an Eulerian–Lagrangian model to analyze the cooling performance of a micropin-fin heat sink, employing both water and nanofluid as the cooling mediums. The nanofluid in question consisted of alumina nanoparticles with varying volume fractions. The primary focus of their analysis was a heat sink featuring staggered circular micropin-fins, with a keen emphasis on evaluating key parameters such as heat transfer coefficients, enhancement ratios, and thermal resistance. Notably, their findings underscored the superior performance of the nanofluid at higher particle concentrations, with the potential to achieve a substantial 16% enhancement in the heat transfer coefficient, all while maintaining a limited pressure drop. This research thus illuminates the significant advantages of nanofluid-based cooling solutions for micropin-fin heat sinks, offering improved thermal performance with minimal trade-offs in terms of pressure drop [111].

Ambreen et al. [128] embarked on a comprehensive study focusing on entropy generation and hydrothermal properties within heat sinks, utilizing water–borne multi-walled carbon nanotubes (MWCNT) and MWCNT/Nnanofluids of graphene nanoplatelets (GNP) nanofluids and exploring various fin-equipped configurations. Employing a multiphase Eulerian–Lagrangian model enriched with slip mechanisms and nanolayering effects, their analysis provided valuable insights into performance optimization. Notably, the study highlighted the exceptional performance of rhombus and hydrofoil pin-fin designs for both nanofluids, with triangular pin-fins also demonstrating commendable results. However, it was discerned that the streamlined shapes of hydrofoil and rhombus fins could impede coolant distribution, influencing heat dissipation. Among the array of configurations explored, the rhombus pin-fin heat sink emerged as the standout choice, boasting superior performance criteria and concurrently offering lower pressure drop, underscoring its potential as an efficient heat dissipation solution in the realm of hydrothermal applications [128].

Ambreen et al. [129] delved into a meticulous exploration of the thermohydraulic characteristics inherent to a hybrid nanofluid within a micro pin-fin heat sink. Employing a multiphase Lagrangian–Eulerian approach, their study was characterized by a comprehensive analysis that encompassed various fin configurations and slip mechanisms. Utilizing temperature differences and Nusselt numbers as performance indicators, the research underscored the remarkable enhancements in thermal efficiency achieved with different types of fins and their specific configurations when nanofluids were employed as the cooling medium. This research, therefore, serves as an invaluable resource, providing profound insights into the optimization of heat sink performance through the strategic implementation of hybrid nanofluids in tandem with tailored fin arrangements [129].

8.6 Conclusion and outlook

The application of nanofluids within microchannel heat sinks presents a promising avenue for revolutionizing thermal management across diverse engineering applications. Nanofluids, created by blending nanoparticles into conventional heat transfer fluids, introduce remarkable heat transfer properties, predominantly heightened thermal conductivity, and improved convective heat transfer characteristics. This distinctive combination positions nanofluids as an appealing solution for optimizing the performance of microchannel heat sinks.

Numerical simulations serve as indispensable tools for comprehending and optimizing the intricate behavior of nanofluids within microchannels. Computational methodologies, exemplified by CFD, have opened windows to profound insights into the complex fluid dynamics and heat transfer processes occurring within these tiny channels. Such simulations offer a cost-effective and efficient means to explore a gamut of parameters, encompassing microchannel geometries, flow conditions, and nanoparticle concentrations, allowing for the fine-tuning of microchannel heat sink designs tailored to precise cooling requirements.

The trajectory of nanofluids in microchannel heat sinks holds tremendous promise, driven by the ever-expanding demand for efficient thermal management in electronics, microelectronics, and various engineering systems. Several key domains of development and research are set to shape the future of this field:

1. *Nanoparticle selection*: Researchers will embark on a broader exploration of nanoparticles, emphasizing those endowed with superior thermal conductivity, steadfast stability, and harmonious compatibility with diverse base fluids. This exploration is likely to yield nanofluids characterized by heightened efficiency and long-term stability.
2. *Enhanced simulation techniques*: The continuous evolution of computational techniques will further refine simulation accuracy and efficiency. This will enable the modeling of nanofluid behavior in microchannels with increasing complexity and detail. Machine learning and AI-driven simulations are poised to play a pivotal role in optimizing heat sink designs.
3. *Experimental validation*: As the practical use of nanofluids gains momentum, there will be a heightened emphasis on empirical validation to corroborate simulation results. This validation process will be instrumental in establishing the reliability and safety of nanofluid-based cooling systems.
4. *Customized heat sink design*: The trend toward customizing microchannel heat sink designs to align with specific applications and thermal requisites is set to intensify. This entails a meticulous optimization of channel geometries, nanoparticle concentrations, and fluid flow dynamics, ultimately culminating in the maximization of heat dissipation.
5. *Environmental and safety considerations*: Environmental sustainability and safety will occupy a central role as nanofluid technologies advance. Researchers and industries will need to address concerns related to the responsible use and disposal of nanofluids to ensure their eco-friendly and secure integration into industrial processes.
6. *Commercialization*: With the maturation of technology and growing recognition of its benefits, the commercial adoption of nanofluid-based microchannel heat sinks is poised for expansion across diverse sectors. This encompasses electronics cooling, automotive, aerospace, and renewable energy systems, among others.

In summation, the integration of nanofluids into microchannel heat sinks, propelled by advanced simulation techniques, represents a promising solution for addressing contemporary thermal management challenges in engineering. As research and innovation continue to unfold, with sustainability at the core, nanofluid-based cooling solutions are poised to assume a pivotal role in the future landscape of thermal management technology, promising enhanced efficiency and sustainability across a myriad of applications.

References

[1] J. Mathew, S. Krishnan, A review on transient thermal management of electronic devices, J. Electron. Packaging 144 (2021).

[2] H.-S.P. Wong, K. Akarvardar, D. Antoniadis, J. Bokor, C. Hu, T.-J. King-Liu, et al., A density metric for semiconductor technology [point of view], Proc. IEEE 108 (2020) 478–482.
[3] P. Ye, T. Ernst, M.V. Khare, The last silicon transistor: nanosheet devices could be the final evolutionary step for Moore's law, IEEE Spectr. 56 (2019) 30–35.
[4] A. Bar-Cohen, P. Wang, Thermal management of on-chip hot spot, J. Heat Transfer. 134 (5) (2012) 051017.
[5] T. Wei, H. Oprins, V. Cherman, E. Beyne, M. Baelmans, Low-cost energy-efficient on-chip hotspot targeted microjet cooling for high-power electronics, IEEE Trans. Compon. Packag. Manuf. Technol. 10 (2019) 577–589.
[6] P. Wang, P. McCluskey, A. Bar-Cohen, Two-phase liquid cooling for thermal management of IGBT power electronic module, J. Electron. Packag. 135 (2013) 021001.
[7] F. Zhou, S.N. Joshi, Y. Liu, E.M. Dede, Near-junction cooling for next-generation power electronics, Int. Commun. Heat Mass Transf. 108 (2019) 104300.
[8] M. Pecht, P. Lall, E.B. Hakim, The influence of temperature on integrated circuit failure mechanisms, Qual. Reliab. Eng. Int. 8 (1992) 167–176.
[9] M. Pecht, J. Gu, Physics-of-failure-based prognostics for electronic products, Trans. Inst. Meas. Control. 31 (2009) 309–322.
[10] E.M. Alawadhi, C.H. Amon, Thermal analyses of a PCM thermal control unit for portable electronic devices: experimental and numerical studies, in: ITherm 2002. Eighth Intersociety Conference on Thermal and Thermomechanical Phenomena in Electronic Systems (Cat. No. 02CH37258), IEEE, 2002, pp. 466–475.
[11] S.S. Murshed, C.N.D. Castro, A critical review of traditional and emerging techniques and fluids for electronics cooling, Renew. Sustain. Energy Rev. 78 (2017) 821–833.
[12] K.P. Drummond, D. Back, M.D. Sinanis, D.B. Janes, D. Peroulis, J.A. Weibel, et al., A hierarchical manifold microchannel heat sink array for high-heat-flux two-phase cooling of electronics, Int. J. Heat Mass Transf. 117 (2018) 319–330.
[13] L. Du, W. Hu, An overview of heat transfer enhancement methods in microchannel heat sinks, Chem. Eng. Sci. 280 (2023) 119081.
[14] C. Li, G. Peterson, Parametric study of pool boiling on horizontal highly conductive microporous coated surfaces, J. Heat Transfer. 129 (11) (2007) 1465–1475.
[15] I. Mudawar, D. Bharathan, K. Kelly, S. Narumanchi, Two-phase spray cooling of hybrid vehicle electronics, IEEE Trans. Compon. Packag. Technol. 32 (2009) 501–512.
[16] C. Green, P. Kottke, X. Han, C. Woodrum, T. Sarvey, P. Asrar, et al., A review of two-phase forced cooling in three-dimensional stacked electronics: technology integration, J. Electron. Packag. 137 (2015) 040802.
[17] C.H. Amon, S.-C. Yao, C.-F. Wu, C.-C. Hsieh, Microelectromechanical system-based evaporative thermal management of high heat flux electronics, J. Heat Transf. 127 (2005) 66–75.
[18] D.B. Tuckerman, R.F.W. Pease, High-performance heat sinking for VLSI, IEEE Electron. Device Lett. 2 (1981) 126–129.
[19] D. Deng, L. Chen, W. Wan, T. Fu, X. Huang, Flow boiling performance in pin fin-interconnected reentrant microchannels heat sink in different operational conditions, Appl. Therm. Eng. 150 (2019) 1260–1272.
[20] D. Deng, X. Chen, L. Chen, Y. Lian, T. Fu, Preparation of porous structures on copper microchannel surfaces by laser writing, Sci. China Technol. Sci. 62 (2019) 2261–2270.

[21] D. Deng, G. Pi, W. Zhang, P. Wang, T. Fu, Numerical study of double-layered microchannel heat sinks with different cross-sectional shapes, Entropy 21 (2018) 16.
[22] S.G. Kandlikar, W.J. Grande, Evolution of microchannel flow passages—thermohydraulic performance and fabrication technology, Heat Transf. Eng. 24 (2003) 3–17.
[23] S. Mehendale, A. Jacobi, R. Shah, Fluid flow and heat transfer at micro- and meso-scales with application to heat exchanger design, Appl. Mech. Rev. 53 (7) (2000) 175–193.
[24] N. Liu, Z. Yu, T. Zhu, X. Yin, H. Zhang, Effects of microstructured surface and mixed surfactants on the heat transfer performance of pulsed spray cooling, Int. J. Therm. Sci. 158 (2020) 106530.
[25] L.E. O'Neill, I. Mudawar, Review of two-phase flow instabilities in macro-and micro-channel systems, Int. J. Heat Mass Transf. 157 (2020) 119738.
[26] P. Rosa, T. Karayiannis, M. Collins, Single-phase heat transfer in microchannels: the importance of scaling effects, Appl. Therm. Eng. 29 (2009) 3447–3468.
[27] A. Alfaryjat, H. Mohammed, N.M. Adam, M. Ariffin, M.I. Najafabadi, Influence of geometrical parameters of hexagonal, circular, and rhombus microchannel heat sinks on the thermohydraulic characteristics, Int. Commun. Heat Mass Transf. 52 (2014) 121–131.
[28] N. Tran, Y.-J. Chang, C.-C. Wang, Optimization of thermal performance of multi-nozzle trapezoidal microchannel heat sinks by using nanofluids of Al_2O_3 and TiO_2, Int. J. Heat Mass Transf. 117 (2018) 787–798.
[29] Y. Zhang, S. Wang, P. Ding, Effects of channel shape on the cooling performance of hybrid micro-channel and slot-jet module, Int. J. Heat Mass Transf. 113 (2017) 295–309.
[30] J. Song, F. Liu, Y. Sui, D. Jing, Numerical studies on the hydraulic and thermal performances of trapezoidal microchannel heat sink, Int. J. Therm. Sci. 161 (2021) 106755.
[31] M. Khoshvaght-Aliabadi, S. Hassani, S.H. Mazloumi, M. Nekoei, Effects of nooks configuration on hydrothermal performance of zigzag channels for nanofluid-cooled microelectronic heat sink, Microelectron. Reliab. 79 (2017) 153–165.
[32] F. Alnaimat, D. Varghese, B. Mathew, Investigation of thermal and hydraulic performance of MEMS heat sinks with zig-zag microchannels, Int. J. Thermofluids 16 (2022) 100213.
[33] N. Tiwari, M.K. Moharana, Conjugate effect on flow boiling instability in wavy microchannel, Int. J. Heat Mass Transf. 166 (2021) 120791.
[34] A.S. El-dean, O. Hassan, H. Shafey, Heat transfer characteristics of two-phase flow in a double-layer microchannel heat sink, Int. Commun. Heat Mass Transf. 132 (2022) 105899.
[35] K. Derakhshanpour, R. Kamali, M. Eslami, Improving performance of single and double-layered microchannel heat sinks by cylindrical ribs: a numerical investigation of geometric parameters, Int. Commun. Heat Mass Transf. 126 (2021) 105440.
[36] N. Hu, Q. Wang, S. Liu, J. Gu, L. Li, J. Lyu, A narrow shape double-layer micro-channel heat sink (DL-MCHS) designed for high-power laser crystal, Appl. Therm. Eng. 211 (2022) 118456.
[37] Y. Wang, G.-F. Ding, Numerical analysis of heat transfer in a manifold microchannel heat sink with high efficient copper heat spreader, Microsyst. Technol. 14 (2008) 389–395.

[38] D. Kong, Y. Kim, M. Kang, E. Song, Y. Hong, H.S. Kim, et al., A holistic approach to thermal-hydraulic design of 3D manifold microchannel heat sinks for energy-efficient cooling, Case Stud. Therm. Eng. 28 (2021) 101583.

[39] F. Brighenti, N. Kamaruzaman, J.J. Brandner, Investigation of self-similar heat sinks for liquid cooled electronics, Appl. Therm. Eng. 59 (2013) 725–732.

[40] D. Sreehari, A.K. Sharma, On thermal performance of serpentine silicon microchannels, Int. J. Therm. Sci. 146 (2019) 106067.

[41] N.H. Naqiuddin, L.H. Saw, M.C. Yew, F. Yusof, T.C. Ng, M.K. Yew, Overview of micro-channel design for high heat flux application, Renew. Sustain. Energy Rev. 82 (2018) 901–914.

[42] M. Jamshidmofid, M. Olfati, A. Arabkoohsar, A. Abbassi, Entropy generation and exergy destruction in two types of wavy microchannels working with various aqueous nanofluids using a multi-phase mixture model, J. Therm. Anal. Calorim. 148 (2023) 3711–3729.

[43] T.-C. Hung, W.-M. Yan, W.-P. Li, Analysis of heat transfer characteristics of double-layered microchannel heat sink, Int. J. Heat Mass Transf. 55 (2012) 3090–3099.

[44] F. Pourfattah, A.A. Abbasian Arani, M.R. Babaie, H.M. Nguyen, A. Asadi, On the thermal characteristics of a manifold microchannel heat sink subjected to nanofluid using two-phase flow simulation, Int. J. Heat Mass Transf. 143 (2019) 118518.

[45] M. Bahiraei, M. Jamshidmofid, S. Heshmatian, Entropy generation in a heat exchanger working with a biological nanofluid considering heterogeneous particle distribution, Adv. Powder Technol. 28 (2017) 2380–2392.

[46] Z. Soleymani, M. Rahimi, M. Gorzin, Y. Pahamli, Performance analysis of hotspot using geometrical and operational parameters of a microchannel pin-fin hybrid heat sink, Int. J. Heat Mass Transf. 159 (2020) 120141.

[47] M.E. Polat, F. Ulger, S. Cadirci, Multi-objective optimization and performance assessment of microchannel heat sinks with micro pin-fins, Int. J. Therm. Sci. 174 (2022) 107432.

[48] G. Lu, J. Yang, X. Zhai, X. Wang, Hotspot thermal management in microchannel heat sinks with vortex generators, Int. J. Therm. Sci. 161 (2021) 106727.

[49] H.-l Liu, H.-y Guo, Z.-l Xie, L. Sang, Numerical investigations for optimizing a novel micro-channel sink with perforated baffles and perforated walls, Int. Commun. Heat Mass Transf. 126 (2021) 105342.

[50] M.S. Lori, K. Vafai, Heat transfer and fluid flow analysis of microchannel heat sinks with periodic vertical porous ribs, Appl. Therm. Eng. 205 (2022) 118059.

[51] K. Derakhshanpour, R. Kamali, M. Eslami, Effect of rib shape and fillet radius on thermal-hydrodynamic performance of microchannel heat sinks: a CFD study, Int. Commun. Heat Mass Transf. 119 (2020) 104928.

[52] A. Husain, K.-Y. Kim, Thermal optimization of a microchannel heat sink with trapezoidal cross section, (2009).

[53] Q. Zhu, H. Xia, J. Chen, X. Zhang, K. Chang, H. Zhang, et al., Fluid flow and heat transfer characteristics of microchannel heat sinks with different groove shapes, Int. J. Therm. Sci. 161 (2021) 106721.

[54] T. Hou, Y. Chen, Pressure drop and heat transfer performance of microchannel heat exchanger with different reentrant cavities, Chem. Eng. Process. Process Intensif. 153 (2020) 107931.

[55] X. Jiang, S. Waqar Ali Shah, J. Liu, Y. Li, S. Zhang, Z. Wang, et al., Design of micro-nano structures for counter flow diverging microchannel heat sink with extraordinarily high energy efficiency, Appl. Therm. Eng. 209 (2022) 118229.

[56] M.I. Hasan, H.L. Tbena, Using of phase change materials to enhance the thermal performance of micro channel heat sink, Eng. Sci. Technol. 21 (2018) 517–526.

[57] X.-Y. Shen, M. Hekmatifar, M.Y.A. Shukor, A.A. Alizadeh, Y.-L. Sun, D. Toghraie, et al., Molecular dynamics simulation of water-based Ferro-nanofluid flow in the microchannel and nanochannel: effects of number of layers and material of walls, J. Mol. Liq. 338 (2021) 116924.

[58] A.H. Al-Zaidi, M.M. Mahmoud, T.G. Karayiannis, Flow boiling in copper and aluminium microchannels, Int. J. Heat Mass Transf. 194 (2022) 123101.

[59] M.T. Sarowar, Performance comparison of microchannel heat sink using boron-based ceramic materials, Adv. Mater. Res. 1163 (2021) 73–88.

[60] M.T. Sarowar, Numerical analysis of a liquid metal cooled mini channel heat sink with five different ceramic substrates, Ceram. Int. 47 (2021) 214–225.

[61] P. Gunnasegaran, H.A. Mohammed, N.H. Shuaib, R. Saidur, The effect of geometrical parameters on heat transfer characteristics of microchannels heat sink with different shapes, Int. Commun. Heat Mass Transf. 37 (2010) 1078–1086.

[62] M. Khoshvaght-Aliabadi, S.M. Hassani, S.H. Mazloumi, Enhancement of laminar forced convection cooling in wavy heat sink with rectangular ribs and Al_2O_3/water nanofluids, Exp. Therm. Fluid Sci. 89 (2017) 199–210.

[63] M. Khoshvaght-Aliabadi, S.M. Hassani, S.H. Mazloumi, Performance enhancement of straight and wavy miniature heat sinks using pin-fin interruptions and nanofluids, Chem. Eng. Process. Process Intensif. 122 (2017) 90–108.

[64] M. Khoshvaght-Aliabadi, S.M. Hassani, S.H. Mazloumi, Comparison of hydrothermal performance between plate fins and plate-pin fins subject to nanofluid-cooled corrugated miniature heat sinks, Microelectron. Reliab. 70 (2017) 84–96.

[65] M. Khoshvaght-Aliabadi, E. Ahmadian, O. Sartipzadeh, Effects of different pin-fin interruptions on performance of a nanofluid-cooled zigzag miniature heat sink (MHS), Int. Commun. Heat Mass Transf. 81 (2017) 19–27.

[66] M. Khoshvaght-Aliabadi, O. Sartipzadeh, S. Pazdar, M. Sahamiyan, Experimental and parametric studies on a miniature heat sink with offset-strip pins and Al_2O_3/water nanofluids, Appl. Therm. Eng. 111 (2017) 1342–1352.

[67] M. Khoshvaght-Aliabadi, S. Deldar, S.M. Hassani, Effects of pin-fins geometry and nanofluid on the performance of a pin-fin miniature heat sink (PFMHS), Int. J. Mech. Sci. 148 (2018) 442–458.

[68] M. Khoshvaght-Aliabadi, M. Sahamiyan, Performance of nanofluid flow in corrugated minichannels heat sink (CMCHS), Energy Convers. Manage. 108 (2016) 297–308.

[69] A. Muhammad, D. Selvakumar, A. Iranzo, Q. Sultan, J. Wu, Comparison of pressure drop and heat transfer performance for liquid metal cooled mini-channel with different coolants and heat sink materials, J. Therm. Anal. Calorim. 141 (2020) 289–300.

[70] H.M. Park, Y.H. Jeong, S. Heo, Effect of heater material and coolant additives on CHF for a downward facing curved surface, Nucl. Eng. Des. 278 (2014) 344–351.

[71] M. Jamshidmofid, A. Abbassi, M. Bahiraei, Efficacy of a novel graphene quantum dots nanofluid in a microchannel heat exchanger, Appl. Therm. Eng. 189 (2021) 116673.

[72] G. Sriharan, S. Harikrishnan, H.F. Oztop, A review on thermophysical properties, preparation, and heat transfer enhancement of conventional and hybrid nanofluids utilized in micro and mini channel heat sink, Sustain. Energy Technol. Assess. 58 (2023) 103327.

[73] S. Kumar, A. Kumar, A. Darshan Kothiyal, M. Singh Bisht, A review of flow and heat transfer behaviour of nanofluids in micro channel heat sinks, Therm. Sci. Eng. Prog. 8 (2018) 477–493.

[74] S. Rashidi, F. Hormozi, N. Karimi, W. Ahmed, Chapter 14 – Applications of nanofluids in thermal energy transport, in: W. Ahmed, M. Booth, E. Nourafkan (Eds.), Emerging Nanotechnologies for Renewable Energy, Elsevier, 2021, pp. 345–368.

[75] M. Gupta, V. Singh, R. Kumar, Z. Said, A review on thermophysical properties of nanofluids and heat transfer applications, Renew. Sustain. Energy Rev. 74 (2017) 638–670.

[76] Y. Xuan, Q. Li, Heat transfer enhancement of nanofluids, Int. J. Heat Fluid Flow 21 (2000) 58–64.

[77] G. Sriharan, S. Harikrishnan, H.F. Oztop, Performance improvement of the mini hexagonal tube heat sink using nanofluids, Therm. Sci. Eng. Prog. 34 (2022) 101390.

[78] I. Wole-Osho, H. Adun, M. Adedeji, E.C. Okonkwo, D. Kavaz, M. Dagbasi, Effect of hybrid nanofluids mixture ratio on the performance of a photovoltaic thermal collector, Int. J. Energy Res. 44 (2020) 9064–9081.

[79] Z. Haddad, C. Abid, H.F. Oztop, A. Mataoui, A review on how the researchers prepare their nanofluids, Int. J. Therm. Sci. 76 (2014) 168–189.

[80] M. Wagener, B. Murty, B. Günther, Preparation of metal nanosuspensions by high-pressure DC-sputtering on running liquids, MRS Online Proc. Lib. (OPL) 457 (1996) 149–154.

[81] A.R.I. Ali, B. Salam, A review on nanofluid: preparation, stability, thermophysical properties, heat transfer characteristics and application, SN Appl. Sci. 2 (2020) 1636.

[82] A. Karakaş, S. Harikrishnan, H.F. Öztop, Preparation of EG/water mixture-based nanofluids using metal-oxide nanocomposite and measurement of their thermophysical properties, Therm. Sci. Eng. Prog. 36 (2022) 101538.

[83] M.-C. Lu, C.-H. Huang, Specific heat capacity of molten salt-based alumina nanofluid, Nanoscale Res. Lett. 8 (2013) 1–7.

[84] P.N. Nwosu, J. Meyer, M. Sharifpur, A review and parametric investigation into nanofluid viscosity models, J. Nanotechnol. Eng. Med. 5 (2014) 031008.

[85] M. Sharifpur, S. Yousefi, J.P. Meyer, A new model for density of nanofluids including nanolayer, Int. Commun. Heat Mass Transf. 78 (2016) 168–174.

[86] X. Zhang, H. Gu, M. Fujii, Effective thermal conductivity and thermal diffusivity of nanofluids containing spherical and cylindrical nanoparticles, J. Appl. Phys. 100 (2006).

[87] S. Chakraborty, P.K. Panigrahi, Stability of nanofluid: a review, Appl. Therm. Eng. 174 (2020) 115259.

[88] Y. Hu, Y. He, Z. Zhang, D. Wen, Enhanced heat capacity of binary nitrate eutectic salt-silica nanofluid for solar energy storage, Sol. Energy Mater. Sol. Cell 192 (2019) 94–102.

[89] H. Zhu, C. Zhang, S. Liu, Y. Tang, Y. Yin, Effects of nanoparticle clustering and alignment on thermal conductivities of Fe_3O_4 aqueous nanofluids, Appl. Phys. Lett. 89 (2006).

[90] M. Almahfood, B. Bai, The synergistic effects of nanoparticle-surfactant nanofluids in EOR applications, J. Pet. Sci. Eng. 171 (2018) 196–210.

[91] A.S. Habeeb, A.A. Karamallah, S. Aljabair, Review of computational multi-phase approaches of nano-fluids filled systems, Therm. Sci. Eng. Prog. 28 (2022) 101175.

[92] M. Aramesh, F. Pourfayaz, A. Kasaeian, Numerical investigation of the nanofluid effects on the heat extraction process of solar ponds in the transient step, Sol. Energy 157 (2017) 869–879.

[93] M.I. Abdulwahab, S. Thahab, A.H. Dhiaa, Experimental study of thermophysical properties of TiO_2 nanofluid, Iraqi J. Chem. Pet. Eng. 17 (2016) 1–6.

[94] F. Garoosi, B. Rohani, M.M. Rashidi, Two-phase mixture modeling of mixed convection of nanofluids in a square cavity with internal and external heating, Powder Technol. 275 (2015) 304–321.

[95] A. Rezaei Gorjaei, R. Rahmani, Numerical simulation of nanofluid flow in a channel using Eulerian–Eulerian two-phase model, Int. J. Thermophys. 42 (2021) 1–16.

[96] M. Jamshidmofid, M. Bahiraei, Hydrothermal performance of single and hybrid nanofluids in left-right and up-down wavy microchannels using two-phase mixture approach, Int. Commun. Heat Mass Transf. 129 (2021) 105752.

[97] M. Kalteh, A. Abbassi, M. Saffar-Avval, J. Harting, Eulerian–Eulerian two-phase numerical simulation of nanofluid laminar forced convection in a microchannel, Int. J. Heat Fluid Flow 32 (2011) 107–116.

[98] M. Jamshidmofid, M. Bahiraei, Thermohydraulic assessment of a novel hybrid nanofluid containing cobalt oxide-decorated reduced graphene oxide nanocomposite in a microchannel heat sink with sinusoidal cavities and rectangular ribs, Int. Commun. Heat Mass Transf. 131 (2022) 105769.

[99] S. Göktepe, K. Atalık, H. Ertürk, Comparison of single and two-phase models for nanofluid convection at the entrance of a uniformly heated tube, Int. J. Therm. Sci. 80 (2014) 83–92.

[100] M. Bahiraei, M. Jamshidmofid, M. Goodarzi, Efficacy of a hybrid nanofluid in a new microchannel heat sink equipped with both secondary channels and ribs, J. Mol. Liq. 273 (2019) 88–98.

[101] N.A.C. Sidik, M.N.A.W.M. Yazid, S. Samion, M.N. Musa, R. Mamat, Latest development on computational approaches for nanofluid flow modeling: Navier–Stokes based multiphase models, Int. Commun. Heat Mass Transf. 74 (2016) 114–124.

[102] M.R. Safaei, A. Jahanbin, A. Kianifar, S. Gharehkhani, A.S. Kherbeet, M. Goodarzi, et al., Mathematical modeling for nanofluids simulation: a review of the latest works, Model. Simul. Eng. Sci. (2016) 189–220.

[103] S. Rashidi, S. Akar, M. Bovand, R. Ellahi, Volume of fluid model to simulate the nanofluid flow and entropy generation in a single slope solar still, Renew. Energy 115 (2018) 400–410.

[104] A. Alimoradi, M. Olfati, M. Maghareh, Numerical investigation of heat transfer intensification in shell and helically coiled finned tube heat exchangers and design optimization, Chem. Eng. Process. Process. Intensif. 121 (2017) 125–143.

[105] M. Bahiraei, Particle migration in nanofluids: a critical review, Int. J. Therm. Sci. 109 (2016) 90–113.

[106] N.M. Noh, A. Fazeli, N.C. Sidik, Numerical simulation of nanofluids for cooling efficiency in microchannel heat sink, J. Adv. Res. Fluid Mech. Therm. Sci. 4 (2014) 13–23.

[107] M. Amani, P. Amani, A. Kasaeian, O. Mahian, W.-M. Yan, Two-phase mixture model for nanofluid turbulent flow and heat transfer: effect of heterogeneous distribution of nanoparticles, Chem. Eng. Sci. 167 (2017) 135–144.
[108] M. Bahiraei, M. Jamshidmofid, M. Amani, R. Barzegarian, Investigating exergy destruction and entropy generation for flow of a new nanofluid containing graphene–silver nanocomposite in a micro heat exchanger considering viscous dissipation, Powder Technol. 336 (2018) 298–310.
[109] M. Bahiraei, S.M. Naghibzadeh, M. Jamshidmofid, Efficacy of an eco-friendly nanofluid in a miniature heat exchanger regarding to arrangement of silver nanoparticles, Energy Convers. Manage. 144 (2017) 224–234.
[110] M. Bahiraei, M. Jamshidmofid, M. Dahari, Second law analysis of hybrid nanofluid flow in a microchannel heat sink integrated with ribs and secondary channels for utilization in miniature thermal devices, Chem. Eng. Process. Process Intensif. 153 (2020) 107963.
[111] T. Ambreen, A. Saleem, C.W. Park, Numerical analysis of the heat transfer and fluid flow characteristics of a nanofluid-cooled micropin-fin heat sink using the Eulerian–Lagrangian approach, Powder Technol. 345 (2019) 509–520.
[112] V. Kumar, J. Sarkar, Numerical and experimental investigations on heat transfer and pressure drop characteristics of Al_2O_3–TiO_2 hybrid nanofluid in minichannel heat sink with different mixture ratio, Powder Technol. 345 (2019) 717–727.
[113] V. Kumar, J. Sarkar, Two-phase numerical simulation of hybrid nanofluid heat transfer in minichannel heat sink and experimental validation, Int. Commun. Heat Mass Transf. 91 (2018) 239–247.
[114] M. Bahiraei, N. Mazaheri, M.R. Daneshyar, A. Mwesigye, Two-phase simulation of irreversibilities for Ag–water nanofluid flow inside an elliptical pin-fin heat sink: entropy generation and exergy considerations, Powder Technol. 409 (2022) 117723.
[115] A. Sarlak, A. Ahmadpour, M.R. Hajmohammadi, Thermal design improvement of a double-layered microchannel heat sink by using multi-walled carbon nanotube (MWCNT) nanofluids with non-Newtonian viscosity, Appl. Therm. Eng. 147 (2019) 205–215.
[116] Q. Wang, S. Zhang, Y. Zhang, J. Fu, Z. Liu, Enhancing performance of nanofluid mini-channel heat sinks through machine learning and multi-objective optimization of operating parameters, Int. J. Heat Mass Transf. 210 (2023) 124204.
[117] J. Mohammadpour, F. Salehi, A. Lee, L. Brandt, Nanofluid heat transfer in a microchannel heat sink with multiple synthetic jets and protrusions, Int. J. Therm. Sci. 179 (2022) 107642.
[118] S. Bazkhane, I. Zahmatkesh, Taguchi–based sensitivity analysis of hydrodynamics and heat transfer of nanofluids in a microchannel heat sink (MCHS) having porous substrates, Int. Commun. Heat Mass Transf. 118 (2020) 104885.
[119] M.K. Moraveji, R.M. Ardehali, CFD modeling (comparing single and two-phase approaches) on thermal performance of Al_2O_3/water nanofluid in mini-channel heat sink, Int. Commun. Heat Mass Transf. 44 (2013) 157–164.
[120] Z.Y. Ghale, M. Haghshenasfard, M.N. Esfahany, Investigation of nanofluids heat transfer in a ribbed microchannel heat sink using single-phase and multiphase CFD models, Int. Commun. Heat Mass Transf. 68 (2015) 122–129.
[121] Y.-T. Yang, K.-T. Tsai, Y.-H. Wang, S.-H. Lin, Numerical study of microchannel heat sink performance using nanofluids, Int. Commun. Heat Mass Transf. 57 (2014) 27–35.

[122] M. Bahiraei, N. Mazaheri, M.R. Daneshyar, Employing elliptical pin-fins and nanofluid within a heat sink for cooling of electronic chips regarding energy efficiency perspective, Appl. Therm. Eng. 183 (2021) 116159.
[123] T. Ambreen, A. Saleem, C.W. Park, Analysis of hydro-thermal and entropy generation characteristics of nanofluid in an aluminium foam heat sink by employing Darcy–Forchheimer–Brinkman model coupled with multiphase Eulerian model, Appl. Therm. Eng. 173 (2020) 115231.
[124] A. Radwan, M. Ahmed, Thermal management of concentrator photovoltaic systems using microchannel heat sink with nanofluids, Sol. Energy 171 (2018) 229–246.
[125] B. Fani, A. Abbassi, M. Kalteh, Effect of nanoparticles size on thermal performance of nanofluid in a trapezoidal microchannel-heat-sink, Int. Commun. Heat Mass Transf. 45 (2013) 155–161.
[126] M. Kalteh, A. Abbassi, M. Saffar-Avval, A. Frijns, A. Darhuber, J. Harting, Experimental and numerical investigation of nanofluid forced convection inside a wide microchannel heat sink, Appl. Therm. Eng. 36 (2012) 260–268.
[127] P. Naphon, S. Wiriyasart, T. Arisariyawong, L. Nakharintr, ANN, numerical and experimental analysis on the jet impingement nanofluids flow and heat transfer characteristics in the micro-channel heat sink, Int. J. Heat Mass Transf. 131 (2019) 329–340.
[128] T. Ambreen, A. Saleem, M. Tanveer, K. Anirudh, S.A. Shehzad, C.W. Park, Irreversibility and hydrothermal analysis of the MWCNTs/GNPs-based nanofluids for electronics cooling applications of the pin-fin heat sinks: multiphase Eulerian–Lagrangian modeling, Case Stud. Therm. Eng. 31 (2022) 101806.
[129] T. Ambreen, A. Saleem, H.M. Ali, S.A. Shehzad, C.W. Park, Performance analysis of hybrid nanofluid in a heat sink equipped with sharp and streamlined micro pin-fins, Powder Technol. 355 (2019) 552–563.

CHAPTER

9 Optimizing fluid flow efficiency: third-grade hybrid nanofluid flow with electro-magneto-hydrodynamics in confined vertical spaces

M.M. Bhatti[1], Munawwar Ali Abbas[2] and Saima Muhammad[2]

[1]*Material Science Innovation and Modelling (MaSIM) Research Focus Area, North-West University (Mafikeng Campus), Mmabatho, South Africa*

[2]*Department of Mathematics, University of Baltistan, Skardu, Pakistan*

9.1 Introduction

The method of augmenting heat transfer has significant importance across several industrial, technological, and medicinal domains due to its cost-effectiveness and energy efficiency. The present advancements in science and technology have fostered a growing need for miniature electronic gadgets that possess exceptional attributes, including superior performance, extended durability, and precise functionality. A group of researchers from several international institutions has developed a novel kind of fluid known as hybrid nanofluids (HNFs), which exhibits a superior heat transfer capability in comparison to traditional fluids [1–3]. HNFs may be produced by the integration of several types of nanoparticles into the base fluid medium. The term "hybrid nanofluid" is often used to describe a mixture of several nanomaterials spread inside a base fluid. The dimensions of these nanoparticles often fall within the nanoscale range, typically ranging from 1 to 100 nm. The objective of incorporating nanoparticles into a given base fluid is to augment the desired characteristics tailored for certain applications. Various categories of nanoparticles, including ceramic, metallic, and carbon-based nanoparticles, are often blended together to generate HNFs that possess a desired amalgamation of distinct characteristics. The aforementioned qualities may exhibit enhanced electrical conductivity, thermal conductivity, and rheological behavior in comparison to the base fluid in isolation. The precise composition of a HNF is contingent

Nanofluids. DOI: https://doi.org/10.1016/B978-0-443-13625-2.00012-7

upon the individual application and the desired qualities that need to be achieved. The researchers have developed HNFs with the aim of achieving targeted performance in several domains, including electronic cooling, lubrication, and heat transfer improvement, among others.

Recent research has discovered that hybrid nanoparticles, which are especially effective at high temperatures, may effectively replace traditional cooling. In comparison to mono nanofluids, HNFs exhibit improved thermal properties, making them desirable for heating applications like [4–7] heat sinks, thermal energy storage, automobile cooling systems, and solar thermal systems. As a result, several researchers have already studied the behavior of HNFs in a variety of structures. In their study, Tian et al. [8] used the support vector machine (SVM) approach to evaluate the apparent viscosity and shear rate of a non-Newtonian HNF flow. Following this, a sensitivity analysis was performed. The SVM approach was used in the researcher's study. The researcher's discoveries made a significant contribution to the understanding and management of multiphase non-Newtonian fluids. The present mathematical and statistical model has a unique capacity to forecast the shear stress, shear rate, and apparent viscosity of a non-Newtonian HNF composed of TiO_2/ZnO/EG. The findings of the study indicate that the proposed formulation and novel approach are suitable for evaluating the apparent viscosity and shear rate in non-Newtonian HNFs. Cheng et al. [9] conducted a study on the thermal characteristics of HNFs, using many fluid models such as the Jeffrey, Maxwell, and Oldroyd-B models. To achieve the intended objective, the researchers used several generalized non-Newtonian rheological models, namely the Maxwell model, Jeffrey model, and Oldroyd-B model. The findings of this work indicate that the memory effects seen in Oldroyd-B HNFs are more pronounced compared to those shown in Maxwell and Jeffrey HNFs. Furthermore, it was noted that the occurrence of viscous dissipation was comparatively less pronounced in relation to other models. Additionally, the rate of heat energy transmission was determined to be greater in the Jeffery HNF when compared to both the Oldroyd model and Maxwell model. The significance of magnetic nanofluids arises from their capacity to be externally manipulated and their capability to modify physical attributes in response to changes in volume fraction and magnetic field intensity. The inspiration for NS Wahid's work stemmed from the development of HNF, which represents a notable improvement over conventional nanofluid. The researcher investigated the flow characteristics at the stagnation point of a HNF comprising magnetite-cobalt ferrite particles dispersed in water. The results of their study indicate that Jeffrey nanofluids have a higher heat transmission capability when compared to the Oldroyd-B and Maxwell models. Furthermore, it was shown that the Oldroyd-B models had superior memory effects in comparison to the Jeffrey and Maxwell models. In their study, Wahid et al. [10] investigated the phenomenon of dual solutions and conducted an analysis of the characteristics of HNF flow across a shrinking Riga plate. This analysis results in the identification of two solutions as a consequence of the counteracting flow originating from the diminishing Riga plate. The solution that met the boundary requirements was determined to be the physically stable solution. Moreover, it was noted that the augmentation of the electrohydrodynamic effect and the velocity slip parameter led to enhanced heat transfer efficiency, hence emphasizing the significance of these factors in enhancing the cooling performance of the heat transfer fluid. The numerical solutions were obtained using the bvp4c built-in tool in MATLAB®. It was found that the initial solutions are both physically stable and satisfy the boundary conditions. A numerical investigation of the HNFs employed in the double-tube heat exchangers using carbon nanotubes (CNT) has been reported by Anitha and Pichumani [11]. The HNF consisted of CNT as nanoparticles blended with either Newtonian or non-Newtonian fluids as coolants. The findings of this study revealed the efficacy of the whole heat exchanger system and also examined its pumping power and pressure drop. They discovered that utilizing Newtonian HNFs increased the heat exchanger's coefficient of heat transfer by up to 25%. For Newtonian models, the performance index has grown to 67% in the turbulent zone. Hemmat Esfe et al. [12] reported an experimental investigation with a novel correlation on non-Newtonian (oil) HNFs containing zinc oxide and multiwalled CNT. It has been discovered that the viscosity of nanofluids has an inverse relationship with temperature, with higher temperatures resulting in lower viscosity values. This suggests that nanofluids have non-Newtonian fluid behavior at lower temperatures.

The assessment conducted by Modi et al. [13] offers a thorough review of the effectiveness of mono-nanofluid and HNF in improving the performance of solar stills. According to their study findings, water is found in plenty on Earth in the form of oceans. However, it is noted that the normal composition of seawater includes a total dissolved solid concentration that falls within the range of 35,000 to 45,000 parts per million (ppm). To address the need for potable water, it becomes imperative to engage in the process of desalination, which involves the removal of salt and impurities from salty water sources, such as saltwater or other contaminated water bodies. Solar stills are a kind of solar energy technology used to desalinate salty water. Nevertheless, solar stills often encounter obstacles about their limited production. The use of water-based nanofluids has emerged as a potentially effective strategy to improve the operational efficiency of solar stills. Furthermore, the researchers directed their attention to investigating the impact of many factors, including the kind of nanofluid, concentration of nanoparticles, and depth of nanofluid, on the efficacy of solar stills. This investigation was based on a comprehensive analysis of contemporary scholarly works. Furthermore, the authors engage in a comprehensive examination of current progressions about the utilization of nanofluids within the context of solar stills, encompassing a diverse range of augmentations. The results of their study are presented in a tabular style. In addition, this study delves into the examination of HNFs and their comparative benefits in comparison to mono-nanofluids, drawing upon a comprehensive analysis of pertinent scholarly works. Furthermore, an evaluation of their suitability in solar stills is conducted. In their article, Chu et al. [14] conducted a comprehensive investigation using a model-based approach to analyze the behavior of unsteady magnetohydrodynamic (MHD) HNF flow between two parallel infinite plates. The research specifically focused on examining the influence of particle shape on the flow characteristics. The researchers directed their attention to the examination of the impacts arising from the presence of unstable viscous flow in a compressing channel. The present study introduces HNF particles, consisting of a mix of silver and gold, exhibiting diverse morphologies, into the host fluid, namely engine oil. The present inquiry aims to explore the dynamics of flow and heat transfer in the presence of MHDs inside a system consisting of two parallel plates of infinite extent. To evaluate the effects of various forms of hybrid nanoparticles (blade, platelet, cylinder, and brick), thermal conductivity models, including the Yamada–Ota and Hamilton–Crosser models, are used for analysis. The authors have devised a mathematical model including a set of very nonlinear partial differential equations (PDEs). These equations are then turned into a system of linked and extremely nonlinear ordinary differential equations (ODEs) by the use of a boundary layer approximation. The numerical solution for the coupled system is computed with the bvp4c solver software in MATLAB®. Graphical representations are used to analyze the impact of different factors on physical attributes. Furthermore, the authors provide a comparative analysis with previously documented outcomes to substantiate the validity of their discoveries. In their study, Alqahtani et al. [15] conducted research aimed at developing a computational approach to address the challenges associated with a HNF including silver and gold nanoparticles. The fluid in question is characterized by its electrical conductivity and rotational motion and is situated between two parallel surfaces. The model developed by the researchers has attracted much interest and recognition from the scientific community. Traditional fluids, including vegetable oils, water, ethylene glycol (EG), and transformer oil, are of paramount importance in many industrial applications. Nevertheless, the limited thermal conductivity shown by these foundational fluids poses considerable obstacles in certain industrial processes. As a result, there has been an increasing focus on the advancement of nanotechnology to tackle these concerns. Nanoscience plays a significant role in improving heat transfer processes in diverse heating and transmission apparatus. Furthermore, the authors provide a clear definition of the MHD spinning flow of a HNF between two permeable surfaces. The HNF under consideration consists of silver (Ag) and gold (Au) nanoparticles that are distributed inside EG. The governing equations have been rendered nondimensional and then converted into a system of ODEs by use of similarity substitutions. The model utilizes the parametric continuation method as the numerical technique for solving first-order ODEs. The current research examined the importance of velocity and energy profiles in relation to different physical factors, and the findings are given via the use of tables and figures. The convective-radiative heat transport of a HNF in an open trapezoidal enclosure was modeled by Atashafrooz et al. [16]. Numerical simulations are used to examine the confluence of convective and radiative heat transport phenomena inside an open trapezoidal enclosure under the influence of MHDs. The enclosure is characterized by the presence of two diagonal side walls and a horizontal wall. The embedded boundary technique properly represents the sloping side walls of the open enclosure. The Roseland approximation is used to do calculations pertaining to the thermal radiation component inside the energy equation. The researchers used a nanofluid in their study, which consists of a combination of hybrid nanoparticles and a base fluid. To analyze the thermal characteristics, the study focuses on the spatial distribution of dimensionless temperature inside the enclosure and the changes in Nusselt numbers along the heated bottom wall. The experiments are carried out over a range of values for the radiative parameter, volume percentages of hybrid nanoparticles, and magnetic force. Arif et al. [17] investigated the heat transfer properties of a fractional model of pair stress Casson trihybrid system including nanoparticles of various shapes in blood. The study also discussed the possible biological uses of this system. The issue was defined by expressing it via the use of momentum and energy equations, which were represented as PDEs. These equations were accompanied by appropriate physical starting and boundary conditions to provide realism. Furthermore, the classical model has been converted into its fractional equivalents by the use of the Atangana–Baleanu time-fractional operator. The velocity and temperature equations have been solved using the Laplace and Fourier transforms. Additionally, the impact of physical factors on velocity and temperature profiles has been examined using graphical analysis using MATHCAD.

The study of fluid behavior in the presence of electric conduction has garnered significant attention from scholars because of its wide-ranging applications in several disciplines such as geophysics, astronomy, and engineering. Electromagnetic-hydrodynamic (EMHD), also referred to as EMHDs, is an interdisciplinary domain that integrates the fundamental principles of fluid dynamics with electromagnetics. HNFs exhibiting EMHD effects have been used across diverse technical domains such as heat pipes, solar collectors, blood flow analysis, thermal conductivity enhancement, heat exchangers, and radiators, among others. Jakeer and Reddy [18] employed the homotopy perturbation approach to study HNFs under the influence of EMHD and entropy production. Higher electric field values tend to boost the velocity field, but the magnetic field resists fluid motion. They also discovered that magnetic fields, temperature variations, and Eckert numbers boost entropy output. Khashi'ie et al. [19] assessed the effects of EMHD on copper and aluminum oxide nanoparticles floating in water-based HNFs. The HNFs flow over elastic Riga surfaces. They discovered that in the presence of significant EMHD, form factor, and thermal radiation, the heat transfer rate rises. Furthermore, they discovered that blade-shaped nanoparticles achieve greater thermal transfer rates, whereas sphere-shaped nanoparticles achieve the lowest thermal rates. Souayeh et al. [20] explored the effects of EMHD on gold and silver nanoparticles suspended in blood flow. Peristaltic flow with a convection mechanism is postulated. The thermophoretic force and the Grashof number were discovered to maximize the fluid velocity closer to the right wall. The thermal profile, on the other hand, reduces the Schmidt number and the radiation parameter. Mkhatshwa and Khumalo [21] demonstrated temperature-dependent EMHD effects on the Carreau HNF flow. The study revealed that the thermal transmission of the Carreau HNF was enhanced by the use of temperature fluctuations and thermal radiation. In their study, Ghazwani [22] conducted an evaluation of the effects of EMHD, viscosity dissipation, and thermal radiation on the flow of a HNF over a moving Riga plate. The researchers obtained data via the use of computational methodology and found that the variable of unsteadiness has a significant impact on both the Nusselt number and the profile of skin friction. In a study conducted by Gamachu et al. [23], the author examined the dynamics of nonlinear convection in the stagnation zone of a revolving sphere. The controlling equations are originally expressed as partial differential equations (DEs) and are afterward converted into ordinary DEs by the use of a similarity transformation approach. The research utilizes the numerical methodology offered by the bvp4c solver in MATLAB®. The findings also indicate that an augmentation in factors such as thermal radiation limitation, temperature ratio, magnetic field limitation, electric field limitation, and other variables results in an escalation in the temperature dispersion of both the HNF composed of nanoparticles and methanol, as well as the nanofluid consisting of nanoparticles and methanol. Furthermore, the heat transfer rate experiences the most significant enhancement when the volumetric fraction comprises 4% of nanoparticles. This leads to a notable increase of 5.13% and 4.37%, respectively, compared to a conventional fluid. The implications of these studies are relevant to a range of practical applications, such as the implementation of nanocoating spin processing in the field of chemical engineering, the use of agricultural chemical spray applications, the reduction of drag in pipeline flows, and the manufacturing of home cleaning products. The study performed by Kumar et al. [24] assessed the impact of the Arrhenius energy model and nonuniform heat sources on the flow behavior of an EMHD HNF across a rotating and deformable disk. Additionally, the researchers evaluated the implications of nonlinear radiation in their analysis. The analysis of the flow problem was conducted by using the modified Buongiorno model and considering the thermophysical properties of a nanoliquid consisting of $Cu - Fe_3O_4$ nanoparticles dispersed in a water medium. To enhance the credibility of this study, several factors such as the passive manipulation of nanoparticles, hydrodynamic slip phenomena, and convective boundary conditions have been thoroughly taken into account. Moreover, this study investigates engineering factors, including the moment coefficient and pumping efficiency of the disk, thereby augmenting the novelty of the research. The model's governing equations are converted into a system of first-order ODEs using suitable similarity transformations. The numerical solution of these equations is obtained using the bvp5c algorithm, which is based on the finite-difference method. It is worth noting that a one-unit increase in the electric parameter leads to a significant decrease of 41.75% in the skin friction coefficient, as well as a notable rise of 15.31% in the heat transfer rate. Moreover, it has been shown that the velocity of entrainment exhibits a direct relationship with fluctuations in the electric field parameter while displaying an inverse relationship with alterations in the volume fraction of copper and magnetite nanoparticles. Saleem et al. [25] conducted a computational study to investigate the unstable mixed convection of nanofluid flow at a stagnation point over a movable EMHD Riga plate. The nanocomposites are formed by combining TiO_2 nanostructures with two unique base fluids, namely H_2O and $C_2H_6O_2$. The flow problem encompasses the fundamental equations that regulate the dynamics of motion and energy, in addition to a distinct model that characterizes viscosity and thermal conductivity. The use of comparable elements is applied to streamline these computations for model problems. The simulation outcomes are acquired via the use of the Runge–Kutta (RK-4) technique and are visually shown in graphical form as well as given in a tabular manner. The flow and temperature profiles of the nanofluid are calculated and studied for both basic fluid theories being considered. Based on the study results, it has been shown that the heat exchange rate of the $C_2H_6O_2$ model exhibits a statistically significant increase in comparison to the H_2O model. Moreover, with an increase in the volume % of nanoparticles, the velocity field experiences a decline, whilst the temperature distribution exhibits enhancement. Moreover, the analysis of their results reveals that $TiO_2/C_2H_6O_2$ has the best heat coefficient for higher acceleration parameters, whereas TiO_2/H_2O displays the highest skin friction coefficient. The primary finding of this study is that the nanofluid based on $C_2H_6O_2$ has somewhat superior performance compared to the nanofluid based on H_2O. Khashi'ie et al. [26] provided an analysis of the unsteady separated stagnation point flow of HNFs in the presence of an EMHD Riga plate. Under the impact of electro-osmosis and MHDs. The primary aim of their study is to assess and compare the numerical solutions pertaining to the unstable separated stagnation point flow that is caused by a Riga plate. This study utilizes two different HNFs, namely copper–alumina/water and graphene–alumina/water. The presence of the Riga plate induces EMHD effects, which are anticipated to result in a postponement of the separation of the boundary layer. The mathematical representation of the flow and energy equations is derived by using boundary layer assumptions, followed by the use of similarity variables to simplify these equations. The numerical outcomes are acquired using the bvp4c function and are shown via visual and tabular depictions. It is important to acknowledge that this particular model had several limitations, including its dependence on a Riga plate as the test surface and water as the underlying fluid medium. The outcomes may exhibit variability when various types of wall surfaces or base fluids are taken into account. One additional constraint pertains to the use of the Takabi and Salehi correlation at the computational stage for HNFs. The outcomes of the research demonstrate the presence of two distinct solutions, with the first solution exhibiting stability, as verified by an investigation of its stability. It is worth noting that the composite material consisting of graphene–alumina/water displays the most elevated skin friction coefficient. Conversely, the composite material including copper–alumina/water shows the greatest thermal coefficient when subjected to increased acceleration parameters. Maraj et al. [27] did a study whereby they analyzed the peristaltic flow of a nanofluid that consisted of menthol inside a channel of uniform shape, which also included a shape element. The inquiry focused on the analysis of the peristaltic transport of electrolytes containing menthol, which were subjected to modification by the application of an external electric field. The phenomenon in question leads to the generation of an overall electric charge on the surface, which in turn causes the attraction of counterions and the repulsion of coions inside the nanofluid containing menthol. As a result, the aforementioned process gives rise to the creation of an electrical double layer, which is a distinct area distinguished by an overall charge in the immediate vicinity of the surface. This research is especially focused on examining the effects of electroosmotic body forces on the base fluid, with a particular emphasis on the presence and absence of copper nanoparticles. The selection of menthol-based fluid as the focal point of investigation is motivated by its significance within the domains of medical and taste enhancement. Furthermore, this scholarly study investigates the collective impacts of electro-osmotically regulated magneto-hydrodynamic peristaltic flow inside a homogeneous conduit, a subject that has not been examined in previous scholarly works. Streamlines are used to visually represent and study how the interaction between electroosmosis and peristalsis impacts the entrapment of a fluid volume, such as a bolus, inside the channel. Bhatti et al. [28] conducted parametric research on the investigation of heat transfer in the context of EMHD. Specifically, they used the Darcy–Brinkman–Forchheimer model to analyze the flow of an Oldroyd 6-constant EMHD fluid between parallel micro-plates. The research conducted by the author primarily centered on the investigation of electro-magneto-hydrodynamics, specifically pertaining to the flow of a fluid characterized by the Oldroyd 6-constants model, inside a configuration of parallel micro-plates. Notably, the author placed significant attention on the examination of heat transfer phenomena within this context. The area located between the micro-plates is described as having a porous nature, and the modeling process includes the use of the Darcy–Brinkman–Forchheimer model to accommodate this porous medium. The shooting technique is used for numerical calculations to solve the nonlinear equations that arise. The results are visually shown using graphical representations, showcasing the velocity profile, Nusselt number, and temperature distribution. The authors further clarified that the study of this matter involves a significant emphasis on the analysis of various fluid and heat transport characteristics. The parameters included in this study are the Hartmann number, Darcy–Brinkman–Forchheimer parameter, Brinkmann number, viscous dissipation, Darcy parameter, and Joule heating coefficient.

The HNFs that flow across permeable media are also vital for a variety of applications [29] such as nuclear engineering, material sciences, solar sciences, and other fields. Furthermore, HNFs passing through porous media may be used in drilling procedures, water purification, improved oil recovery, and electronic device cooling. However, major obstacles emerge, such as nanofluid stability owing to the substantial effect of nanoparticles on the solid matrix and pressure loss, particularly in the case of forced convection. As a result, many researchers investigated HNFs in a porous media. For instance, the thermal nonequilibrium of a HNF flow using a porous cavity was explored by Chamkha et al. [30]. They evaluated the solutions using finite element analysis and found that increasing the translucent section significantly boosted heat transmission because HNFs circulated more effectively. A unique application of multibanding on HNF by means of a porous medium with magnetic phenomena was reported by Manna et al. [31]. They discovered that the multibanding method uses less energy than the whole domain's magnetic field. Kumar et al. [32] employed the Darcy–Brinkman–Forchheimer model to study non-Newtonian HNFs. They took into account the Casson nanofluid movement through a spinning disk. They discovered that the suction–injection parameter is affected similarly by the disk's upward and downward motion, but that the downward motion of the disk has a negative influence on velocity and thermal profile. A thorough examination of the flow of a HNF across a medium that is porous with magnetic effects was provided by Talebi Rostami et al. [33]. The radial basis function approach was used to determine the solutions. Slip and magnetic effects on HNF flow across a porous media have been examined by Eid and Nafe [34]. They also studied the effect of generating heat and varied thermal conductivity. They discovered that iron oxide nanoparticles maximize heat transmission when the surface shrinks while minimizing it when it stretches. The Refs. [35–39] list some notable works on HNF flow with EMHD effects across a porous media.

The primary objective of the present investigation is to analyze the behavior of iron oxide and gold nanoparticles that are dispersed in a nanofluid composed of sodium alginate (SA). This inquiry is motivated by the widespread use of HNFs in many engineering fields and industrial sectors. In situations when they are located at a finite distance in vertical spaces, the movement of fluid occurs inside a porous media that exhibits nonDarcian behavior and is influenced by convection effects. Both the purported fluid and the nanoparticles are considered to possess electrical conductivity. The energy equation includes the implications of joule heating and viscous dissipation as well. The use of equations pertaining to momentum and energy is a common practice in the realm of mathematical modeling. The shooting method is used as an algorithm on computers for solving systems of nonlinear coupled differential equations. A solution, in essence, refers to a systematic approach used to resolve a certain issue that arises from a problem. The act of seeking a solution to a problem includes effectively addressing the issue or achieving the desired goal. When addressing the challenges at hand, we are presented with two distinct categories of answers: analytical solutions and numerical solutions. Analytical solutions are known for their higher degree of accuracy, yet their applicability is limited to comparatively less complex issues. These approaches are often deemed unfeasible in real-world circumstances and are not attainable when dealing with more intricate challenges. When faced with the task of modeling physical events, it is common to encounter nonlinear PDEs. Analytical solutions are often unavailable for the majority of PDEs, with finding theoretical solutions for nonlinear PDEs proving to be notably challenging.

Researchers have devised many numerical algorithms to get approximate answers to address this challenge. Numerical solutions provide a convenient approach for effectively addressing complicated and nonlinear problems, particularly those that often emerge in the mathematical modeling of physical systems. Approximations are often utilized in numerical methods to get numerical solutions. To approach a numerical solution, iterative procedures are used. Due to the inherent limitations of approximation, the results obtained may not be entirely correct. However, by allocating more processing resources, it is possible to enhance the accuracy of the outcomes. Obtaining a comprehensive curve of findings may be a time-consuming and expensive process. Therefore, to avoid the occurrence of divergent or unsuitable outcomes, it is advisable to use these procedures while taking into account their stability and convergence. Perturbation methods are applicable to problems characterized by the presence of small or large parameters, which are often referred to as perturbation quantities. Semiexact methods may be used to address nonlinear problems that lack perturbation quantities, hence offering a solution that surpasses the constraints associated with perturbation approaches. The homotopy perturbation method [40–42] is a technique that has been developed to address the aforementioned issue. Additional techniques that may be used in this context include the homotopy analysis approach [43–45], the differential transformation method [46–48], and the Adomian decomposition approach [49–51]. The optimal homotopy asymptotic method (OHAM) is a potent approach for addressing nonlinear problems that do not possess perturbation parameters [52–54]. The authors used the OHAM in the cited literature to get approximate solutions for a specific category of Volterra integral equations characterized by weakly singular kernels. The researchers demonstrated that the OHAM method exhibits both reliability and efficiency in the context of solving weakly singular Volterra integral equations. The exploration of OHAM's applicability to various issue domains has also been investigated. To effectively address nonlinear differential equations, it is necessary to linearize and discretize them. However, this process introduces errors associated with discretization. The direct application of this approach to nonlinear differential equations, without resorting to linearization and discretization techniques, enables us to get the desired level of accuracy. In their study, Sheikholeslami et al. [55] investigate the issue of laminar viscous flow in a semiporous channel under the influence of a transverse magnetic field. The authors demonstrated that the optimum homotopy asymptotic technique is a very effective methodology for the solution of nonlinear differential equations. Additionally, the researchers investigated the impact of significant factors to determine the influence of these parameters on the behavior of nanofluids. There are three ways that may be used to get the OHAM's constant parameters. One often-used technique is the least-square method (LSM) [56]. Typically, researchers use this particular technique to address the constant parameters. However, it has been noted that the central processing unit requires a significant amount of time to execute this strategy, intending to minimize the residual of the OHAM. The use of the Levenberg–Marquardt algorithm was employed by the author to address the fractional intergrow differential equation. The researchers' results suggest that the application of the LSM to each cell, by dividing it by the nth cell, may provide an approximate solution in the form of a kth-order polynomial. The researchers conducted an analysis of the condition number of their matrix pertaining to the linear system. They then presented significant findings on the stability and originality of their approach. To address this issue via the reduction of the residual, some publications, including Li et al. [57], propose the use of the Galerkin method (GM) as an alternative approach. The researchers have investigated the behavior of the classical element-free GM, which is well-recognized as a meshless technique. The authors attempted to directly demonstrate the overall numerical inaccuracy in the calculated results, which can be divided into two main components: spatial discretization and temporal discretization. The third approach presented by Yang et al. [58] is the collocation approach (CM), which is used for obtaining the constant parameters. The researchers engaged in a discussion on the use of the space-time sinc collecting technique for the resolution of a fourth-order nonlocal heat model that emerges in the field of viscoelasticity. To address the initial singularity of the solution in the temporal domain, the author used the sinc-collocation method with a single exponential transformation. This approach represents a noteworthy technique for handling the singularity inherent in the equation. The experimental findings demonstrate the high level of precision shown by the novel sinc-collocation approach, as well as the accuracy of the theoretical forecast for this particular nonlocal heat problem.

9.2 Mathematical and physical structure of the hybrid nanofluid model

Let us explore a scenario where the fluid flow of SA of third grade is assumed to propagate between a pair of parallel vertical walls that are positioned at a limited distance of $2d$ as shown in Fig. 9.1. The geometrical structure is assumed to be porous and modeled with the help of nonDarcian law. The cartesian coordinate is proposed for the geometrical structure i.e., such that the walls of the channel are located at $x = -d$ and $x = d$. The base fluid is electrically conductive and is surrounded by an electric and magnetic field. The vertical walls are filled with a HNF containing a combination of iron oxide and gold nanoparticles (Fe_3O_4-Au-NPS). It is postulated that the nanoparticles and SA are in a state of thermal equilibrium, resulting in the absence of any slip between them. Moreover, both plates are held at a constant temperature T_2 and T_1, where $T_1 > T_2$. Therefore, the mathematical modeling for the continuity because of the proposed assumptions is expressed as:

$$\nabla \cdot \underline{u} = 0, \tag{9.1}$$

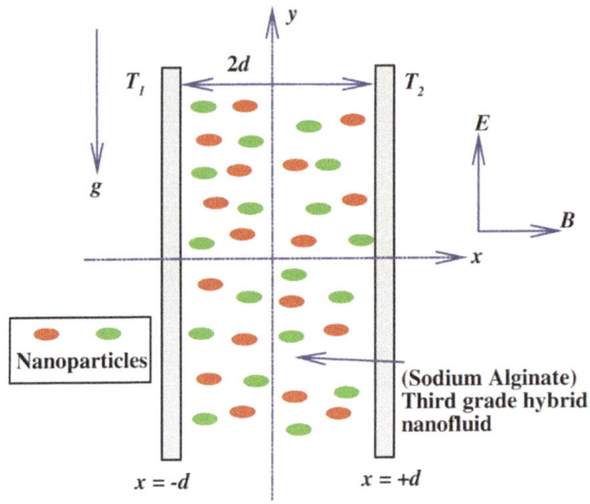

FIGURE 9.1

Orientation of the hybrid third-grade nanofluid flow between vertical walls.

The equation that represents the momentum equation including Darcy resistance and Ohm's law may be stated as follows [59–61]:

$$\rho_{hnf}\left(\underline{u}\cdot\nabla\underline{u}+\frac{\partial\underline{u}}{\partial t}\right)=-\nabla\cdot p+\nabla\underline{\tau}+\underline{\mathrm{J}}\times\underline{\mathrm{B}}+\boldsymbol{R}+(\rho\beta)_{hnf}g(T-T_m), \tag{9.2}$$

where the third-grade fluid stress tensor is denoted by $\underline{\tau}$, $\boldsymbol{R}$ Darcy's resistance, pressure is expressed by p, g expressed as the gravity force, β the thermal expansion coefficient, T is the temperature, the mean temperature is $T_m=\frac{T_1+T_2}{2}$, density is indicated by ρ, hnf in subscript represents the HNF; t is time, B the magnetic field, and J expressed as local ion current density vector, it is expressed as [62,63]:

$$\underline{J}=\sigma_{hnf}\left(\underline{u}\times\underline{B}+\underline{E}\right), \tag{9.3}$$

where $\underline{E}$ the electrical field, and σ expressed as electrical conductivity.

The following statement presents the formal description of the basic equation governing the behavior of a third-grade Reiner–Rivlin differential viscoelastic fluid:

$$\underline{\tau}=\mu_{hnf}\underline{h}_1+\tau_1\underline{h}_2+\tau_2\underline{h}_1^2+\overline{\tau}_1\underline{h}_3+\overline{\tau}_2\left(\underline{h}_1\underline{h}_2+\underline{h}_2\underline{h}_1\right)+\overline{\tau}_3\left(\mathrm{tr}\underline{h}_1^2\right)\underline{h}_1, \tag{9.4}$$

where μ expressed as the viscosity of the third-grade fluid, the material constants are described as τ_1, τ_2, $\overline{\tau}_1$, $\overline{\tau}_2$, $\overline{\tau}_3$, and the kinematical tensors $\underline{h}_1$, $\underline{h}_2$, $\underline{h}_3$ are expressed as

$$\left[\begin{array}{l}\underline{h}_1=\mathrm{grad}\underline{u}+\mathrm{grad}\underline{u}', \\ \underline{h}_n=\dfrac{\mathrm{d}\underline{h}_{n-1}}{\mathrm{d}t}+\mathrm{grad}\underline{u}'\underline{h}_{n-1}+\underline{h}_{n-1}\mathrm{grad}\underline{u}, \quad n=2,3,\ldots\end{array}\right. \tag{9.5}$$

If the model exhibits thermodynamic compatibility, meaning that all fluid motions adhere to the Clausius–Duhem inequality, and if the hypothesis of the specific Helmholtz free energy being minimized in equilibrium holds, then the further prerequisites are established:

$$\mu \geq 0, \quad \overline{\tau}_1 = \overline{\tau}_2 = 0, \quad \tau_1 \geq 0, \quad |\tau_1 + \tau_2| \leq \sqrt{24\mu\overline{\tau}_3}, \quad \overline{\tau}_3 \geq 0. \tag{9.6}$$

The thermal energy equation is formulated to include the implications of viscous dissipation and Joule heating:

$$(\rho c_p)_{hnf} \frac{\mathrm{d}T}{\mathrm{d}t} = \kappa_{hnf} \mathrm{grad}\ T + \underline{\tau}:\mathrm{grad}\underline{u} + \frac{\underline{J} \cdot \underline{J}}{\sigma_{hnf}} - \boldsymbol{R}\underline{u}, \tag{9.7}$$

where κ the is thermal conductivity and c_p represents the specific heat.

We are going to explore the velocity component and temperature field in a particular form for the presented problem:

$$\underline{u} = [0, v(x), 0], T = T(x). \tag{9.8}$$

The Darcy resistance R is defined as:

$$\boldsymbol{R} = -\left[\mu_{hnf} + 2\overline{\tau}_3\left(\frac{\mathrm{d}v}{\mathrm{d}x}\right)^2\right] v\frac{}{\underline{k}} - \rho_{hnf} C_F \frac{v^2}{\underline{k}^{1/2}}, \tag{9.9}$$

where $\underline{k}$ is the permeability of the porosity.

We derive the following set of equations by substituting Eq. (9.8) into governing Eq. (9.2):

$$\frac{\partial p}{\partial x} = (2\tau_1 + \tau_2)\frac{\mathrm{d}}{\mathrm{d}x}\left[\left(\frac{\mathrm{d}v}{\mathrm{d}x}\right)^2\right], \tag{9.10}$$

$$\begin{aligned} \frac{\partial p}{\partial y} &= \mu_{hnf}\frac{\mathrm{d}^2 v}{\mathrm{d}x^2} + 6\overline{\tau}_3\frac{\mathrm{d}^2 v}{\mathrm{d}x^2}\left(\frac{\mathrm{d}v}{\mathrm{d}x}\right)^2 - \sigma_{hnf}B^2 v + \sigma_{hnf}BE \\ &- 1\frac{}{\underline{k}}\left[\mu_{hnf} + 2\overline{\tau}_3\left(\frac{\mathrm{d}v}{\mathrm{d}x}\right)^2\right] v - \frac{\rho_{hnf}C_F}{\underline{k}^{1/2}}v^2 + g(\rho\beta)_{hnf}(T - T_m), \end{aligned} \tag{9.11}$$

$$\frac{\partial p}{\partial z} = 0. \tag{9.12}$$

The modified pressure is described as follows:

$$\underline{p} = p - (\tau_2 + 2\tau_1)\left(\frac{\mathrm{d}v}{\mathrm{d}x}\right)^2, \tag{9.13}$$

Eqs. (9.8–9.11) may be simplified to the following form in conjunction with Eq. (9.12):

$$\frac{\partial p}{\partial x} = 0, \tag{9.14}$$

$$\begin{aligned}\frac{\partial p}{\partial y} &= \mu_{hnf}\frac{d^2v}{dx^2} + 6\overline{\tau}_3\frac{d^2v}{dx^2}\left(\frac{dv}{dx}\right)^2 - \sigma_{hnf}B^2v + \sigma_{hnf}BE \\ &- \frac{1}{\underline{k}}\left[\mu_{hnf} + 2\overline{\tau}_3\left(\frac{dv}{dx}\right)^2\right]v - \frac{\rho_{hnf}C_F}{\underline{k}^{1/2}}v^2 + g(\rho\beta)_{hnf}(T - T_m),\end{aligned} \tag{9.15}$$

$$\frac{\partial p}{\partial z} = 0. \tag{9.16}$$

The preceding equations show that the pressure function is constant. We acquire:

$$\begin{aligned}&\mu_{hnf}\frac{d^2v}{dx^2} + 6\overline{\tau}_3\frac{d^2v}{dx^2}\left(\frac{dv}{dx}\right)^2 - \sigma_{hnf}B^2v + \sigma_{hnf}BE \\ &- \frac{1}{\underline{k}}\left[\mu_{hnf} + 2\overline{\tau}_3\left(\frac{dv}{dx}\right)^2\right]v - \frac{\rho_{hnf}C_F}{\underline{k}^{1/2}}v^2 + (\rho\beta)_{hnf}g(T - T_m) = 0,\end{aligned} \tag{9.17}$$

The following are the velocity boundary conditions at the vertical walls:

$$v(\pm d) = 0, \tag{9.18}$$

The energy equation for the electro-magnetic hydrodynamic fluid may be expressed as follows [64–66]:

$$\begin{aligned}&\kappa_{hnf}\frac{\partial^2 T}{\partial x^2} + \frac{1}{\underline{k}}\left[\mu_{hnf} + 2\overline{\tau}_3\left(\frac{dv}{dx}\right)^2\right]v^2 + \frac{\rho_{hnf}C_F}{\underline{k}^{1/2}}v^3 + \sigma_{hnf}\left(B^2v^2 + E^2 - 2BEv\right) + \mu_{hnf}\left(\frac{dv}{dx}\right)^2 \\ &+ 2\overline{\tau}_3\left(\frac{dv}{dx}\right)^4 = 0.\end{aligned} \tag{9.19}$$

The last components in the aforementioned equation describe the heat production caused by viscous dissipation and Joule heating.

According to the geometrical structure, the thermal boundary conditions are as follows:

$$T(-d) = T_1, \quad T(+d) = T_2, \tag{9.20}$$

The physical and thermodynamic characteristics of HNFs are characterized by the relationships presented [67–69]:

- Density:

$$\rho hnf = \rho_1^{np}\psi_1 + \rho_2^{np}\psi_2 + \rho f(1 - \psi hnf),\ \psi hnf = \psi_1 + \psi_{2,} \tag{9.21}$$

- Heat capacity:

$$\rho hnf = \rho_1^{np}\psi_1 + \rho_2^{np}\psi_2 + \rho f(1 - \psi hnf),\ \psi hnf = \psi_1 + \psi_{2,} \tag{9.22}$$

- Dynamic viscosity:

$$\mu hnf = \frac{\mu_f}{\left(1-\psi_{hnf}\right)^{205,}} \tag{9.23}$$

- Thermal conductivity:

$$\kappa_{hnf} = \kappa_f \times \left[\frac{2\kappa_f + \left(\psi_1\kappa_1^{np} + \psi_2\kappa_2^{np}\right) + 2\left(\psi_1\kappa_1^{np} + \psi_2\kappa_2^{np}\right) - 2\psi_{hnf}\kappa_f}{2\kappa_f + \left(\psi_1\kappa_1^{np} + \psi_2\kappa_2^{np}\right) - \left(\psi_1\kappa_1^{np} + \psi_2\kappa_2^{np}\right) + \psi_{hnf}\kappa_f}\right] \tag{9.24}$$

- Thermal expansion coefficient:

$$(\rho\beta)_{hnf} = \left(1-\psi_{hnf}\right)(\rho\beta)_f + (\rho\beta)_1^{np}\psi_1 + (\rho\beta)_2^{np}\psi_{2,} \tag{9.25}$$

- Electric conductivity:

$$\sigma_{hnf} = \left[1 + \frac{3(\underline{\sigma}-1)\psi_{hnf}}{(\underline{\sigma}+2) - \psi_{hnf}(\underline{\sigma}-1)}\right] \times \sigma_f, \quad \sigma_{hnp} = \frac{\sigma_1^{np}\psi_1 + \sigma_2^{np}\psi_2}{\psi_{hnf}}, \quad \underline{\sigma} = \frac{\sigma_{hnp}}{\sigma_f}. \tag{9.26}$$

Here ψ_1, ψ_2 denotes the nanoparticle volume fraction of Fe_3O_4 and Au nanoparticles.

Both Takabi and Salehi [68] and Tayebi and Chamkha [69] effectively used the aforementioned model to an Al_2O_3–Cu/water HNF in a variety of geometrical formats.

The subsequent procedure involves converting the previously described equations into a dimensionless structure, using the variables enumerated below:

$$\overline{v} = \frac{v}{v_0}, \quad \theta = \frac{T-T_m}{T_1-T_2}, \quad \overline{x} = \frac{x}{d}, \tag{9.27}$$

And the reference velocity is expressed as v_0.

By using Eq. (9.27), Eqs. (9.17) and (9.20) may be expressed in a dimensionless form as:

$$\begin{aligned} &\frac{\mu_{hnf}}{\mu_f}\frac{d^2v}{dx^2} + 6\chi\left(\frac{dv}{dx}\right)^2\frac{d^2v}{dx^2} - \frac{\sigma_{hnf}}{\sigma_f}\gamma^2 v + \frac{\sigma_{hnf}}{\sigma_f}\xi - \gamma_1\left[\frac{\mu_{hnf}}{\mu_f} + 2\chi\left(\frac{dv}{dx}\right)^2\right]v^2 \\ &- \gamma_2\frac{\rho_{hnf}}{\rho_f}v^3 + \Gamma\frac{(\rho\beta)_{hnf}}{(\rho\beta)_f}T = 0, \end{aligned} \tag{9.28}$$

$$\begin{aligned} &\frac{\kappa_{hnf}}{\kappa_f}\frac{d^2T}{dx^2} + \delta\frac{\mu_{hnf}}{\mu_f}\left(\frac{dv}{dx}\right)^2 + 2\delta\chi\left(\frac{dv}{dx}\right)^4 + \frac{\sigma_{hnf}}{\sigma_f}\left(\delta\gamma^2v^2 - \beta_1 v + \beta_2\right) \\ &+ \gamma_1 v^2\delta\left[\frac{\mu_{hnf}}{\mu_f} + 2\overline{\tau}_3\left(\frac{dV}{dX'}\right)^2\right] + \frac{\rho_{hnf}}{\rho_f}\gamma_1\delta v^3 = 0, \end{aligned} \tag{9.29}$$

The boundary conditions (9.18) and (9.20) are now reformulated in a dimensionless manner as:

$$v=0, \quad T=\frac{1}{2}\bigg|_{x=-1}, \quad v=0, \quad T=-\frac{1}{2},\bigg|_{x=+1} \tag{9.30}$$

And

$$\begin{aligned} &Gr=\frac{g(\rho\beta)_f d^3(T_1-T_2)}{\mu_f \upsilon_f}, \quad Re=\frac{v_0 d}{\upsilon_f}, \quad \xi=\frac{\sigma_f BEd^2}{\mu_f v_0}, \quad \chi=\frac{v_0^2 \overline{T}_3}{\mu_f \lambda^2}, \\ &\gamma_1=\frac{d^2}{k}, \quad \gamma_2=\frac{C_F v_0 d^2}{\upsilon_f \sqrt{k}}, \quad \Gamma=\frac{Gr}{Re}, \quad \beta_1=\frac{2d^2 v_0 \sigma_f BE}{\kappa_f(T_1-T_2)}, \quad \beta_2=\frac{E^2 d^2 \sigma_f}{(T_1-T_2)\kappa_f}, \\ &\gamma=B\lambda\sqrt{\frac{\sigma_f}{\mu_f}}, \quad \delta=\frac{\mu_f v_0^2}{\kappa_f(T_1-T_2)}. \end{aligned} \tag{9.31}$$

Here, Gr *is* the Grashof number, Re the Reynolds number, ξ the electrical strength, χ the third-grade fluid parameter, γ_1 the Darcy number, γ_2 the Darcy–Brinkman–Forchheimer, γ the Hartmann number, δ the Brinkman number, Γ the buoyancy convection parameter, β_1 the amount of heat generated through electrical and magnetic field coupling to heat conduction, and β_2 the ratio of Joule heating to heat conduction.

The dimensionless forms of the skin friction coefficient and the Nusselt number on the left plate are defined as:

$$S\sim\frac{\mu_{hnf}}{\mu_f}\frac{dv}{dx}\bigg|_{x=-1}+2\chi\left(\frac{dv}{dx}\right)^3\bigg|_{x=-1}, N\sim\frac{\kappa_{hnf}}{\kappa_f}\frac{dT}{dx}\bigg|_{x=-1}.\Big\} \tag{9.32}$$

9.3 Shooting method

The shooting method [70–74] is a computational approach used for solving the solution of boundary value problems (BVPs), with a special emphasis on differential equations, within the realm of academic research. The aforementioned method has significant efficacy in facilitating the estimation of solutions for issues characterized by disparate situations at two distinct sites. Nevertheless, similar to other numerical methods, the shooting technique has both benefits and downsides. This article will comprehensively examine the pros and downsides at hand.

The shooting approach exhibits notable adaptability as a primary asset. The use of this method extends to a diverse set of boundary value issues, making it a very important instrument in several domains such as physics, engineering, and finance. The shooting approach has a versatile nature as it does not need the issue to be constrained to a particular equation or domain, hence broadening its use.

The implementation of the approach is quite basic, particularly in comparison to other numerical techniques such as finite difference or finite element methods.

An additional benefit of the shooting approach lies in its capacity to effectively address situations characterized by intricate boundary conditions. In situations with mixed boundary conditions, when the conditions are prescribed at distinct locations inside the domain, the shooting technique remains a viable and efficient approach. The software provides the user with the capability to change the shooting parameter or an initial condition estimate to get precise answers that meet the requirements of both sets of boundary conditions.

Moreover, the firing technique offers a significant level of precision when executed accurately. Through the manipulation of shooting parameters, it is possible to enhance the precision of the solution until a desired degree of accuracy is attained. The degree of precision management has significant value in addressing issues with stringent criteria or in examining delicate systems.

Furthermore, the shooting method is a resilient strategy capable of effectively addressing nonlinear situations. Nonlinear boundary value issues often manifest in several practical applications, and the shooting technique proves to be an excellent approach for addressing such scenarios. This feature makes it a beneficial numerical technique for a diverse array of practical situations.

Nevertheless, it is important to acknowledge that the shooting approach, despite its merits, nevertheless has some drawbacks that need careful consideration. The shooting technique necessitates an initial estimation of the unknown boundary condition to start the iterative process. The process of selecting a suitable starting condition may be a challenging endeavor because an imprecise initial estimate has the potential to result in a solution that either diverges or fails to converge.

Additionally, the use of the shooting approach might result in significant computing costs, particularly when dealing with issues that need a high degree of precision or include a substantial number of unknown variables. The computing cost may be significantly increased due to the iterative nature of the procedure and the need to predict the beginning condition for each iteration. The efficiency and usability of the firing approach may be constrained in some instances.

A further drawback pertains to the potential nonconvergence of the shooting approach, particularly in scenarios where the issue showcases several solutions, or the solution space demonstrates significant oscillations. In circumstances of this kind, it may be necessary to use supplementary methodologies or make adjustments to the shooting approach to get a viable answer.

Moreover, it should be noted that the shooting technique might exhibit sensitivity to the selection of the firing parameter. The convergence and accuracy of the solution may be influenced by the range and step size of the firing parameter. The process of determining suitable values may include a series of experimental attempts, which may be a laborious and time-intensive endeavor.

In summary, it should be noted that the shooting approach may not be applicable to problems characterized by discontinuous or unique solutions. The shooting technique may encounter difficulties when dealing with these particular solutions because it heavily depends on the properties of continuity and differentiability shown by the solution.

9.3.1 Steps in implementing the shooting method

The following are the fundamental procedures for implementing the shooting approach in solving a nonlinear BVP [75–77]:

Step 1. The differential equations and boundary conditions that characterize the issue should be provided. It is important to ensure that the BVP is accurately described, including the appropriate domain and boundary conditions at both ends.
Step 2. The conversion of a BVP into an initial value problem (IVP) may be achieved by the incorporation of an additional parameter or unknown variable. This parameter is often known in academic literature as the "shooting parameter" or "shooting variable." By manipulating this parameter, the IVP solutions may be modified to meet the specified boundary conditions.
Step 3. Choose initial approximations for the shooting parameter(s) and the accompanying initial conditions for the IVP. The beginning values should be carefully chosen to increase the likelihood of the IVP solutions fulfilling the given boundary conditions.
Step 4. Utilize a numerical method, such as the Runge–Kutta approach, to solve the IVP by using the selected initial approximations. The numerical solution for the IVP may be obtained using this method.
Step 5. The resultant solution generated from the fourth step should be examined and compared to the required boundary conditions. Iteratively modify the shooting parameter(s) and/or the starting projections until the boundary requirements are fulfilled with a desired level of accuracy.
Step 6. The process of repeating steps 4 and 5 should be continued until the answer sufficiently meets the boundary criteria. This may include enhancing the numerical technique, modifying the increment magnitude, or altering the initial approximations.
Step 7. After the shooting approach has reached convergence and produced a solution that meets the boundary criteria, the numerical solution obtained may be regarded as an approximation to the nonlinear BVP.
Step 8. It is noteworthy that the shooting approach may need a certain degree of experimentation, especially when it comes to the selection of suitable starting approximations and shooting settings. Furthermore, more sophisticated methods such as bisection or Newton's method may be used to enhance the precision of the shooting parameter modifications.

9.3.2 Results and discussion

In this section, we will examine the graphical and numerical outcomes related to the different formulation parameters. In the previous section, a numerical method known as the shooting approach was used to get numerical outcomes for the velocity field and temperature field. The results are shown for the skin friction profile, Nusselt number, velocity, and temperature. The numerical results were obtained based on the numerical data shown in Table 9.1. However, Table 9.2 presents the Nusselt number profile and skin friction number data. The computation of both profiles is accomplished by using Eq. (9.25). In this table, we can see that the skin friction coefficient has been significantly enhanced due to the strong impact of the third-grade fluid parameter χ, Brinkman number δ, electric strength ξ, buoyancy convection parameter Γ, and nanoparticle volume fraction of gold ψ_1 and iron oxide ψ_2 nanoparticles. On the other hand, the behavior is converse on the skin friction number for Hartmann number γ, Darcy number γ_1, Darcy–Brinkman–Forchheimer parameter γ_2, and heat generation parameter β_1. Moreover, we observed that the Nusselt number diminishes due to the increment in the third-grade fluid parameter χ, Hartmann number γ, Darcy number γ_1, Darcy–Brinkman–Forchheimer parameter γ_2, nanoparticle volume fraction of gold nanoparticles ψ_2, and heat generation parameter β_1. While it increases due to the impact of the brinkman number δ, electric strength ξ, buoyancy convection parameter Γ, and nanoparticle volume fraction of iron oxide nanoparticles ψ_1.

9.3.3 Velocity profile

Figs. 9.2–9.8 show the consequences of all the parameters on the velocity profile. It can be seen in Fig. 9.2 that the third-grade fluid parameter opposes the fluid flow along the whole channel. Third-grade fluid is viscoelastic, which means it can deform and resist deformation. Moreover, the results for Newtonian fluid reduce when third-grade fluid approaches zero ($\chi \rightarrow 0$). Fig. 9.3 shows that the velocity field has been enhanced as a result of the buoyancy convection parameter's substantial impact. Due to the increased temperature difference between the fluid and its surroundings, which tends to improve heat transfer in the fluids, the velocity field grows as a consequence of the rise in the buoyant convection parameter. The variation of the Hartmann number on the velocity profile is

Table 9.1 Thermophysical properties of sodium alginate third-grade fluid, Iron oxide, and gold nanoparticles [78,79].

Physical properties	Sodium alginate	Au	Fe_3O_4
ρ (kg/m^3)	989	19,300	5180
κ (W/m·K)	0.6376	320	9.7
c_p (J/kg·K)	4175	129.1	670
β (1/K)	0.00099	0.0000142	0.000014
σ (S/m)	2.64×10^{-4}	4.5×10^{7}	2.5×10^{4}

Table 9.2 Numerical values of skin friction coefficient and the local Nusselt number against all the pertinent parameters.

χ				
	0	**0.5**	**1**	**1.5**
S	1.265382404	1.309168085	1.335736549	1.354841561
N	0.030365102	0.028935127	0.027956376	0.027194536
δ				
	0.5	**1**	**1.5**	**2**
S	1.335736549	1.35022853	1.365605941	1.381973066
N	0.027956376	0.213744006	0.409915392	0.617667411
ξ				
	1	**1.5**	**2**	**2.5**
S	1.335736549	1.918226945	2.511077533	3.112483186
N	0.027956376	0.139649699	0.294303478	0.483927243
γ				
	0	**1**	**2**	**3**
S	1.382309083	1.214130265	0.906786735	0.671916699
N	0.028711737	0.025875044	0.019986768	0.015097503
Γ				
	0	**1**	**2**	**3**
S	1.014345174	1.214130265	1.419097185	1.630021544
N	0.005381773	0.025875044	0.051094655	0.081161031
β_1				
	0	**0.4**	**0.7**	**1**
S	1.228022252	1.216875694	1.208687387	1.200643776
N	0.170589768	0.054497583	-0.030897651	-0.114890921
γ_1				
	0.5	**1**	**2**	**3**
S	1.399599523	1.298879117	1.141926762	1.025589242
N	0.028875578	0.027294119	0.024602829	0.022436292
γ_2				
	0	**1**	**2**	**3**
S	1.268277116	1.214130265	1.170363079	1.133857831
N	0.026953011	0.025875044	0.024980297	0.024224577
ψ_1				
	0.0	**0.05**	**0.1**	**0.15**
S	0.999450206	1.098781288	1.214130265	1.349031415
N	0.024656982	0.024771524	0.025875044	0.028303068

(Continued)

Table 9.2 Numerical values of skin friction coefficient and the local Nusselt number against all the pertinent parameters. *Continued*

ψ_2				
	0.0	**0.05**	**0.1**	**0.15**
S	1.022149150	1.107693066	1.214130265	1.341415655
N	0.056070491	0.027251996	0.025875044	0.025415652

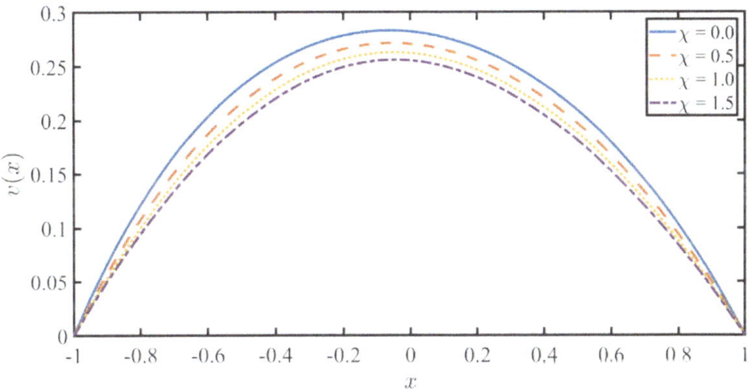

FIGURE 9.2

Consequences of third-grade fluid parameter χ on velocity profile.

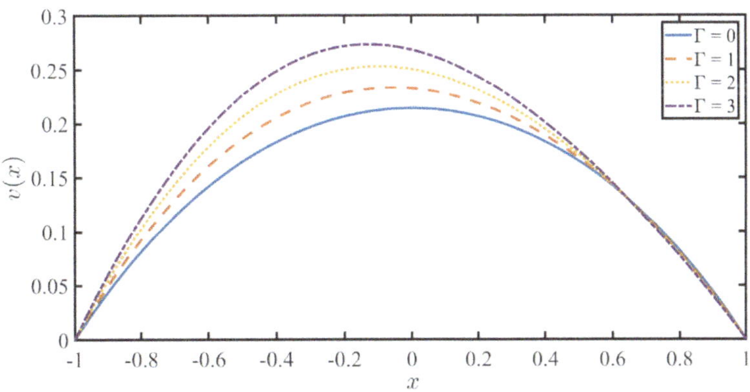

FIGURE 9.3

Consequences of buoyancy convection parameter Γ on velocity profile.

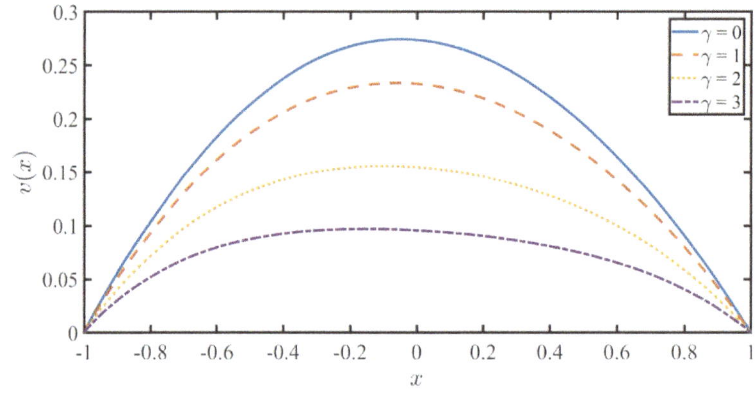

FIGURE 9.4

Consequences of Hartmann number γ on velocity profile.

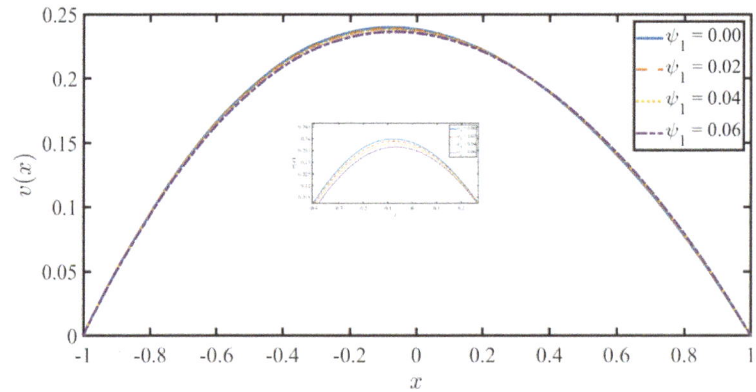

FIGURE 9.5

Consequences of iron oxide nanoparticles ψ_1 on velocity profile.

apparent in Fig. 9.4. It should be noted that the magnetic field's existence significantly increases the resistance that fluid motion experiences since the magnetic field results in the Lorentz force, which works against fluid motion. The velocity field is apparent to be decreasing in Figs. 9.5 and 9.6 as a result of an upsurge in the volume fraction of iron oxide and gold nanoparticles. The findings demonstrate the behavior of mono nanofluids with $\psi_1 = 0$ or $\psi_2 = 0$. The viscosity of the fluid tends to grow dramatically as the value of nanoparticles increases, which tends to slow the fluid's velocity. Fig. 9.7 illustrates that the velocity field falls as the Darcy number goes up. The permeability of the porous media is inversely correlated with the Darcy number, as is well known; nevertheless, this is different from the

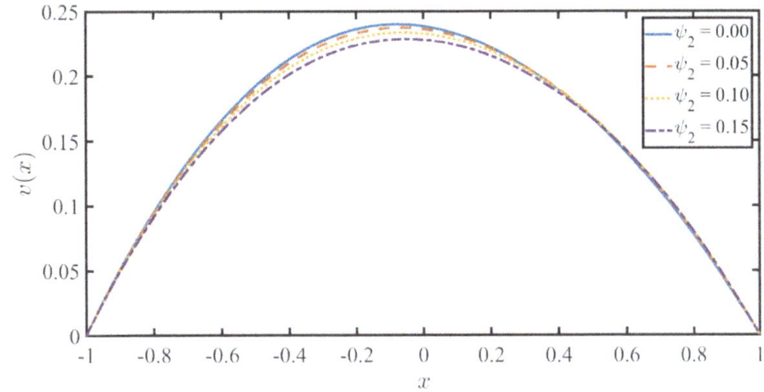

FIGURE 9.6

Consequences of gold nanoparticles ψ_2 on velocity profile.

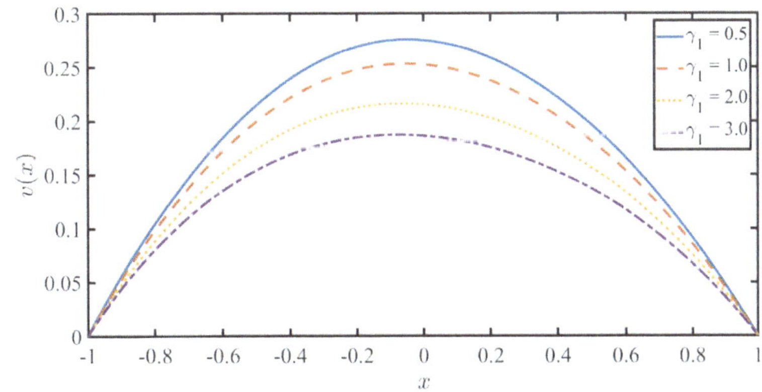

FIGURE 9.7

Consequences of Darcy number γ_1 on velocity profile.

conventional Darcy number. As the Darcy number rises, permeability decreases, slowing the flow across the whole regime. The Darcy–Brinkman–Forchheimer parameter's change on the velocity profile is demonstrated in Fig. 9.8, and it is discovered that this parameter functions as an opposing force. The findings for the conventional Darcian flow are found when $\gamma_2 = 0$.

9.3.4 Temperature profile

The variation of the temperature profile for the nanoparticle volume fractions of gold and iron oxide is depicted in Figs. 9.9 and 9.10. Here, it can be observed

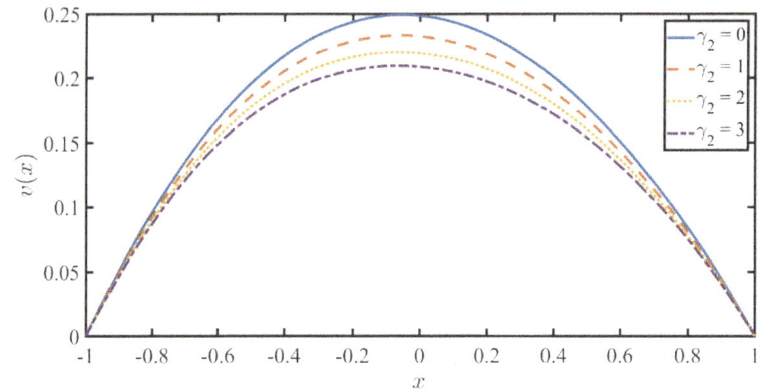

FIGURE 9.8

Consequences of Darcy–Brinkmann–Forchheimer γ_2 on velocity profile.

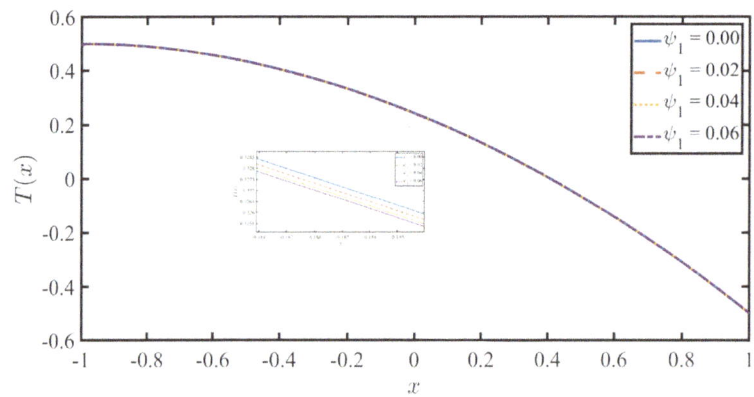

FIGURE 9.9

Consequences of iron oxide nanoparticles ψ_1 on temperature profile.

that the findings show a minimal effect. The temperature profile is reduced as the volume percentage of nanoparticles rises, although the effects are inconsequential. Fig. 9.11 demonstrates the way the heat generation parameter declines the temperature profile throughout the whole channel. However, $\beta_1 = 0$ shows that the influence of heat generated on thermal conduction caused by the combined EMHD effects has vanished and temperatures have reached their maximum. This parameter has a cooling effect on the flow regime because as it rises, the thermal field decreases. Figs. 9.12 and 9.13 display that both the Darcy number and the Darcy–Brinkman–Forchheimer parameter slightly optimize the thermal profile. Because an increase in the thermal profile with a higher Darcy number indicates

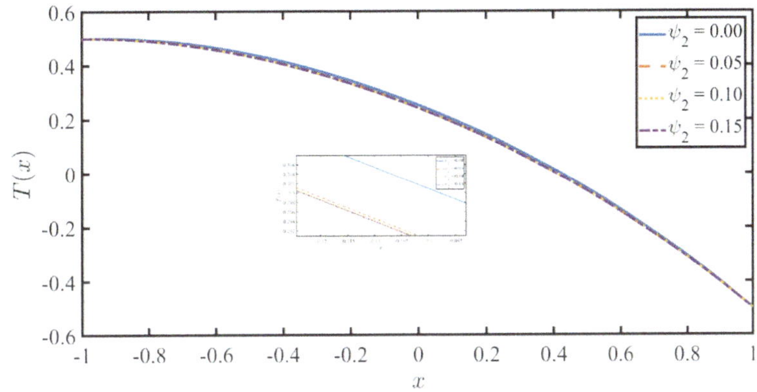

FIGURE 9.10

Consequences of gold nanoparticles ψ_2 on temperature profile.

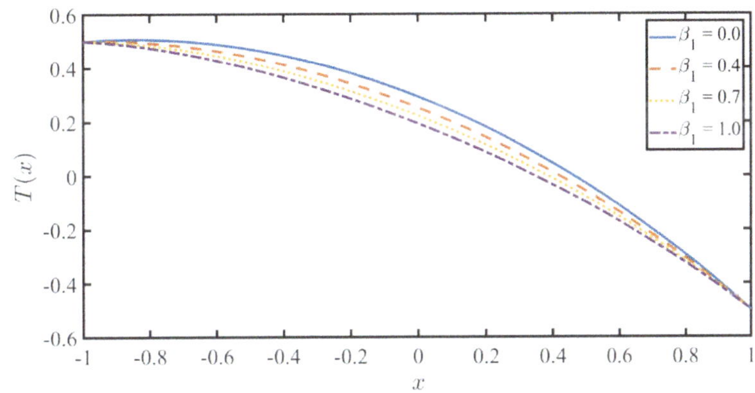

FIGURE 9.11

Consequences of heat generation (due to electrical and magnetic field) parameter β_1 on temperature profile.

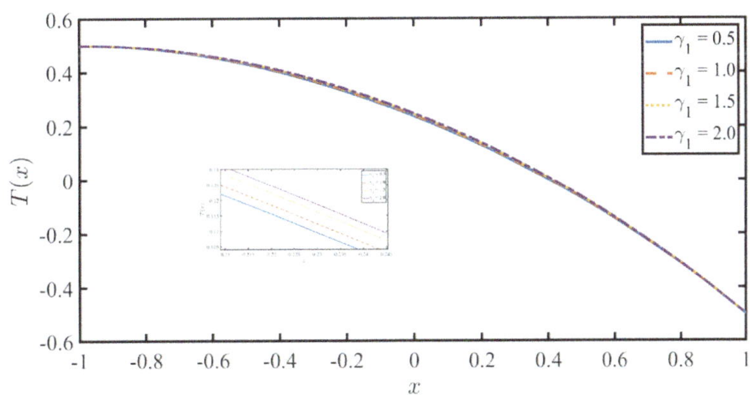

FIGURE 9.12

Consequences of Darcy number γ_1 on temperature profile.

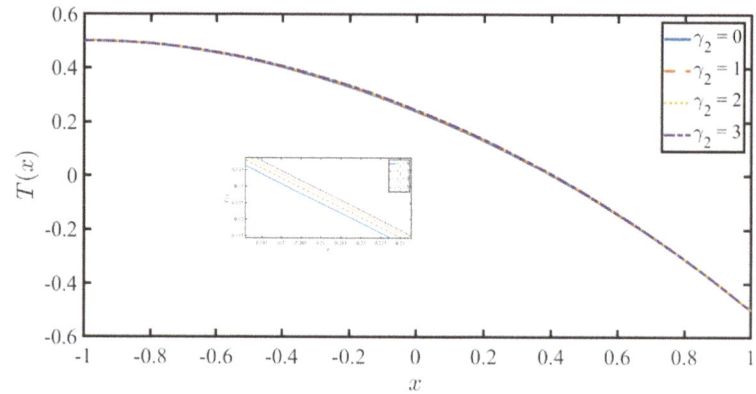

FIGURE 9.13

Consequences of Darcy–Brinkman–Forchheimer γ_2 on temperature profile.

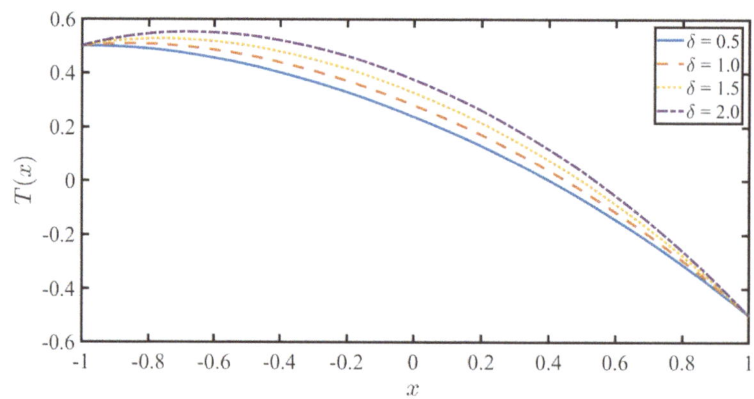

FIGURE 9.14

Consequences of Brinkman number δ on temperature profile.

that permeability has been lowered and there are fewer solid matrices in the flow regime, both parameters are related to viscous dissipation due to the impacts of the Brinkman number. Fig. 9.14 indicates that the Brinkman number enhances the thermal profile in the whole regime. Due to the increment in the Brinkman number, the viscous forces increase, which causes a rise in the thermal profile. Fig. 9.15 indicates that the third-grade fluid parameter increases the thermal profile exclusively not much. As the fluid moves, it encounters varying velocities and deformation rates, causing the fluid to undergo varying amounts of shear and internal friction heating. Because of the fluid's non-Newtonian nature, the

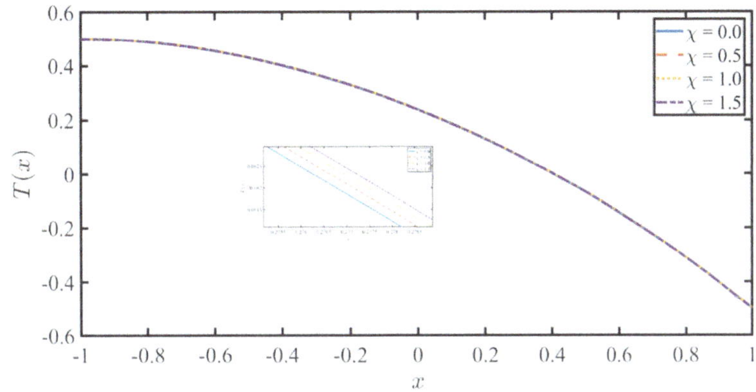

FIGURE 9.15

Consequences of third-grade fluid parameter δ on temperature profile.

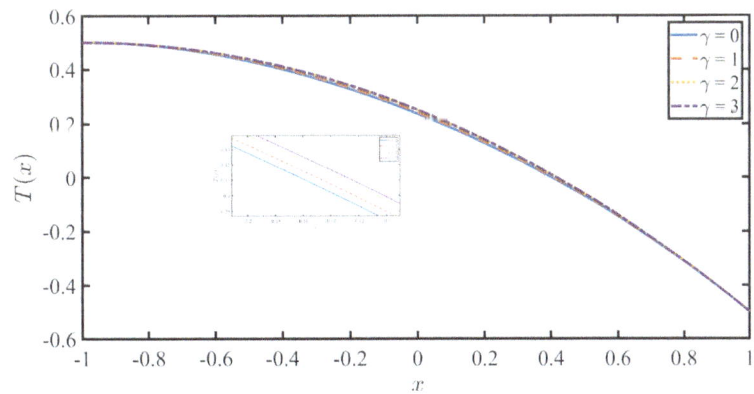

FIGURE 9.16

Consequences of Hartmann number γ on temperature profile.

viscosity is substantially lower in areas of high shear rate compared to regions of low shear rate, resulting in differential heating across the fluid. As a consequence, the temperature field in third-grade fluid is raised owing to internal frictional heating caused by the fluid's non-Newtonian actions, causing differential heating throughout the fluid as it travels through different levels of shear. It is noticed in Fig. 9.16 that increasing the Hartmann number improves the thermal profile. Again, the impacts are insignificant. As the Hartmann number progresses, fluid motion decreases, heat transmission becomes less efficient, and temperature rises as less heat is transmitted from the afflicted region.

9.4 Conclusions and outlooks

In this work, we examined the characteristics of hybrid third-grade nanofluids as they propagate through a pair of vertical walls. When subjected to an applied electric and magnetic field, the third-grade fluid, which is composed of SA, exhibits electrical conductivity. The medium of the geometrical structure is purportedly permeable. The mathematical formulation of momentum and energy equations was established via the use of the Darcy–Brinkman–Forchheimer model in combination with Ohm's law. The shooting method is used for the numerical solution of nonlinear coupled differential equations. The shooting approach has several benefits, including its adaptability, capacity to handle intricate boundary conditions, precision, and resilience in addressing nonlinear issues. Nevertheless, the method also has some drawbacks, such as the need for a suitable starting estimate, high processing costs, convergence challenges, sensitivity to the shooting parameter, and limits in dealing with solutions that are discontinuous or single. Notwithstanding these constraints, the shooting method continues to be a powerful numerical approach for resolving boundary value issues and has shown effective application across many scientific and engineering domains. The present research presents the main discoveries:

1. The mere existence of hybrid nanoparticles is apparent in suppressing the velocity and temperature profiles.
2. The Darcy number and Darcy–Brinkman–Forchheimer coefficients are shown to strengthen the velocity profile while having little impact on the thermal profile.
3. It is discovered that the Hartmann number and third-grade fluid characteristics impede fluid flow, but these variables improve the thermal profile.
4. The simulations show that the buoyancy convection parameter optimizes the velocity profile.

References

[1] A. Tripathy, K. Regulagadda, C.W.E. Lam, M.A. Donati, A. Milionis, C.S. Sharma, E. Mitridis, T.M. Schutzius, D. Poulikakos, Ultrathin durable organic hydrophobic coatings enhancing dropwise condensation heat transfer, Langmuir 38 (2022) 11296–11303.

[2] A. Tripathy, C.W.E. Lam, D. Davila, M. Donati, A. Milionis, C.S. Sharma, D. Poulikakos, Ultrathin lubricant-infused vertical graphene nanoscaffolds for high-performance dropwise condensation, ACS Nano 15 (2021) 14305–14315.

[3] S. Rashidi, M. Eskandarian, O. Mahian, S. Poncet, Combination of nanofluid and inserts for heat transfer enhancement: gaps and challenges, J. Therm. Anal. Calorim. 135 (2019) 437–460.

[4] S. Suresh, K.P. Venkitaraj, P. Selvakumar, M. Chandrasekar, Effect of Al_2O_3–Cu/water hybrid nanofluid in heat transfer, Exp. Therm. Fluid Sci. 38 (2012) 54–60.

[5] M. Baghbanzadeh, A. Rashidi, D. Rashtchian, R. Lotfi, A. Amrollahi, Synthesis of spherical silica/multiwall carbon nanotubes hybrid nanostructures and investigation of thermal conductivity of related nanofluids, Thermochim. Acta 549 (2012) 87–94.

[6] Y. Ma, R. Mohebbi, M.M. Rashidi, Z. Yang, MHD convective heat transfer of Ag-MgO/water hybrid nanofluid in a channel with active heaters and coolers, Int. J. Heat Mass Transfer 137 (2019) 714–726.
[7] M.M. Rashidi, M.A. Nazari, I. Mahariq, M.E.H. Assad, M.E. Ali, R. Almuzaiqer, A. Nuhait, N. Murshid, Thermophysical properties of hybrid nanofluids and the proposed models: An updated comprehensive study, Nanomaterials (Basel) 11 (2021) 3084.
[8] Z. Tian, H. Arasteh, A. Parsian, A. Karimipour, M.R. Safaei, T.K. Nguyen, Estimate the shear rate & apparent viscosity of multi-phased non-Newtonian hybrid nanofluids via new developed support vector machine method coupled with sensitivity analysis, Phys. A: Stat. Mech. Appl. 535 (2019) 122456.
[9] Cheng, M. Nawaz, H. Kaneez, M.K. Alaoui, A. Selmi, C. Li, H. Assilzadeh, Flow and heat transfer analysis of elastoviscoplastic generalized non-Newtonian fluid with hybrid nano structures and dust particles, Int. Commun. Heat Mass Transfer 126 (2021) 105275.
[10] N.S. Wahid, N.M. Arifin, N.S. Khashi'ie, I. Pop, N. Bachok, M.E.H. Hafidzuddin, Hybrid nanofluid stagnation point flow past a slip shrinking Riga plate, Chin. J. Phys. 78 (2022) 180–193.
[11] S. Anitha, M. Pichumani, Numerical analysis on heat transfer performance of industrial double-tube heat exchanger using CNT: Newtonian/non-Newtonian hybrid nanofluids, J. Therm. Anal. Calorim. 147 (2022) 9603–9624.
[12] M. Hemmat Esfe, D. Toghraie, S. Alidoust, Experimental analysis on the rheological characteristics of MWCNT-ZnO (50:50)/5W30 oil non-Newtonian hybrid nanofluid to obtain a new correlation, Powder Technol. 407 (2022) 117595.
[13] K.V. Modi, P.R. Patel, S.K. Patel, Applicability of mono-nanofluid and hybrid-nanofluid as a technique to improve the performance of solar still: a critical review, J. Clean. Prod. 387 (2023) 135875.
[14] Y.-M. Chu, S. Bashir, M. Ramzan, M.Y. Malik, Modelntechn comparative study of magnetohydrodynamics unsteady hybrid nanofluid flow between two infinite parallel plates with particle shape effects, Math. Methods Appl. Sci. 46 (2023) 11568–11582.
[15] A.M. Alqahtani, M. Bilal, A. Ali, T.R. Alsenani, S.M. Eldin, Numerical solution of an electrically conducting spinning flow of hybrid nanofluid comprised of silver and gold nanoparticles across two parallel surfaces, Sci. Rep. 13 (2023) 7180.
[16] M. Atashafrooz, H. Sajjadi, A. Amiri Delouei, Simulation of combined convective-radiative heat transfer of hybrid nanofluid flow inside an open trapezoidal enclosure considering the magnetic force impacts, J. Magn. Magn. Mater. 567 (2023) 170354.
[17] M. Arif, L. Di Persio, P. Kumam, W. Watthayu, A. Akgül, Heat transfer analysis of fractional model of couple stress Casson tri-hybrid nanofluid using dissimilar shape nanoparticles in blood with biomedical applications, Sci. Rep. 13 (2023) 4596.
[18] S. Jakeer, P.B.A. Reddy, Entropy generation on EMHD stagnation point flow of hybrid nanofluid over a stretching sheet: homotopy perturbation solution, Phys. Scr. 95 (2020) 125203.
[19] N.S. Khashi'ie, N.M. Arifin, M. Sheremet, I. Pop, Shape factor effect of radiative $Cu-Al_2O_3/H_2O$ hybrid nanofluid flow towards an EMHD plate, Case Stud. Therm. Eng. 26 (2021) 101199.
[20] B. Souayeh, K. Ramesh, N. Hdhiri, E. Yasin, M.W. Alam, K. Alfares, A. Yasin, Heat transfer attributes of gold-silver-blood hybrid nanomaterial flow in an EMHD peristaltic channel with activation energy, Nanomaterials (Basel) 12 (2022) 1615.

[21] M.P. Mkhatshwa, M. Khumalo, Irreversibility scrutinization on EMHD Darcy–Forchheimer slip flow of Carreau hybrid nanofluid through a stretchable surface in porous medium with temperature-variant properties, Heat Transfer 52 (2023) 395–429.
[22] H.A. Ghazwani, Time-dependent mixed stagnation point hybrid nanofluid flow with radiative heat flux and viscous dissipation effects over a movable EMHD Riga plate, Int. J. Mod. Phys. B (2023) 2450090.
[23] D. Gamachu, W. Ibrahim, L.K. Bijiga, Nonlinear convection unsteady flow of electro-magnetohydrodynamic Sutterby hybrid nanofluid in the stagnation zone of a spinning sphere, Results Phys. 49 (2023) 106498.
[24] P. Kumar, H. Poonia, A.S. Sujesh Areekara, A. Sabu, L. Mathew, Significance of irregular heat source and Arrhenius energy on electro-magnetohydrodynamic hybrid nanofluid flow over a rotating stretchable disk with nonlinear radiation, Numer. Heat Transfer, A: Appl. (2023) 1–23.
[25] A.S. Saleem, T. Berrouk, I. Gul, W. Zari, I. Alghamdi, Unsteady mix convectional stagnation point flow of nanofluid over a movable electro-magnetohydrodynamics Riga plate numerical approach, Sci. Rep. 13 (2023).
[26] N. Khashiie, Insight into unsteady separated stagnation point flow of hybrid nanofluids subjected to an electro-magnetohydrodynamics Riga plate, Magnetochemistry 9 (2023) 46.
[27] E.N. Maraj, N.S. Akbar, I. Zehra, A.W. Butt, H. Ahmed Alghamdi, Electro-osmotically modulated magneto hydrodynamic peristaltic flow of menthol based nanofluid in a uniform channel with shape factor, J. Magn. Magn. Mater. 576 (2023) 170774.
[28] M.M. Bhatti, E.E. Michaelides, Oldroyd 6-constant electro-magneto-hydrodynamic fluid flow through parallel micro-plates with heat transfer using Darcy–Brinkman–Forchheimer model: a parametric investigation, Math. Eng. 5 (2023) 1–19.
[29] K.K. Al-Chlaihawi, H.H. Alaydamee, A.E. Faisal, K. Al-Farhany, M.A. Alomari, Newtonian and non-Newtonian nanofluids with entropy generation in conjugate natural convection of hybrid nanofluid-anoflu enclosures: a review, Heat Transfer 51 (2022) 1725–1745.
[30] A.J. Chamkha, S. Sazegar, E. Jamesahar, M. Ghalambaz, Thermal non-equilibrium heat transfer modeling of hybrid nanofluids in a structure composed of the layers of solid and porous media and free nanofluids, Energies 12 (2019) 541.
[31] N.K. Manna, M.K. Mondal, N. Biswas, A novel multi-banding application of magnetic field to convective transport system filled with porous medium and hybrid nanofluid, Phys. Scr. 96 (2021) 065001.
[32] R.N. Kumar, R.J.P. Gowda, B.J. Gireesha, B.C. Prasannakumara, Non-Newtonian hybrid nanofluid flow over vertically upward/downward moving rotating disk in a Darcy–Forchheimer porous medium, Eur. Phys. J.: Spec. Top. 230 (2021) 1227–1237.
[33] H. Talebi Rostami, M. Fallah Najafabadi, K. Hosseinzadeh, D.D. Ganji, Investigation of mixture-based dusty hybrid nanofluid flow in porous media affected by magnetic field using RBF method, Int. J. Ambient Energy (2022) 1–11.
[34] M.R. Eid, M.A. Nafe, Thermal conductivity variation and heat generation effects on magneto-hybrid nanofluid flow in a porous medium with slip condition, Waves Random Complex. Media 32 (2022) 1103–1127.
[35] J. Wang, Y.-P. Xu, R. Qahiti, M. Jafaryar, M.A. Alazwari, N.H. Abu-Hamdeh, A. Issakhov, M.M. Selim, Simulation of hybrid nanofluid flow within a microchannel

heat sink considering porous media analyzing CPU stability, J. Pet. Sci. Eng. 208 (2022) 109734.
[36] S. Jakeer, S.R.R. Reddy, A.M. Rashad, M.L. Rupa, C. Manjula, Nonlinear analysis of Darcy−Forchheimer flow in EMHD ternary hybrid nanofluid (Cu-CNT-Ti/water) with radiation effect, Forces Mech. (2023) 100177.
[37] M.M. Bhatti, F. Ishtiaq, R. Ellahi, S.M. Sait, Novel aspects of cilia-driven flow of viscoelastic fluid through a non-Darcy medium under the influence of an induced magnetic field and heat transfer, Mathematics 11 (2023) 2284.
[38] T. Siva, S. Jangili, B. Kumbhakar, Analytical solution to optimise the entropy generation in EMHD flow of non-Newtonian fluid through a microchannel, Pramana 96 (2022) 168.
[39] A. Zaib, R.U. Haq, A.J. Chamkha, M.M. Rashidi, Impact of partial slip on mixed convective flow towards a Riga plate comprising micropolar TiO_2-kerosene/water nanoparticles, Int. J. Numer. Methods Heat Fluid Flow 29 (2019) 1647−1662.
[40] J.- He, C.-H. Huan, A.A. He, A good initial guess for approximating nonlinear oscillators by the homotopy perturbation method, Facta Univ. Ser.: Mech. Eng. 21 (2023) 21−029.
[41] N.H. Aljahdaly, A.M. Alweldi, On the modified Laplace homotopy perturbation method for solving damped modified Kawahara equation and its application in a fluid, Symmetry 15 (2023) 394.
[42] M.K. Kolawole, M.O. Olayiwola, A.I. Alaje, H.O. Adekunle, K.A. Odeyemi, Conceptual analysis of the combined effects of vaccination, therapeutic actions, and human subjection to physical constraint in reducing the prevalence of COVID-19 using the homotopy perturbation method, Beni-Suef Univ. J. Basic. Appl. Sci. 12 (2023) 10.
[43] K.K. Ali, M. Maneea, Optical solitons using optimal homotopy analysis method for time-fractional (1 + 1)-dimensional coupled nonlinear Schrodinger equations, Opt. Int. J. Light Electron Opt. 283 (2023) 170907.
[44] W. Fafa, Z. Odibat, N. Shawagfeh, The homotopy analysis method for solving differential equations with generalized Caputo-type fractional derivatives, J. Comput. Nonlinear Dyn. 18 (2023) 1−22.
[45] C.N. Guled, J.V. Tawade, P. Kumam, S. Noeiaghdam, I. Maharudrappa, S.M. Chithra, V. Govindan, The heat transfer effects of MHD slip flow with suction and injection and radiation over a shrinking sheet by optimal homotopy analysis method, Results Eng. 18 (2023) 101173.
[46] K. Shah, T. Abdeljawad, F. Jarad, Q. Al-Mdallal, On nonlinear conformable fractional order dynamical system via differential transform method, Comput. Model. Eng. Sci. 136 (2023) 1457−1472.
[47] M. Derikvand, F. Farhatnia, D.H. Hodges, Functionally graded thick sandwich beams with porous core: buckling analysis via differential transform method, Mech. Based Des. Struct. Mach. (2021) 1−28.
[48] F. Hıra, On-differential transform method, J. Phys. A: Math. Theor. 56 (2023) 325202.
[49] M. Mallikarjuna, R. Senthamarai, An amperometric biosensor and its steady state current in the case of substrate and product inhibition: Taylors series method and Adomian decomposition method, J. Electroanal. Chem. 946 (2023) 117699.

[50] H. Fatoorehchi, S. Djilali, Stability analysis of linear time-invariant dynamic systems using the matrix sign function and the Adomian decomposition method, Int. J. Dyn. Control. 11 (2023) 593–604.

[51] R. Ali, U.N. Ghosh, L. Mandi, P. Chatterjee, Application of Adomian decomposition method to study collision effect in dusty plasma in the presence of polarization force, Indian J. Phys. 97 (2023) 2209–2216.

[52] H. Jan, H. Ullah, M. Ullah, I. Fiza, A. Khan, A. Mohamed, A. Allah, Modification of optimal homotopy asymptotic method for multi-dimensional time-fractional model of Navier–Stokes equation, Fractals 31 (2023) 2340021.

[53] Z. Ayati, S. Pourjafar, A new optimal multistep optimal homotopy asymptotic method to solve nonlinear system of two biological species, Nonlinear Eng. 12 (2023) 20220230.

[54] A. Jameel, D.J. Fareed, N. Hashim, O. Anakira, A. Ababneh, A.-K. Qazza, K.S.A. Alomari, Application of the optimal homotopy asymptotic approach for solving two-point fuzzy ordinary differential equations of fractional order arising in physics, Axioms 12 (2023) 387.

[55] M. Sheikholeslami, D.D. Ganji, Magnetohydrodynamic flow in a permeable channel filled with nanofluid, Sci. Iran. B 21 (2014) 203–212.

[56] Y. Jia, M. Xu, Y. Lin, D. Jiang, An efficient technique based on least-squares method for fractional integro-differential equations, Alex. Eng. J. 64 (2023) 97–105.

[57] Y. Li, C. Liu, W. Li, Y. Chai, Numerical investigation of the element-free Galerkin method (EFGM) with appropriate temporal discretization techniques for transient wave propagation problems, Appl. Math. Comput. 442 (2023) 127755.

[58] X. Yang, L. Wu, H. Zhang, A space-time spectral order sinc-collocation method for the fourth-order nonlocal heat model arising in viscoelasticity, Appl. Math. Comput. 457 (2023) 128192.

[59] L. Wang, Y. Jian, Q. Liu, F. Li, L. Chang, Electromagnetohydrodynamic flow and heat transfer of third grade fluids between two micro-parallel plates, Colloids Surf. A: Physicochem. Eng. Asp. 494 (2016) 87–94.

[60] C. Yang, Y. Jian, Z. Xie, F. Li, Electromagnetohydrodynamic electroosmotic flow and entropy generation of third-grade fluids in a parallel microchannel, Micromachines (Basel) 11 (2020) 418.

[61] A. Ahlawat, M.K. Sharma, M.M. Rashidi, M.A. Sheremet, Entropy production minimization and heat transfer enhancement in a cavity filled with micropolar hybrid nanofluid under an influence of discrete heaters and uniform magnetic field, J. Therm. Anal. Calorim. 148 (2023) 10803–10820.

[62] R.V. Polovia, V.P. Demutskii, 1990th ed Fundamentals of Magnetohydrodynamics, Kluwer Academic/Plenum, New York, NY, 1990.

[63] F. Baharifard, K. Parand, M.M. Rashidi, Novel solution for heat and mass transfer of a MHD micropolar fluid flow on a moving plate with suction and injection, Eng. Comput. 38 (2022) 13–30.

[64] B.K. Sharma, A. Kumar, R. Gandhi, M.M. Bhatti, N.K. Mishra, Entropy generation and thermal radiation analysis of EMHD Jeffrey nanofluid flow: applications in solar energy, Nanomaterials (Basel) 13 (2023) 544.

[65] J. Kayalvizhi, A.G. Vijaya, Kumar, Entropy analysis of EMHD hybrid nanofluid stagnation point flow over a porous stretching sheet with melting heat transfer in the presence of thermal radiation, Energies 15 (2022) 8317.

[66] M.M. Bhatti, R. Ellahi, M. Hossein Doranehgard, Numerical study on the hybrid nanofluid (Co_3O_4−Go/H_2O) flow over a circular elastic surface with non-Darcy medium: application in solar energy, J. Mol. Liq. 361 (2022) 119655.

[67] M. Ibrahim, M.M. Rashidi, T.A. Assiri, N. Aamir, T. Saeed, V. Ali, Influence of dimension variations of a fin and wall emissivity on the nanofluids flow inside a square cavity using the two-phase Lattice Boltzmann method, Eng. Anal. Bound. Elem. 157 (2023) 148−156.

[68] B. Takabi, S. Salehi, Augmentation of the heat transfer performance of a sinusoidal corrugated enclosure by employing hybrid nanofluid, Adv. Mech. Eng. 6 (2014) 147059.

[69] T. Tayebi, A.J. Chamkha, Entropy generation analysis during MHD natural convection flow of hybrid nanofluid in a square cavity containing a corrugated conducting block, Int. J. Numer. Methods Heat Fluid Flow 30 (2020) 1115−1136.

[70] H.A. El-Dawy, M.F. El-Amin, Z.A. Raizah, Joule heating and viscous dissipation effects on a stretching/shrinking cannel filled by micropolar hybrid nanofluid in presence thermal/solar radiation, J. Nanofluids 12 (2023) 738−744.

[71] M. Alrehili, The flow of a thermo nanofluid thin film inside an unsteady stretching sheet with a heat flux effect, Energies 16 (2023) 1160.

[72] H.U. Rasheed, Zeeshan, S. Islam, B. Ali, Q. Shah, R. Ali, Implementation of shooting technique for Buongiorno nanofluid model driven by a continuous permeable surface, Heat Transfer 52 (2023) 3119−3134.

[73] G.-M. Gie, C.-Y. Jung, H. Lee, Semi-analytic shooting methods for Burgers' equation, J. Comput. Appl. Math. 418 (2023) 114694.

[74] H. Feng, X. Yue, X. Wang, Z. Zhang, Decoupling and quasi-linearization methods for boundary value problems in relative orbital mechanics, Nonlinear Dyn. 111 (2023) 199−215.

[75] S.N. Ha, A nonlinear shooting method for two-point boundary value problems, Comput. Appl. Math. 42 (2001) 1411−1420.

[76] B.S. Attili, M.I. Syam, Efficient shooting method for solving two point boundary value problems, Chaos, Solitons Fractals 35 (2008) 895−903.

[77] R.J. Punith Gowda, F.S. Al-Mubaddel, R. Naveen Kumar, B.C. Prasannakumara, A. Issakhov, M. Rahimi-Gorji, Y.A. Al-Turki, Computational modelling of nanofluid flow over a curved stretching sheet using Koo−Kleinstreuer and Li (KKL) correlation and modified Fourier heat flux model, Chaos, Solitons Fractals 145 (2021) 110774.

[78] F.A. Alwawi, H.T. Alkasasbeh, A.M. Rashad, R. Idris, MHD natural convection of sodium alginate Casson nanofluid over a solid sphere, Results Phys. 16 (2020) 102818.

[79] F.M. Abbasi, M. Gul, S.A. Shehzad, Effectiveness of temperature-dependent properties of Au, Ag, FeO, Cu nanoparticles in peristalsis of nanofluids, Int. Commun. Heat Mass Transfer 116 (2020) 104651.

CHAPTER

10 Optimal homotopy asymptotic method with Caputo fractional derivatives: a new approach for solving time-fractional Navier-Stokes equation

Faranak Rabiei[1,2], Zeeshan Ali[2], Kamyar Hosseini[3] and M.M. Bhatti[4]

[1]*Department of Mathematics, Texas A&M University Kingsville, Kingsville, TX, United States*
[2]*School of Engineering, Monash University Malaysia, Bandar Sunway, Selangor, Malaysia*
[3]*Department of Mathematics, Near East University TRNC, Mersin, Turkey*
[4]*Material Science Innovation and Modelling (MaSIM) Research Focus Area, North-West University (Mafikeng Campus), Mmabatho, South Africa*

10.1 General introduction

Partial differential equations (PDEs) serve as indispensable mathematical instruments employed to elucidate and tackle an extensive array of physical phenomena that entail multiple functions contingent on various variables. Within the domain of fluid mechanics, one of the foundational PDEs is the Navier–Stokes (NS) equation, which serves as a fundamental framework for modeling the motion of incompressible fluids while accommodating the intricate interplay of viscosity and pressure. The NS equation was initially presented by Claude–Louis Navier (1785–1836) in a pair of his autobiographies [1,2]. These publications marked a significant milestone as they formally incorporated friction into the equations governing fluid motion. Before this, all theoretical advancements concerning fluid motion had treated fluids as having ideal fluidity and lacked viscosity considerations. Since the renowned Euler equation for nonviscous fluids was published in 1755, the equations describing fluid motion have essentially been bound to ideal flows [3].

Despite the long-standing recognition of fluid friction as the primary source of experimental deviations from theory, there had been relatively limited efforts in incorporating viscosity effects into the equations governing fluid motion. Leonhard Euler (1707–83) was among the few who attempted this, introducing a theoretical development [4] in which he mistakenly assumed that fluid friction,

Nanofluids. DOI: https://doi.org/10.1016/B978-0-443-13625-2.00009-7

akin to solid friction, was directly proportional to pressure [5]. It seems that Navier was the one motivated to methodically approach this task and finally succeed after Euler's research. His dedication to this subject was not coincidental but rather driven by his established expertise in hydraulics, as evidenced by his editing of Belidor's renowned treatise on "Architecture Hydraulique."

Between 1807 and 1820, Navier played a pivotal role in elevating mathematical analysis into a crucial tool for civil engineers. During this period, he also laid the groundwork for the emerging idea of mechanical work within the field of mechanical science. As an outcome, Navier had the necessary background knowledge to develop what is now known as the NS equation, sometimes referred to as the NS equation. These prerequisites encompassed both mathematical precision and practical engineering insight, effectively embodying the qualities of an engineer within a scientist. Subsequent to Navier's initial formulation, various authors independently derived the NS equation using different methodologies. However, Navier never got the credit he deserved for his ground-breaking work because neither his successors nor specialist magazines in the area gave his ideas any attention.

Curiously, it seems that every person who discovered the NS equation after Navier was aware of his earlier work but decided to ignore it. We have opted to reexamine Navier's methods despite the fact that a number of other writers have subsequently deduced the NS equation, frequently in a more exacting way. This choice was made in recognition of the fact that, despite initially relying on shaky techniques to account for the molecular forces that cause friction, Navier's pioneering efforts not only laid the groundwork but also unexpectedly produced the first extensive equation for the flow of real viscous liquids.

The equations regulating the equilibrium and motion of an elastic solid were initially laid out by Navier in his memoir, Sur les Lois de l'Equilibre et du Mouvement des Corps Solides Elastiques in the year 1821, marking the beginning of the current theory of elasticity. This seminal work, presented to the Academy on May 14, 1821, and later published in 1827 in the "Mémoires de L'Académie Royale des Sciences de L'Institut de France," marked a significant milestone. Navier initially formulated these equations for isotropic elastic solids with a single elastic constant. These equations may be expanded to represent various continuous media, as Navier quickly recognized. In his first memoir, he developed the equation explaining the flow of viscous liquids by building on the formulae for elastic solids. In this derivation, he replaced the displacements of elastic solids with the velocities of fluid particles and substituted the elastic solid constant with the fluid viscosity constant (referred to as the 'adherence constant' by Navier).

The work of Navier certainly served as an inspiration for other scientists, including Cauchy, Poisson, and Saint-Venant, as they developed equations for the movement of viscous liquids. By taking into account the scenario of a "nonelastic body" and supposing that the stress tensor is proportional to the rate of deformation tensor, Cauchy [6], for example, was able to derive the equation of motion for viscous liquids from his formulae of elasticity. The NS equation was created as a consequence of Poisson's [7] relationship between fluid stresses and rate of deformation, which is analogous to the relationship between solid stresses and strain. Saint–Venant [8] developed a stress tensor that produced the differential equations of Navier, Cauchy, and Poisson, all with a single constant parameter, by concentrating on the transverse force operating on moving particles of liquid.

These 18th-century scientists, including Bernoullis, d'Alembert, Euler, and Lagrange, sought to close the gap in understanding the rational fluid mechanics of ideal nonviscous fluids and the real-world actions of viscous liquids in hydraulic structures. Stokes' interest in pendulum studies that suggested viscosity may affect the behavior of ideal pendulum motion sparked his participation in the equation of motion for viscous flows, generally known as the NS equation. To examine how actual fluids differ from ideal ones, Stokes originally tried to develop motion situations utilizing oscillating spheres and tubes to approximate pendulum components [9].

He acknowledged that friction was a cause of deviation but not the sole one, pointing out other factors such as flow discontinuity and instabilities leading to turbulence. Stokes recognized the need for accurate experiments and eventually shifted his strategy, incorporating internal fluid friction into the fundamental equations of hydrodynamics. Employing methods akin to those of Cauchy and Poisson, he derived the NS equation, emphasizing its applicability to various practical scenarios, such as the motion of water in pipes, tidal and wave friction calculations, and similar problems.

Stokes became associated with the NS equation because he extensively compared theoretical predictions with experimental data for various setups, including cylindrical rods, spheres, pendula oscillating in air and water, and more. Despite some initial misconceptions, such as assuming viscosity's proportionality to density and neglecting certain terms in the equation, Stokes achieved good agreement between theory and experiments, particularly for oscillating pendula. By systematically comparing theory and experiments, Stokes, in contrast to other NS equation writers, had a clear aim to use his work realistically. This led to his affiliation with Navier as the coauthor of the formula that describes motion for viscosity fluids.

The development of the NS equation is a topic of great significance in the history of mechanics, and it has been thoroughly explored by historians of the field. René Dugas' "A History of Mechanics" [10] is a notable publication that provides a description of Navier's second memoir and a short summary of his research on elasticity. The "Convolutions in French Mathematics" [11] by Grattan-Guinness provides a thorough analysis of Navier's memoirs on fluid dynamics, as well as his ideas related to elastic rods, surfaces, and bodies. Darrigol's "Worlds of Flow" [12] gives a great overview and context for all of Navier's studies on fluids, highlighting their effect on and often ignored by other well-known individuals such as Cauchy, Saint-Venant, Poisson, and Stokes. The history of the NS equation that was previously discussed may be uncovered in Ref. [13].

The fundamental PDE in fluid dynamics known as the NS equation has the following generic form:

$$\frac{\partial \underline{\xi}(\overline{\varepsilon},\omega)}{\partial \omega} + \left(\underline{\xi}(\overline{\varepsilon},\omega)\cdot\nabla\right)\underline{\xi}(\overline{\varepsilon},\omega) = -\frac{1}{\rho}\nabla P + \eta\nabla^2\underline{\xi}(\overline{\varepsilon},\omega), \tag{10.1}$$

where ρ is the density, $\overline{\varepsilon}$ is the spatial variable, ξ is the velocity, ω is the time, η is the kinematic viscosity, P is the pressure, and ∇ denotes the gradient differential operator [13].

Fractional differential equations have gained popularity as a practical tool for simulating a variety of real-world events. Notably, under long-time scaling limits, time-fractional derivatives play a vital role as infinitesimal creators of time evolution. The applications of the aforementioned equations are found in various fields, including biology [14], blood flow [15], capacitor theory [16], aerodynamics [17], viscoelasticity [18], thermodynamics [19], control theory [20], fitting of experimental data [21], electrodynamics of complex media [22], electrochemistry [23], electrical circuits [24], and polymer technology and physics [25]. They provide effective descriptions for numerous essential phenomena in these diverse domains. The ideas of ordinary calculus are extended into the field of mathematics known as fractional calculus (FC), which has a rich history spanning more than three centuries. FC was born out of a large correspondence between eminent mathematicians Leibniz and L'Hospital around 1695. During this correspondence, Leibniz posed a fundamental question that laid the foundation for the development of FC.

"Is it possible to apply the definition of derivatives with integer orders to derivatives with noninteger orders?" L'Hopital expressed some interest in the aforementioned subject and, in response, asked Leibniz a new, simpler question: "What if the order would be 1/2?" In a letter dated September 30, 1695, Leibniz said, "It will eventually lead to a paradox, from which one day useful implications will be drawn." The exact birthday of FC is commonly acknowledged to be September 30, 1695. The question posed by Leibniz concerning fractional derivatives, or more precisely, semiderivatives, continued to captivate mathematicians in the decades that followed [26,27]. Following L'Hopital and Leibniz's original investigation, FC continued to be a field of research that attracted the best mathematicians in Europe. Euler [27], for instance, in 1730, remarked: "When 'n' is a positive integer, and 'Θ' is a function of 'ω,' denoted as '$\Theta(\omega)$,' the ratio of $\frac{d^{n}\Theta}{d\omega^{n}}$ can always be described algebraically. However, what happens when 'n' is a fraction? What kind of ratio can be formulated then?"

Subsequent references to fractional derivatives emerged from the works of several notable mathematicians throughout history. This esteemed group comprises Lagrange (1772), Lacroix (1819), Laplace (1812), Riemann (1847), Fourier (1822), Green (1859), Grunwald (1867), Holmgren (1865), Letnikov (1868), Sonini (1869), Laurent (1884), Krug (1890), Nekrassov (1888), Weyl (1919), and numerous additional researchers [28–30]. Throughout the 19th century, the theory of FC primarily developed thanks to the profound insights and exceptional abilities of these eminent mathematicians. In particular, Lacroix [31] presented an earlier accurate answer to the conundrum that Leibniz and L'Hospital had first presented in 1819. He asserted that for a fractional derivative, specifically $\frac{d^{\frac{1}{2}}\omega}{d\omega^{\frac{1}{2}}} = 2\sqrt{\frac{\omega}{\pi}}$. Lacroix's monumental work, a 700-page treatise on calculus published in 1819, also encompassed the development of formulas for the n-th derivative of $y = \omega^m$, with 'm' being a positive integer:

$$\mathbf{D}^n y = \frac{d^n}{d\omega^n}(\omega^m) = \frac{m!}{(m-n)!}\omega^{m-n}, m \geq n. \tag{10.2}$$

He created an equation for the fractional derivative of a power function by replacing the factorial sign with the gamma function.

$$\mathbf{D}^\nu \omega^\beta = \frac{\Gamma(\beta+1)}{\Gamma(\beta-\nu+1)}\omega^{\beta-\nu}. \tag{10.3}$$

Within this context, ν and β are fractional values, while the gamma function $\Gamma(z)$ is precisely defined for $z > 0$ as follows:

$$\Gamma(z) = \int_0^\infty e^{-\omega}\omega^{z-1}d\omega.$$

Lacroix estimated in particular

$$\mathbf{D}^{\frac{1}{2}}\omega = \frac{\Gamma(2)}{\Gamma\left(\frac{3}{2}\right)}\omega^{\frac{1}{2}} = 2\sqrt{\frac{\omega}{\pi}}.$$

It is interesting to note that the fractional derivative of a constant function, ($\beta = 0$), produces a nonzero result using the previous definition as

$$\mathbf{D}^\nu 1 = \mathbf{D}^\nu \omega^0 = \frac{1}{\Gamma(1-\nu)}\omega^{-\nu} \neq 0.$$

By methodically differentiating an analytic function's power series expansion term by term, the technique developed by Lacroix, which takes advantage of the linearity of fractional derivatives, may be applied to any kind of analytic function. However, it is worth noting that this method's applicability is limited due to its narrow scope, which prevents it from being considered truly general.

It is interesting to note that with these early theoretical advancements, the first real-world uses of FC appeared. One noteworthy instance of this was the work of Abel in 1823 [32,33]. Abel started working on a tautochrone problem-related integral equation. He made a crucial discovery: an integral transform that could be written as a semiderivative might be used to get the answer. To be more specific, Abel took into account the following integral transform:

$$K = \int_0^{\omega} (\bar{\varepsilon} - \omega)^{-\frac{1}{2}} \Theta(\bar{\varepsilon}) d\bar{\varepsilon}, K = \text{const.} \tag{10.4}$$

A fractional derivative of order 1/2 was employed by Abel to write the right side of (10.4):

$$\sqrt{\pi}\left(\frac{d^{-\frac{1}{2}}}{d\omega^{-\frac{1}{2}}}(\Theta(\omega))\right).$$

Joseph Liouville was intrigued by Abel's answer and began the first substantial investigation of FC [34–37]. Around 1832, Liouville started a comprehensive research of FC and made the most significant advancements in this subject by effectively applying his results to issues with potential theory. Liouville initiated his theoretical journey by building upon the well-established results for derivatives of integer order '$\mathscr{w}$'.

$$\mathbf{D}^{\mathscr{w}} e^{a\omega} = a^{\mathscr{w}} e^{a\omega}. \tag{10.5}$$

Expression (10.5) can be readily extended in a formal manner to encompass noninteger values of '$\mathscr{w}$', thereby yielding:

$$\mathbf{D}^{\nu} e^{a\omega} = a^{\nu} e^{a\omega}. \tag{10.6}$$

By mixing several functions as a superposition of complex exponentials, Fourier expansion makes it feasible to express a wide variety of different functions.

$$\Theta(\omega) = \sum_{\mathscr{w}=0}^{\infty} c_{\mathscr{w}} \exp(a_{\mathscr{w}}\omega), \text{Re } a_{\mathscr{w}} > 0,$$

Liouville once again proposed the subsequent equation for determining the derivative of order by referencing the linearity of the fractional derivative ν:

$$\mathbf{D}^{\nu}\Theta(\omega) = \sum_{\mathscr{w}=0}^{\infty} c_{\mathscr{w}} a_{\mathscr{w}}^{\nu} e^{a_{\mathscr{w}}\omega}. \tag{10.7}$$

It is acknowledged that formula (10.7) is Liouville's first formula for a fractional derivative [34,35]. Nevertheless, owing to its restricted application, it cannot be considered a general definition of the fractional derivative, much as Lacroix's formula. Liouville set out to create a second definition in order to overcome this restriction. He began by using a definite integral, which is closely related to the gamma function:

$$\mathbf{I} = \int_0^{\infty} u^{\beta-1} e^{-\omega u} du, \beta > 0, \omega > 0.$$

He continued his research and came up with what is now referred to as the second Liouville formula.

$$\mathbf{D}^{\nu}\omega^{-\beta} = (-1)^{\nu}\frac{\Gamma(\nu+\beta)}{\Gamma(\beta)}x^{-\nu-\beta}, \beta > 0.$$

None of the aforementioned definitions of a fractional derivative proved to be appropriate as a general definition. In the years that followed, a number of similar formulations appeared. For instance, Greer [38] used (10.6) in the following way to create equations for the fractional derivatives of trigonometric functions:

$$\mathbf{D}^{\nu}e^{ia\omega} = a^{\nu}\left(\cos\frac{\pi\nu}{2} + i\sin\frac{\pi\nu}{2}\right)(\cos a\omega + i\sin a\omega).$$

Joseph Fourier [39] derived integral representations for both $\Theta(\omega)$ and its derivative, as follows:

$$\mathbf{D}^{n}\Theta(\omega) = \frac{1}{2\pi}\int_{-\infty}^{+\infty}\Theta(\mathfrak{P})d\mathfrak{P}\int_{-\infty}^{+\infty}t^{n}\cos\left[t(\omega-\mathfrak{P}) + \frac{n\pi}{2}\right]dt,$$

By formally substituting the integer 'n' with any arbitrary real value 'ν,' he derived the following:

$$\mathbf{D}^{\nu}\Theta(\omega) = \frac{1}{2\pi}\int_{-\infty}^{+\infty}\Theta(\mathfrak{P})d\mathfrak{P}\int_{-\infty}^{+\infty}t^{\nu}\cos\left[t(\omega-\mathfrak{P}) + \frac{\nu\pi}{2}\right]dt,$$

A work written by G. F. Bernhard Riemann [40], while still a student, is responsible for one of the most important developments in the development of FC. Unfortunately, this work was not published until 1892, after he had passed away. In 1853, while attempting to extend the concept of a Taylor series, Riemann crafted an alternative definition that hinged on a definite integral. This new definition proved to be applicable to power series featuring noninteger exponents.

$$\mathbf{D}^{-\nu}\Theta(\omega) = \frac{1}{\Gamma(\nu)}\int_{c}^{\omega}(\omega-t)^{\alpha-1}\Theta(\omega)dt + \Psi(\omega). \tag{10.8}$$

Indeed, the expression obtained stands as the most commonly employed contemporary definition of fractional integration. To address the uncertainty associated with the lower limit of integration 'c,' Riemann introduced a "complementary" function $\Psi(\omega)$ to his definition. However, the modern definition of fractional integration has eliminated the need for this potentially troublesome complementary function. Because neither Riemann nor Liouville was able to answer the problem of the complementary function, the historical process that resulted in the development of the current Riemann-Liouville formulation is of special significance.

The initial efforts that laid the foundation for what is now known as the Riemann–Liouville definition can be traced back to the work of N. Ya. Sonin in 1869 [41]. In his paper, Sonin embarked on a journey that began with Cauchy's integral formula and led to the concept of differentiation with an arbitrary index. Shortly thereafter, in 1872, A. V. Letnikov [42] further expanded upon Sonin's ideas [43]. Both Sonin and Letnikov attempted to establish definitions for fractional derivatives by employing closed contours as their starting point. They built upon Cauchy's integral formula, which was originally formulated for derivatives of integer order and can be expressed as:

$$\Theta^{(n)}(z) = \frac{n!}{2\pi i}\int_C \frac{\Theta(t)}{(t-z)^{n+1}}\,dt. \tag{10.9}$$

By using Euler's gamma function in place of the factorial sign, which is expressed as the expansion to the fractional world may be accomplished as $\nu! = \Gamma(1+\nu)$. However, when attempting a direct expansion to noninteger values of α, a significant challenge emerges: the integrand in Eq. (10.9) contains a branching point. Addressing this challenge would necessitate an appropriate contour that includes a branch cut—a crucial element not accounted for in the analysis of Sonin and Letnikov. Ultimately, Laurent [44] introduced a novel approach. Laurent employed an open circuit, known as a Laurent loop, as opposed to Sonin and Letnikov's closed circuit. The contemporary explanation of the Riemann-Liouville fractional integral that we employ currently was created using this ground-breaking technique.

$$\mathbf{D}^{-\nu}\Theta(\omega) = \frac{1}{\Gamma(\nu)}\int_c^{\omega} (\omega-\overline{\varepsilon})^{\nu-1}\Theta(\overline{\varepsilon})d\overline{\varepsilon},\ \mathrm{Re}(\nu)>0. \tag{10.10}$$

In Eq. (10.10), Riemann's formula (10.8) is immediately recognizable; however, it is without the troublesome complementary function. In contemporary terminology, when setting the lower terminal value, c, to negative infinity, expression (10.10) is referred to as the Liouville fractional integral. Choosing $c=0$ simplifies expression (10.10) to the well-known Riemann fractional integral, and when c is arbitrary, it is called the Riemann-Liouville fractional integral. The fractional integration operator is expressed most often in the application by employing Eq. (10.10). By choosing $c=0$ in (10.10), Riemann's formula (10.8) is produced without the troublesome complimentary function $\Psi(\omega)$, and by choosing $c=-\infty$, Liouville's original formulation (10.7) is made comparable. The designation of Eq. (10.10) as the Riemann-Liouville fractional integral is explained by these two facts.

Although the sign of the parameter α in formula (10.10) is the primary difference between the symbols for fractional integration and differentiation, the change from fractional integration to differentiation cannot be achieved immediately by replacing a minus v on the right-hand side of (10.10). The problem is caused by the fact that when working with negative integration orders, the integral on the right side of (10.10) diverges. However, by continuing the analysis, it can be shown that:

$$\begin{aligned}\mathbf{D}^{\nu}\Theta(\omega) &= \mathbf{D}^{n-\beta}\Theta(\omega) = \mathbf{D}^{n}\Theta(\omega)\mathbf{D}^{-\beta}\Theta(\omega)\\ &= \frac{d^{n}}{d\omega^{n}}\left(\frac{1}{\Gamma(\beta)}\int_c^{\omega}(\omega-\overline{\varepsilon})^{\beta-1}\Theta(\overline{\varepsilon})d\overline{\varepsilon}\right),\end{aligned} \tag{10.11}$$

provides what is now understood to be the Riemann-Liouville fractional derivative formulation. In Eq. (10.11), the value $n = \nu$ denotes the smallest integer greater than ν, where $0 < \beta = n - \nu < 1$. When either $c = 0$ or $c = \infty$, the integral in (10.11) transforms into the beta-integral, allowing for straightforward evaluation across a wide range of functions.

In close proximity to each other, Grunwald and Letnikov established the foundation for another widely used formulation of fractional derivatives [45]. Grunwald (1867) started with the idea of a derivative as the limit of a difference quotient after becoming frustrated with the shortcomings of Liouville's methodology. Riemann's definite integral should be interpreted with a finite lower limit, he underlined. He also created definite-integral formulae for regular derivatives and showed that Liouville's formulation, which lacked a clear lower limit, basically corresponded to a lower limit of negative infinity. Formally referred to as:

$$\mathbf{D}^{\nu}\Theta(\omega) = \lim_{h \to 0} \frac{(\Delta_h^{\nu}\Theta)}{h^{\nu}}(\omega) = \lim_{h \to 0} \frac{\sum_{k=0}(-1)^k \binom{\nu}{k} \Theta(\omega - kh)}{h^{\nu}}, \nu > 0. \quad (10.12)$$

The Grunwald-Letnikov fractional derivative is the term used to describe this in modern times. In definition (10.12), the symbol $\binom{\nu}{k}$ represents the generalized binomial coefficient, where the traditional factorials have been replaced with Euler's gamma function. Notably, Letnikov [42] showed that definition (10.12) agrees with the definitions proposed by Riemann and Liouville under certain very moderate situations.

Presently, the Grunwald-Letnikov definition primarily finds utility in the development of various numerical methods. These methods leverage formula (10.12) with finite sums to approximate fractional derivatives accurately. It is also important to acknowledge O. Heaviside's work [46] and the developments in FC regarding the end of the 19th century. The development of Heaviside's operational calculus, which was first created to address particular issues in electromagnetic theory, was a crucial step in the use of generalized derivatives. The connection to FC is evident in Heaviside's use of arbitrary powers of 'p' often $\sqrt{p}$, to arrive at solutions for various engineering challenges.

In 1917, Weyl [47] and Hardy [48,49] delved into the study of some distinct yet intrinsic characteristics of different integrals applied to functions within the Lebesgue and Lipschitz classes. Furthermore, Weyl established that the fractional integrals that follow may be stated for values between 0 and 1, provided that the integrals described in (10.13) converge over an infinite range:

$$\begin{cases} \mathbf{I}_{+}^{\nu}\Theta(\omega) = \dfrac{1}{\Gamma(\nu)} \displaystyle\int_{-\infty}^{\omega} (\omega - \overline{\varepsilon})^{\alpha-1} \Theta(\overline{\varepsilon}) d\overline{\varepsilon}, \\ \mathbf{I}_{-}^{\nu}\Theta(\omega) = \dfrac{1}{\Gamma(\nu)} \displaystyle\int_{\omega}^{\infty} (\overline{\varepsilon} - \omega)^{\alpha-1} \Theta(\overline{\varepsilon}) d\overline{\varepsilon}. \end{cases} \quad (10.13)$$

Particularly noteworthy is the Riemann-Liouville fractional integral definition provided in (10.10) with a lower limit set to $c = -\infty$. This particular form, equivalent to Liouville's proposed definition of a fractional integral, is frequently denoted as the Weyl fractional integral. Modern nomenclature distinguishes between left-sided and right-sided forms of fractional operators. The left and right Liouville fractional integrals, as stated in (10.13), are other names for the Weyl operators.

Later, in 1927, Marchaud [50] developed an integral-based definition of fractional derivatives that used the Grunwald-Letnikov concept (10.12), utilizing:

$$
\begin{aligned}
{}^{M}\mathbf{D}^{\nu}\Theta(\omega) &= \frac{\nu}{\Gamma(-\nu+1)}\int_{0}^{\infty}\frac{\left(\Delta_{\tau}^{l}\Theta\right)(\omega)}{\tau^{\nu+1}}d\tau \\
&= \frac{\nu}{\Gamma(-\nu+1)}\int_{0}^{\infty}\frac{\Theta(\omega)-\Theta(\omega-\tau)}{\tau^{\nu+1}}d\tau, \nu>0.
\end{aligned}
$$

This expression now represents the Marchaud fractional derivative, which characterizes the fractional derivative of a given function 'Θ.' The symbol $\left(\Delta_{\tau}^{l}\Theta\right)(\omega)$ represents a finite difference of order 'l' greater than ν, with 'c' serving as a normalizing constant. Importantly, under certain circumstances, this definition agrees with the Riemann-Liouville definition and has a relationship to the Grunwald-Letnikov definition. Starting in 1938 [51,52], M. Riesz made substantial contributions, with the integral serving as the primary subject of his research:

$$
{}^{R}\mathbf{I}^{\nu}\Theta = \frac{1}{2\Gamma(\nu)\cos\left(\frac{\nu\pi}{2}\right)}\int_{-\infty}^{\infty}\frac{\varphi(\tau)}{|\tau-\omega|^{1-\nu}}d\tau, \ \mathrm{Re}\,\nu>0, \nu\neq 1,3,5,\ldots
$$

today known as Riesz potential. This integral, particularly its extension to $n-$dimensional Euclidean space, is closely related to Weyl fractional integrals (10.13), and hence to Riemann-Liouville fractional integrals, via the subsequent relationship:

$$
{}^{R}\mathbf{I}^{\nu} = \left(\mathbf{I}_{+}^{\nu}+\mathbf{I}_{-}^{\nu}\right)\left(2\cos\left(\frac{\nu\pi}{2}\right)\right)^{-1}.
$$

In 1949, Riesz [52] further expanded the theory of fractional integration to encompass functions involving multiple variables. This extension necessitated a change to the Riemann-Liouville definition of fractional integrals, which is represented as follows:

$$
\frac{2\omega^{-2(\nu+\varrho)}}{\Gamma(\nu)}\int_{0}^{\omega}\left(\omega^{2}-\tau^{2}\right)^{\nu-1}\tau^{2\varrho+1}\Theta(\tau)d\tau,
$$

$$
\frac{2\omega^{2\varrho}}{\Gamma(\nu)}\int_{0}^{\omega}\left(\omega^{2}-\tau^{2}\right)^{\nu-1}\tau^{1-2\nu-2\varrho}\Theta(\tau)d\tau,
$$

was initially presented by Erdelyi et al. in their work [53], where it found utility across a range of applications. Although these concepts are closely related to the fractional differentiation of functions such as ϖ^{2} and $\sqrt{\varpi}$, a task pioneered by

Liouville in 1832, it is noteworthy that Erdelyi and Kober employed Mellin's transform to derive their results.

M. Caputo made an important modern addition to FC in 1967 [54]. When used for fractional differential equations, the Riemann-Liouville formulation of the fractional derivative has a significant disadvantage in that it requires an unusual set of beginning conditions. To make it easier to solve the fractional differential equation, specific fractional integrals and derivatives should be supplied at the first time instant. In order to use standard beginning conditions—such as those required for integer-order differential equations—Caputo reformulated the Riemann-Liouville fractional derivative. Caputo defined the fractional derivative with respect to a function Θ with $(n - 1)$ completely continuous integer-order derivatives as follows:

$$ {}^{c}\mathbf{D}^{\nu}\Theta(\omega) = \frac{1}{\Gamma(n-\nu)}\int_{0}^{\omega}(\omega-\overline{\varepsilon})^{n-\nu-1}\left(\frac{\mathrm{d}}{\mathrm{d}\overline{\varepsilon}}\right)^{n}\Theta(\overline{\varepsilon})\mathrm{d}\overline{\varepsilon}. \tag{10.14}$$

The derivative presented in Eq. (10.14) bears a close relationship to the Riemann-Liouville fractional derivative and has become a common choice in contemporary applications. Interestingly, Rabotnov actually introduced the same differential operator in Russian viscoelastic research a year before Caputo's paper was published. Despite this, it is noteworthy that in current literature, this operator is predominantly referred to as the Caputo operator. The entire literature about FC mentioned above can be found in Ref. [55].

Due to the aforementioned applications and the importance of FC, Eq. (10.1) can be reformulated using a time-fractional derivative of order $\nu \in (0, 1]$ as follows:

$$\frac{\partial^{\nu}\underline{\xi}(\overline{\varepsilon},\omega)}{\partial\omega^{\nu}} + \left(\underline{\xi}(\overline{\varepsilon},\omega)\cdot\nabla\right)\underline{\xi}(\overline{\varepsilon},\omega) = -\frac{1}{\rho}\nabla P + \eta\nabla^{2}\underline{\xi}(\overline{\varepsilon},\omega). \tag{10.15}$$

Here, $\frac{\partial^{\nu}}{\partial\omega^{\nu}}$ represents the Caputo fractional derivative (CFD). As a result of its nonlinear nature, there is currently no universally recognized approach for obtaining an exact solution to this Eq. (10.15). Only a few specific cases allow for exact solutions, which can be used to make predictions about the fluid's state and consider simplified flow patterns. For more details, refer to Ref. [56].

Nonlinear differential equations that characterize physical phenomena often pose formidable challenges when exact solutions are sought. Consequently, a plethora of analytical and numerical methodologies have been prominently employed in their resolution. Over the past few decades, perturbation techniques rooted in Poincare's homotopy theory within the realm of topology have emerged as valuable tools in this endeavor. In 1992, Liao [57] pioneered the development of the Homotopy Analysis Method (HAM), laying the foundation for subsequent advancements in the field. Building upon Liao's seminal work, He [58] introduced the Homotopy Perturbation Method (HPM) in 1998, thereby expanding the analytical capabilities of this methodology. Both HAM and HPM have demonstrated remarkable efficacy in addressing complex technical and scientific challenges.

To demonstrate the effectiveness of these techniques, Ganji and Rafei [59] utilized the HPM to effectively solve the nonlinear Hirota-Satsuma coupled Korteweg-de Vries (KdV) problem. In a similar manner, Lu and Liu [60] utilized the HAM to solve the KdV-Burgers equation, demonstrating the adaptability of these methods in various problem domains. In addition, Siddiqui et al. [61] used HPM to study the complex flow structures of a viscous magnetohydrodynamic fluid restricted among a pair of parallel surfaces. This study further showcases the broad applicability and efficacy of homotopy-based addresses in solving nonlinear differential equations.

Marinca and Herisanu [62] proposed a unique computational approach called the optimal homotopy asymptotic approach (OHAM), which is a revised and improved iteration of the HAM with built-in convergence criteria. This novel computational technique stands out by offering superior flexibility in comparison to its precursor, HAM. The efficacy and reliability of OHAM have been substantiated through its successful application in resolving a diverse spectrum of scientific challenges. The researchers have successfully applied OHAM to solve various scientific problems, demonstrating its reliability and effectiveness.

OHAM has been strategically employed to scrutinize an array of complex physical phenomena. Noteworthy applications encompass the investigation of electrical machine dynamics [63], the analysis of particle oscillations [64], the exploration of systems governed by fractional power restoring forces [65], heat transfer [66], and steady flow of a fourth-grade fluid [67]. The method's value is underscored by its intrinsic capability to conveniently regulate and manipulate the convergence region during problem-solving endeavors. Therefore, OHAM, with its integrated convergence criteria and inherent adaptability, has emerged as a valuable and potent computational tool for addressing a wide array of scientific challenges characterized by nonlinear differential equations. Its successful applications in various scientific domains attest to its robustness and effectiveness as a problem-solving technique.

The authors of Ref. [49] explored the unstable one-dimensional motion of a fluid that is viscous in a tube. The NS equations in cylindrical dimensions regulate the flow field in the tube. The Eq. (10.15) was converted into operator representation as follows:

$$\frac{\partial^{\nu}\underline{\xi}(\overline{\varepsilon},\omega)}{\partial\omega^{\nu}} = \mathcal{P} + \eta\left(\frac{\partial^{2}\underline{\xi}(\overline{\varepsilon},\omega)}{\partial\overline{\varepsilon}^{2}} + \frac{1}{\overline{\varepsilon}}\frac{\partial\underline{\xi}(\overline{\varepsilon},\omega)}{\partial\overline{\varepsilon}}\right) \tag{10.16}$$

The equation involves a fractional order derivative where $0 < \nu \leq 1$, and $\mathcal{P} = -\frac{\partial p}{\rho\partial z}$. When $\nu = 1$, the standard NS equation is obtained. The analytical solution for Eq. (10.16) was successfully derived through the utilization of the Adomian decomposition method. In tackling this equation, a rich spectrum of effective analytical methodologies has been employed within the scientific community. Noteworthy among these techniques are He's HPM and the variational iteration method, as extensively discussed by previous research endeavors [68].

Additionally, researchers have explored various innovative approaches to address Eq. (10.16) with great success. Among these techniques are the Elzaki transform combined with the homotopy perturbation technique [69], the natural homotopy perturbation approach [70], the iterative Elzaki transform approach [71], the new homotopy perturbation transform approach [72], the Q-homotopy analysis transform scheme [73], and the modified Laplace decomposition approach [74]. These methodologies have demonstrated their efficacy in providing analytical solutions to complex problems akin to Eq. (10.16). However, it is noteworthy that none of the aforementioned techniques have hitherto employed a semianalytical approach in addressing the specific challenges posed by Eq. (10.16). This underscores the novelty and potential significance of the current research effort, which introduces a novel dimension to the analytical toolbox for solving this particular equation.

In addition, we use the least squares technique (LST) and residuals. LST is a mathematical marvel spanning a spectrum of disciplines from statistics to machine learning, intricately weaves its influence through the optimization landscape, finding optimal fits by meticulously minimizing the sum of squared differences between observed and predicted values. Its historical lineage, dating back to the 18th century with the celestial musings of Carl Friedrich Gauss, took a profound turn with Adrien-Marie Legendre's formalization in the 19th century, setting the stage for its ubiquity across fields. LST's evolution unfolds against the backdrop of its theoretical foundation, grounded in optimization principles that seek to unravel the mysteries of data relationships by sculpting a model that best aligns with observed phenomena. This methodological elegance extends its reach into regression analysis, where LST emerges as the linchpin, carving precise lines through data points in simple linear regression and gracefully navigating multidimensional landscapes in multiple linear regression, providing a robust framework for predictive modeling. Yet, LST's prowess transcends the confines of classical statistics, infiltrating realms as diverse as finance, engineering, and medical research, where its adaptability and reliability in handling intricate problems find resonance. As we delve into the intricate tapestry of LST, its significance becomes palpable—an indispensable tool fostering optimal solutions by delicately minimizing the squared discrepancies between theoretical models and empirical observations. The journey of LST, a historical odyssey rooted in celestial contemplations, unfolds into a contemporary narrative where its theoretical underpinnings and practical applications converge, giving rise to a methodology that stands resilient against the complexities of real-world data analysis.

In the realm of statistical modeling, particularly within the context of the LST, the computation of residuals plays a pivotal role in assessing the similarity between observed and predicted values. Residuals, in essence, represent the differences between the actual data points and those predicted by a chosen model. The pursuit of an optimal fit, characteristic of LST, hinges on minimizing these residuals.

To delve into the computation of residuals, it is essential to understand their nature. Residuals are the vertical distances between each observed data point and the corresponding point predicted by the model. In the simplest terms, they are the errors or the discrepancies that the model introduces. In the context of LST, where the primary objective is to minimize the sum of squared residuals, understanding how these residuals are computed becomes paramount. The computation of residuals involves subtracting the predicted value from the observed value for each data point. Mathematically, the residual (e) for a particular observation (i) can be expressed as:

$$e_i = y_i - \hat{y}_i$$

Here, y_i represents the observed value, $\hat{y}_i$ is the predicted value, and e_i denotes the residual for the i-th observation. The residual represents the vertical distance between the observed point and the point predicted by the model along the vertical axis.

The significance of computing residuals within the framework of LST lies in evaluating how well the chosen model captures the underlying patterns in the data. A perfectly fitting model would yield residuals of zero for every observation, indicating that the model precisely predicts each data point. However, in the real world, achieving zero residuals is often impractical due to inherent variability and noise in the data. The process of minimizing residuals is, in fact, a process of optimizing the model parameters to reduce the overall discrepancy between observed and predicted values. This is where the LST exhibits its strength. By minimizing the sum of squared residuals, LST seeks to find the optimal parameters that result in the best-fitting line or surface through the dataset.

The similarity between observed and predicted values is gauged by the magnitude and distribution of residuals. Small residuals suggest a close match between the model and the data, indicating a high degree of predictive accuracy. Conversely, larger residuals signify areas where the model deviates from observed values, highlighting potential areas for improvement or further model refinement.

In summary, the computation of residuals within the context of LST is a crucial step in assessing the similarity between observed and predicted values. The process involves calculating the vertical differences between each observed data point and its corresponding prediction. By minimizing the sum of squared residuals, LST optimizes model parameters to achieve the best possible fit, ultimately unveiling the underlying patterns in the data and providing a quantitative measure of the model's predictive accuracy.

In this Chapter, we embark on a comprehensive exploration of the application of the OHAM to address a specific time-fractional derivative of the NS Eq. (10.16). Our objective is to introduce a semianalytical solution for this equation, utilizing OHAM, and assess its accuracy and effectiveness. To achieve this, we have structured the chapter into several sections, each contributing to a deeper understanding of the problem at hand.

Section 10.2 provides a foundational overview of the definitions and characteristics of fractional order operators, setting the stage for our subsequent analysis. In Section 10.3, we delve into the basic formulation of OHAM, explaining its principles and mechanics. Building upon this understanding, Section 10.4 demonstrates the practical application of OHAM to time-fractional Navier-Stokes (TFNS)-type equations. Our findings are presented through tables and plots, offering a quantitative assessment of OHAM's accuracy and effectiveness in solving these complex equations.

Finally, in Section 10.5, we summarize our research findings, highlighting the contributions and implications of our work in the broader context of solving nonlinear fractional differential equations. By following this structured approach, we aim to provide readers with a clear and organized presentation of our research, facilitating a deeper understanding of the capabilities of OHAM in solving challenging problems in fluid mechanics and related fields.

10.2 Preliminaries

This section delves into the introduction and discussion of several key definitions and properties that hold significant importance in the context of the chapter.

Definition 10.1: A real function $\Theta(\omega), \omega > 0$ is said to belong to the space $C_{\overline{\varepsilon}}(\overline{\varepsilon} > 0)$ if it can be expressed as $\Theta(\omega) = \omega^{p}\psi(\omega)$ for certain $p > \overline{\varepsilon}$ where $\psi(\omega) \in C[0, \infty)$, and it is required to be in the space $C^{k}_{\overline{\varepsilon}} \Leftrightarrow \Theta^{k} \in C_{\overline{\varepsilon}}, k \in N$.

Definition 10.2: [30] The $v \geq 0$ order integral operator, in the Riemann-Liouville sense, for $\Theta \in C_{\overline{\varepsilon}}, \overline{\varepsilon} \geq -1$ is expressed as follows:

$$\mathbf{I}^{v}\Theta(\omega) = \frac{1}{\Gamma(v)} \int_{0}^{\omega} \frac{\Theta(\overline{\varepsilon})}{(\omega - \overline{\varepsilon})^{1-v}} \, d\overline{\varepsilon}$$

Let $\Theta \in C_{\overline{\varepsilon}}, \overline{\varepsilon} \geq -1, v, \gamma \geq 0$, and $\mu > -1$. The properties [30] of the function are as follows:

- $\mathbf{I}^{v}\mathbf{I}^{\gamma}\Theta(\omega) = \mathbf{I}^{v+\gamma}\Theta(\omega)$,
- $\mathbf{I}^{v}\mathbf{I}^{\gamma}\Theta(\omega) = \mathbf{I}^{\gamma}\mathbf{I}^{v}\Theta(\omega)$,
- $\mathbf{I}^{v}\omega^{\mu} = \frac{\Gamma(\mu+1)}{\Gamma(\mu+v+1)}\omega^{\mu+v}$.

Definition 10.3: [30] The operator that represents the $v > 0$ order Caputo derivative for a function Θ belonging to the space $C^{k}_{-1}, k \in N$, is defined as:

$${}^{c}\mathbf{D}^{v}\Theta(\omega) = \frac{1}{\Gamma(\nu - v)} \int_{0}^{\omega} \frac{\Theta^{(\nu)}(\overline{\varepsilon})}{(\omega - \overline{\varepsilon})^{v-\nu+1}} \, d\overline{\varepsilon}, \omega > 0.$$

In the given context, where $[v]$ indicates the integer part of v and $\mathcal{n} = [v] + 1$, certain properties of functions Θ and ϕ belonging to the space $C_{\overline{\varepsilon}}$ (where $\overline{\varepsilon} \geq -1$) can be established. These properties, as provided in [30], are as follows:

- ${}^{c}\mathbf{D}^{v}(\alpha\Theta(\omega) + \beta\phi(\omega)) = \alpha{}^{c}\mathbf{D}^{v}\Theta(\omega) + \beta{}^{c}\mathbf{D}^{v}\phi(\omega), \alpha, \beta \in R,$
- ${}^{c}\mathbf{D}^{v}\mathbf{I}^{v}\Theta(\omega) = \Theta(\omega),$
- $\mathbf{I}^{v}[{}^{c}\mathbf{D}^{v}]\Theta(\omega) = \Theta(\omega) - \sum_{j=0}^{k-1} \Theta^{(j)}(0)\frac{x_j}{j!}.$

10.3 Formulation of the OHAM

To solve PDEs with their respective boundary conditions using OHAM, the procedure outlined in [75] is followed:

$$L\left(\underline{\xi}(\overline{\varepsilon}, \omega)\right) + N\left(\underline{\xi}(\overline{\varepsilon}, \omega)\right) + \hbar^{*}(\overline{\varepsilon}, \omega) = 0, \mathcal{B}\left(\underline{\xi}, \frac{\partial \underline{\xi}}{\partial \omega}\right) = 0, \overline{\varepsilon} \in \Omega, \tag{10.17}$$

The problem involves a linear operator is designated as L and a nonlinear operator is designated as N. The unknown function is represented by the function $\underline{\xi}(\overline{\varepsilon}, \omega)$, and the boundary operator is denoted by $\mathcal{B}$. The spatial and time variables are represented by $\overline{\varepsilon}$ and ω, respectively. Additionally, it is expected that the domain of this problem can be expressed by Ω and that the function $\hbar^{*}(\overline{\varepsilon}, \omega)$ exists.

Through OHAM, the homotopy function $w(\overline{\varepsilon}, \omega; \varphi): \Omega \times \Upsilon \rightarrow R$ is constructed, where $\Upsilon = [0, 1]$. This homotopy function possesses the following properties:

$$(1 - \varphi)\left\{L(w(\overline{\varepsilon}, \omega; \varphi)) + \hbar^{*}(\overline{\varepsilon}, \omega)\right\} = H(\varphi)\left\{L(w(\overline{\varepsilon}, \omega; \varphi)) + N(w(\overline{\varepsilon}, \omega; \varphi)) + \hbar^{*}(\overline{\varepsilon}, \omega)\right\}, \tag{10.18}$$

where the parameter φ belongs to the interval Υ and serves as an embedding parameter. The auxiliary function $H(\varphi)$ is nonzero for $\varphi \neq 0$, and it satisfies $H(0) = 0$. Eq. (10.18) is considered the optimal homotopy equation. Evidently, the following holds:

$$\varphi = 0 \Rightarrow h^{*}(\overline{\varepsilon}, \omega) + L(w(\overline{\varepsilon}, \omega; \varphi)) = 0$$
$$\varphi = 1 \Rightarrow N(w(\overline{\varepsilon}, \omega; \varphi)) + h^{*}(\overline{\varepsilon}, \omega) + L(w(\overline{\varepsilon}, \omega; \varphi)).$$

For $\varphi = 0$, we can obtain $w(\overline{\varepsilon}, \omega; 0) = \underline{\xi}_0(\overline{\varepsilon}, \omega)$ and for $\varphi = 1$, we can get $w(\overline{\varepsilon}, \omega; 1) = \underline{\xi}(\overline{\varepsilon}, \omega)$. Consequently, as φ ranges from 0 to 1, $w(\overline{\varepsilon}, \omega; \varphi)$ transitions from $\underline{\xi}_0(\overline{\varepsilon}, \omega)$ to $\underline{\xi}(\overline{\varepsilon}, \omega)$, with $\underline{\xi}_0(\overline{\varepsilon}, \omega)$ being obtained from Eq. (10.18) when $\varphi = 0$:

$$L\left(\underline{\xi}_0(\overline{\varepsilon}, \omega; \varphi)\right) + \hbar^{*}(\overline{\varepsilon}, \omega) = 0, \mathcal{B}\left(\underline{\xi}_0, \frac{\partial \underline{\xi}_0}{\partial \omega}\right) = 0. \tag{10.19}$$

Then, we express the auxiliary function $H(\varphi)$ as follows:

$$H(\varphi) = \varphi K_1 + \varphi^2 K_2 + \varphi^3 K_3 + \cdots + \varphi^{R} K_{\mathcal{L}}. \tag{10.20}$$

To obtain a numerical solution, Taylor's series expansion is used around φ, and $w(\overline{\varepsilon},\omega;\varphi,K_\sigma)$ is expanded as follows:

$$w(\overline{\varepsilon},\omega;\varphi,K_\sigma)=\underline{\xi}_0(\overline{\varepsilon},\omega)+\sum_{\tau=1}^{\infty}\underline{\xi}_\tau(\overline{\varepsilon},\omega;K_\sigma)\varphi^\tau,\sigma=1,2,3,\ldots \tag{10.21}$$

The zeroth-order problem, as given in Eq. (10.19), can be acquired by equating the coefficients of φ powers after inserting Eq. (10.20) into Eq. (10.18). Similarly, the first-order problem and second-order problem can be obtained using Eqs. (10.22) and (10.23), respectively. The general governing equations for $\underline{\xi}_\tau(\overline{\varepsilon},\omega)$ are represented by Eq. (10.24).

$$L\left(\underline{\xi}_1(\overline{\varepsilon},\omega)\right)=K_1N_0\left(\underline{\xi}_0(\overline{\varepsilon},\omega)\right),B\left(\underline{\xi}_1,\frac{\partial\underline{\xi}_1}{\partial\omega}\right)=0, \tag{10.22}$$

$$\begin{aligned}L\left(\underline{\xi}_2(\overline{\varepsilon},\omega)\right)-L\left(\underline{\xi}_1(\overline{\varepsilon},\omega)\right)&=K_2N_0\left(\underline{\xi}_0(\overline{\varepsilon},\omega)\right)+K_1\left[L\left(\underline{\xi}_1(\overline{\varepsilon},\omega)\right)\right.\\&\left.+N_1\left(\underline{\xi}_0(\overline{\varepsilon},\omega),\underline{\xi}_1(\overline{\varepsilon},\omega)\right)\right],B\left(\underline{\xi}_2,\frac{\partial\underline{\xi}_2}{\partial\omega}\right)=0,\end{aligned} \tag{10.23}$$

$\vdots$

$$\begin{aligned}&L\left(\underline{\xi}_\tau(\overline{\varepsilon},\omega)\right)-L\left(\underline{\xi}_{\tau-1}(\overline{\varepsilon},\omega)\right)=K_\tau N_0\left(\underline{\xi}_0(\overline{\varepsilon},\omega)\right)\\&+\sum_{\sigma=1}^{\tau-1}K_\sigma\left[L\left(\underline{\xi}_{\tau-\sigma}(\overline{\varepsilon},\omega)\right)+N_{\tau-\sigma}\left(\underline{\xi}_0(\overline{\varepsilon},\omega),\ \underline{\xi}_1(\overline{\varepsilon},\omega),\ \underline{\xi}_2(\overline{\varepsilon},\omega),\ \ldots,\underline{\xi}_{\tau-\sigma}(\overline{\varepsilon},\omega)\right)\right],\ B\left(\underline{\xi}_\tau,\frac{\partial\underline{\xi}_\tau}{\partial\omega}\right),\tau=2,3,\ldots,\end{aligned} \tag{10.24}$$

where $N_{\tau-\sigma}\left(\underline{\xi}_0(\overline{\varepsilon},\omega),\underline{\xi}_1(\overline{\varepsilon},\omega),\ldots,\underline{\xi}_{\tau-\sigma}(\overline{\varepsilon},\omega)\right)$ is the coefficient of $\varphi^{\tau-\sigma}$ in the expansion of $N(w(\overline{\varepsilon},\omega;\varphi))$ about the embedding parameter φ.

$$N(w(\overline{\varepsilon},\omega;\varphi,K_\sigma))=N_0\left(\underline{\xi}_0(\overline{\varepsilon},\omega)\right)+\sum_{\tau\geq1}N_\tau\left(\underline{\xi}_0,\underline{\xi}_1,\ldots,\underline{\xi}_\tau\right)\varphi^\tau.$$

The task of solving the linear equations with linear boundary conditions for $\underline{\xi}_\tau$, where $\tau\geq0$, can be accomplished with ease in this context.

If the series in Eq. (10.21) is convergent at $\varphi=1$, it implies a dependence on $K_1,K_2,\ldots$.

$$\tilde{\underline{\xi}}(\overline{\varepsilon},\omega;K_\sigma)=\underline{\xi}_0(\overline{\varepsilon},\omega)+\sum_{\tau=1}^{E}\underline{\xi}_\tau(\overline{\varepsilon},\omega,K_\sigma). \tag{10.25}$$

By substituting Eq. (10.25) into Eq. (10.17), the residual expression can be obtained in the subsequent structure:

$$R(\overline{\varepsilon},\omega;K_\sigma)=L\left(\tilde{\underline{\xi}}(\overline{\varepsilon},\omega;K_\sigma)\right)+N\left(\tilde{\underline{\xi}}(\overline{\varepsilon},\omega;K_\sigma)\right)+\hbar^*(\overline{\varepsilon},\omega). \tag{10.26}$$

In the case where $\mathcal{R}(\overline{\varepsilon}, \omega; \mathcal{K}_\sigma) = 0$, the solution $\tilde{\underline{\xi}}(\overline{\varepsilon}, \omega; \mathcal{K}_\sigma)$ would be the exact solution. It should be emphasized that, generally speaking, this condition cannot always be satisfied in nonlinear issues. To calculate the coefficients $\mathcal{K}_\sigma$, where $\sigma = 1, 2, \ldots, \mathcal{L}$, one can employ the LST, as outlined below:

$$\psi(\mathcal{K}_\sigma) = \int_0^\omega \int_\Omega \mathcal{R}^2(\overline{\varepsilon}, \omega; \mathcal{K}_\sigma) d\overline{\varepsilon} d\omega,$$

where $\mathcal{R}$ denotes the residual as defined by Eq. (10.26), and

$$\frac{\partial \psi}{\partial \mathcal{K}_1} = \frac{\partial \psi}{\partial \mathcal{K}_2} = \frac{\partial \psi}{\partial \mathcal{K}_3} = \cdots = \frac{\partial \psi}{\partial \mathcal{K}_\mathcal{L}} = 0. \tag{10.27}$$

Using Eq. (10.27), the convergence, which is determined by the coefficients $\mathcal{K}_1, \mathcal{K}_2, \mathcal{K}_3, \ldots$, can be effectively identified and minimized.

10.4 Numerical examples

In this portion, the utilization of fractional OHAM is implemented to arrive at the solution to the TFNS equations [76].

Example 10.1: Certainly, let us consider the TFNS equation provided below:

$$\frac{\partial^\nu \underline{\xi}(\overline{\varepsilon}, \omega)}{\partial \omega^\nu} = \mathcal{P} + \frac{\partial^2 \underline{\xi}(\overline{\varepsilon}, \omega)}{\partial \overline{\varepsilon}^2} + \frac{1}{\overline{\varepsilon}} \frac{\partial \underline{\xi}(\overline{\varepsilon}, \omega)}{\partial \overline{\varepsilon}}, \tag{10.28}$$

with the initial condition:

$$\underline{\xi}(\overline{\varepsilon}, 0) = 1 - \overline{\varepsilon}^2. \tag{10.29}$$

In Eq. (10.28), the term $\frac{\partial^\nu}{\partial \omega^\nu}$ represents the *CFD*, where $0 < \nu \leq 1$. The exact solution to Eq. (10.28), as provided by [56], is given below:

$$\underline{\xi}(\overline{\varepsilon}, \omega) = 1 - \overline{\varepsilon}^2 + (\mathcal{P} - 4)\omega. \tag{10.30}$$

Based on the information provided in Section 10.3, the construction of the homotopy is as follows:

$$(1 - \varphi) \frac{\partial^\nu w(\overline{\varepsilon}, \omega; \varphi)}{\partial \omega^\nu} = \mathcal{H}(\varphi, \mathcal{K}_\sigma) \left[\frac{\partial^\nu w(\overline{\varepsilon}, \omega; \varphi)}{\partial \omega^\nu} - \mathcal{P} - \frac{\partial^2 w(\overline{\varepsilon}, \omega; \varphi)}{\partial \overline{\varepsilon}^2} - \frac{1}{\overline{\varepsilon}} \frac{\partial w(\overline{\varepsilon}, \omega; \varphi)}{\partial \overline{\varepsilon}} \right], \tag{10.31}$$

where

$$w(\overline{\varepsilon}, \omega; \varphi) = \underline{\xi}_0(\overline{\varepsilon}, \omega) + \sum_{\tau=1}^{\infty} \underline{\xi}_\tau(\overline{\varepsilon}, \omega; \mathcal{K}_\sigma) \varphi^\tau, \sigma = 1, 2, 3, \ldots \tag{10.32}$$

$$\mathcal{H}(\varphi, \mathcal{K}_\sigma) = \varphi \mathcal{K}_1 + \varphi^2 \mathcal{K}_2 + \varphi^3 \mathcal{K}_3 + \varphi^4 \mathcal{K}_4 + \ldots \tag{10.33}$$

The simplified problems can be derived by inserting Eqs. (10.32) and (10.33) into Eq. (10.31) and subsequently equating the coefficients of the corresponding powers of φ.

Zero-order problem:

$$\frac{\partial^{v}\underline{\xi}_{0}(\overline{\varepsilon},\omega)}{\partial\omega^{v}}=0,\underline{\xi}_{0}(\overline{\varepsilon},0)=-\overline{\varepsilon}^{2}+1. \tag{10.34}$$

First-order problem:

$$\begin{aligned}\frac{\partial^{v}\underline{\xi}_{1}(\varepsilon,\omega;\mathsf{K}_{1})}{\partial\omega^{v}}&=(1+\mathsf{K}_{1})\frac{\partial^{v}\underline{\xi}_{0}(\overline{\varepsilon},\omega)}{\partial\omega^{v}}-\mathsf{K}_{1}\mathsf{P}-\mathsf{K}_{1}\frac{\partial^{2}\underline{\xi}_{0}(\overline{\varepsilon},\omega)}{\partial\overline{\varepsilon}^{2}}\\&-\frac{\mathsf{K}_{1}}{\overline{\varepsilon}}\frac{\partial\underline{\xi}_{0}(\overline{\varepsilon},\omega)}{\partial\overline{\varepsilon}},\underline{\xi}_{1}(\overline{\varepsilon},0)=0.\end{aligned} \tag{10.35}$$

After applying fractional integration and considering the initial condition, we obtain the solutions to Eqs. (10.34) and (10.35) as follows:

$$\begin{aligned}\underline{\xi}_{0}(\overline{\varepsilon},\omega)&=1-\overline{\varepsilon}^{2}\\\underline{\xi}_{1}(\overline{\varepsilon},\omega;\mathsf{K}_{1})&=-\frac{\mathsf{K}_{1}(\mathsf{P}-4)\omega^{v}}{\Gamma(1+v)}.\end{aligned}$$

By following the calculations, we arrive at the following expression:

$$\begin{aligned}\tilde{\underline{\xi}}(\overline{\varepsilon},\omega)&=\underline{\xi}_{0}(\overline{\varepsilon},\omega)+\underline{\xi}_{1}(\overline{\varepsilon},\omega;\mathsf{K}_{1})+\underline{\xi}_{2}(\overline{\varepsilon},\omega;\mathsf{K}_{1},\mathsf{K}_{2})+\ldots\\&=-\overline{\varepsilon}^{2}+1-\frac{\mathsf{K}_{1}(\mathsf{P}-4)\omega^{v}}{\Gamma(1+v)}.\end{aligned} \tag{10.36}$$

Upon implementing the least squares method and computing the residual, we have successfully determined the auxiliary constant value for $v=1$ as $\mathsf{K}_1=-1$. By substituting this obtained value of K_1 into Eq. (10.36), the resulting expression becomes:

$$\tilde{\underline{\xi}}(\overline{\varepsilon},\omega)=1+\frac{(\mathsf{P}-4)\omega^{v}}{\Gamma(1+v)}-\overline{\varepsilon}^{2}. \tag{10.37}$$

The obtained result in Eq. (10.37) is in complete agreement with references [56,68–74]. Figs. 10.1–10.3 depict the semianalytical solution outcome assessment for example (10.28) with $\mathsf{P}=2$. Fig. 10.1B showcases the dynamics of the solution obtained using OHAM for different noninteger order Brownian motions, along with the standard motion where $v=1$. As shown in Fig. 10.1B, the solution obtained using this technique decreases rapidly as $\overline{\varepsilon}$ increases in example (10.28).

Fig. 10.1A provides evidence of the suggested method's effectiveness. Moreover, Fig. 10.3A illustrates the solution behavior for $v=0.3$, while Fig. 10.3B demonstrates the solution behavior for $v=0.6$. Additionally, the exact solution has been obtained using OHAM.

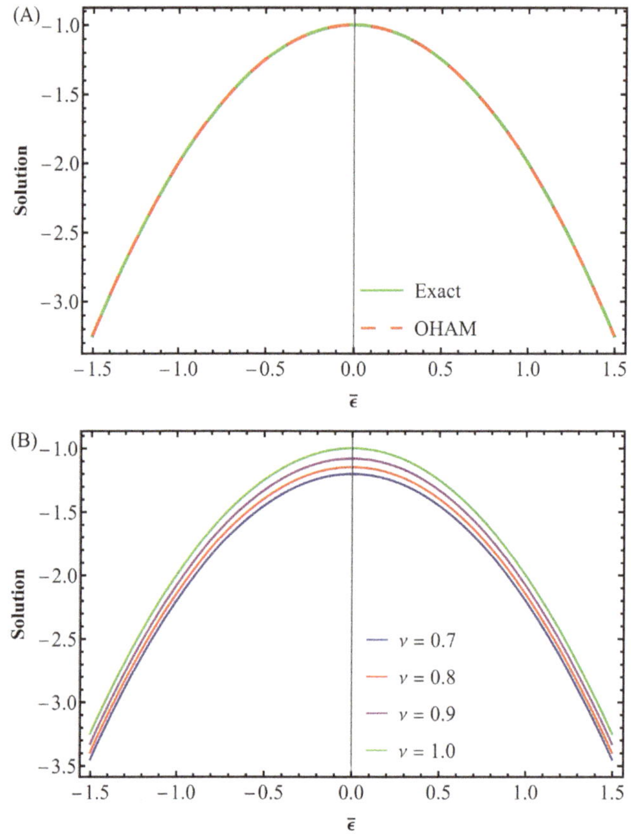

FIGURE 10.1

(A) Exact vs OHAM solution and (B) solution behavior with varying ν values.

Example 10.2: Let us consider the TFNS equation given below:

$$\frac{\partial^{\nu}\underline{\xi}(\overline{\varepsilon},\omega)}{\partial\omega^{\nu}}=\frac{\partial^{2}\underline{\xi}(\overline{\varepsilon},\omega)}{\partial\overline{\varepsilon}^{2}}+\frac{1}{\overline{\varepsilon}}\frac{\partial\underline{\xi}(\overline{\varepsilon},\omega)}{\partial\overline{\varepsilon}}, \tag{10.38}$$

with the initial condition

$$\underline{\xi}(\overline{\varepsilon},0)=\overline{\varepsilon}, \tag{10.39}$$

The CFD, denoted by $\frac{\partial^{\nu}}{\partial\omega^{\nu}}$, is used in the TFNS equation, where $0<\nu\leq 1$. According to Ref. [56], the exact solution to Eq. (10.38) can be expressed as a series, and the first four terms of this series are listed below:

$$\underline{\xi}(\overline{\varepsilon},\omega)=\overline{\varepsilon}+\frac{\omega}{\overline{\varepsilon}}+\frac{\omega^{2}}{2\overline{\varepsilon}^{3}}+\frac{9\omega^{3}}{6\overline{\varepsilon}^{5}}. \tag{10.40}$$

(A)

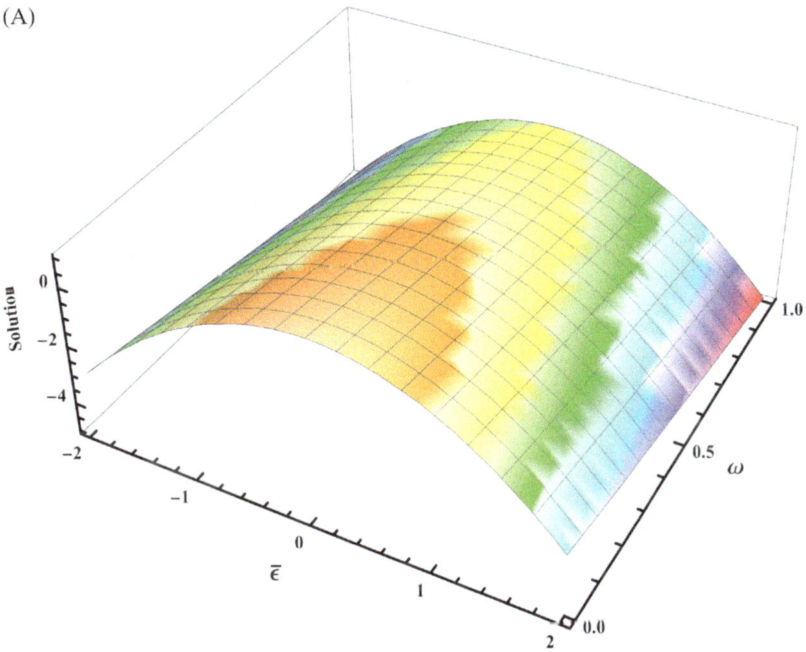

(B)

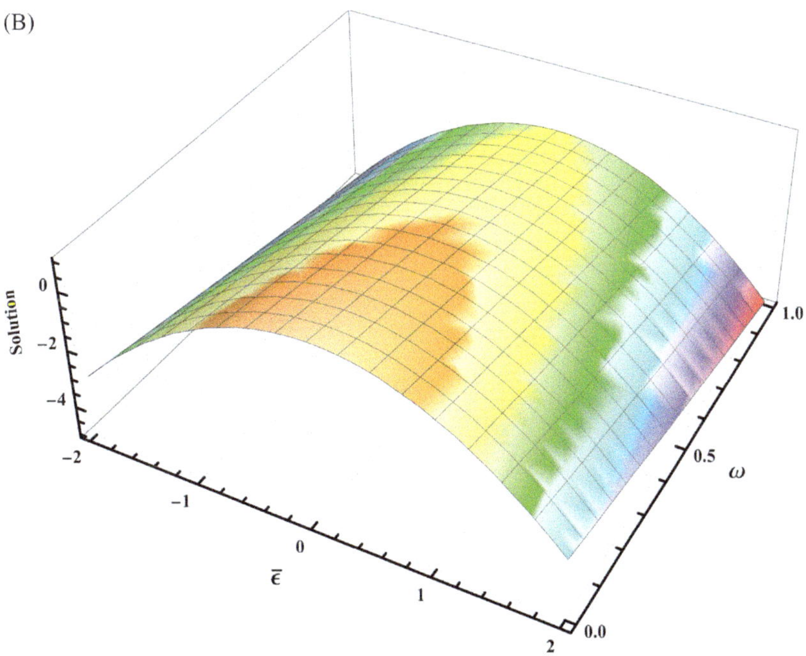

FIGURE 10.2

(A) Exact solution and (B) OHAM solution.

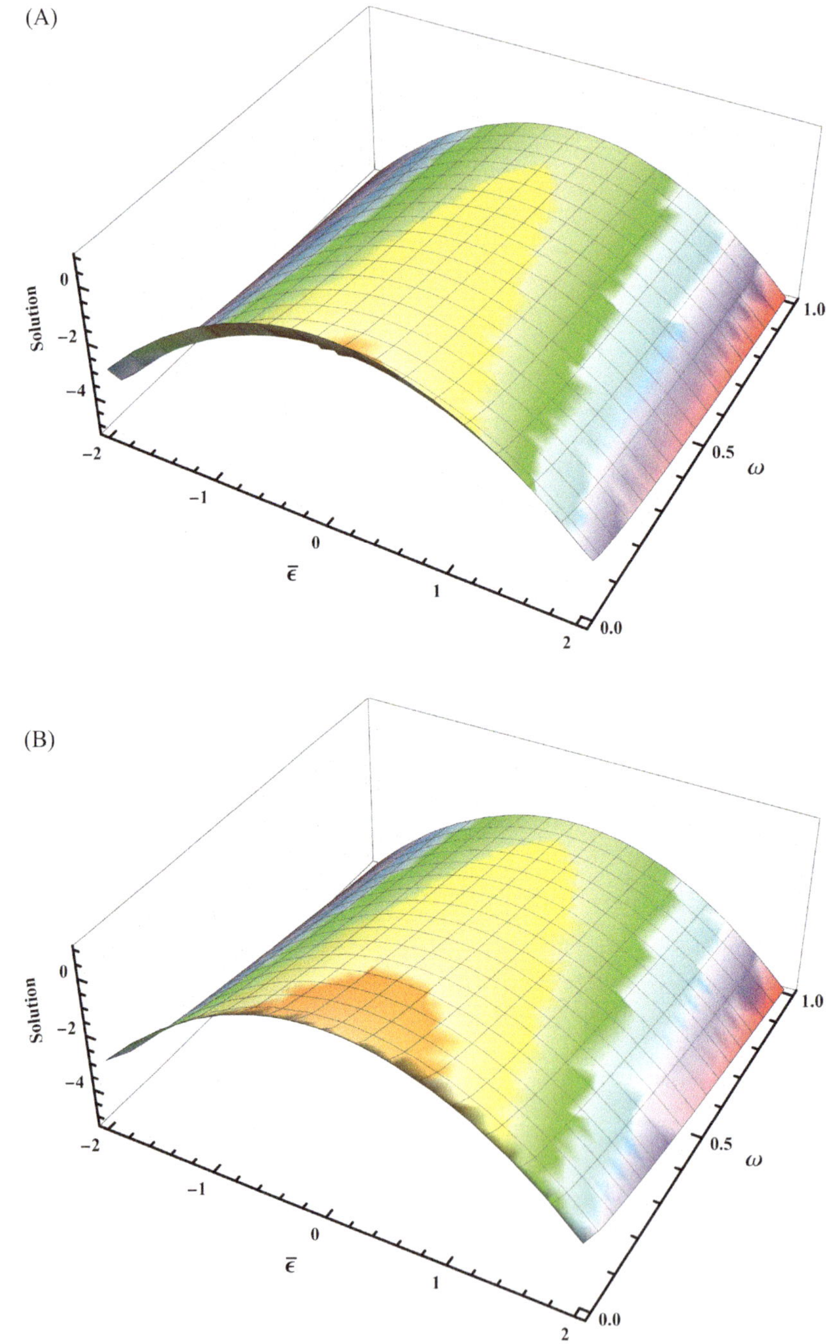

FIGURE 10.3

(A) Solution for $v = 0.3$ and (B) solution for $v = 0.6$.

Following the procedure detailed in Section 10.3, the homotopy can be developed as follows:

$$(1-\varphi)\frac{\partial^{v} w(\overline{\varepsilon},\omega;\varphi)}{\partial\omega^{v}} = H(\varphi,K_{\sigma})\left[\frac{\partial^{v} w(\overline{\varepsilon},\omega;\varphi)}{\partial\omega^{v}} - \frac{\partial^{2} w(\overline{\varepsilon},\omega;\varphi)}{\partial\overline{\varepsilon}^{2}} - \frac{1}{\overline{\varepsilon}}\frac{\partial w(\overline{\varepsilon},\omega;\varphi)}{\partial\overline{\varepsilon}}\right], \quad (10.41)$$

where

$$w(\overline{\varepsilon},\omega;\varphi) = \underline{\xi}_{0}(\overline{\varepsilon},\omega) + \sum_{\tau=1}^{\infty}\underline{\xi}_{\tau}(\overline{\varepsilon},\omega;K_{\sigma})\varphi^{\tau}, \sigma = 1,2,3,\ldots \quad (10.42)$$

$$H(\varphi,K_{\sigma}) = \varphi K_{1} + \varphi^{2}K_{2} + \varphi^{3}K_{3} + \varphi^{4}K_{4} + \ldots \quad (10.43)$$

Comparing the parameters of the relevant powers of φ after inserting Eqs. (10.42) and (10.43) into Eq. (10.41) will result in the simplified problems shown below:

Zero-order problem:

$$\frac{\partial^{v}\underline{\xi}_{0}(\overline{\varepsilon},\omega)}{\partial\omega^{v}} = 0, \underline{\xi}_{0}(\overline{\varepsilon},0) = \overline{\varepsilon}. \quad (10.44)$$

First-order problem:

$$\frac{\partial^{v}\underline{\xi}_{1}(\overline{\varepsilon},\omega;K_{1})}{\partial\omega^{v}} = (1+K_{1})\frac{\partial^{v}\underline{\xi}_{0}(\overline{\varepsilon},\omega)}{\partial\omega^{v}} - K_{1}\frac{\partial^{2}\underline{\xi}_{0}(\overline{\varepsilon},\omega)}{\partial\overline{\varepsilon}^{2}} - \frac{K_{1}}{\overline{\varepsilon}}\frac{\partial\underline{\xi}_{0}(\overline{\varepsilon},\omega)}{\partial\overline{\varepsilon}}, \underline{\xi}_{1}(\overline{\varepsilon},0) = 0. \quad (10.45)$$

Second-order problem:

$$\frac{\partial^{v}\underline{\xi}_{2}(\overline{\varepsilon},\omega;K_{1},K_{2})}{\partial\omega^{v}} = (1+K_{1})\frac{\partial^{v}\underline{\xi}_{1}(\overline{\varepsilon},\omega;K_{1})}{\partial\omega^{v}} - K_{1}\frac{\partial^{2}\underline{\xi}_{1}(\overline{\varepsilon},\omega;K_{1})}{\partial\overline{\varepsilon}^{2}} - \frac{K_{1}}{\overline{\varepsilon}}\frac{\partial\underline{\xi}_{1}(\overline{\varepsilon},\omega;K_{1})}{\partial\overline{\varepsilon}}$$

$$+K_{2}\frac{\partial^{v}\underline{\xi}_{0}(\overline{\varepsilon},\omega)}{\partial\omega^{v}} - K_{2}\frac{\partial^{2}\underline{\xi}_{0}(\overline{\varepsilon},\omega)}{\partial\overline{\varepsilon}^{2}} - \frac{K_{2}}{\overline{\varepsilon}}\frac{\partial\underline{\xi}_{0}(\overline{\varepsilon},\omega)}{\partial\overline{\varepsilon}}, \underline{\xi}_{2}(\overline{\varepsilon},0) = 0. \quad (10.46)$$

Third-order problem:

$$\frac{\partial^{v}\underline{\xi}_{3}(\overline{\varepsilon},\omega;K_{1},K_{2},K_{3})}{\partial\omega^{v}}$$

$$= (1+K_1)\frac{\partial^v \xi_2(\bar{\varepsilon},\omega;K_1,K_2)}{\partial \omega^v} - K_1\frac{\partial^2 \xi_2(\bar{\varepsilon},\omega;K_1,K_2)}{\partial \bar{\varepsilon}^2} - \frac{K_1}{\bar{\varepsilon}}\frac{\partial \xi_2(\bar{\varepsilon},\omega;K_1,K_2)}{\partial \bar{\varepsilon}}$$
$$+ K_2\frac{\partial^v \xi_1(\bar{\varepsilon},\omega;K_1)}{\partial \omega^v} - K_2\frac{\partial^2 \xi_1(\bar{\varepsilon},\omega;K_1)}{\partial \bar{\varepsilon}^2} - \frac{K_2}{\bar{\varepsilon}}\frac{\partial \xi_1(\bar{\varepsilon},\omega;K_1)}{\partial \bar{\varepsilon}} + K_3\frac{\partial^v \xi_0(\bar{\varepsilon},\omega)}{\partial \omega^v} \quad (10.47)$$
$$- K_3\frac{\partial^2 \xi_0(\bar{\varepsilon},\omega)}{\partial \bar{\varepsilon}^2} - \frac{K_3}{\bar{\varepsilon}}\frac{\partial \xi_0(\bar{\varepsilon},\omega)}{\partial \bar{\varepsilon}}, \xi_3(\bar{\varepsilon},0) = 0.$$

After applying fractional integration and considering the initial conditions, we obtain the solutions to Eqs. (10.44)–(10.47) as follows:

$$\xi_0(\bar{\varepsilon},\omega) = \bar{\varepsilon},$$
$$\xi_1(\bar{\varepsilon},\omega;K_1) = -\frac{K_1\omega^v}{\bar{\varepsilon}\Gamma(1+v)},$$
$$\xi_2(\bar{\varepsilon},\omega;K_1,K_2) = -\frac{(K_1+K_1^2+K_2)\omega^v}{\bar{\varepsilon}\Gamma(1+v)} + \frac{K_1^2\omega^{2v}}{\bar{\varepsilon}^3\Gamma(1+2v)},$$
$$\xi_3(\bar{\varepsilon},\omega;K_1,K_2,K_3) = -\frac{(K_1+2K_1^2+K_1^3+2K_1K_2+K_2+K_3)\omega^v}{\bar{\varepsilon}\Gamma(1+v)}$$
$$+\frac{2K_1(K_1+K_1^2+K_2)\omega^{2v}}{\bar{\varepsilon}^3\Gamma(1+2v)} - \frac{9K_1^3\omega^{3v}}{\bar{\varepsilon}^5\Gamma(1+3v)}.$$

The solutions to the problems of higher orders that follow can be calculated by following a similar process. This leads us to the final expression:

$$\tilde{\xi}(\bar{\varepsilon},\omega) = \xi_0(\bar{\varepsilon},\omega) + \xi_1(\bar{\varepsilon},\omega;K_1) + \xi_2(\bar{\varepsilon},\omega;K_1,K_2) + \xi_3(\bar{\varepsilon},\omega;K_1,K_2,K_3) + \ldots$$
$$= \bar{\varepsilon} - \frac{(3K_1+3K_1^2+K_1^3+2K_2+2K_1K_2+K_3)\omega^v}{\bar{\varepsilon}\Gamma(1+v)} \quad (10.48)$$
$$+\frac{K_1(3K_1+2K_1^2+2K_2)\omega^{2v}}{\bar{\varepsilon}^3\Gamma(1+2v)} - \frac{9K_1^3\omega^{3v}}{\bar{\varepsilon}^5\Gamma(1+3v)}\ldots$$

After obtaining the residual, the least square method was used to determine the auxiliary constant value for v. Table 10.1 presents the values of the auxiliary constant corresponding to different v values. Additionally, Table 10.2 displays the absolute errors of both solutions.

Upon substituting the values of K_1, K_2, and K_3 for $v = 1$ into Eq. (10.49), we obtain the following expression:

$$\tilde{\xi}(\bar{\varepsilon},\omega) = \bar{\varepsilon} + \frac{\omega^v}{\bar{\varepsilon}\Gamma(1+v)} + \frac{1.00008\omega^{2v}}{\bar{\varepsilon}^3\Gamma(1+2v)} + \frac{8.99873\omega^{3v}}{\bar{\varepsilon}^5\Gamma(1+3v)}. \quad (10.49)$$

Figs. 10.4–10.6 demonstrate the outcomes of the examination of the TFNS equation's answer behavior employing the OHAM. Particularly, the resulting curves in Fig. 10.4B exhibit a sharp decline for higher fractional orders before

Table 10.1 K_1, K_2, K_3 for various values of v.

v	K_1	K_2	K_3
0.7	− 1.036692817879	− 0.001240278601	-4.24687×10^{-5}
0.8	− 1.020417062606	− 0.000376864638	-7.02931×10^{-6}
0.9	− 1.010550021386	− 0.000101578844	-9.88500×10^{-7}
1.0	− 0.999953117565	− 0.000040256679	3.30467×10^{-8}

Table 10.2 Comparing the accuracy of the exact and OHAM solutions for the given problem.

$\bar{\varepsilon}$	**Exact**	**OHAM**	**Abs Error**
1.0	4.0000000	3.9998292	1.70739×10^{-4}
1.4	2.5754031	2.5753785	2.45769×10^{-5}
1.8	2.5206726	2.5206683	4.27894×10^{-6}
2.2	2.7306083	2.7306079	3.26557×10^{-7}
2.6	3.0256880	3.0256885	5.03320×10^{-7}
3.0	3.3580246	3.3580253	6.12886×10^{-7}

converging to the typical fluid motion at $v = 1$. We evaluate the dynamics of the solution in the TFNS problem employing the OHAM. The findings are shown in Figs. 10.4–10.6. Particularly, in Fig. 10.4B, the lines representing the solution drop quickly when the fractional orders get higher. They eventually settle into the regular fluid motion when v is 1. This is interesting because it suggests a clear trend in how the solution changes with different fractional orders. The graphs not only show these patterns but also make it easier for researchers and others to understand what is happening in the TFNS equation under different circumstances. This way of looking at things helps us see how the fluid dynamics react to different levels of complexity, adding to our overall knowledge of the system.

Fig. 10.4A shows how well our proposed technique works. It demonstrates that our method is efficient and reliable for solving complex fluid mechanics problems involving fractional orders. This demonstrates how effective the OHAM is at researching issues of this kind.

Figs. 10.6A and B offer a more nuanced exploration of the OHAM by delving into its performance under specific fractional orders, such as $v = 0.3$ and $v = 0.6$. These focused analyses showcase OHAM's remarkable adaptability and efficacy across varying degrees of complexity and nonlinearity within fractional order systems. The graphical representations provide a visual narrative of OHAM's ability to navigate and comprehend the intricate dynamics inherent in fluid mechanics problems, offering insights that transcend a generic approach and are tailored to the specific challenges posed by fractional orders.

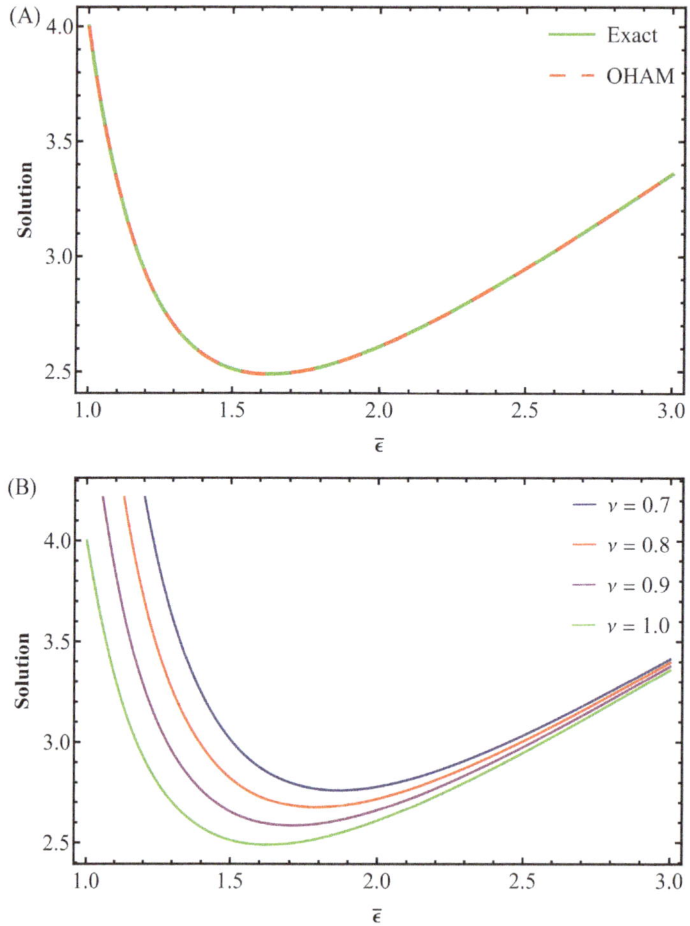

FIGURE 10.4

(A) Exact vs OHAM solution and (B) solution behavior with varying v.

Zooming in on Fig. 10.6A and its depiction of OHAM's response to a fractional order $v = 0.3$, we observe a harmonious correspondence between the predicted outcomes and the actual behavior of the system. This alignment underscores OHAM's precision and reliability in capturing the subtleties associated with a specific fractional order, reaffirming its utility in scenarios where fractional orders play a pivotal role in shaping fluid mechanics problems. The clear insights derived from this figure illuminate OHAM's ability to discern and interpret nuances at a granular level, contributing to a more profound understanding of the system's behavior.

Fig. 10.6B, which examines OHAM's performance under a different fractional order, $v = 0.6$, extends this narrative by showcasing the method's consistency and versatility. Once again, the graph attests to OHAM's adeptness in addressing the

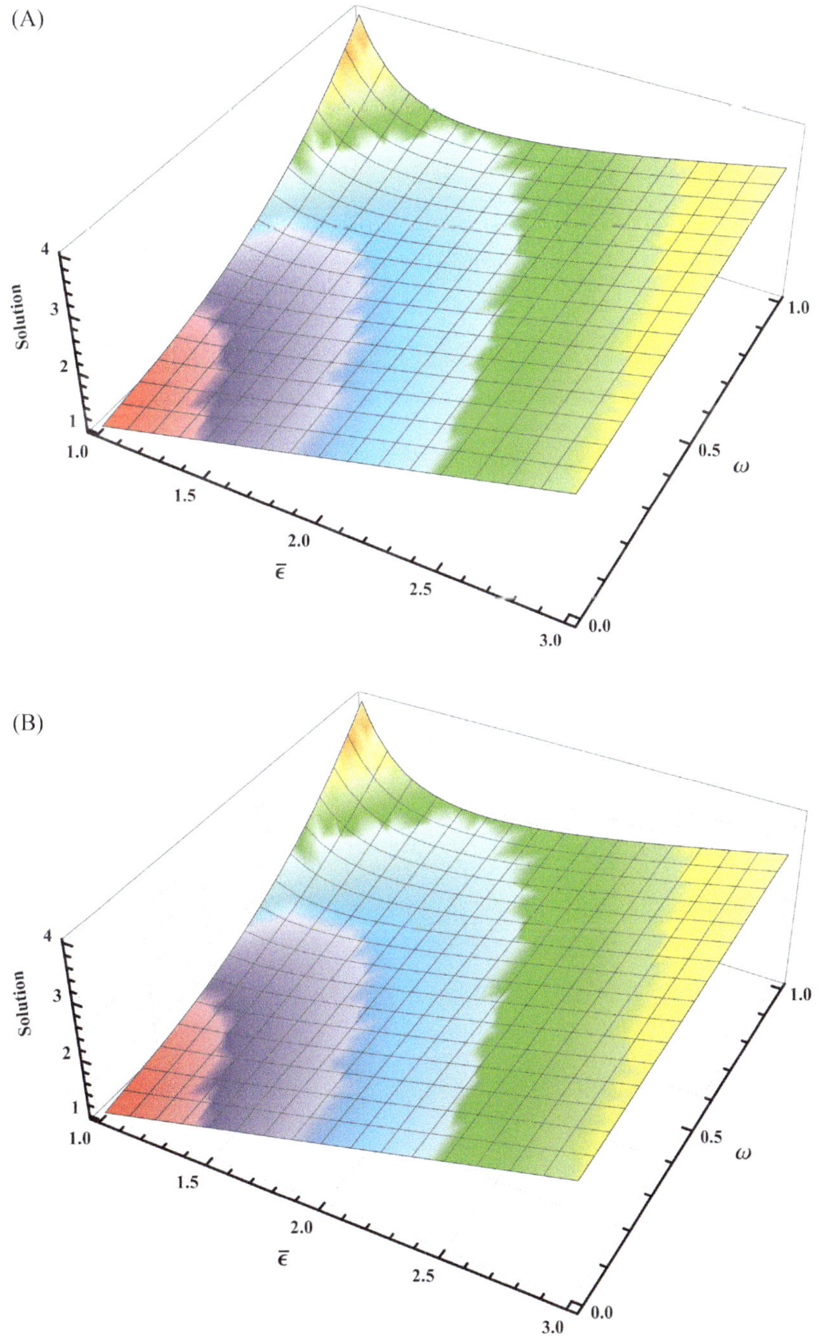

FIGURE 10.5

(A) Exact solution and (B) OHAM solution.

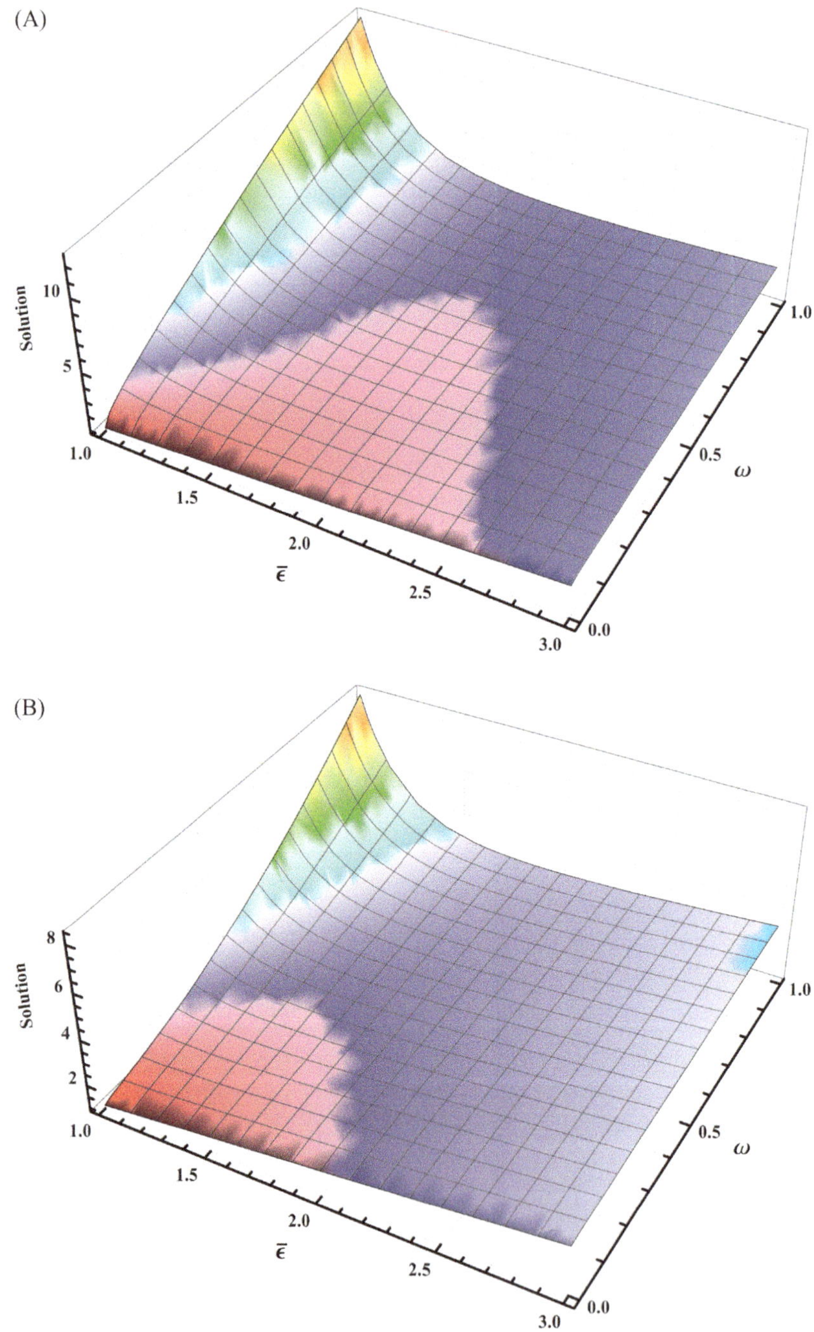

FIGURE 10.6

(A) Solution for $v = 0.3$ and (B) solution for $v = 0.6$.

challenges posed by varying degrees of complexity. The adaptability demonstrated in these visual representations transcends the limitations of a one-size-fits-all solution, emphasizing OHAM's capacity to cater to the specific demands of fluid mechanics problems characterized by distinct fractional orders.

Connecting these graphical analyses with the findings presented in Table 10.2 reinforces the robustness of OHAM as a comprehensive analytical tool. The quantitative data in the table, which encompasses various scenarios and fractional orders, aligns seamlessly with the qualitative insights derived from the visual representations. The agreement between the pictures in Figs. 10.4A and 10.6A and B and the corresponding data in Table 10.2 serves as a powerful convergence of evidence, collectively affirming that OHAM is not only efficient and accurate but also remarkably versatile in solving fractional order fluid mechanics problems.

Moreover, the synergy between the graphical representations and the tabular data strengthens the generalizability of OHAM's applicability. The method's success is not confined to a singular instance but extends across a spectrum of fractional orders, reflecting its universal competence in unraveling the complexities embedded in fluid mechanics problems. This universality underscores OHAM as a go-to methodology for researchers and practitioners confronted with a diverse array of scenarios, providing a robust framework for gaining profound insights into the behavior of fluid systems governed by fractional orders.

In conclusion, Figs. 10.6A and B deepen our understanding of OHAM by spotlighting its performance under specific fractional orders. These illustrations not only underscore the method's adaptability and efficacy in handling complexity and nonlinearity but also provide a nuanced view of its capabilities. The coherence between the graphical representations and the data in Table 10.2 reinforces OHAM's efficiency, accuracy, and versatility, establishing it as a valuable and dependable tool for advancing our comprehension of fractional order fluid mechanics problems.

10.5 Conclusion

In the ever-evolving landscape of scientific and engineering research, the fractional-order OHAM has emerged as an extraordinary and indispensable analytical tool. This comprehensive study has delved deep into OHAM's capabilities and its successful application in tackling the formidable TFNS problem. The findings underscore OHAM's remarkable versatility, precision, and reliability, establishing it as a cornerstone in the toolkit of researchers and professionals across a multitude of disciplines. The conducted study has revealed that the fractional-order OHAM stands out as an effective and dependable semianalytical approach for obtaining optimal solutions to a broad spectrum of fractional-order linear and nonlinear problems, commonly encountered in the domains of science and engineering. The successful application of OHAM in resolving the TFNS problem serves as a testament to both its versatility and precision. Furthermore, the consistency observed between OHAM solutions and exact solutions, coupled with the demonstrated convergence through numerical approximation, underscores the efficacy of the proposed technique. This finding instills confidence in researchers and professionals within the field, encouraging them to employ OHAM with assurance in tackling a diverse array of challenges spanning the realms of science and technology. OHAM, with its demonstrated capabilities, holds the potential to be a valuable tool for addressing complex problems in these domains.

OHAM's journey from theory to practical application illuminates its adaptability and prowess in addressing complex, real-world problems. The versatility of OHAM stands out as one of its most significant attributes. It possesses the remarkable ability to address a wide spectrum of problems, both linear and nonlinear, making it an invaluable asset in diverse scientific and engineering domains. The TFNS problem, a quintessential example of nonlinear PDEs governing fluid dynamics, serves as a testament to OHAM's aptitude for handling intricate equations. OHAM excels in accommodating complexities, proving itself as a versatile tool for resolving real-world challenges.

Precision is another hallmark of OHAM's success story. The consistency observed between OHAM-derived solutions and exact solutions underscores its reliability. In the context of the TFNS problem, OHAM solutions are closely aligned with the exact solutions, reaffirming the method's accuracy and dependability. Researchers and professionals can place their trust in OHAM with confidence, knowing that it offers a robust approach to solving intricate problems with a high degree of precision.

However, OHAM's contributions extend far beyond the TFNS problem. Its potential ramifications span across a multitude of domains, each benefiting from OHAM's analytical acumen:

- Fluid dynamics and heat transfer: OHAM's promise in addressing intricate fluid dynamics and heat transfer problems is evident. Complex systems, such as turbulent flows or heat conduction in heterogeneous media, often defy straightforward analytical solutions. OHAM, with its capability to handle nonlinear equations with precision, simplifies the simulation and analysis of such phenomena. In the context of fluid dynamics, this can revolutionize fields such as aviation, automotive design, and environmental engineering, where a deeper understanding of complex fluid behavior is crucial.
- Electrical engineering and circuit design: The realm of electrical engineering frequently encounters fractional-order elements in circuits, introducing complexity to analysis and design. OHAM's proficiency in providing analytical solutions for circuits containing these elements streamlines the design process. This simplification holds the potential to reshape the development of electronic devices and systems.

- Biomedical engineering and mathematical biology: OHAM's applications in biomedical engineering and mathematical biology are burgeoning. It facilitates the modeling of physiological processes, population dynamics, and drug diffusion in biological tissues. With OHAM, researchers can gain deeper insights into intricate biological systems, opening doors to improved medical treatments and more accurate predictions in epidemiology.
- Materials science and engineering: OHAM's analytical skills are very important in the field of materials science and engineering. It makes it possible to develop materials suited to particular applications by assisting in the prediction of the mechanical and thermal characteristics of composite materials. OHAM's precision further extends to nanomaterials, where it aids in deciphering the unique properties of materials at the nanoscale, with implications for industries ranging from electronics to healthcare.
- Environmental engineering and sustainability: Environmental challenges such as pollution control and sustainable energy generation require innovative solutions. OHAM can model pollutant dispersion with precision, aiding in pollution control measures. In the domain of renewable energy, OHAM's analytical solutions contribute to the design and optimization of sustainable energy systems, a pivotal step toward a more sustainable future.
- Beyond traditional fields: Interdisciplinary applications: OHAM's potential transcends traditional scientific and engineering disciplines. Its relevance extends into interdisciplinary areas where FC is emerging as a powerful tool.
- Financial mathematics: In finance, where risk assessment, portfolio optimization, and derivative pricing are paramount, OHAM's precision can aid in quantifying risks and optimizing investment strategies.
- Medical research: OHAM's analytical prowess is applicable in modeling biological processes, drug delivery, and disease spread, leading to improved medical research and more effective healthcare strategies.

As we gaze into the future, OHAM holds the promise of even greater contributions to the scientific and engineering communities. Several avenues for further research and development beckon. Algorithmic improvements: While OHAM has demonstrated remarkable precision, ongoing research can focus on refining computational algorithms associated with the method. Enhancing OHAM's computational efficiency can render it an even more attractive option for addressing large-scale problems, making it accessible to an even broader audience. Multiscale modeling: Many real-world problems span multiple scales, from the molecular to the macroscopic. OHAM's analytical solutions could potentially be extended to multiscale modeling, facilitating a comprehensive understanding of complex systems that exhibit behaviors across various scales. Experimental validation: Although OHAM solutions have consistently aligned with exact solutions, experimental validation remains a critical step in many applications. Future research endeavors can focus on conducting experiments to corroborate the accuracy of OHAM predictions in diverse scenarios, bridging the gap between theoretical modeling and practical reality. Integration with computational tools: To further enhance its accessibility and usability, OHAM can be seamlessly integrated into existing computational tools and software. Developing user-friendly interfaces and libraries that incorporate OHAM can expand its reach, empowering a broader community of researchers and professionals to leverage its capabilities.

In a world characterized by the escalating complexity of scientific and engineering challenges, OHAM stands as a beacon of analytical precision and reliability, guiding researchers and professionals toward innovative solutions. Its profound impact resonates across multiple domains, offering a pathway to deeper insights and more effective problem-solving.

OHAM's remarkable journey, from a mathematical concept to a practical problem-solving tool, underscores the transformative power of innovative mathematical techniques. Its unparalleled versatility, precision, and reliability make it an invaluable asset, simplifying the intricacies of modeling, analysis, and optimization in complex systems.

The implications of OHAM's capabilities reverberate through scientific and engineering disciplines, as well as interdisciplinary realms. OHAM empowers researchers and professionals to explore the intricacies of fluid dynamics, revolutionize electrical circuit design, gain deeper insights into biological processes, predict materials' properties, and address pressing environmental and sustainability challenges. It finds application in financial modeling, medical research, and myriad other fields, shaping the trajectory of research and development in unprecedented ways.

As we embark on the journey ahead, the potential of OHAM remains largely untapped. Algorithmic enhancements promise to make OHAM even more efficient and accessible, amplifying its impact. Multiscale modeling holds the key to unlocking new dimensions of understanding in systems that span diverse scales. Experimental validation will solidify OHAM's reputation as a reliable problem-solving tool, bridging the gap between theory and reality. Integration with computational tools will democratize its use, making OHAM an indispensable component of the researcher's toolkit.

In conclusion, the fractional-order OHAM represents not only a remarkable achievement in mathematical modeling but also a profound enabler of scientific and engineering progress. Its journey from theory to application illustrates its ability to adapt and thrive in the face of complex challenges. OHAM's versatility, precision, and reliability make it a guiding star for researchers and professionals, illuminating a path toward innovation, discovery, and progress.

As we embrace the boundless opportunities and possibilities that OHAM offers, we embark on a collective journey of exploration and transformation. OHAM, with its unwavering analytical precision, will continue to shape the future of science, engineering, and interdisciplinary research. It stands as a testament to the enduring capacity of human ingenuity to forge new pathways, solve intricate problems, and create a brighter, more interconnected world.

This comprehensive tool, OHAM, invites us to push the boundaries of our understanding and explore the uncharted territories of science and engineering. With each new discovery and application, OHAM brings us closer to solving the grand challenges of our time and unlocking the potential of tomorrow's innovations. It is a testament to the relentless pursuit of knowledge, the power of mathematical insight, and the indomitable spirit of human inquiry. OHAM is not merely a method, it is a symbol of our capacity to conquer complexity, navigate uncertainty, and chart a course toward a better future.

References

[1] C. Navier, Sur les Lois des Mouvement des Fluides, en Ayant Egard a L'adhesion des, Mol. Ann. Chimie 19 (1821) 244–260.

[2] C. Naiver, Sur les lois du mouvement des fluides, Memoires del. Academie Royale des. Sci. 6 (1827) 389–440.

[3] L. Euler, Principes généraux du mouvement des fluides, Mémoires de.. l'académie des. Sci. de Berl. 11 (1757) 274–315.

[4] L. Euler, Tentamen theoriae de frictione fluidorum, Novi Commentarii Academiae Scientiarum Petropolitanae, Forgotten Books, 1761, pp. 338–388.

[5] S.R. Bistafa, Euler's friction of fluids theory and the estimation of fountain jet heights, Eur. Phys. J. H. 40 (2015) 375–384.

[6] A.L.B. Cauchy, Exercices de mathématiques, De Bure Frères, 1828.

[7] S. Masuda, Poisson' "Sloughs" in His Final Works in Life, 2018. uni-miskolc.hu.

[8] M. Ersoy, O. Lakkis, P. Townsend, A. Saint-Venant, model for overland flows with precipitation and recharge, Math. Comput. Appl. 26 (2020) 1.

[9] G.G. Stokes, Mathematical and Physical Papers, Cambridge University Press, Cambridge, UK, 1901.

[10] R. Dugas, A History of Mechanics, Courier Corporation, 2012.

[11] I. Grattan-Guinness, Convolutions in French Mathematics, 1800–1840: From the Calculus and Mechanics to Mathematical Analysis and Mathematical Physics, Birkhäuser, 2017.

[12] O. Darrigol, Worlds of Flow: A History of Hydrodynamics from the Bernoullis to Prandtl, Oxford University Press, 2005.

[13] S.R. Bistafa, On the development of the Navier-Stokes equation by Navier, Rev. Brasileira de.. Ensino de Física 40 (2017) e2603.

[14] F. Liu, K. Burrage, Novel techniques in parameter estimation for fractional dynamical models arising from biological systems, Comput. Math. Appl. 62 (2011) 822–833.

[15] N.A. Shah, D. Vieru, C. Fetecau, Effects of the fractional order and magnetic field on the blood flow in cylindrical domains, J. Magn. Magn. Mater. 409 (2016) 10–19.

[16] Y. Jiang, B. Zhang, X. Shu, Z. Wei, Fractional-order autonomous circuits with order larger than one, J. Adv. Res. 25 (2020) 217–225.

[17] O.W. Abdulwahhab, N.H. Abbas, A new method to tune a fractional-order PID controller for a twin rotor aerodynamic system, Arab. J. Sci. Eng. 42 (2017) 5179–5189.

[18] F. Meral, T. Royston, R. Magin, Fractional calculus in viscoelasticity: an experimental study, Commun. Nonlinear Sci. Numer. Simul. 15 (2010) 939−945.
[19] R. Meilanov, R. Magomedov, Thermodynamics in fractional calculus, J. Eng. Phys. Thermophys. 87 (2014) 1521−1531.
[20] R. Matušů, Application of fractional order calculus to control theory, Int. J. Math. Model. Methods Appl. Sci. 5 (2011) 1162−1169.
[21] N. Can, H. Jafari, M. Ncube, Fractional calculus in data fitting, Alex. Eng. J. 59 (2020) 3269−3274.
[22] V.E. Tarasov, Fractional Dynamics: Applications of Fractional Calculus to Dynamics of Particles, Fields and Media, Springer Science & Business Media, 2011.
[23] K.B. Oldham, Fractional differential equations in electrochemistry, Adv. Eng. Softw. 41 (2010) 9−12.
[24] S. Das, Application of generalized fractional calculus in electrical circuit analysis and electromagnetics, Funct. Fract. Calculus (2011) 387−436.
[25] R. Hilfer, Applications of Fractional Calculus in Physics, World Scientific, Singapore, 2000, 472 pp.
[26] G. Pertz, C. Gerhardt, Leibnizens gesammelte Werke, Lebinizens mathematische Schriften, Erste Abtheilung, Band II, 301−302, Dritte Folge Mathematik (Erster Band). A. Asher & Comp., Briefwechsel zwischen Leibniz, Hugens van Zulichem und dem Marquis de l'Hospital, 1849.
[27] L. Euler, De progressionibus transcendentibus seu quarum termini generales algebraice dari nequeunt, Commentarii Academiae Scientiarum Petropolitanae 5 (1738) 36−57.
[28] K. Oldham, J. Spanier, The Fractional Calculus Theory and Applications of Differentiation and Integration to Arbitrary Order, Elsevier, 1974.
[29] K.S. Miller, B. Ross, An Introduction to the Fractional Calculus and Fractional Differential Equations, John Wiley & Sons, New York, NY, USA, 1993.
[30] A.A. Kilbas, H.M. Srivastava, J.J. Trujillo, Theory and Applications of Fractional Differential Equations, Elsevier, 2006.
[31] S.F. Lacroix, Traité du calcul différentiel et du calcul intégral, JBM Duprat, 1797.
[32] N.H. Abel, Auflösung einer mechanischen Aufgabe, Journal für die reine und angewandte, Mathematik 1 (1826) 153−157.
[33] R. Gorenflo, S. Vessella, Abel integral equations: analysis and applications, Lecture Notes in Mathematics, Springer, Berlin Heidelberg, 1991.
[34] J. Liouville, Mémoire sur quelques questions de géométrie et de mécanique et sur un nouveau genre de calcul pour résoudre ces équations, Éc. Polytechnique 13 (1832) 71−162.
[35] J. Liouville, Mémoire sur le calcul des différentielles à indices quelconques, Journal de l'École Polytechnique, Paris 13 (1832) 71−162.
[36] J. Liouville, Mémoire sur le théorème des fonctions complémentaires, Journal für die reine und angewandte Mathematik 11 (1834) 1−19.
[37] J. Liouville, Memoire sur l'integration des equations differentielles a indices fractionnaires, Journal de l'École Polytechnique 15 (1837) 55−84.
[38] H.R. Greer, On fractional differentiation, Quart. J. Math. 3 (1858) 327−330.
[39] J.B.J. Fourier, Théorie analytique de la chaleur, Gauthier-Villars et fils, 1888.
[40] B. Riemann, Versuch einer allgemeinen Auffassung der Integration und Differentiation, Gesammelte Werke 62 (1876) 331−344.

[41] N.Y. Sonin, On differentiation with arbitrary index, Mosc. Matem. Sb. 6 (1869) 1–38.
[42] A. Letnikov, Theory of differentiation with an arbitrary index, Sb. Math. 3 (1868) 1–66 (Russian).
[43] A. Letnikov, An explanation of the concepts of the theory of differentiation of arbitrary index, Mosc. Matem. Sb. 6 (1872) 413–445.
[44] H. Laurent, Sur le calcul des dérivées à indices quelconques, Nouvelles annales de. mathématiques: J. des. candidats aux. écoles polytechnique et. normale 3 (1884) 240–252.
[45] A.K. Grunwald, Uber "begrente", Derivationen und deren Anwedung, Zangew Math. und Phys. 12 (1867) 441–480.
[46] O. Heaviside, Electrical Papers, Cambridge University Press, 2011.
[47] H. Weyl, Bemerkungen zum begriff des differentialquotienten gebrochener ordnung, Vierteljschr. Naturforsch. Gesellsch. Zur. 62 (1917) 296–302.
[48] G.H. Hardy, J.E. Littlewood, Some properties of fractional integrals. I, Mathematische Z. 27 (1928) 565–606.
[49] G.H. Hardy, J.E. Littlewood, Some properties of fractional integrals. II, Mathematische Z. 34 (1932) 403–439.
[50] A. Marchaud, Sur les dérivées et sur les différences des fonctions de variables réelles, J. de. Mathematiques Pures et. Appliquees 6 (1927) 337–425.
[51] M. Riesz, Intégrales de Riemann-Liouville et potentiels, Acta Sci. Math. Szeged. 9 (1938) 1–42.
[52] M. Riesz, L'intégrale de Riemann-Liouville et le problème de Cauchy, Acta Mathematica 81 (1949) 1–222.
[53] A. Erdélyi, On fractional integration and its application to the theory of Hankel transforms, Q. J. Math. 44 (1940) 293–303.
[54] M. Caputo, Linear models of dissipation whose Q is almost frequency independent—II, Geophys. J. Int. 13 (1967) 529–539.
[55] M.P. Lazarević, M.R. Rapaić, T.B. Šekara, V. Mladenov, N. Mastorakis, Introduction to fractional calculus with brief historical background, Advanced topics on applications of fractional calculus on control problems, system stability and modeling, WSEAS Press, 2014, pp. 3–16.
[56] S. Momani, Z. Odibat, Analytical solution of a time-fractional Navier–Stokes equation by Adomian decomposition method, Appl. Math. Comput. 177 (2006) 488–494.
[57] S. Liao, On the Proposed Homotopy Analysis Technique for Nonlinear Problems and Its Applications, Shanghai Jiao Tong University, 1992.
[58] J. He, An approximate solution technique depending on an artificial parameter: a special example, Commun. Nonlinear Sci. Numer. Simul. 3 (1998) 92–97.
[59] D. Ganji, M. Rafei, Solitary wave solutions for a generalized Hirota–Satsuma coupled KdV equation by homotopy perturbation method, Phys. Lett. A 356 (2006) 131–137.
[60] D. Lu, J. Liu, Application of the homotopy analysis method for solving the variable coefficient KdV-Burgers equation, Abstract and Applied Analysis, Hindawi, 2014.
[61] A.M. Siddiqui, S. Irum, A.R. Ansari, Unsteady squeezing flow of a viscous MHD fluid between parallel plates, a solution using the homotopy perturbation method, Math. Model. Anal. 13 (2008) 565–576.

[62] V. Marinca, N. Herişanu, Application of optimal homotopy asymptotic method for solving nonlinear equations arising in heat transfer, Int. Commun. Heat. Mass. Transf. 35 (2008) 710–715.
[63] N. Herişanu, V. Marinca, T. Dordea, G. Madescu, A new analytical approach to nonlinear vibration of an electrical machine, Proc. Rom. Acad. Ser. A 9 (2008) 229–236.
[64] V. Marinca, N. Herişanu, Determination of periodic solutions for the motion of a particle on a rotating parabola by means of the optimal homotopy asymptotic method, J. Sound. Vib. 329 (2010) 1450–1459.
[65] N. Herişanu, V. Marinca, Accurate analytical solutions to oscillators with discontinuities and fractional-power restoring force by means of the optimal homotopy asymptotic method, Comput. Math. Appl. 60 (2010) 1607–1615.
[66] K. Muhammad, T. Hayat, Alsaedi, OHAM analysis of Newtonian heating in mixed convective flow of CNTs over a stretched cylinder, Alex. Eng. J. 61 (2022) 3697–3707.
[67] V. Marinca, N. Herişanu, C. Bota, B. Marinca, An optimal homotopy asymptotic method applied to the steady flow of a fourth-grade fluid past a porous plate, Appl. Math. Lett. 22 (2009) 245–251.
[68] N.A. Khan, A. Ara, S.A. Ali, A. Mahmood, Analytical study of Navier-Stokes equation with fractional orders using He's homotopy perturbation and variational iteration methods, Int. J. Nonlinear Sci. Numer. Simul. 10 (2009) 1127–1134.
[69] B.K. Albuohimad, Analytical technique of the fractional Navier-Stokes model by Elzaki transform and homotopy perturbation method, AIP Conference Proceedings, AIP Publishing LLC, 2019, p. 050002.
[70] S. Maitama, Analytical solution of time-fractional Navier–Stokes equation by natural homotopy perturbation method, Prog. Fract. Differ. Appl. 4 (2018) 123–131.
[71] K. Wang, S. Liu, Analytical study of time-fractional Navier-Stokes equation by using transform methods, Adv. Difference Equ. 2016 (2016) 1–12.
[72] D. Kumar, J. Singh, S. Kumar, A fractional model of Navier–Stokes equation arising in unsteady flow of a viscous fluid, J. Assoc. Arab. Univ. Basic Appl. Sci. 17 (2015) 14–19.
[73] A. Prakash, D.G. Prakasha, P. Veeresha, A reliable algorithm for time-fractional Navier-Stokes equations via Laplace transform, Nonlinear Eng. 8 (2019) 695–701.
[74] S. Kumar, D. Kumar, S. Abbasbandy, M. Rashidi, Analytical solution of fractional Navier–Stokes equation by using modified Laplace decomposition method, Ain Shams Eng. J. 5 (2014) 569–574.
[75] S. Iqbal, M. Idrees, A.M. Siddiqui, A.R. Ansari, Some solutions of the linear and nonlinear Klein–Gordon equations using the optimal homotopy asymptotic method, Appl. Math. Comput. 216 (2010) 2898–2909.
[76] Z. Ali, S.N. Nia, F. Rabiei, K. Shah, M.K. Tan, A semi-analytical approach for the solution of time-fractional Navier-Stokes equationmethod, Adv. Math. Phys. (2021) 13. Article ID: 5547804.

CHAPTER

Numerical simulations of nanofluids heat and mass transfer (lattice Boltzmann methods and applications) 11

Rasul Mohebbi[1], Yuan Ma[2] and Molka El Maghraoui[3]
[1]*School of Engineering, Damghan University, Damghan, Iran*
[2]*Department of Mechanical Engineering, The Hong Kong Polytechnic University, Kowloon, Hong Kong, P.R. China*
[3]*Faculty of Science of Tunis, University of Tunis El Manar, Tunis, Tunisia*

11.1 General introduction

Thermal transfer mechanisms are of critical importance in various engineering systems, ranging from power plants and cooling systems to electronic devices and chemical reactors. Efficient heat transfer not only enhances the performance and reliability of these systems but also contributes to reducing their size, weight, cost, and environmental impact. In comparison to solid materials, conventional working fluids such as water and ethylene glycol demonstrate limited thermal conductivity and specific heat capacity. Consequently, multiple approaches have been devised to boost heat transfer efficiency by altering fluid properties or flow characteristics.

In the quest for enhanced heat transfer capabilities, one particularly promising approach involves the use of nanofluids (NFs). Nanofluids consist of colloidal suspensions of metallic or ceramic nanoparticles, finely blended with a base fluid, typically in the nanometer size range. These nanofluids have shown great potential in significantly augmenting heat transfer performance. This chapter delves into the realm of nanofluids and their impact on heat transfer. By exploring diverse types of nanofluids, researchers have sought to harness their unique properties to improve heat transfer capabilities in various applications [1]. As a result, numerous investigations have been conducted in diverse geometries, examining scenarios of natural, forced, and mixed convection of nanofluids.

Through a systematic analysis of existing research, this chapter offers a valuable understanding of the application of nanofluids and their numerical simulations. We focus on utilizing the LBM for numerical simulations, as it provides an efficient means to assess the movement of the fluid and heat exchange behavior of nanofluids. Specifically, we investigate nanofluidic convection heat transfer,

Nanofluids. DOI: https://doi.org/10.1016/B978-0-443-13625-2.00005-X

seeking to understand and optimize its behavior for practical engineering applications. Possessing an extensive grasp of the heat and mass transfer within nanofluids, the aims of this chapter are to the advancement of engineering systems by offering innovative and efficient solutions to heat transfer challenges.

Numerous researchers have extensively investigated the heat transfer phenomenon, focusing on the manipulation of various parameters, including nanofluid properties and geometric configurations. Khanafer et al. [2] were among the first to study this phenomenon in a rectangular cavity. Zhang et al. [3] conducted a study on heat transfer in porous media by using nanofluids, such as Ag, Cu, and Al_2O_3. Arya and colleagues [4] examined the effect of nanofluids on heat transfer within a heat exchanger.

The Ag-MgO [5], CuO [6], and MWCNT-Fe_3O_4 [7] have also been studied by scientists. Ma et al. [8] investigated the convective behavior of MWCNTs/H_2O nanofluids within a U-shaped cavity. In a similar vein, Izadi et al. [9] utilized MWCNT-Fe_3O_4/water hybrid nanofluid to enhance thermal transmission within a shaped enclosure. On the other hand, Dogonchi et al. [10] examined the energy transport of Cu/H_2O nanofluid in a cavity featuring a semicircular bottom wall. Peyman et al. [11] studied the effect of nanofluid on improving the heat transfer performance in lithium-ion batteries. They investigated the different parameters and found that the nanofluid is a very useful method for enhancing the cooling efficiency.

Izadi et al. [12] investigated the impact of a magnetic field on nanofluid behavior in a porous medium. The research findings demonstrate that at low Ra, the Nu rises with an increment in the magnetic numbers. Mohebbi et al. [13] delved into nanofluid heat transfer within an F-shaped cavity using the LBM. Mahmoodi [14] scrutinized the thermal behavior of an L-shaped cavity utilizing a Cu/water nanofluid, employing the finite volume method (FVM). Sidik et al. [15] conducted a numerical assessment of fin-based thermal performance within a channel, employing the LBM. They examined the effect of adding fins and nanofluid on heat transfer performance. Kefayati et al. [16] used the LBM to study the natural convection heat transfer of a nanofluid in different geometries. They studied the different parameters and found that their method is in excellent agreement with other numerical methods. Mohebbi and Rashidi [17] employed LBM to investigate Al_2O_3/water nanofluid heat transfer within an L-shaped cavity. The research carried out by these authors contributes to the expanding knowledge base on the utilization of nanofluids for augmenting heat transfer. Qi et al. [18] investigated the heat transfer of Al_2O_3-H_2O and (Al_2O_3/Cu)-H_2O nanofluids in a cavity employing a two-stage LBM model, while Ma and Yang [19] employed the LBM approach to conduct simulations of heat transfer involving a hybrid nanofluid in a cavity with a heated obstacle. Also, a numerical investigation performed by Ma et al. [20] focused on forced convection heat transfer of a hybrid nanofluid, composed of multiwall carbon nanotubes and iron oxide nanoparticles suspended in water (MWCNT-Fe_3O_4/water hybrid nanofluid), within a partially heated T-shaped.

Izadi et al. [21] used Al_2O_3 nanoparticles to study the heat transfer rate. Hussein et al. [22] studied the turbulent forced convection nanofluid in tubes. Akhtari et al. [23] examined nanofluid forced convection in heat exchangers using FVM. Moraveji et al. [24] utilized CFD techniques to explore the heat performance of nanofluids in a tube. The outcomes from experiments were obtained under various conditions. Rashidi et al. [25] explored the heat transfer behavior of Cu-water nanofluids within a corrugated channel. Mohebbi et al. [26] employed LBM to examine the heat transfer characteristics of nanofluids in channels equipped with hot blocks. They found that adding nanoparticles and using the block on the walls can enhance heat transfer. Furthermore, they observed up to a 39.04% increase in the Nu. In their research, Mohebbi and colleagues [27] utilized a two-dimensional LBM to examine heat transfer in a channel with fins on two horizontal walls, considering the use of three distinct nanofluids.

Sheikholeslami et al. [28] utilized LBM to study the behavior of CuO-H_2O nanofluid inside an enclosure equipped with a hot block while considering the influence of a magnetic field. Their findings indicated that an increment in the Ha resulted in a reduction in the thermal plume, while the convective mode improved with increasing Da and Re. Izadi et al. [29] employed a numerical code to investigate the impact of using nanofluid with spindle-shaped turbulators in a heat exchanger. The proven ability of nanofluids to improve heat transfer has sparked significant interest in studying their behavior under forced convection. Consequently, a multitude of studies have been dedicated to this area to examine the heat transfer characteristics of nanofluids under such conditions [30–34].

Numerous researchers have dedicated their efforts to improving the heat transfer of nanofluids under mixed convection scenarios [35]. Izadi and their colleagues have carried out comprehensive research on nanofluids mixed convection in various geometries, as evidenced by their published works [36–38]. Based on their discoveries, a rise in the volume fraction leads to an improvement in heat transfer. Ogban and Naterer [39] introduced a novel approach aimed at reducing the entropy production rate in various flow configurations involving nanofluids. Their investigation demonstrated that augmenting the nanoparticle volume fraction resulted in a decline in entropy production rate during natural convection, while for mixed convection, an optimal concentration of nanoparticles was identified that minimized overall entropy generation. They proposed that managing the nanoparticle distribution and adjusting the cavity aspect ratio (AR) can efficiently decrease entropy production rates. Esfe and colleagues [40] employed a numerical simulation to investigate mixed convection in a cavity filled with nanofluids. According to their results, lowering the Ri resulted in an increased average Nu. Sebdani et al. [41] explored the mixed convection of Al_2O_3 nanofluid. They observed that shifting the heat source towards the side wall increased the rate of heat transfer.

Most above numerical simulations of fluid mechanics problems involving nanofluids have relied on semiempirical models and conventional CFD techniques [42,43] like finite difference method (FDM) [44,45], FVM [46,47], and finite element method (FEM) [48–52]. Nonetheless, a thorough review and analysis of the existing literature suggests that the LBM is a promising alternative that has been effectively used in various fluid dynamics problems and has the potential to produce precise results [53–55]. The LBM is based on kinetic theory and has numerous advantages over traditional methods, including explicitness, ease of implementation, and efficient parallel computation [56,57]. The LBM has already demonstrated successful applications in isothermal hydrodynamic problems, positioning itself as a formidable competitor to conventional CFD techniques [58–61]. As a result, the LBM has gained attention as a promising alternative due to its various advantages over traditional methods. With proper boundary settings, the LBM can efficiently handle complex domain geometries [62–64].

Most of the references cited employ the single-phase approach, which uses the uniform model to examine the thermal properties of nanofluids. This particular methodology relies on the assumption that the velocity disparity between the phase and nanoparticles are considered negligible, an acceptable assumption when both the volume fraction and the size of the particles are limited [65]. However, micro-factors such as nanoparticle radius and interparticle forces should be considered when simulating buoyancy-driven convection in nanofluids. Additionally, the two-phase LBM is particularly useful for investigating interparticle forces due to its simplicity. Qi et al. [66] employed a dual-phase LBM to examine the convection thermal transfer of Cu–Ga nanofluid with the incorporation of interaction forces involved in the nanoparticles. They explored the influence of distinct parameters concerning nanoparticle distribution and thermal transfer properties, including nanoparticle size, interaction forces, geometry AR, nanoparticle concentration, and Rayleigh number (Ra). Their findings showed that the use of smaller nanoparticle size augmented thermal transfer significantly, and that the temperature contrast which resulted in driving force had a maximal impact on nanoparticle location. Moreover, the average Nu grew with elevation of AR, concentration of nanoparticles, and Ra. In other works, Qi et al. [67] expanded their utilization of the two-phase lattice Boltzmann model to assess the consequences of three distinct categories of nanoparticles and two different base fluids (H_2O and Ga) on innate convection thermal transfer. Saberi and Sidik et al. [68] used the LBM to investigate he heat transfer of Cu-water nanofluid in a two-dimensional channel. Ma et al. [69] applied Buongiorno's two-phase method to examine the heat transfer characteristics of nanofluids in a triangular solar collector possessing an undulating wall. Their findings suggest that increasing the concentration of nanofluids (φ) below a certain critical value can improve the heat transfer enhancement ratio. Nevertheless, exceeding a critical value of solid volume fraction in the nanofluid, augmenting the φ could result in a reduction of the heat transfer efficiency. Ren et al. [70] proposed a two-phase LBM to investigate the heat transfer characteristics of CuO-H_2O nanomagnetic fluids subjected to nonuniform and variable magnetic fields. While two phases approach was also adopted by Kafayati and Bassom [71,72].

Abdollahzadeh et al. [65] used a two-phase scheme for simulating the flow of Al_2O_3/water nanofluid in a rectangular channel. The aforementioned approach took into account the drift velocity of each phase and was validated through simulations performed at low Re. The findings revealed that taking into consideration the drift velocity only holds a notable influence in low Reynolds numbers, and the model proved to be suitable for both engineering and scientific purposes. Kefayati and Bassom [71] employed the LBM to examine the behavior of both Newtonian and non-Newtonian nanofluids. The suggested method exhibits the capability to integrate any form of additional tensor directly, and it is not reliant on relaxation time, rendering it suitable for investigating non-Newtonian nanofluids. The efficacy of this method was assessed by benchmarking it against prior investigations, using a natural convection problem in a chamber.

Therefore the objective of this chapter is to conduct a numerical study based on the LBM to analyze the augmentation of heat transport in various configurations containing nanofluids, considering different convection scenarios including natural, forced, and mixed convection. Consequently, a thorough examination of flow patterns, temperature distributions, and heat transfer performance is carried out.

11.2 Nanofluids modeling methods

The incorporation of nanoparticles, which typically range from 1 to 100 nm, into the base fluid represents a relatively recent and innovative technique employed to significantly improve the thermal conductivity of the fluid [72,73]. It was Choi [74] who originally proposed the idea of nanofluids, revolutionizing the field of heat transfer [75]. These nanofluids exhibit the potential for substantially higher heat transfer rates when compared to conventional fluids such as water [76], enabling more effective heating or cooling processes and contributing to a reduction in overall energy consumption.

Nanofluids have attracted considerable attention due to their distinctive attributes, encompassing heightened thermal conductivity, improved stability, and intriguing rheological characteristics. The integration of nanoparticles within nanofluids results in a significantly amplified effective surface area, which leads to intensified convective heat transfer, ultimately elevating the overall thermal efficiency of the fluid. The dynamic Brownian motion of nanoparticles within nanofluids plays a pivotal role in shaping their heat transfer performance, presenting opportunities for precise tuning of their thermal behavior. To this end, several equations have been formulated and made accessible to facilitate precise simulation and prediction of nanofluid properties.

11.2.1 Nanofluid single-phase method

The nanofluid single-phase technique, an innovative approach in nanofluid research, entails the direct dispersion of nanoparticles into the base fluid without introducing any additional phase or interface differentiation. Employing the nanofluid single-phase approach allows researchers to delve into intricate fluid dynamics and explore the synergistic effects of nanoparticles, all without grappling with the intricacies linked to two-phase systems. Consequently, the nanofluid single-phase method introduces distinct challenges in precisely characterizing nanofluid behavior, especially at the nanoscale, prompting the need for inventive measurement techniques to achieve comprehensive analysis.

Certainly, nanofluid parameters play a crucial role in characterizing the behavior of nanofluids in simulations. These parameters help describe the properties of the fluid and how they interact with the nanoparticles. Here is a general outline of the types of parameters described [77]:

Density:

$$\rho_{nf} = (1-\phi)\rho_f + \phi\rho_p \tag{11.1}$$

Heat capacity:

$$(\rho c_p)_{nf} = (1-\phi)(\rho c_p)_f + \phi(\rho c_p)_p \tag{11.2}$$

Dynamic viscosity:

$$\frac{\mu_{nf}}{\mu_f} = (1 + 2.5\phi + 6.5\phi^2) \tag{11.3}$$

Thermal conductivity by [78]:

$$\frac{k_{nf}-k_f}{k_f} = \frac{k_p}{k_f}\left(1 + \mathrm{c}\frac{u_p d_p}{\alpha_f}\right)\frac{d_f}{d_p}\frac{\phi}{1-\phi} \tag{11.4}$$

Thermal diffusivity:

$$\alpha_{nf} = \frac{k_{nf}}{(\rho c_p)_{nf}} \tag{11.5}$$

Prandtl number:

$$\mathrm{Pr}_{nf} = \frac{(\mu c_p)_{nf}}{k_{nf}} \tag{11.6}$$

Here $\boldsymbol{c}$ is 25,000 and (u_p) is:

$$u_p = \frac{2k_B T}{\pi\mu_f d_p^2} \tag{11.7}$$

These parameters collectively define the characteristics of the nanofluid and provide insights into how it behaves in various flow and heat transfer scenarios. It is important to consider these parameters carefully when setting up nanofluid simulations to accurately capture the behavior of the system.

Koo and Kleinstreuer [79] introduced a bicomponent thermal conductivity, a model that considers the influences of both the base liquid and the nanoparticles within a nanofluid. This model offers an approach to estimating the effective thermal conductivity of the nanofluid, which is a crucial parameter in understanding its heat transfer characteristics. The specific form of the model, as described in reference [80]

$$k_{\text{eff}} = k_{\text{static}} + k_{\text{Brownian}} \tag{11.8}$$

$$\frac{k_{\text{static}}}{k_f} = 1 + \frac{3\left(\frac{k_s}{k_f} - 1\right)\phi}{\left(\frac{k_s}{k_f} + 2\right) - \left(\frac{k_s}{k_f} - 1\right)\phi} \tag{11.9}$$

$$k_{\text{Brownian}} = 5 \times 10^4 \beta\phi\rho_f c_{p,f}\sqrt{\frac{k_b T}{\rho_s d_s}} f(T,\phi) \tag{11.10}$$

Recently, there has been a growing recognition of the significance of thermal interfacial resistance between the base liquid and nanoparticles [78,80]. Li [81] introduced a model for thermal interfacial resistance $R_f = 4 \times 10^{-8}\text{kw}^2/\text{W}$, and the original k_s in Eq. (11.9) was depicted by $k_{s,eff}$ as below:

$$R_f + \frac{d_p}{k_s} = \frac{d_p}{k_{s,eff}} \tag{11.11}$$

Furthermore, he integrated the model proposed by Koo and Kleinstreuer [82] with the empirical functions β and f to formulate a novel functiong'. This function examines the influence of temperature, particle diameter, and nanoparticle volume fraction. For CuO/water nanofluids, the function is expressed as follows:

$$g'(T,\phi,d_p) = \left(a_1 + a_2\ln(d_p) + a_3\ln(\phi) + a_4\ln(\phi)\ln(d_p) + a_5\ln(d_p)^2\right)\ln(T) + \left(a_6 + a_7\ln(d_p)\right.$$
$$\left. + a_8\ln(\phi) + a_9\ln(\phi)\ln(d_p) + a_{10}\ln(d_p)^2\right) \tag{11.12}$$

where $a_i(i = 0\text{–}10)$ are shown in Table 11.1.

Thus k_{Brownian} can be computed by Koo–Kleinstreuer–Li (KKL) model:

$$k_{\text{Brownian}} = 5 \times 10^4 \beta\phi\rho_f c_{p,f}\sqrt{\frac{k_b T}{\rho_s d_p}} g'(T,\phi,d_p) \tag{11.13}$$

The effective viscosity is:

$$\mu_{\text{eff}} = \mu_{\text{static}} + \mu_{\text{Brownian}} = \mu_{\text{static}} + \frac{k_{\text{Brownian}}}{k_f} \times \frac{\mu_f}{\text{Pr}_f} \tag{11.14}$$

where $\mu_{\text{static}} = \mu_f/(1-\phi)^{2.5}$ is the viscosity of the nanofluid.

11.2.2 Nanofluid two-phase method

The nanofluid two-phase method stands as a crucial and pioneering technique in nanofluid research, involving the introduction of nanoparticles into the base fluid as a separate phase, distinct from the continuous fluid phase. This unique two-phase configuration induces special interactions and behaviors, leading to the formation of dispersed clusters or agglomerates of nanoparticles within the fluid. By

Table 11.1 The coefficient values of CuO/water nanofluid [79].

Coefficient values	CuO/water
a_1	−26.593310846
a_2	−0.403818333
a_3	−33.3516805
a_4	−1.915825591
a_5	6.42185846658E-02
a_6	48.40336955
a_7	−9.787756683
a_8	190.245610009
a_9	10.9285386565
a_{10}	−0.72009983664

Reprinted with permission from A. Rahimi, A. Kasaeipoor, E.H. Malekshah, A.S. Far, M. Sepehr. Heat transfer intensification using CuO-water nanofluid in a finned capsule-shaped heat exchanger using lattice Boltzmann method. Chem. Eng. Process. Process Intensif. 127 (2018) 17–27.

employing the nanofluid two-phase method, researchers gain an invaluable platform to investigate intricate phenomena, such as nanoparticle migration, coalescence, and sedimentation, all while considering varying flow conditions and external influences. This method allows for a comprehensive exploration of the impact of interfacial forces and dynamic interactions between nanoparticles and the surrounding fluid, which play a critical role in understanding the overall heat transfer enhancement and stability attributes of nanofluids. Nevertheless, the successful application of the nanofluid two-phase method demands sophisticated experimental setups and advanced numerical simulations to accurately capture the intricacies of the multiphase system. Despite the challenges, this approach holds immense promise for advancing nanofluid research and unlocking innovative applications in various fields, ranging from electronics' thermal management to energy-intensive industries' advanced cooling systems.

It is essential to take into account various nanoparticle volume fractions when studying the thermal system. A seminal study by Buongiorno [83] in 2006 explored and proposed a two-phase model for studying nanofluids, which has since gained widespread adoption among researchers investigating nanofluid heat transfer.

The nanoparticle transport equation is solved by the FDM. The equation is expressed as follows

$$\frac{\phi_i^{n+1}-\phi_i^n}{\delta t}+u\frac{\phi_E^n-\phi_W^n}{\delta x}+v\frac{\phi_N^n-\phi_S^n}{\delta y}=\frac{1}{Le}\left[\frac{\phi_{i+1,j}^n-2\phi_{i,j}^n+\phi_{i-1,j}^n}{\delta x^2}+\frac{\phi_{i,j+1}^n-2\phi_{i,j}^n+\phi_{i,j-1}^n}{\delta y^2}\right]$$
$$+\frac{N_T}{LeN_B}\left[\frac{T_{i+1,j}^n-2T_{i,j}^n+T_{i-1,j}^n}{\delta x^2}+\frac{T_{i,j+1}^n-2T_{i,j}^n+T_{i,j-1}^n}{\delta y^2}\right] \tag{11.15}$$

The ϕ_E, ϕ_W, ϕ_N, and ϕ_S can be calculated with a limiter function $\psi(r_i)$ [84]. As an example, ϕ_E^n can be defined as:

$$\phi_E^n = \phi_i^n + \frac{1}{2}\psi(r_i)\left(\phi_{i+1}^n - \phi_i^n\right), r_i = \frac{\phi_i^n - \phi_{i-1}^n}{\phi_{i+1}^n - \phi_i^n}. \tag{11.16}$$

The $\psi(r_i)$ [85] is defined as:

$$\psi(r_i) = \max\left[0, \min\left(2r_i, \min\left(\frac{1+2r_i}{3}, 2\right)\right)\right]. \tag{11.17}$$

The k and μ of a nanofluid are expressed as follows [86]:

$$k_{nf} = k_f\left[1 + 4.4\left(\frac{2\rho_f k_b T}{\pi\mu_f^2 d_f}\right)^{0.4} \mathrm{Pr}^{0.66}\left(\frac{T}{T_{fr}}\right)^{10}\left(\frac{k_s}{k_f}\right)^{0.03}\phi^{0.66}\right], \tag{11.18}$$

$$\mu_{nf} = \frac{\mu_f}{1 - 34.87\left(\frac{d_p}{d_f}\right)^{-0.3}\phi^{1.03}}, \tag{11.19}$$

$$d_f = 0.1\frac{6M}{N\pi\rho_{f0}}^{1/3}. \tag{11.20}$$

11.3 Heat transfer simulations of nanofluids

11.3.1 LBM

The LBM is a numerical technique that has gained significant traction in recent years as a popular numerical technique for solving a diverse array of intricate fluid flow challenges encountered in both scientific research and engineering applications. Unlike traditional numerical methods that rely on solving partial differential equations, the LBM is grounded on kinetic theory, enabling more efficient and accurate simulation of fluid dynamics.

The LBM has demonstrated success in numerous fluid mechanics problems, including viscous flow and single and multiphase fluid flows. These are problems that are often difficult to solve using traditional methods due to their complex nature. Moreover, the LBM has been applied in simulating fluid dynamics in porous media, which is crucial for understanding a range of geological and environmental processes.

In summary, the LBM is a robust and powerful numerical tool that can be applied in various scientific and engineering disciplines to solve complex fluid flow problems. Its efficiency and accuracy make it a valuable tool for advancing our understanding of fluid dynamics and for developing new technologies that rely on fluid flow.

11.3.1.1 LBM equations

In the context of simulating nanofluid flow within a cavity using the LBM, the method employs two distribution functions. These functions are denoted as "f" and "g," and they respectively represent the flow field and the temperature field. The LBM offers a versatile approach that allows for the separate consideration and simulation of these two distinct fields, enabling a comprehensive analysis of both fluid flow and heat transfer phenomena within the system.

For this simulation, the D2Q9 model is commonly employed. The D2Q9 model is characterized by having 9 discrete velocity directions within a 2D lattice. The subscripts δ_x and δ_y likely denote the lattice spacing in the x and y directions, respectively. The equality $\delta_x = \delta_y$ indicates that the lattice is uniform in both directions. The e_i terms are related to the discrete velocity directions within the lattice. The e_i is given as [87]:

$$e_i \begin{cases} (0,0)i = 0 \\ (\cos\left[(i-1)\pi/2\right], \sin[(i-1)\pi/2]) \cdot ci = 1, \ldots, 4 \\ \sqrt{2}\left(\cos\left[(i-5)\pi/2 + \pi/4\right]\right), \sin\left[(i-5)\pi/2 + \pi/4\right] \cdot ci = 5, \ldots, 8 \end{cases} \tag{11.21}$$

Certainly, when simulating the flow of nanofluids within a lattice using the LBM, the lattice speed c is standardized to a value of 1. In this context, the distribution functions, denoted as f and g, are obtained through the solution of the lattice Boltzmann equation (LBE). This equation describes the evolution of these distribution functions over discrete time steps within the lattice.

Incorporating the Bhatnagar–Gross–Krook (BGK) approximation, the LBE can be expressed for the flow field with an external force in the i-direction F_i as follows:

$$f_i(x + e_i\Delta t, t + \Delta t) = f_i(x,t) + \frac{\Delta t}{\tau_v}\left[f_i^{eq}(x,t) - f_i(x,t)\right] + \Delta t e_i F_i \tag{11.22}$$

In the context of simulating the temperature field within the lattice using the LBM, particularly for nanofluid flow, the evolution of the thermal distribution function g can be described by a similar LBE framework as used for the thermal field. For discrete time steps, where Δt is equal to 1, the LBE for the temperature field g can be expressed as:

$$g_i(x + e_i\Delta t, t + \Delta t) = g_i(x,t) + \frac{\Delta t}{\tau_c}\left[g_i^{eq}(x,t) - g_i(x,t)\right] \tag{11.23}$$

f_i^{eq} and g_i^{eq} represent the equilibrium distribution functions and τ_v and τ_c represent the lattice relaxation times. Additionally, the v and α defined in terms of the relaxation times, τ_v and τ_c, which are:

$$\tau_v = 0.5 + v\frac{\delta t}{c_s^2} \tag{11.24}$$

$$\tau_c = 0.5 + \alpha\frac{\delta t}{c_s^2} \tag{11.25}$$

where $c_s = c/\sqrt{3}$ represents the speed of sound. It is crucial to note here that $\tau > 0.5$ should be met for the two relaxation times to guarantee positive values of viscosity and thermal diffusivity. The f_i^{eq} and g_i^{eq} can be expressed as:

$$f_i^{eq} = w_i \rho \left[1 + \frac{e_i u}{c_s^2} + \frac{1}{2}\frac{(e_i u)^2}{c_s^4} - \frac{1}{2}\frac{u^2}{c_s^2} \right] \tag{11.26}$$

$$g_i^{eq} = w_i T \left[1 + \frac{c_i u}{c_s^2} \right] \tag{11.27}$$

where the weight function w_i is $w_0 = 4/9$, $w_{1-4} = 1/9$, $w_{5-8} = 1/36$. In order to include buoyant, the force term in Eq. (11.22) is calculated in the Y-direction:

$$F_i = 3 w_i \rho g_y \beta (T - T_m) \tag{11.28}$$

Certainly, the simulation of fluid flow and heat transfer using the LBM involves deriving macroscopic variables such as density, momentum, and temperature from the distribution functions. These variables are crucial for understanding the behavior of the fluid system on a larger scale. The relationships between the distribution functions and macroscopic variables can be expressed using equations like the following

$$\begin{aligned} &\text{Flow density:} && \rho = \sum_i f_i \\ &\text{Momentum:} && \rho u = \sum_i e_i f_i \\ &\text{Temperature:} && T = \sum_i g_i \end{aligned} \tag{11.29}$$

These equations allow the conversion from the microscopic scale (distribution functions) to the macroscopic scale (density, momentum, and temperature). They are essential for obtaining information about the bulk behavior of the fluid and temperature fields. The LBM iteratively updates the distribution functions according to the LBE rules, which ultimately influence the evolution of these macroscopic variables over time and space.

11.3.1.2 Nusselt number calculation

The Nusselt number (Nu) is a fundamental dimensionless parameter that holds significant importance in characterizing heat transfer phenomena. It quantifies the enhancement of heat transfer due to convection over heat transfer through conduction alone. The Nu number is often used to assess the effectiveness of various heat transfer mechanisms, particularly in fluid flow scenarios. The Nu number is widely used in heat transfer studies to assess the efficiency of heat exchange mechanisms, design and optimization of heat exchangers, and understanding fluid behavior under varying conditions.

The Nu number is defined by a mathematical expression, which is provided in references [88,89]. The Nu number can be expressed by:

$$Nu_i = \frac{hW}{k_f} \tag{11.30}$$

where h and k_f are according to the following [90]:

$$h = \frac{q_w}{T_h - T_c} \tag{11.31}$$

$$k_{nf} = -\frac{q_w}{\partial T/\partial x} \quad \text{for vertical walls} \tag{11.32a}$$

$$k_{nf} = -\frac{q_w}{\partial T/\partial y} \quad \text{for horizontal walls} \tag{11.32b}$$

Using the previous relations, the expression of local Nu for the source wall is:

$$Nu_n = -\left(\frac{k_{nf}}{k_f}\right)\left(\frac{\partial \theta}{\partial X}\right) \quad \text{for vertical walls} \tag{11.33a}$$

$$Nu_n = -\left(\frac{k_{nf}}{k_f}\right)\left(\frac{\partial \theta}{\partial Y}\right) \quad \text{for horizontal walls} \tag{11.33b}$$

Eventually, the average Nu could be achieved by integrating the local Nu along the heat sink.

11.3.1.3 Curved boundary modeling

In addressing the curved surface within the simulation, the methodologies introduced by Mei et al. [91] and Guo et al. [92] were employed. This method is instrumental in effectively handling the intricacies associated with curved surfaces within the simulation framework. Fig. 11.1 provides a visual representation of the specific portion of the curved wall boundary that is of interest within the study. According to this figure, the value of Δ is:

$$\Delta = \left|\frac{\vec{x}_f - \vec{x}_w}{\vec{x}_f - \vec{x}_b}\right| \tag{11.34}$$

In the depicted figure, the different colors of dots serve to categorize and differentiate various nodes based on their locations and characteristics:

Blue dots: The dots colored in blue correspond to nodes located along the border denoted as x_w. These nodes are associated with the boundary of the domain under consideration and are crucial in defining the geometry and shape of the simulation area.
Green dots: Dots colored in green represent nodes situated on the liquid sides, identified as x_f. These nodes belong to the regions within the simulation domain that are occupied by the fluid.
White Dots: The white dots are indicative of nodes positioned within the solid part, referred to as x_b. These nodes correspond to the solid boundaries or

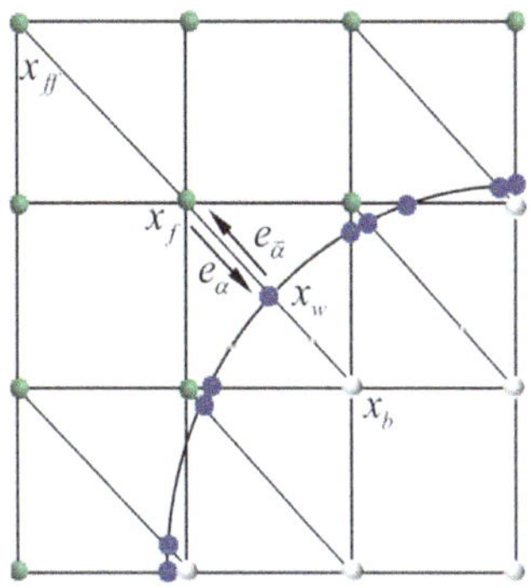

FIGURE 11.1

Schematic of the curved wall boundary [93].

Reprinted with permission from M.A. Abchouyeh, O.S. Fard, R. Mohebbi, M.A. Sheremet, Enhancement of heat transfer of nanofluids in the presence of sinusoidal side obstacles between two parallel plates through the LBM, Int. J. Mech. Sci. 156 (2019) 159–169.

obstacles present in the simulation, which influence fluid flow patterns and heat transfer dynamics.

By color-coding the nodes in this manner, the figure effectively visualizes the spatial distribution of different node types across the simulated domain. This classification is essential for accurate representation and simulation of fluid behavior, heat transfer, and other phenomena within the system.

The equation of the distribution function is:

$$\tilde{f}_{\bar{a}}(\vec{x}_b, t + \Delta t) = (1 - \lambda)\tilde{f}_a(\vec{x}_f, t + \Delta t) + \lambda f_a^0(\vec{x}_b, t + \Delta t) - 2\frac{3}{c^2}\omega_a \rho(\vec{x}_f, t + \Delta t)\vec{e}_a \vec{u}_w \tag{11.35}$$

where

$$f_a^0(\vec{x}_b, t + \Delta t) = f_a^{eq}(\vec{x}_f, t + \Delta t) + \frac{3}{c^2}\omega_a \rho(\vec{x}_f, t + \Delta t)\vec{e}_a(\vec{u}_{bf} - \vec{u}_f) \tag{11.36}$$

where u_{bf} and λ are:

$$\begin{cases} \vec{u}_{bf} = \vec{u}_{ff} = (\vec{x}_{ff}, t), \lambda = \dfrac{2\Delta - 1}{\tau - 2} & \text{if } 0 \leq \Delta \leq 0.5 \\ \vec{u}_{bf} = \left(1 - \dfrac{3}{2\Delta}\right)\vec{u}_f + \dfrac{3}{2\Delta}\vec{u}_w, \lambda = \dfrac{2\Delta - 1}{\tau + 0.5} & \text{if } 0.5 \leq \Delta \leq 1.0 \end{cases} \tag{11.37}$$

where u_w symbolizes the velocity of the rigid border, u_{bf} is the interpolation's velocity, and $\vec{e}_{\tilde{a}} = -\vec{e}_a$. For the temperature field, the thermal function is used:

$$g_{\bar{a}}(\vec{x}_b, t + \Delta t) = g_{\bar{a}}^{eq}(\vec{x}_b, t) + g_{\bar{a}}^{neq}(\vec{x}_b, t) \tag{11.38}$$

where

$$\tilde{g}_{\overline{a}}(\vec{x}_b, t+\Delta t) = g_{\overline{a}}^{eq}(\vec{x}_b, t) + \left(1 - \frac{1}{\tau_g}\right) g_{\overline{a}}^{neq}(\vec{x}_b, t) \tag{11.39}$$

$$g_{\overline{a}}^{eq}(\vec{x}_b, t) = \omega_{\overline{\alpha}} T_b^* \left[1 + \frac{3}{c^2} \vec{e}_a \cdot \vec{u}^*{}_b\right] \tag{11.40}$$

$$\begin{cases} T_b^* = T_{b1} & \\ \vec{u}_b{}^* = \left[\vec{u}_w + (\Delta - 1)\vec{u}_f\right] \ /\Delta & \text{if } \Delta \geq 0.75 \\ T_b^* = T_{b1} + (1-\Delta)T_{b2} & \text{if } \Delta < 0.75 \\ \vec{u}^*{}_b = \left[\vec{u}_w + (\Delta - 1)\vec{u}_f\right] + (1-\Delta)\left[2\vec{u}_w + (\Delta - 1)\vec{u}_{ff}\right] \ /(1+\Delta) & \end{cases} \tag{11.41}$$

$$\begin{aligned} T_{b1} &= \left[T_w + (\Delta - 1)T_f\right]/\Delta \\ T_{b2} &= \left[2T_w + (\Delta - 1)T_{ff}\right]/(1+\Delta) \end{aligned} \tag{11.42}$$

$$g_{\overline{a}}^{neq}(\vec{x}_b, t) = \Delta g_{\overline{a}}^{neq}\left(\vec{x}_f, t\right) + (1-\Delta) g_{\overline{a}}^{neq}\left(\vec{x}_{ff}, t\right). \tag{11.43}$$

11.3.1.4 Magnetohydrodynamics modeling

Magnetohydrodynamics (MHD) finds extensive applicability across various engineering domains due to its relevance in systems involving the interaction between magnetic fields and fluid flow. In MHD simulations, the uniform magnetic field $(\vec{B} = B_x\vec{e}_x + B_e\vec{e}_y)$ of constant magnitude $B = \sqrt{B_x^2 + B_y^2}$ can be employed to model the magnetic influence on fluid dynamics. The electric current J is $J = \sigma(\rightarrow V \times \vec{B})$ and the electromagnetic force F is $F = \sigma(\vec{V} \times \vec{B}) \times \vec{B}$.

In the context of Eq. (11.22), the body force associated with MHD plays a crucial role in describing the behavior of the fluid under the influence of both the magnetic field and fluid flow. This body force term encapsulates the combined effects of pressure, velocity, magnetic field, and other pertinent factors that govern the fluid motion in MHD scenarios. The body force in Eq. (11.22) for MHD is as follows:

$$F_i = F_{ix} + F_{iy}, \tag{11.44}$$

$$\begin{aligned} F_{ix} &= 3w_i\rho[A(v\sin\theta\cos\theta) - (u\sin^2(\theta))], \\ F_{iy} &= 3w_i\rho[A(u\sin\theta\cos\theta) - \left(v\cos^2(\theta)\right)], \end{aligned}$$

11.3.1.5 Immersed boundary-simplified TLBM

Chen et al. [94] introduced a novel approach in the realm of numerical simulations, combining the immersed boundary method with the SLBM (Single LBM). This innovative fusion led to the development of a new simulation technique known as IB-SLBM, designed specifically to address the complex flow dynamics of moving plates, circular cylinders, and other geometric shapes. This method stands as a versatile tool for effectively modeling and analyzing fluid behavior around such objects.

Furthermore, the researchers [94] expanded the application of this method to encompass thermal fields. In doing so, they introduced the IB-STLBM (Immersed Boundary-Single Thermal LBM). This extended technique allowed for the simulation of not only fluid flow but also the associated thermal interactions. By incorporating thermal effects, this method offered a comprehensive approach to studying complex systems where both fluid motion and heat transfer play pivotal roles.

The governing equations are based on two steps, predictor and corrector steps:

$$\frac{\partial \rho}{\partial t} + \nabla \cdot (\rho \mathbf{u}) = 0 \tag{11.45}$$

$$\frac{\partial (\rho \mathbf{u})}{\partial t} + \nabla \cdot (\rho \mathbf{u}\mathbf{u}) = -\nabla p + \nu \nabla \cdot \left[\nabla \rho \mathbf{u} + \nabla \rho \mathbf{u}^T\right] + \mathbf{F}_B \tag{11.46}$$

$$\frac{\partial (\rho T)}{\partial t} + \nabla \cdot (\rho T \mathbf{u}) = \alpha \nabla^2 (\rho T) \tag{11.47}$$

The second step will be as:

$$\frac{\partial \rho}{\partial t} = 0 \tag{11.48}$$

$$\frac{\partial (\rho \mathbf{u})}{\partial t} = \mathbf{f} \tag{11.49}$$

$$\frac{\partial (\rho T)}{\partial t} = q \tag{11.50}$$

The predictor step in IB-SLBM is based on the following steps:

Step 1:

$$\rho^* = \sum_i f_i^{eq}(x_k - e_i \Delta t, t - \Delta t) \tag{11.51}$$

$$\rho^* \mathbf{u}^* = \sum_i e_i f_i^{eq}(x_k - e_i \Delta t, t - \Delta t) \tag{11.52}$$

$$\rho^* T^* = \sum_i g_i^{eq}(x_k - e_i \Delta t, t - \Delta t) \tag{11.53}$$

Step 2:

$$\rho(x_k, t) = \rho^* \tag{11.54}$$

$$\rho(x_k, t)\mathbf{u}^{**} = \rho^* \mathbf{u}^* + (\tau_\nu - 1)\sum_i e_i f_i^{eq}(x_k + e_i \Delta t, t) - (\tau_\nu - 1)\rho(x_k, t - \Delta t)\mathbf{u}(x_k, t - \Delta t) + \mathbf{F}_B \Delta t \tag{11.55}$$

$$\rho(x_k, t)T^{**} = \rho^* T^* + (\tau_c - 1)\sum_i g_i^{eq}(x_k + e_i \Delta t, t) - (\tau_c - 1)\rho(x_k, t - \Delta t)T(x_k, t - \Delta t) \tag{11.56}$$

For correcting the velocity and temperature the Eqs. (11.57) and (11.58) can be used:

$$\mathbf{u}^{n+1} = \mathbf{u}^{**} + \Delta\mathbf{u} \tag{11.57}$$

$$T^{n+1} = T^{**} + \Delta T \tag{11.58}$$

The force and source terms are

$$\mathbf{f} = \rho^{n+1}\frac{\Delta\mathbf{u}}{\Delta t} \tag{11.59}$$

$$q = \rho^{n+1}\frac{\Delta T}{\Delta t} \tag{11.60}$$

By defining Eq. (11.61) for abbreviation and using the Dirichlet boundary condition, the corrected variables are as:

$$\varphi^{n+1} = \varphi^{**} + \Delta\varphi \tag{11.61}$$

$$\varphi^{n+1}\left(\mathbf{X}_B^l\right) = \varphi_B\left(\mathbf{X}_B^l\right) \tag{11.62}$$

The variable φ is:

$$\varphi^{n+1}\left(\mathbf{X}_B^l\right) = \sum_j \varphi^{n+1}(\mathbf{r}_j)D\left(\mathbf{r}_j - \mathbf{X}_B^l\right)h^2, l = 1, 2, \ldots, N; j = 1, 2, \ldots, M \tag{11.63}$$

$$D_{jl} = D\left(\mathbf{r}_j - \mathbf{X}_B^l\right) = \delta\left(r_j^x - X_B^l\right)\delta\left(r_j^y - Y_B^l\right) \tag{11.64}$$

$$\delta(r) = \begin{cases} \dfrac{1 + \cos\left(\dfrac{\pi|r|}{2}\right)}{4h}, & |r| \le 2 \\ 0, & |r| > 2 \end{cases} \tag{11.65}$$

$$\Delta\varphi(\mathbf{r}_j) = \sum_j \delta\varphi_B^l D\left(\mathbf{r}_j - \mathbf{X}_B^l\right)\Delta_s^l \tag{11.66}$$

$$\varphi_B\left(\mathbf{X}_B^l\right) = \sum_j \varphi^{**}(\mathbf{r}_j)D\left(\mathbf{r}_j - \mathbf{X}_B^l\right)h^2 + \sum_j\sum_k \delta\varphi_B^k\Delta_s^k D\left(\mathbf{r}_j - \mathbf{X}_B^l\right)D\left(\mathbf{r}_j - \mathbf{X}_B^k\right)h^2 \tag{11.67}$$

$$\mathbf{A}\cdot\mathbf{X} = \mathbf{B} \tag{11.68}$$

where

$$A_{ij} = h^2\sum_{k=1}^{M} D_{ik}D_{jk} \tag{11.69}$$

$$X = \left[\delta\varphi_B^1\Delta_s^1, \delta\varphi_B^2\Delta_s^2, \ldots, \delta\varphi_B^N\Delta_s^N\right]^T \tag{11.70}$$

$$B_i = \varphi_i - \sum_{k=1}^{M} D_{ik}\varphi_k^{**} \tag{11.71}$$

11.3.2 Application samples of LBM

The LBM has been extensively applied to investigate various configurations, as evidenced by studies [95–98]. In Fig. 11.2, a collection of specific geometric configurations is presented. These configurations primarily involve 2D square cavities or channels that are occupied by laminar Newtonian and incompressible nanofluids. The enclosures within these configurations encompass diverse suspensions of nanoparticles within a water medium. These investigations encompass a wide spectrum of scenarios, delving into the intricate behaviors and interactions within nanofluidic systems. The LBM serves as a powerful tool in unveiling the complex fluid dynamics, heat transfer mechanisms, and nanoparticle dispersion patterns that are inherent in such configurations. The choice of 2D square cavities and channels as the fundamental geometries underlines the need to comprehend and model these fundamental scenarios before venturing into more complex and practical applications. The utilization of different nanoparticle-water suspensions further enriches the diversity of the study, enabling the examination of a range of fluid properties and behaviors. By examining these selected geometric configurations and employing the LBM as the analytical framework, researchers aim to deepen their understanding of the fundamental principles governing nanofluidic behavior. This knowledge lays the groundwork for further exploration into more intricate and real-world applications, offering insights into heat transfer enhancement, fluid mixing, and other phenomena that hold promise for various engineering and scientific domains.

In this context, two distinct methods were employed to establish boundary conditions: the standard bounce-back method [105] and Mohamad's method [106]. These methodologies were instrumental in effectively defining the behavior of the fluid near the boundaries within the enclosure under study. Specifically, the choice of boundary conditions was pivotal in ensuring accurate representation and simulation of the fluid flow and heat transfer processes.

In accordance with the standard bounce-back method, the interaction between the fluid particles and the boundary was modeled in a manner that mirrored the physical phenomena of particle rebounding upon collision with the boundary. This approach serves to effectively simulate the no-slip condition at the solid boundaries, which is a common assumption in fluid dynamics.

Concurrently, Mohamad's method was employed to address boundary conditions as well. This method, devised by Mohamad, offers an alternative approach for handling temperature boundaries.

For the remaining walls within the enclosure, an adiabatic boundary condition was implemented. This choice implies that there is no heat exchange across these particular boundaries, contributing to a controlled and consistent representation of the thermal interactions within the system.

Mohebbi et al. [53] conducted a comprehensive investigation involving a two-dimensional (2D) Γ-shaped cavity utilizing the LBM, as illustrated in Fig. 11.2J. The impact of multiple parameters on fluid flow, temperature distribution, and

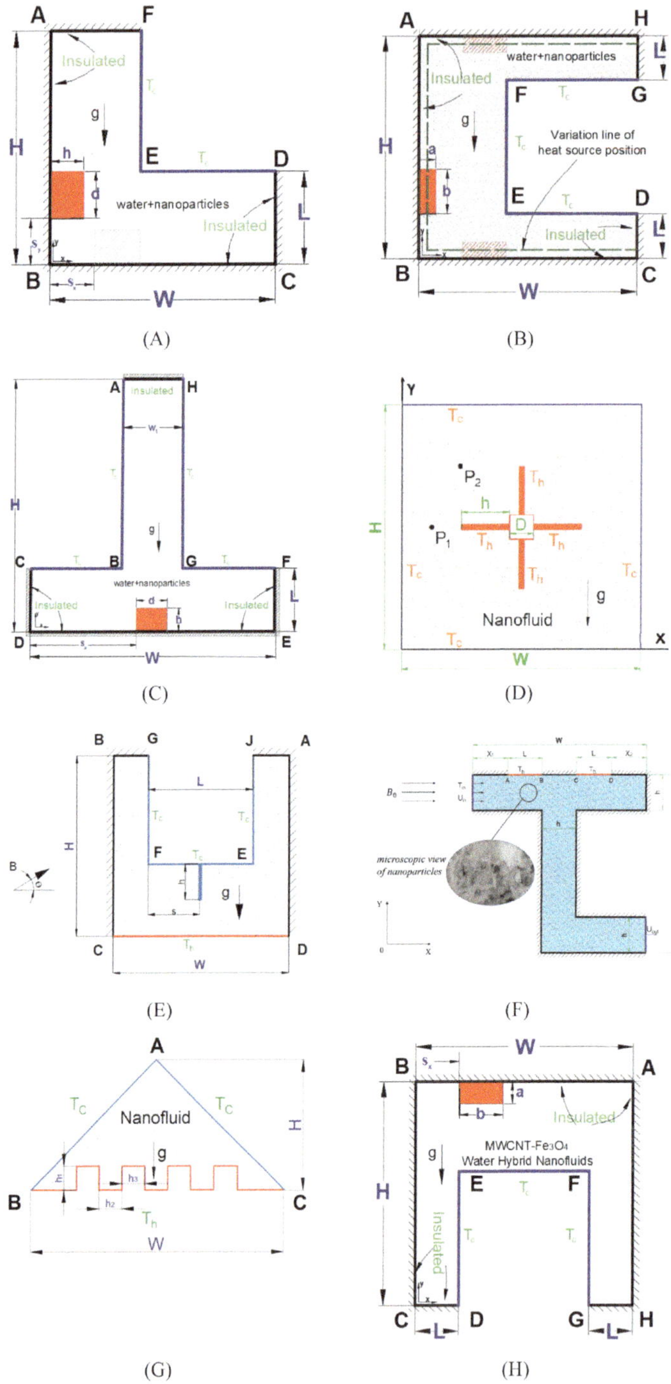

FIGURE 11.2

(*Continued*)

Schematic representation of some applications of LBM. (A) [17]; (B) [55]; (C) [9]; (D) [19]; (E) [99]; (F) [7]; (G) [69]; (H) [100]; (I) [101]; (J) [53]; (K) [102]; (L) [103]; (M) [5]; (N) [31]; (O) [104]; (P) [89].

(A) Reprinted with permission from R. Mohebbi, M.M. Rashidi, Numerical simulation of natural convection heat transfer of a nanofluid in an L-shaped enclosure with a heating obstacle, J. Taiwan Inst. Chem. Eng. 72 (2017) 70–84. https://doi.org/10.1016/J.JTICE0.2017.01.006. (B) Reprinted with permission from R. Mohebbi, M. Izadi, A.J. Chamkha, Heat source location and natural convection in a C-shaped enclosure saturated by a nanofluid, Phys. Fluids. 29 (2017) 122009. (C) Reprinted with permission from M. Izadi, R. Mohebbi, D. Karimi, M.A. Sheremet, Numerical simulation of natural convection heat transfer inside a ⊥ shaped cavity filled by a MWCNT-Fe3O4/water hybrid nanofluids using LBM, Chem. Eng. Process. – Process Intensif. 125 (2018) 56–66. (D) Reprinted with permission from Y. Ma, Z. Yang, Simplified and highly stable thermal lattice Boltzmann method simulation of hybrid nanofluid thermal convection at high Rayleigh numbers, Phys. Fluids. 32 (2020) 012009. (E) Reprinted with permission from Y. Ma, R. Mohebbi, M.M. Rashidi, Z. Yang, M.A. Sheremet. Numerical study of MHD nanofluid natural convection in a baffled U-shaped enclosure. Int. J. Heat Mass Transfer. 130 (2019) 123–34. (F) Reprinted with permission from Y. Ma, R. Mohebbi, M.M. Rashidi, Z. Yang, MHD forced convection of MWCNT–Fe3O4/water hybrid nanofluid in a partially heated τ-shaped channel using LBM. J. Therm. Anal. Calorim. (2018). (G) Reprinted with permission from Y. Ma, M.M. Rashidi, R. Mohebbi, Z. Yang, Nanofluid natural convection in a corrugated solar power plant using the hybrid LBM-TVD method, Energy 199 (2020) 117402. https://doi.org/10.1016/J.ENERGY.2020.117402. (H) Reprinted with permission from A. Matori, R. Mohebbi, Z. Hashemi, et al., Lattice Boltzmann study of multi-walled carbon nanotube (MWCNT)-Fe3O4/water hybrid nanofluids natural convection heat transfer in a Π-shaped cavity equipped by hot obstacle. J. Therm. Anal. Calorim. 136 (2019) 2495–2508. (I) Reprinted with permission from Y. Ma, R. Mohebbi, M.M. Rashidi, Z. Yang, Koo–Kleinstreuer–Li correlation for simulation of nanofluid natural convection in hollow cavity in existence of magnetic field. J. Therm. Anal. Calorim. 137 (2019) 1413–29. (J) Reprinted with permission from R. Mohebbi, M. Izadi, H. Sajjadi, A.A. Delouei, M.A. Sheremet, Examining of nanofluid natural convection heat transfer in a Γ-shaped enclosure including a rectangular hot obstacle using the lattice Boltzmann method, Phys. A Stat. Mech. Appl. 526 (2019) 120831. (K) Reprinted with permission from R. Mohebbi, H. Lakzayi, H. Rasam, Numerical simulation of conjugate heat transfer in a square cavity consisting the conducting partitions by utilizing lattice Boltzmann method, Phys. A: Stat. Mech. Appl. 546 (2020) 123050. (L) Reprinted with permission from Y. Ma, R. Mohebbi, M.M. Rashidi, Z. Yang, M. Sheremet, Nanoliquid thermal convection in I-shaped multiple-pipe heat exchanger under magnetic field influence. Phys. A: Stat. Mech. Appl. 550 (2020) 124028. (M) Reprinted with permission from Y. Ma, R. Mohebbi, M.M. Rashidi, Z. Yang, MHD convective heat transfer of Ag-MgO/water hybrid nanofluid in a channel with active heaters and coolers, Int. J. Heat Mass Transf. 137 (2019) 714–726. 169. (N) Reprinted with permission from Y. Ma, R. Mohebbi, M. Sheremet, Heat transfer and fluid flow around starfish-inspired cylinder by immersed boundary-simplified thermal lattice Boltzmann method, Eng. Anal. Bound. Elem. 146 (2023) 656–667. (O) Reprinted with permission from Y. Ma, R. Mohebbi, Hydro-thermal performance on the two phase nanofluid convection heat transfer using the LBM-FD method. Waves Random Complex Media 22 (2022) 1–6. (P) Reprinted with permission from R. Mohebbi, Y. Ma, Optimal configuration of discrete heat sources in a channel with sudden expansion and contraction by lattice Boltzmann method. J. Therm. Anal. Calorim. 148 (2023) 4553–4566.

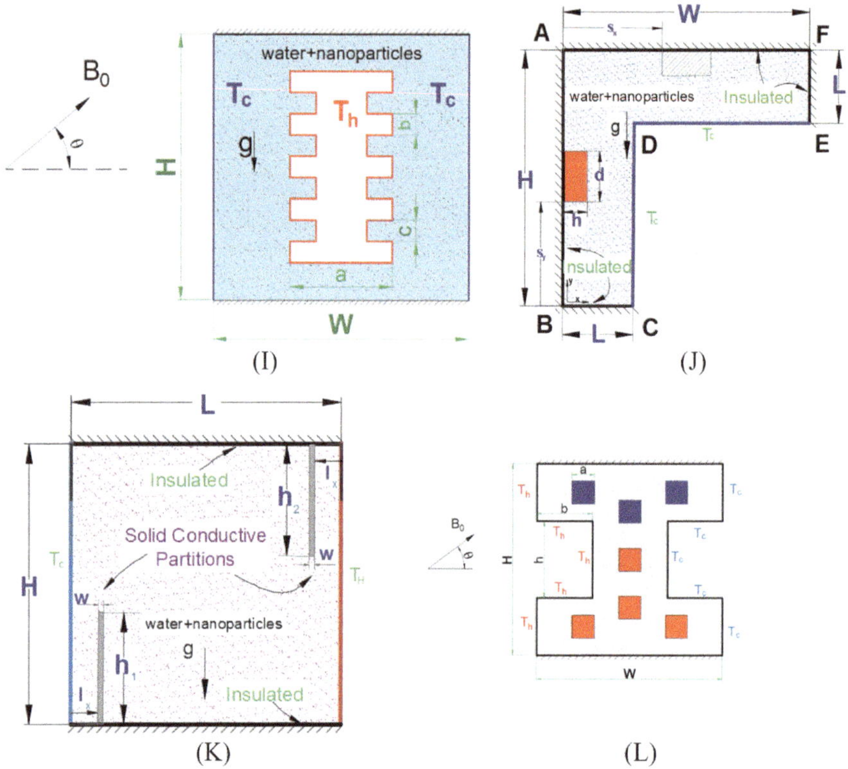

FIGURE 11.2

(Continued).

vortex generation was studied by them. In their study, one of the walls, specifically the CDE wall, was maintained at a temperature (T_c) of 0, while the remaining walls were assumed to be thermally insulating. Within the cavity, a rectangular obstacle was designated as a heat source, characterized by a temperature (T_h) of 1. The entire enclosure was occupied by Al_2O_3 nanoparticles, and the research delved into a comprehensive examination of numerous physical parameters, resulting in the generation of various impactful findings. Presented below are selected outcomes from their research.

Fig. 11.3 portrays the influence of Ra on the structural configuration of streamline and isotherm patterns. It is discernible that at a Ra value of 10^3, the isothermal lines align closely along the vertical walls. As Ra increases

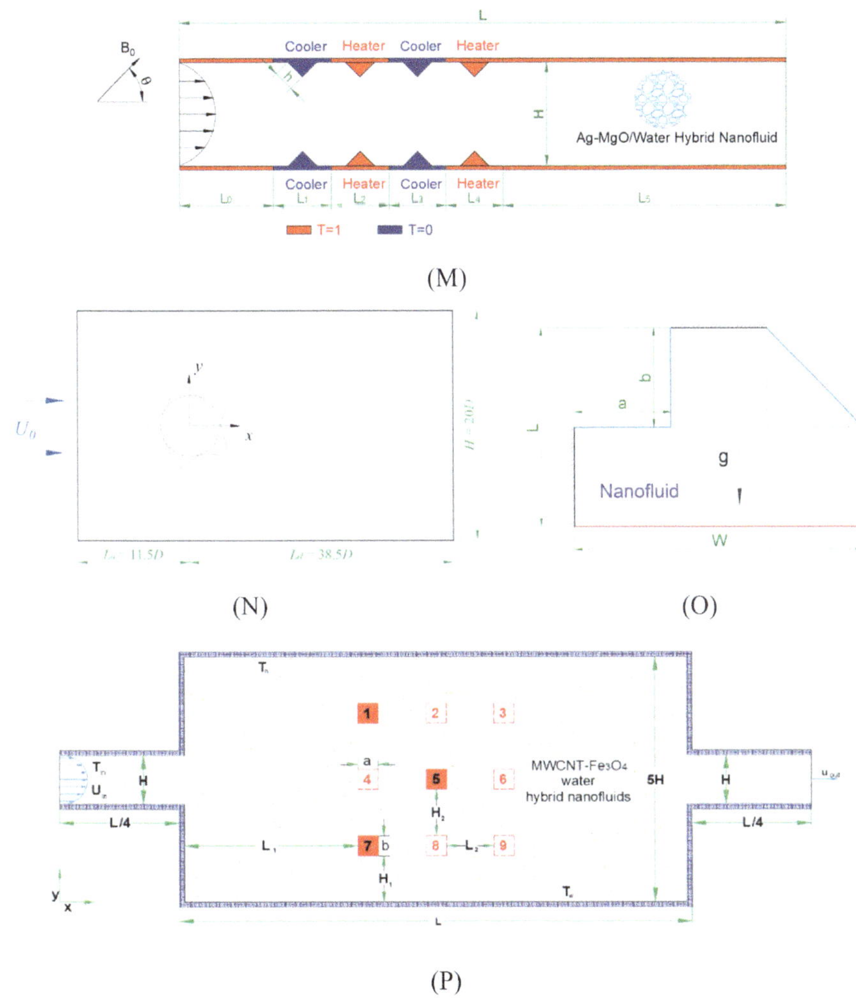

FIGURE 11.2

(Continued).

significantly and proximity to the heated obstacle intensifies, the distribution of isothermal lines becomes denser.

Fig. 11.4 highlights the effect of the cavity's AR at a Ra value of 10^6 on both isotherm and streamline patterns. This exploration is conducted for two distinct scenarios: one where the host liquid's volume fraction (∅) is 0% and the other where ∅ is 5%. Within the enclosure, a clockwise vortex takes form. Notably, the presence of nanoparticles enhances this vortex, which aligns with the structural characteristics of streamlines outlined in Fig. 11.3.

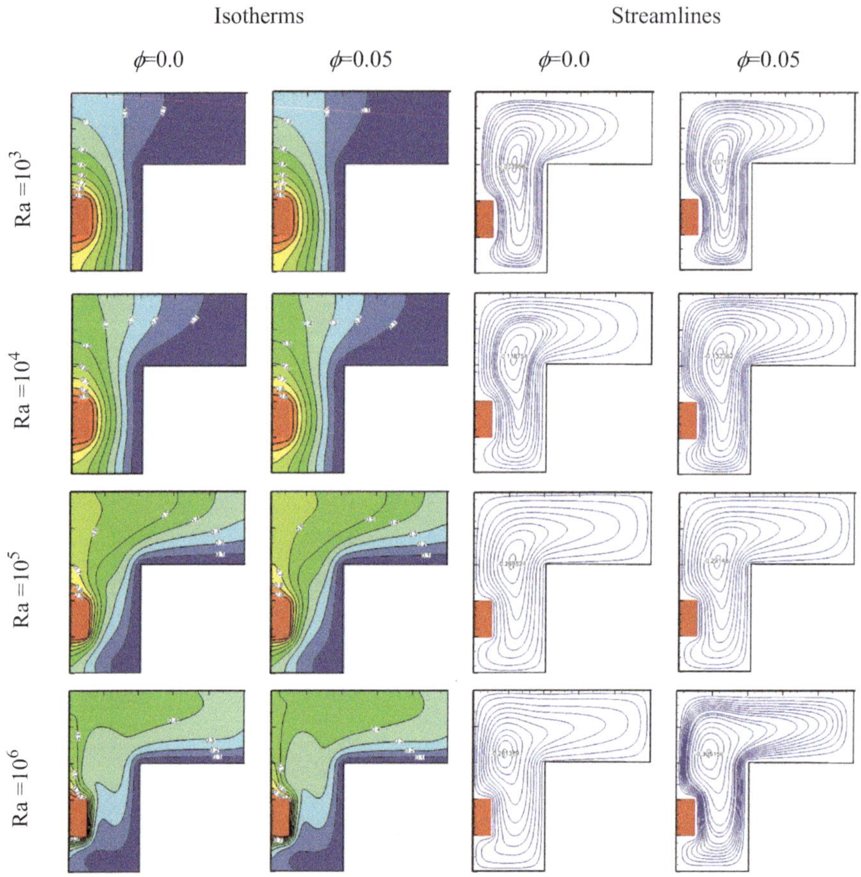

FIGURE 11.3

Effect of Ra and φ on the isotherms and streamlines at h = 0.1H, Sy = 0.2, AR = 0.4 [53].

Reprinted with permission from R. Mohebbi, M. Izadi, H. Sajjadi, A.A. Delouei, M.A. Sheremet, Examining of nanofluid natural convection heat transfer in a Γ-shaped enclosure including a rectangular hot obstacle using the lattice Boltzmann method, Phys. A Stat. Mech. Its Appl. 526 (2019) 120831. https://doi.org/10.1016/J.PHYSA0.2019.04.067.

Turning to Fig. 11.5, the impact of Ra on velocity profiles is showcased. As the AR and Ra values escalate, there is a corresponding increase in the stream function. This trend indicates a heightened circulation of fluid within the system, a result of the compounded effects of increased AR and Ra values.

Mohebbi et al. [55] undertook a comprehensive exploration of nanofluidic heat transfer within a C-shaped cavity, featuring a heated obstruction. The study involved a meticulous consideration of the Ra, which was scrutinized across the range of 10^3–10^6, alongside a variation in the ϕ spanning from

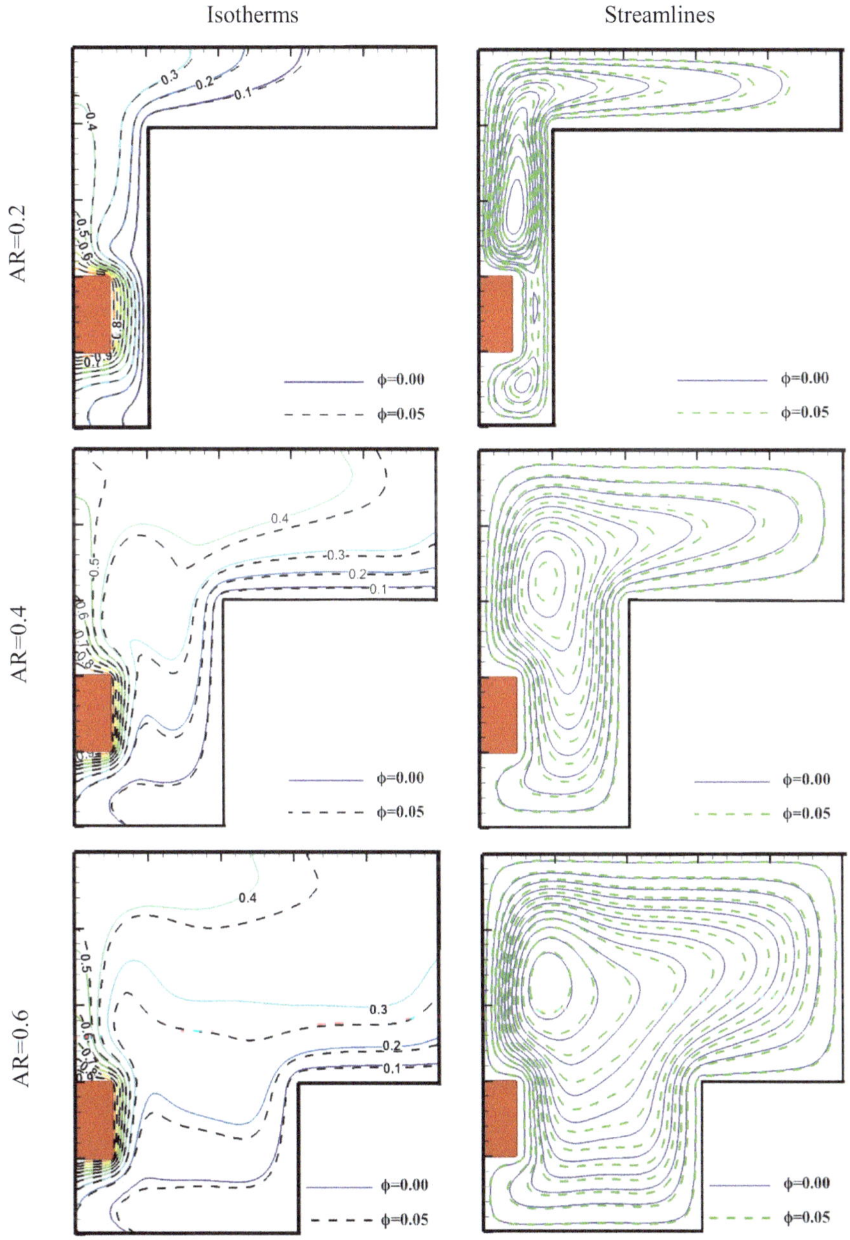

FIGURE 11.4

Isotherms and streamlines at different values of φ and AR ratio for h = 0.1H, Sy = 0.2, Ra = 10^6 [53]. *AR*, Aspect ratio.

Reprinted with permission from R. Mohebbi, M. Izadi, H. Sajjadi, A.A. Delouei, M.A. Sheremet, Examining of nanofluid natural convection heat transfer in a Γ-shaped enclosure including a rectangular hot obstacle using the lattice Boltzmann method, Phys. A Stat. Mech. Its Appl. 526 (2019) 120831. https://doi.org/10.1016/J.PHYSA.2019.04.067.

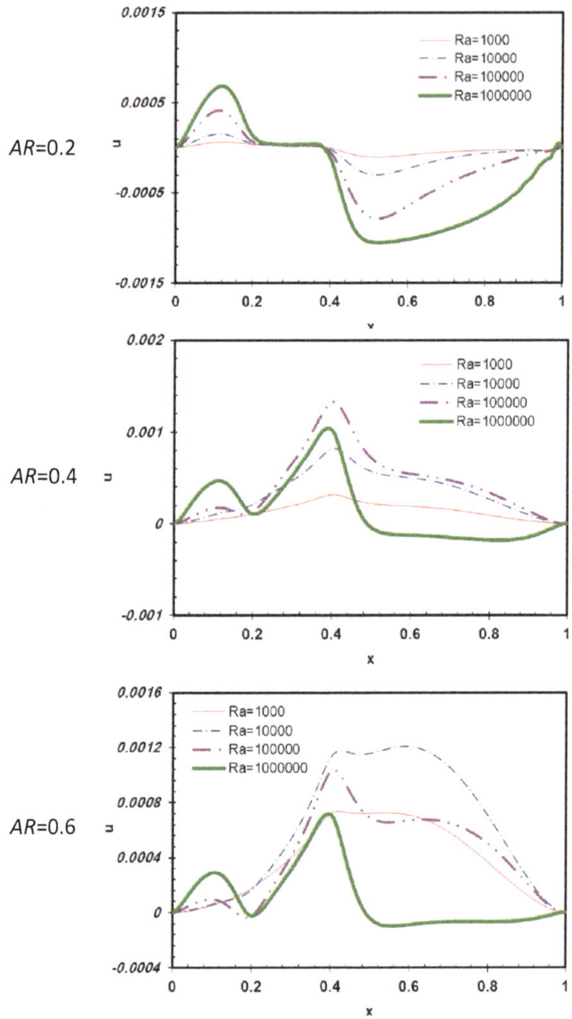

FIGURE 11.5

Effect of AR and Ra on the velocity profiles at y = 0.15, h = 0.1H, Sx = 0.2 [53]. *AR*, Aspect ratio.

Reprinted with permission from R. Mohebbi, M. Izadi, H. Sajjadi, A.A. Delouei, M.A. Sheremet, Examining of nanofluid natural convection heat transfer in a Γ-shaped enclosure including a rectangular hot obstacle using the lattice Boltzmann method, Phys. A Stat. Mech. Its Appl. 526 (2019) 120831. https://doi.org/10.1016/J.PHYSA.2019.04.067.

0% to 5%. The elucidation of evolving vortex patterns and the intricate interplay between heat sources and nanofluidic properties contribute to a more holistic understanding of these phenomena. In Figs. 11.6 and 11.7, the

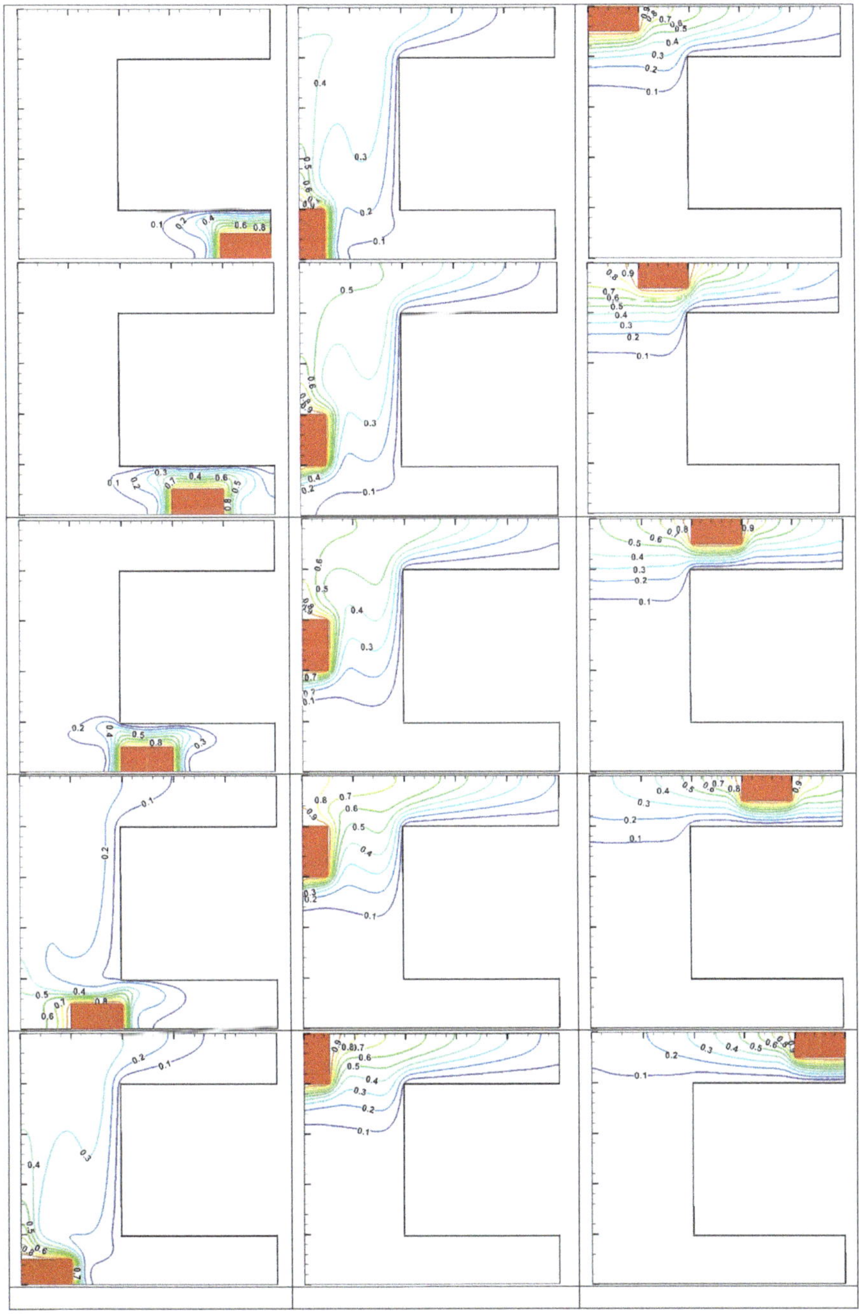

FIGURE 11.6

Isotherms of different cases according to Ref. [55] for Ra = 10^6 and Ø = 0.05.

Reprinted with permission from R. Mohebbi, M. Izadi, A.J. Chamkha, Heat source location and natural convection in a C-shaped enclosure saturated by a nanofluid, Phys. Fluids. 29 (2017) 122009. https://doi.org/10.1063/1.4993866.

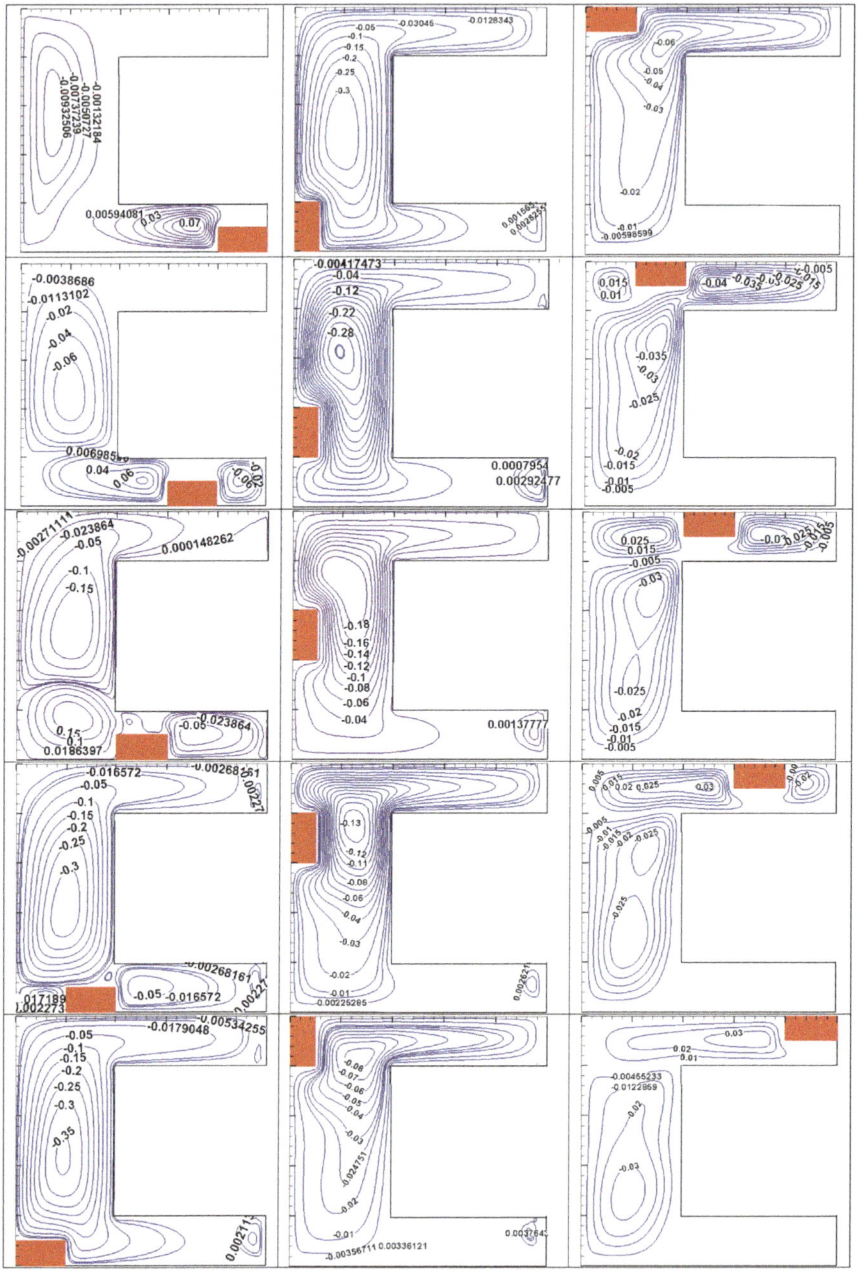

FIGURE 11.7

Streamlines of different cases according to Ref. [55] for Ra = 10^6 and $\varnothing = 0.05$.

Reprinted with permission from R. Mohebbi, M. Izadi, A.J. Chamkha, Heat source location and natural convection in a C-shaped enclosure saturated by a nanofluid, Phys. Fluids. 29 (2017) 122009. https://doi.org/10.1063/1.4993866.

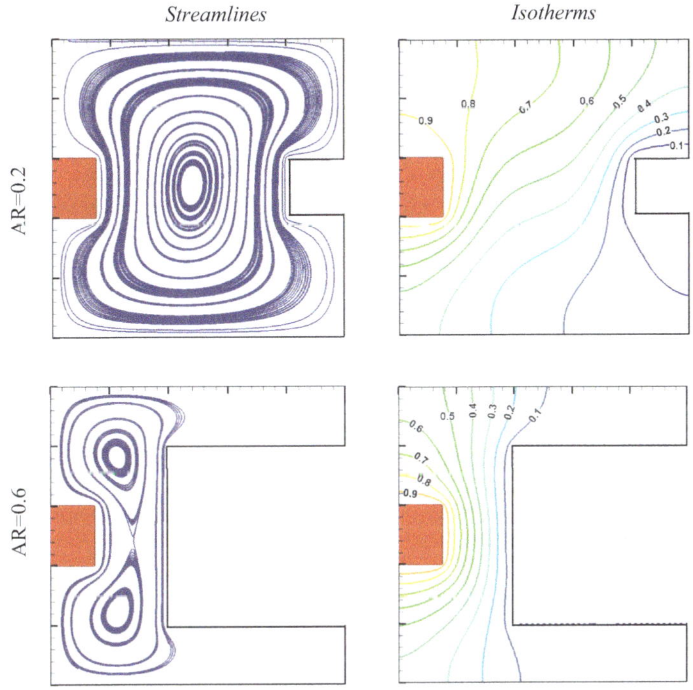

FIGURE 11.8

Streamlines and isotherms inside the cavity at Ra = –0⁴ and ∅ = 0.03 for different ARs of the cavity.

visualization of isothermal contours and streamline patterns is presented for distinct scenarios, particularly focusing on the extreme conditions of $Ra = 10^6$ and (ϕ) of 0.05.

A remarkable observation lies in the deformation and distortion frequently exhibited by the isothermal patterns. Notably, in cases 2–4 and specifically along the right segment of the thermally activated source, an additional secondary vortex emerges. This secondary vortex phenomenon is particularly pronounced within these cases. On progressing to case 5, a distinctive presence of an isolated primary vortex is discernible. Subsequently, as the investigation delves into cases 6–10, the influence of the primary vortex undergoes attenuation, notably diminishing in significance. This attenuation is attributed to alterations in the positioning of the heat source, which has been displaced upward within the vertical section of the enclosure.

Izadi et al. [107] conducted a notable investigation into nanofluid heat transfer within a C-shaped cavity using the LBM. Their study focused on elucidating the impact of the AR between the cavity and the heat source on the rate of heat transfer. Upon perusal of Fig. 11.8, it becomes evident that the formation of more robust vortices becomes pronounced, particularly in the central region of the cavity. This phenomenon is closely correlated with the proximity of streamlines to the heated obstruction, and as the AR increases, this intensification becomes more pronounced. Simultaneously, with an increase in the AR, the efficiency of heat dissipation toward the heat sink experiences an enhancement.

Furthermore, the intricate interplay between heat transfer dynamics and the Ra is discernible in their findings. As the Ra value undergoes an escalation, the characteristics of heat transfer within the system exhibit notable alterations. These outcomes collectively contribute to a comprehensive understanding of the intricate interdependencies governing nanofluidic heat transfer within a C-shaped cavity. The work of Izadi et al. [107] provides valuable insights into the nuanced behavior of nanofluidic systems, offering a compelling perspective on the diverse factors that influence heat transfer phenomena in such complex configurations.

11.4 Conclusion and outlook

In this section, we present a succinct overview of numerical simulations concerning nanofluid flow and heat transfer, with a specific focus on the LBM. The arrangement of our discourse encompasses a comprehensive literature survey, the techniques employed in nanofluid modeling, an in-depth exploration of the LBM, and a meticulously constructed case study. The purpose of this arrangement is to unveil the intricate mathematical principles that underlie the analysis of nanofluid flow and heat transfer phenomena. The integration of nanoparticles into foundational fluids has emerged as a highly promising avenue for augmenting the efficiency of heat transfer processes. The characteristics governing heat transfer in nanofluids within a given domain are influenced by a multitude of intricate factors. These factors encompass the geometric configuration of the domain itself, the volumetric concentration of nanoparticles denoted as ϕ, the Ra, the Reynolds number (Re), the Hartmann number (Ha), and the specific composition of the nanofluid under examination. Moreover, it is imperative to underscore the pivotal role played by the chosen viscosity model in the course of nanofluid computations. The selection of a suitable viscosity model can wield a pronounced influence over the outcomes yielded by heat transfer analyses. Thus the meticulous choice of a viscosity model stands as a critical consideration that can significantly shape the results derived from computations pertaining to heat transfer. In conclusion, this chapter orchestrates a comprehensive journey through the realm of numerical simulations pertaining to nanofluidic phenomena, with a central emphasis on the LBM. By traversing through the landscape of existing literature, dissecting nanofluid modeling methodologies, delving into the intricacies of the

LBM, and culminating in a revealing case study, this chapter aspires to illuminate the fundamental mathematical bedrock underlying the analysis of nanofluidic flow and heat transfer.

References

[1] M. Raja, R. Vijayan, P. Dineshkumar, M. Venkatesan, Review on nanofluids characterization, heat transfer characteristics and applications, Renew. Sustain. Energy Rev. 64 (2016) 163–173. Available from: https://doi.org/10.1016/J.RSER.2016.05.079.

[2] K. Khanafer, K. Vafai, M. Lightstone, Buoyancy-driven heat transfer enhancement in a two-dimensional enclosure utilizing nanofluids, Int. J. Heat. Mass. Transf. 46 (2003) 3639–3653. Available from: https://doi.org/10.1016/S0017-9310(03)00156-X.

[3] C. Zhang, L. Zheng, X. Zhang, G. Chen, MHD flow and radiation heat transfer of nanofluids in porous media with variable surface heat flux and chemical reaction, Appl. Math. Model. 39 (2015) 165–181. Available from: https://doi.org/10.1016/J.APM.2014.05.023.

[4] H. Arya, M.M. Sarafraz, O. Pourmehran, M. Arjomandi, Heat transfer and pressure drop characteristics of MgO nanofluid in a double pipe heat exchanger, Heat. Mass. Transf. Und Stoffuebertragung 55 (2019) 1769–1781. Available from: https://doi.org/10.1007/S00231-018-02554-1/FIGURES/11.

[5] Y. Ma, R. Mohebbi, M.M. Rashidi, Z. Yang, MHD convective heat transfer of Ag-MgO/water hybrid nanofluid in a channel with active heaters and coolers, Int. J. Heat. Mass. Transf. 137 (2019) 714–726. Available from: https://doi.org/10.1016/J.IJHEATMASSTRANSFER.2019.03.169.

[6] Y. Ma, R. Mohebbi, M.M. Rashidi, Z. Yang, Mixed convection characteristics in a baffled U-shaped lid-driven cavity in the presence of magnetic field, J. Therm. Anal. Calorim. 140 (2020) 1967–1984. Available from: https://doi.org/10.1007/S10973-019-08900-7/FIGURES/19.

[7] Y. Ma, R. Mohebbi, M.M. Rashidi, Z. Yang, MHD forced convection of MWCNT–Fe_3O_4/water hybrid nanofluid in a partially heated τ-shaped channel using LBM, J. Therm. Anal. Calorim. 136 (2019) 1723–1735. Available from: https://doi.org/10.1007/S10973-018-7788-4/FIGURES/11.

[8] Y. Ma, R. Mohebbi, M.M. Rashidi, Z. Yang, Numerical simulation of flow over a square cylinder with upstream and downstream circular bar using lattice Boltzmann method, Int. J. Mod. Phys. C 29 (2018). Available from: https://doi.org/10.1142/S0129183118500304.

[9] M. Izadi, R. Mohebbi, D. Karimi, M.A. Sheremet, Numerical simulation of natural convection heat transfer inside a ⊥ shaped cavity filled by a MWCNT-Fe3O4/water hybrid nanofluids using LBM, Chem. Eng. Process. – Process Intensif. 125 (2018) 56–66. Available from: https://doi.org/10.1016/J.CEP.2018.01.004.

[10] A.S. Dogonchi, M.A. Ismael, A.J. Chamkha, D.D. Ganji, Numerical analysis of natural convection of Cu–water nanofluid filling triangular cavity with semicircular bottom wall, J. Therm. Anal. Calorim. 135 (2019) 3485–3497. Available from: https://doi.org/10.1007/S10973-018-7520-4/FIGURES/8.

[11] P. Soleymani, E. Saffarifard, J. Jahanpanah, M. Babaie, A. Nourian, R. Mohebbi, et al., Enhancement of an air-cooled battery thermal management system using liquid

cooling with CuO and Al_2O_3 nanofluids under steady-state and transient conditions, Fluids 8 (2023) 261. Available from: https://doi.org/10.3390/fluids8100261.

[12] M. Izadi, R. Mohebbi, A.A. Delouei, H. Sajjadi, Natural convection of a magnetizable hybrid nanofluid inside a porous enclosure subjected to two variable magnetic fields, Int. J. Mech. Sci. 151 (2019) 154–169. Available from: https://doi.org/10.1016/J.IJMECSCI.2018.11.019.

[13] R. Mohebbi, A. Abbaszadeh, M. Varzandeh, Y. Ma, Natural convection heat transfer of Ag-MgO/water micropolar hybrid nanofluid inside an F-shaped cavity equipped by hot obstacle, J. Heat. Mass. Transf. Res. 8 (2021) 139–150. Available from: https://doi.org/10.22075/JHMTR.2021.21269.1303.

[14] M. Mahmoodi, Numerical simulation of free convection of a nanofluid in L-shaped cavities, Int. J. Therm. Sci. 50 (2011) 1731–1740. Available from: https://doi.org/10.1016/J.IJTHERMALSCI.2011.04.009.

[15] N.A.C. Sidik, M. Khakbaz, L. Jahanshaloo, S. Samion, A.N. Darus, Simulation of forced convection in a channel with nanofluid by the lattice boltzmann method, Nanoscale Res. Lett. 8 (2013) 1–8. Available from: https://doi.org/10.1186/1556-276X-8-178/TABLES/1.

[16] G.H.R. Kefayati, S.F. Hosseinizadeh, M. Gorji, H. Sajjadi, Lattice Boltzmann simulation of natural convection in tall enclosures using water/SiO_2 nanofluid☆, Int. Commun. Heat. Mass. Transf. 38 (2011) 798–805. Available from: https://doi.org/10.1016/j.icheatmasstransfer.2011.03.005.

[17] R. Mohebbi, M.M. Rashidi, Numerical simulation of natural convection heat transfer of a nanofluid in an L-shaped enclosure with a heating obstacle, J. Taiwan. Inst. Chem. Eng. 72 (2017) 70–84. Available from: https://doi.org/10.1016/J.JTICE.2017.01.006.

[18] C. Qi, C. Li, K. Li, D. Han, Natural convection of nanofluids in solar energy collectors based on a two-phase lattice Boltzmann model, J. Therm. Anal. Calorim. 147 (2022) 2417–2438. Available from: https://doi.org/10.1007/S10973-021-10668-8/FIGURES/25.

[19] Y. Ma, Z. Yang, Simplified and highly stable thermal lattice Boltzmann method simulation of hybrid nanofluid thermal convection at high Rayleigh numbers, Phys. Fluids. 32 (2020) 012009. Available from: https://doi.org/10.1063/1.5139092.

[20] Y. Ma, R. Mohebbi, M.M. Rashidi, Z. Yang, MHD forced convection of MWCNT–Fe_3O_4/water hybrid nanofluid in a partially heated τ-shaped channel using LBM, J. Therm. Anal. Calorim. 136 (2019) 1723–1735. Available from: https://doi.org/10.1007/S10973-018-7788-4/METRICS.

[21] M. Izadi, A. Behzadmehr, D. Jalali-Vahida, Numerical study of developing laminar forced convection of a nanofluid in an annulus, Int. J. Therm. Sci. 48 (2009) 2119–2129. Available from: https://doi.org/10.1016/J.IJTHERMALSCI.2009.04.003.

[22] A.M. Hussein, K.V. Sharma, R.A. Bakar, K. Kadirgama, The effect of cross sectional area of tube on friction factor and heat transfer nanofluid turbulent flow, Int. Commun. Heat. Mass. Transf. 47 (2013) 49–55. Available from: https://doi.org/10.1016/J.ICHEATMASSTRANSFER.2013.06.007.

[23] M. Akhtari, M. Haghshenasfard, M.R. Talaie, Numerical and experimental investigation of heat transfer of α-Al_2O_3/water nanofluid in double pipe and shell and tube heat exchangers, Numer. Heat. Transf. Part A Appl. 63 (2013) 941–958. Available from: https://doi.org/10.1080/10407782.2013.772855.

[24] M.K. Moraveji, M. Darabi, S.M.H. Haddad, R. Davarnejad, Modeling of convective heat transfer of a nanofluid in the developing region of tube flow with computational

fluid dynamics, Int. Commun. Heat. Mass. Transf. 38 (2011) 1291–1295. Available from: https://doi.org/10.1016/J.ICHEATMASSTRANSFER.2011.06.011.

[25] M.M. Rashidi, A. Hosseini, I. Pop, S. Kumar, N. Freidoonimehr, Comparative numerical study of single and two-phase models of nanofluid heat transfer in wavy channel, Appl. Math. Mech. 35 (2014) 831–848. Available from: https://doi.org/10.1007/S10483-014-1839-9.

[26] R. Mohebbi, H. Lakzayi, N.A.C. Sidik, W.M.A.A. Japar, Lattice Boltzmann method based study of the heat transfer augmentation associated with Cu/water nanofluid in a channel with surface mounted blocks, Int. J. Heat. Mass. Transf. 117 (2018) 425–435. Available from: https://doi.org/10.1016/J.IJHEATMASSTRANSFER.2017.10.043.

[27] R. Mohebbi, M.M. Rashidi, M. Izadi, N.A.C. Sidik, H.W. Xian, Forced convection of nanofluids in an extended surfaces channel using lattice Boltzmann method, Int. J. Heat. Mass. Transf. 117 (2018) 1291–1303. Available from: https://doi.org/10.1016/J.IJHEATMASSTRANSFER.2017.10.063.

[28] M. Sheikholeslami, T. Hayat, A. Alsaedi, Numerical simulation for forced convection flow of MHD CuO-H_2O nanofluid inside a cavity by means of LBM, J. Mol. Liq. 249 (2018) 941–948. Available from: https://doi.org/10.1016/J.MOLLIQ.2017.10.099.

[29] M. Izadi, H.M. Alshehri, F. Hosseinzadeh, M. Shokri Rad, M.B. Ben Hamida, Numerical study on forced convection heat transfer of TiO_2/water nanofluid flow inside a double-pipe heat exchanger with spindle-shaped turbulators, Eng. Anal Bound. Elem. 150 (2023) 612–623. Available from: https://doi.org/10.1016/J.ENGANABOUND.2023.02.046.

[30] R. Mohebbi, M. Izadi, A. Amiri Delouei, H. Sajjadi, Effect of MWCNT–Fe_3O_4/water hybrid nanofluid on the thermal performance of ribbed channel with apart sections of heating and cooling, J. Therm. Anal. Calorim. 135 (2019) 3029–3042. Available from: https://doi.org/10.1007/S10973-018-7483-5/FIGURES/14.

[31] Y. Ma, R. Mohebbi, M. Sheremet, Heat transfer and fluid flow around starfish-inspired cylinder by immersed boundary-simplified thermal lattice Boltzmann method, Eng. Anal. Bound. Elem. 146 (2023) 656–667. Available from: https://doi.org/10.1016/J.ENGANABOUND.2022.11.017.

[32] P. Soleymani, Y. Ma, E. Saffarifard, R. Mohebbi, M. Babaie, N. Karimi, et al., Numerical investigation on turbulent flow, heat transfer, and entropy generation of water-based magnetic nanofluid flow in a tube with hemisphere porous under a uniform magnetic field, Int. Commun. Heat. Mass. Transf. 137 (2022) 106308. Available from: https://doi.org/10.1016/J.ICHEATMASSTRANSFER.2022.106308.

[33] R. Mohebbi, A.A. Delouei, A. Jamali, M. Izadi, A.A. Mohamad, Pore-scale simulation of non-Newtonian power-law fluid flow and forced convection in partially porous media: thermal lattice Boltzmann method, Phys. A Stat. Mech. Its Appl. 525 (2019) 642–656. Available from: https://doi.org/10.1016/J.PHYSA.2019.03.039.

[34] M. Mahmoodi, Mixed convection inside nanofluid filled rectangular enclosures with moving bottom wall, Therm. Sci. 15 (2011) 889–903. Available from: https://doi.org/10.2298/TSCI101129030M.

[35] B. Ghasemi, S.M. Aminossadati, Mixed convection in a lid-driven triangular enclosure filled with nanofluids, Int. Commun. Heat. Mass. Transf. 37 (2010) 1142–1148. Available from: https://doi.org/10.1016/J.ICHEATMASSTRANSFER.2010.06.020.

[36] M. Izadi, A. Behzadmehr, M.M. Shahmardan, Effects of inclination angle on laminar mixed convection of a nanofluid flowing through an annulus, Chem. Emg. Commun. 202 (2015) 1693–1702. https://doi.org/10.1080/00986445.2014.910770.

[37] M. Izadi, A. Behzadmehr, M.M. Shahmardan, Effects of discrete source-sink arrangements on mixed convection in a square cavity filled by nanofluid, Korean J. Chem. Eng. 31 (2013) 12–19. Available from: https://doi.org/10.1007/S11814-013-0176-7.

[38] M. Izadi, M.M. Shahmardan, M.J. Maghrebi, A. Behzadmehr, Numerical study of developed laminar mixed convection of Al_2O_3/water nanofluid in an annulus, Chem. Eng. Commun. 200 (2013) 878–894. Available from: https://doi.org/10.1080/00986445.2012.723077.

[39] P.U. Ogban, G.F. Naterer, Control volume finite element method for entropy generation minimization in mixed convection of nanofluids, Numer. Heat Transf. B: Fundam. 75 (2019) 363–382. https://doi.org/10.1080/10407790.2019.1627797.

[40] M. Hemmat Esfe, M. Akbari, A. Karimipour, M. Afrand, O. Mahian, S. Wongwises, Mixed-convection flow and heat transfer in an inclined cavity equipped to a hot obstacle using nanofluids considering temperature-dependent properties, Int. J. Heat. Mass. Transf. 85 (2015) 656–666. Available from: https://doi.org/10.1016/J.IJHEATMASSTRANSFER.2015.02.009.

[41] S.M. Sebdani, M. Mahmoodi, S.M. Hashemi, Effect of nanofluid variable properties on mixed convection in a square cavity, Int. J. Therm. Sci. 52 (2012) 112–126. Available from: https://doi.org/10.1016/J.IJTHERMALSCI.2011.09.003.

[42] A. Msaddak, E. Sediki, M. Ben Salah, Assessment of thermal heat loss from solar cavity receiver with lattice Boltzmann method, Sol. Energy 173 (2018) 1115–1125. Available from: https://doi.org/10.1016/j.solener.2018.08.059.

[43] M. Hatami, A.F. Elelamy, D. Jing, Mass transfer modeling in nanofluids: numerical approaches and challenges, Nanofluids Mass. Transf. (2022) 273–296. Available from: https://doi.org/10.1016/B978-0-12-823996-4.00010-0.

[44] S.M. Hussain, W. Jamshed, A comparative entropy based analysis of tangent hyperbolic hybrid nanofluid flow: implementing finite difference method, Int. Commun. Heat. Mass. Transf. 129 (2021) 105671. Available from: https://doi.org/10.1016/J.ICHEATMASSTRANSFER.2021.105671.

[45] M. Irfan, M. Asif Farooq, T. Iqra, A. Mushtaq, Z.H. Shamsi, A simplified finite difference method (SFDM) for EMHD Powell–Eyring nanofluid flow featuring variable thickness surface and variable fluid characteristics, Math. Probl. Eng. 2020 (2020). Available from: https://doi.org/10.1155/2020/8823905.

[46] Z. Boulahia, C. Boulahia, R. Sehaqui, Two-phase computation of free convection and entropy generation inside an enclosure filled by a hybrid Al_2O_3-TiO_2-Cu water nanofluid having a corrugated heat source using the generalized Buongiorno's mathematical model: employment of finite volume method, Mater. Today Proc. 30 (2020) 1056–1067. Available from: https://doi.org/10.1016/J.MATPR.2020.05.523.

[47] N. Muhammad, S. Nadeem, A. Issakhov, Finite volume method for mixed convection flow of Ag–ethylene glycol nanofluid flow in a cavity having thin central heater, Phys. A Stat. Mech. Appl. 537 (2020) 122738. Available from: https://doi.org/10.1016/J.PHYSA.2019.122738.

[48] M. Sheikholeslami, Application of control volume based finite element method (CVFEM) for nanofluid flow and heat transfer, Appl. Control Vol. Based Finite

Elem. Method. Nanofluid Flow. Heat. Transf. (2018) 1−780. Available from: https://doi.org/10.1016/C2017-0-01264-8.

[49] T.K. Nguyen, F.A. Soomro, J.A. Ali, R.U. Haq, M. Sheikholeslami, A. Shafee, Heat transfer of ethylene glycol-Fe_3O_4 nanofluid enclosed by curved porous cavity including electric field, Phys. A Stat. Mech. Appl. 550 (2020) 123945. Available from: https://doi.org/10.1016/J.PHYSA.2019.123945.

[50] M. Izadi, M.A. Sheremet, S.A.M. Mehryan, Natural convection of a hybrid nanofluid affected by an inclined periodic magnetic field within a porous medium, Chin. J. Phys. 65 (2020) 447−458. Available from: https://doi.org/10.1016/J.CJPH.2020.03.006.

[51] F.H. Ali, H.K. Hamzah, K. Egab, M. Arıcı, A. Shahsavar, Non-Newtonian nanofluid natural convection in a U-shaped cavity under magnetic field, Int. J. Mech. Sci. 186 (2020) 105887. Available from: https://doi.org/10.1016/J.IJMECSCI.2020.105887.

[52] A.I. Alsabery, T. Armaghani, A.J. Chamkha, I. Hashim, Two-phase nanofluid model and magnetic field effects on mixed convection in a lid-driven cavity containing heated triangular wall, Alex. Eng. J. 59 (2020) 129−148. Available from: https://doi.org/10.1016/J.AEJ.2019.12.017.

[53] R. Mohebbi, M. Izadi, H. Sajjadi, A.A. Delouei, M.A. Sheremet, Examining of nanofluid natural convection heat transfer in a Γ-shaped enclosure including a rectangular hot obstacle using the lattice Boltzmann method, Phys. A Stat. Mech. Appl. 526 (2019) 120831. Available from: https://doi.org/10.1016/J.PHYSA.2019.04.067.

[54] S.A.M. Mehryan, F.M. Kashkooli, M. Ghalambaz, A.J. Chamkha, Free convection of hybrid Al_2O_3-Cu water nanofluid in a differentially heated porous cavity, Adv. Powder Technol. 28 (2017) 2295−2305. Available from: https://doi.org/10.1016/J.APT.2017.06.011.

[55] R. Mohebbi, M. Izadi, A.J. Chamkha, Heat source location and natural convection in a C-shaped enclosure saturated by a nanofluid, Phys. Fluids. 29 (2017) 122009. Available from: https://doi.org/10.1063/1.4993866.

[56] A.A. Mohamad, Lattice Boltzmann Method: Fundamentals and Engineering Applications with Computer Codes, second ed., 2019, pp. 1−222. Available from: https://doi.org/10.1007/978-1-4471-7423-3.

[57] S. Chen, G.D. Doolen, Lattice Boltzmann method for fluid flows, Annu. Rev. Fluid Mech. 30 (1998) 329−364. Available from: https://doi.org/10.1146/ANNUREV.FLUID.30.1.329.

[58] R. Mohebbi, M. Nazari, M.H. Kayhani, Comparative study of forced convection of a power-law fluid in a channel with a built-in square cylinder, J. Appl. Mech. Tech. Phys. 57 (2016) 55−68. Available from: https://doi.org/10.1134/S0021894416010077/METRICS.

[59] R. Mohebbi, H. Heidari, Lattice Boltzmann simulation of fluid flow and heat transfer in a parallel-plate channel with transverse rectangular cavities, Int. J. Mod. Phys. C 28 (2017). Available from: https://doi.org/10.1142/S0129183117500425.

[60] M. Nazari, R. Mohebbi, M.H. Kayhani, Power-law fluid flow and heat transfer in a channel with a built-in porous square cylinder: lattice Boltzmann simulation, J. Nonnewton. Fluid Mech. 204 (2014) 38−49. Available from: https://doi.org/10.1016/J.JNNFM.2013.12.002.

[61] P. Poureslami, M. Siavashi, H. Moghimi, M. Hosseini, Pore-scale convection-conduction heat transfer and fluid flow in open-cell metal foams: a three-dimensional multiple-relaxation time lattice Boltzmann (MRT-LBM) solution, Int. Commun.

Heat. Mass. Transf. 126 (2021) 105465. Available from: https://doi.org/10.1016/J.ICHEATMASSTRANSFER.2021.105465.

[62] Y. Ma, R. Mohebbi, M.M. Rashidi, Z. Yang, Study of nanofluid forced convection heat transfer in a bent channel by means of lattice Boltzmann method, Phys. Fluids. 30 (2018) 032001. Available from: https://doi.org/10.1063/1.5022060.

[63] H. Sajjadi, A.A. Delouei, M. Atashafrooz, M. Sheikholeslami, Double MRT lattice Boltzmann simulation of 3-D MHD natural convection in a cubic cavity with sinusoidal temperature distribution utilizing nanofluid, Int. J. Heat. Mass. Transf. 126 (2018) 489–503. Available from: https://doi.org/10.1016/J.IJHEATMASSTRANSFER.2018.05.064.

[64] H. Sajjadi, M. Gorji, G.H.R. Kefayati, D.D. Ganji, Lattice Boltzmann simulation of turbulent natural convection in tall enclosures using Cu/water nanofluid, Numer. Heat Transf. B: Fundam. 62 (2012) 512–530. https://doi.org/10.1080/10407782.2012.703054.

[65] Y. Abdollahzadeh, M. Mehrpooya, S.M.A. Mousavian, H. Moqtaderi, Modeling and simulation of nanofluid in low Reynolds numbers using two-phase lattice Boltzmann method based on mixture model, Chem. Eng. Res. Des. 192 (2023) 402–411. Available from: https://doi.org/10.1016/J.CHERD.2023.02.023.

[66] C. Qi, L. Liang, Z. Rao, Study on the flow and heat transfer of liquid metal based nanofluid with different nanoparticle radiuses using two-phase lattice Boltzmann method, Int. J. Heat. Mass. Transf. 94 (2016) 316–326. Available from: https://doi.org/10.1016/J.IJHEATMASSTRANSFER.2015.11.068.

[67] C. Qi, G. Wang, L. Yang, Y. Wan, Z. Rao, Two-phase lattice Boltzmann simulation of the effects of base fluid and nanoparticle size on natural convection heat transfer of nanofluid, Int. J. Heat. Mass. Transf. 105 (2017) 664–672. Available from: https://doi.org/10.1016/J.IJHEATMASSTRANSFER.2016.10.043.

[68] N.A.C. Sidik, M. Khakbaz, L. Jahanshaloo, S. Samion, A.N. Darus, Simulation of forced convection in a channel with nanofluid by the lattice Boltzmann method, Nanoscale Res. Lett. 8 (2013) 178. Available from: https://doi.org/10.1186/1556-276X-8-178.

[69] Y. Ma, M.M. Rashidi, R. Mohebbi, Z. Yang, Nanofluid natural convection in a corrugated solar power plant using the hybrid LBM-TVD method, Energy 199 (2020) 117402. Available from: https://doi.org/10.1016/J.ENERGY.2020.117402.

[70] J. Ren, Z. Jin, X. Huang, S. Belošević, A. Milićević, I. Tomanović, et al., A lattice Boltzmann method for two-phase nanofluid under variable non-uniform magnetic fields, J. Appl. Phys. 132 (2022) 174703. Available from: https://doi.org/10.1063/5.0118137.

[71] G. Kefayati, A.P. Bassom, A lattice Boltzmann method for single- and two-phase models of nanofluids: Newtonian and non-Newtonian nanofluids, Phys. Fluids. 33 (2021) 102008. Available from: https://doi.org/10.1063/5.0067744.

[72] Y. Peng, C. Shu, Y.T. Chew, Simplified thermal lattice Boltzmann model for incompressible thermal flows, Phys. Rev. E. 68 (2003) 026701. Available from: https://doi.org/10.1103/PhysRevE.68.026701.

[73] P.L. Bhatnagar, E.P. Gross, M. Krook, A model for collision processes in gases. I. Small amplitude processes in charged and neutral one-component systems, Phys. Rev. 94 (1954) 511. Available from: https://doi.org/10.1103/PhysRev.94.511.

[74] A.A. Mohamad, A. Kuzmin, A critical evaluation of force term in lattice Boltzmann method, natural convection problem, Int. J. Heat. Mass. Transf. 53 (2010) 990–996. Available from: https://doi.org/10.1016/j.ijheatmasstransfer.2009.11.014.

[75] Z. Guo, B. Shi, N. Wang, Lattice BGK model for incompressible Navier–Stokes equation, J. Comput. Phys. 165 (2000) 288–306. Available from: https://doi.org/10.1006/JCPH.2000.6616.

[76] Y.Y. Yan, Y.Q. Zu, Numerical simulation of heat transfer and fluid flow past a rotating isothermal cylinder—a LBM approach, Int. J. Heat. Mass. Transf. 51 (2008) 2519–2536. Available from: https://doi.org/10.1016/J.IJHEATMASSTRANSFER.2007.07.053.

[77] A. Einstein, A.D. Cowper, Investigations on the Theory of the Brownian Movement, translated by R. Fürth, n.d.

[78] H.E. Patel, T. Sundararajan, T. Pradeep, A. Dasgupta, N. Dasgupta, S.K. Das, A micro-convection model for thermal conductivity of nanofluids, Pramana—J. Phys. 65 (2005) 863–869. Available from: https://doi.org/10.1007/BF02704086.

[79] J. Koo, C. Kleinstreuer, Viscous dissipation effects in microtubes and microchannels, Int. J. Heat. Mass. Transf. 47 (2004) 3159–3169. Available from: https://doi.org/10.1016/j.ijheatmasstransfer.2004.02.017.

[80] A. Einstein, Investigations on the Theory of the Brownian Movement, Courier Corporation, 1956.

[81] J. Li, Computational Analysis of Nanofluid Flow in Microchannels with Applications to Micro-Heat Sinks and Bio-MEMS, NC State University, Raleigh, NC, 2008.

[82] J. Koo, C. Kleinstreuer, Laminar nanofluid flow in microheat-sinks, Int. J. Heat. Mass. Transf. 48 (2005) 2652–2661. Available from: https://doi.org/10.1016/j.ijheatmasstransfer.2005.01.029.

[83] J. Buongiorno, Convective transport in nanofluids, J. Heat. Transfer. 128 (2006) 240–250. Available from: https://doi.org/10.1115/1.2150834.

[84] W.M.H.K. Versteeg, Introduction to Computational Fluid Dynamics, An: The Finite Volume Method, second ed., Pearson Education, 2007.

[85] D. Kuzmin, On the design of general-purpose flux limiters for finite element schemes. I. Scalar convection, J. Comput. Phys. 219 (2006) 513–531. Available from: https://doi.org/10.1016/j.jcp.2006.03.034.

[86] M. Corcione, Empirical correlating equations for predicting the effective thermal conductivity and dynamic viscosity of nanofluids, Energy Convers. Manage. 52 (2011) 789–793. Available from: https://doi.org/10.1016/j.enconman.2010.06.072.

[87] M. El Maghraoui, R. Mohebbi, M. Sheremet, M. Hazami, Simulation of conductive fins effect on mixed convection heat transfer in a two-dimensional inclined channel using the lattice Boltzmann method, Numer. Heat. Transf. Part A Appl. 85 (2023) 1130–1156. Available from: https://doi.org/10.1080/10407782.2023.2197266.

[88] G.H.R. Kefayati, Natural convection of ferrofluid in a linearly heated cavity utilizing LBM, J. Mol. Liq. 191 (2014) 1–9. Available from: https://doi.org/10.1016/J.MOLLIQ.2013.11.021.

[89] R. Mohebbi, Y. Ma, Optimal configuration of discrete heat sources in a channel with sudden expansion and contraction by lattice Boltzmann method, J. Therm. Anal. Calorim. 148 (2023) 4553–4566. Available from: https://doi.org/10.1007/S10973-023-12020-8/FIGURES/12.

[90] Y. Ma, R. Mohebbi, Z. Yang, M. Sheremet, Thermal performance of an environmentally friendly nanoliquid in a cabinet with two I-shaped heaters: application for electronic cooling, Int. J. Numer. Methods Heat. Fluid Flow. 32 (2022) 3637–3656. Available from: https://doi.org/10.1108/HFF-10-2021-0656.

[91] R. Mei, D. Yu, Force evaluation in the lattice Boltzmann method involving curved geometry, Phys. Rev. E, 65 (2002) 041203.

[92] Z. Guo, B. Shi, C. Zheng, A coupled lattice BGK model for the Boussinesq equations, Int. J. Numer. Methods Fluids. 39 (2002) 325–342. Available from: https://doi.org/10.1002/fld.337.

[93] M.A. Abchouyeh, O.S. Fard, R. Mohebbi, M.A. Sheremet, Enhancement of heat transfer of nanofluids in the presence of sinusoidal side obstacles between two parallel plates through the lattice Boltzmann method, Int. J. Mech. Sci. 156 (2019) 159–169. Available from: https://doi.org/10.1016/j.ijmecsci.2019.03.035.

[94] Z. Chen, C. Shu, D. Tan, Immersed boundary-simplified lattice Boltzmann method for incompressible viscous flows, Phys. Fluids 30 (2018). Available from: https://doi.org/10.1063/1.5028353/907872.

[95] A.A. Mohamad, Heat transfer enhancements in heat exchangers fitted with porous media. Part I: Constant wall temperature, Int. J. Therm. Sci. 42 (2003) 385–395. Available from: https://doi.org/10.1016/S1290-0729(02)00039-X.

[96] H.R. Ashorynejad, A.A. Mohamad, M. Sheikholeslami, Magnetic field effects on natural convection flow of a nanofluid in a horizontal cylindrical annulus using lattice Boltzmann method, Int. J. Therm. Sci. 64 (2013) 240–250. Available from: https://doi.org/10.1016/j.ijthermalsci.2012.08.006.

[97] A.A. Mohamad, M. El-Ganaoui, R. Bennacer, Lattice Boltzmann simulation of natural convection in an open ended cavity, Int. J. Therm. Sci. 48 (2009) 1870–1875. Available from: https://doi.org/10.1016/j.ijthermalsci.2009.02.004.

[98] S. Karimnejad, A. Amiri Delouei, M. Nazari, M.M. Shahmardan, A.A. Mohamad, Sedimentation of elliptical particles using immersed boundary – lattice Boltzmann method: a complementary repulsive force model, J. Mol. Liq. 262 (2018) 180–193. Available from: https://doi.org/10.1016/j.molliq.2018.04.075.

[99] Y. Ma, R. Mohebbi, M.M. Rashidi, Z. Yang, M.A. Sheremet, Numerical study of MHD nanofluid natural convection in a baffled U-shaped enclosure, Int. J. Heat. Mass. Transf. 130 (2019) 123–134. Available from: https://doi.org/10.1016/j.ijheatmasstransfer.2018.10.072.

[100] A. Matori, R. Mohebbi, Z. Hashemi, Y. Ma, Lattice Boltzmann study of multi-walled carbon nanotube (MWCNT)-Fe_3O_4/water hybrid nanofluids natural convection heat transfer in a Π-shaped cavity equipped by hot obstacle, J. Therm. Anal. Calorim. 136 (2019) 2495–2508. Available from: https://doi.org/10.1007/s10973-018-7881-8.

[101] Y. Ma, R. Mohebbi, M.M. Rashidi, Z. Yang, Koo–Kleinstreuer–Li correlation for simulation of nanofluid natural convection in hollow cavity in existence of magnetic field, J. Therm. Anal. Calorim. 137 (2019) 1413–1429. Available from: https://doi.org/10.1007/s10973-019-08038-6.

[102] R. Mohebbi, H. lakzayi, H. Rasam, Numerical simulation of conjugate heat transfer in a square cavity consisting the conducting partitions by utilizing lattice Boltzmann method, Phys. A Stat. Mech. Appl. 546 (2020) 123050. Available from: https://doi.org/10.1016/J.PHYSA.2019.123050.

[103] Y. Ma, R. Mohebbi, M.M. Rashidi, Z. Yang, M. Sheremet, Nanoliquid thermal convection in I-shaped multiple-pipe heat exchanger under magnetic field influence, Phys. A Stat. Mech. Appl. 550 (2020) 124028. Available from: https://doi.org/10.1016/j.physa.2019.124028.

[104] Y. Ma, R. Mohebbi, Hydro-thermal performance on the two phase nanofluid convection heat transfer using the LBM-FD method, Waves Random Complex. Media (2022) 1–16. Available from: https://doi.org/10.1080/17455030.2022.2089371.

[105] Q. Zou, X. He, On pressure and velocity boundary conditions for the lattice Boltzmann BGK model, Phys. Fluids. 9 (1998) 1591. Available from: https://doi.org/10.1063/1.869307.

[106] S. Succi, Applied lattice Boltzmann method for transport phenomena, momentum, heat and mass transfer. A. A. Mohamad Sure Printing, Calgary, AB April 2007, Can. J. Chem. Eng. 85 (2008) 946–947. Available from: https://doi.org/10.1002/cjce.5450850617.

[107] M. Izadi, R. Mohebbi, A. Chamkha, I. Pop, Effects of cavity and heat source aspect ratios on natural convection of a nanofluid in a C-shaped cavity using lattice Boltzmann method, Int. J. Numer. Methods Heat. Fluid Flow. 28 (2018) 1930–1955. Available from: https://doi.org/10.1108/HFF-03-2018-0110.

CHAPTER

Parametric influences on nanofluid-jet cooling heat transfer

12

Emmanuel O. Atofarati[1], Sharifpur Mohsen[1,2,3] and Josua P. Meyer[1,4]

[1]*Department of Mechanical and Aeronautical Engineering, University of Pretoria, Hatfield, Gauteng, South Africa*

[2]*School of Mechanical, Industrial and Aeronautical Engineering, University of the Witwatersrand, Gauteng, South Africa*

[3]*Department of Medical Research, China Medical University Hospital, China Medical University Taichung, Taichung, Heilongjiang, Taiwan*

[4]*Department of Mechanical and Mechatronic Engineering, Stellenbosch University, Stellenbosch, Western Cape, South Africa*

12.1 General Introduction

As the need for optimized energy consumption increases in cooling systems, researchers have been on the move to discover novel approaches to diminish this challenge. Several means of heat transfer have been explored, and in some cases, the combination of two or more of these methods has been utilized. The methods include active and passive methods. Active methods require additional energy for heat transfer; examples are forced convection, vortex generation, pulsation, and jet impingement. Passive methods do not need additional energy sources for their operations; examples are porous media [1,2], macro- and micro-channels [3,4], extended surfaces (corrugations, pin-fins, dimples, and ribbed walls) [5,6], natural convection techniques [7,8], and multiphase cooling [9].

Jet impingement is one of the renowned effective methods of heat transfer. It has been applied to both heating and cooling processes. In this review, we will be focusing on jet impingement cooling (JIC). According to Chupp et al. [10], cooling with impinged jet gained enormous research interest in the mid-20th century after World War II when reciprocating prime movers were replaced with gas turbines. Metzger [11] and Walz [12], in their technical report titled "spot cooling and heating of surfaces with high-velocity impinging air jets," presented a pioneering report on JIC using slot jets and circular jets, respectively. Their study experimentally measures the heat transfer coefficient of high-speed air jet impinged on the surface of flat and curved copper targets. Further research by Metzger et al. [13] presented the optimum interjet spacing for an array of inline jets used for cooling concave surfaces. Since then, JIC has remained the most

Nanofluids. DOI: https://doi.org/10.1016/B978-0-443-13625-2.00014-0

effective means of cooling gas turbine aerofoil-blades leading edge, vane, and mid-chord. The turbine leading is usually approximated to a concave surface, while the vane and the mid-chord are considered a flat surface [14].

Nanofluids are specially engineered suspensions formed from the dispersion of nanoparticles (nanoscaled particles [$<$ 100 mm] of metals, metal-oxides, or nonmetals) into conventional fluids. Generally, nanofluids have been proven to have good thermophysical properties required for effective heat transfer either for cooling or heating processes [15–17]. Nanofluids constitute a combination of regular working fluids and nanoscaled materials, resulting in a two-phase fluid. This amalgamation grants nanofluids remarkable thermophysical properties crucial for augmenting heat transfer. The initial introduction of the concept of nanofluids and their industrial applications can be attributed to Choi and Eastman [18]. They reported that nanofluids possess higher thermal conductivity when matched to conventional heat transfer fluids. The enhanced thermal conductivity of nanofluids was also foreseen to facilitate the future design of more compact heat exchangers. Generally, nanofluids are classified as non-Newtonian fluids, with some being shear-thinning fluids and others shear-thickening fluids. This is due to dependence of their viscosity on the shear rate at a constant temperature. However, some nanofluids behave as Newtonian fluid at low shear rate, low concentration of the nanoparticle, and certain temperature of the nanofluids. Also, nanofluids are practically two-phase fluids (liquid-solid or gaseous-solid); however, in most numerical studies, they are often treated as single-phase fluids [19–22]. The special rheological and thermophysical properties of nanofluids make their study broad and unlimited in scope and application.

Nanofluid-jet impingement cooling (NJIC) is a novel method of impingement cooling that utilizes nanofluids as a coolant for JIC. Introducing this special thermal transport medium "nanofluids" into systems utilizing jet impingement for cooling has consistently improved the heat transfer rate in such thermal systems. This review examines the various parameters that impact the heat transfer effectiveness of cooling through nanofluid-jet impingement, explores its industrial applications, and introduces recent correlations for performance prediction. Section 12.2 presents a brief bibliographic analysis of NJIC, Section 12.3 presents the preparation of nanofluids and the parameters that influence its performance as a heat transfer fluid, Section 12.4 presents a general overview of JIC, and Section 12.5 amplifies the NJIC and the parameters affecting their performance.

12.2 Brief bibliographic overview of nanofluid-jet impingement cooling

The bibliographic data presented in Fig. 12.1. shows the paper population and prolific authors in NJIC. The data were extracted from the SCOPUS database on 20th January, 2023. According to SCOPUS, there are 1993 published documents on the subject area of "NJIC," 2620 published documents on the subject area of "JIC," and 91 published documents on the subject area of "Pulsating Jet Impingement

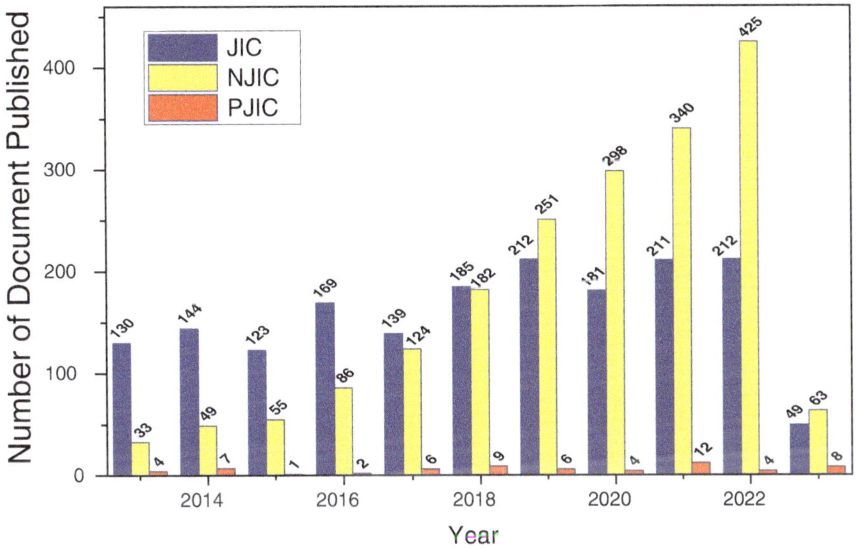

(A)

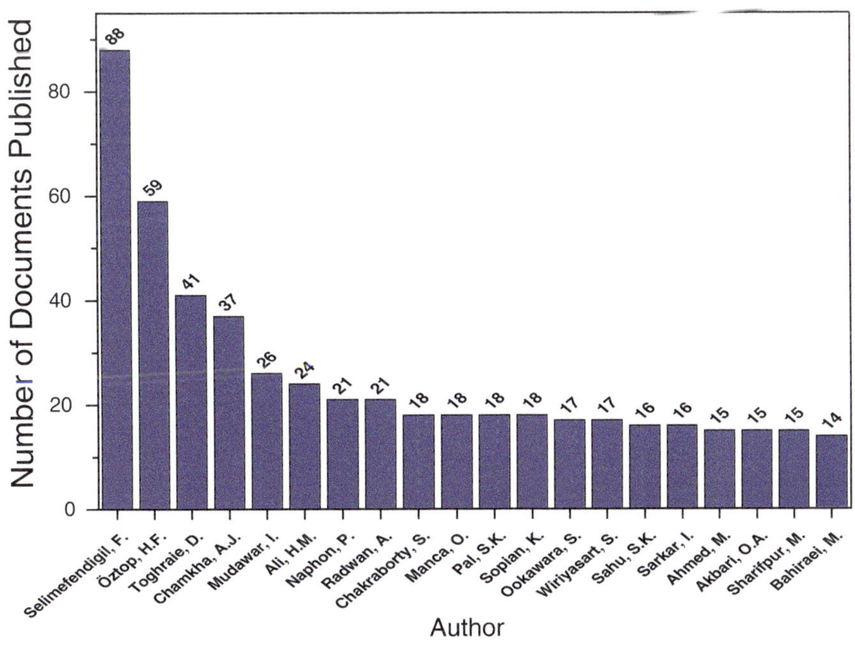

(B)

FIGURE 12.1

Trends of nanofluids jet impingement cooling: (A) publication population and (B) 20 most prolific authors.

Cooling (PJIC)" from 2002 till 2013. Among these, there are 1588 journal articles, 227 review papers, 122 conference papers, 31 book chapters, 17 books, and 8 other types of documents. The trend of research in NJIC has spontaneously been growing since 2013 till date as evident in Fig. 12.1A. An average of 25% increase in the publication per year on NJIC has been recorded since 2013. Most studies in this subject area concluded that nanofluid jets have better heat transfer enhancement than conventional fluid jets. Ravikumar et al. [23] investigated the thermal performance of Al_2O_3-H_2O nanofluid used as a coolant in high heat-load systems. They discussed the effectiveness of this nanofluid with and without surfactants (using SDS and Tween 20) and compared their performance with water. They concluded that the Al_2O_3-Tween 20, Al_2O_3-SDS, and Al_2O_3 nanofluids had about 32%, 19%, and 10% enhancement in thermal performance compared to water. This was validated in the research conducted by Sun et al. [24], as they reported a 120.53% increase in heat transfer coefficient for swirling jets and a 116.67% enhancement for conventional jets using 0.05% (wt.) of Ag/MWCNT hybrid nanofluids. Further discussion on nanofluid and its hybrid's heat transfer enhancement is discussed in the comprehensive review conducted by Aksoy et al. [25], Muneeshwaran et al. [15], Liang and Mudawar [26], and Tyagi et al. [27].

The first 20 most prolific and relevant authors in the NJIC subject area are shown in Fig. 12.1B. Selimefendigil and Oztop are the most remarkable authors, with most of their works being numerical studies. Selimefendigil and Oztop had simulated several complex jet impingement scenarios that have not been experimented with, that is, "Pulsating nanofluids JIC of a heated horizontal surface." This work is their most cited document on the subject, with 106 citations, making them one of the most relevant authors on the subject.

12.3 Nanofluids and hybrid nanofluids for jet impingement cooling

Nanofluids had been classified based on the number of nanoparticles: single-nanofluids, binary-nanofluids, or ternary-nanofluids hybrid. Single-nanofluids are nanofluids of an individual nanoparticle and a base fluid. The utilization of single-nanofluids encounters significant hurdles due to their inherent limitations, characterized by their less stability and elevated viscosity. Also, nanofluids have been classified based on the type of nanoparticle molecules, namely, metallic, metallic oxide, nonmetallic oxide nanofluids, and carbon-based nanofluids [28]. In this section, we will be considering the preparation of nanofluids, parameters influencing their stability and thermophysical properties, and their application.

12.3.1 Preparation of nanofluids and hybrid nanofluids

The preparation of nanofluids can be accomplished through two fundamental techniques, namely, one-step and two-step techniques.

12.3.1.1 One-step technique

The one-step nanofluids production technique involves direct preparation without any transitional steps such as nanoparticle dehydration, storage, and dispersion in fluid [29]. The one-step technique of nanofluid preparation can be by chemical method or by physical method. The chemical methods are sol-gel [30,31], chemi cal reduction [32], microemulsion [33], radiolytic process [34], and polyol [35]. At the same time, the physical methods are laser ablation [36,37], dual plasma synthesis, spray dying process, pulse wire evaporation [38,39], vacuum-based submerged arc [40], and magnetron sputter deposition [41].

12.3.1.2 Two-step technique

The two-step techniques involve ultrasonication and chemical vapor deposition, solvothermal and mechanical processes (milling or stirring, ball milling). The one-step techniques have been proven to produce more stable nanofluids compared to the two-step techniques. However, they are not suitable for large-scale production. Most studies on NJIC synthesize their nanofluids using the two-step techniques, as large volumes with different concentrations and compositions were required [42]. A diagrammatic analogy of the most common two-step technique (chemical vapor deposition) is shown in Fig. 12.2. In this technique, the appropriate weight of the nanoparticle is measured using the chemical scale, and the nanoparticle is added to the equivalent volume of the base fluid. The magnetic stirrer is used to mix them together with the optimized quantity of dispersant. The final mixture is then sonicated using an ultrasonic sonicator or homogenizer.

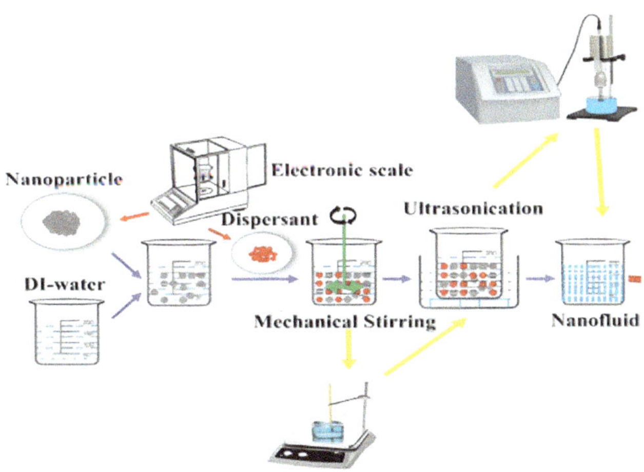

FIGURE 12.2

Fundamental procedure of nanofluid synthesis.

12.3.2 Parametric factors influencing nanofluids stability and thermophysical properties

The suspension nanoparticles have engineered nanofluids in a base fluid that can either be stable or unstable. Stable nanofluids are nanofluids whose nanoparticles are well dispersed with approximately no agglomeration tendency after their preparation [42]. Several parametric factors affect nanofluid's stability and directly affect their thermophysical properties. Some of these parameters include surfactant presence, temperature, pH of the nanoparticles, nanoparticle shape, size, concentration, and the method of preparation.

12.3.2.1 Influence of surfactants

Surfactants or dispersants are chemical agents that promote uniform nanoparticle distribution in the base fluid during nanofluid synthesis. Surfactants can be categorized into anionic, nonanionic, cationic, and amphoteric surfactants. The common surfactants used in nanofluid studies include sodium alkyl benzene sulfonate (SDBS), sodium lauryl sulfate (SLS or SDS), polyvidone (PVP), lauryl betaine (LB), cetrimonium bromide (CTAB), gum arabic (GA), oleic acid (OA), tweens, spans, and oleylamine. Generally, surfactants are amphiphilic particles with features of a hydrophilic head that exhibits an affinity for water molecules, along with a hydrophobic tail that repels water molecules. This unique dual structure allows such molecules to interact effectively with both aqueous and nonaqueous environments. They prevent the agglomeration, coagulation, or sedimentation of the nanoparticle in the base fluid, and optimum quantity increases fluid stability. It has been experimentally proven that increased surfactant concentration in nanofluids usually typically leads to a decrease in the thermal conductivity of the nanofluid and elevates its viscosity.

The addition of an optimized concentration of surfactant (dispersant) to nanofluids has been to enhance its heat transfer capability, as it increases stability, improves thermal conductivity, and reduces agglomeration or coagulation of the nanoparticles for a long period [43–48]. Borode et al. [47] examined the impact of various surfactants on certain thermophysical properties of nanofluids containing graphene nanoplatelets. In their work, the influence of Tween-80, sodium alkyl benzene sulfonate (SDBS), gum arabic (GA), and sodium lauryl sulfate (SDS) surfactants were considered at different mixing ratios. The result shows that the thermal and electrical conductivity of the nanofluid was improved at elevated temperatures with the surfactants. Tween-80, as a surfactant, showed the best thermal conductivity with increased viscosity. Most studies on NJIC prepare their nanofluids without surfactant, while others use it and report its effect on thermal enhancement. The study conducted by Tie et al. [49] revealed that the addition of SDBS surfactant in Cu-water nanofluid reduced the heat transfer effectiveness of the nanofluid during impingement cooling. Their study further concluded that the heat transferred had a reverse correlation to the concentration of SDBS in the nanofluid. Generally, the ratio of the weight of the surfactant to the weight of the nanoparticle used must defined using the dispersion factor and the appropriate surfactant should be selected for the respective nanofluid.

12.3.2.2 Influence of nanofluids pH

The implication of pH to the zeta potential of nanofluids has been extensively studied in the literature [50–52]. Nanofluids' pH value is the level of their acidity or basicity. The pH value at which a nanosuspension has a neutral net charge, which corresponds to the zeta potential tending to zero, is called the isoelectric point (IEP) [42]. Wang et al. [52] explained that high stability can be attained by increasing surface charge density (increasing the pH away from the IEP), increasing the interactive repulsive energy, and improving stability. The pH value of nanofluids depends on the chemical properties of the base fluid and the suspended nanoparticle. Also, Huang et al.[53] and Giwa et al. [54] investigated the impact of pH on nanofluid stability. They reported that the increased pH value of nanofluid increases the zeta potential, which directly enhances the nanofluid's stability and then boosts its thermophysical properties.

12.3.2.3 Influence of sonicating temperature

Adam et al. [55] measured the stability of SiO_2-H_2O nanosuspension by investigating the influence of sonication temperature. Evidence suggests that increasing sonicating temperature increased the nanoparticles' agglomeration, thereby reducing the nanofluid's stability. Chang et al. [56] also confirm this proof while examining temperature's effect on CuO nanofluid. Their findings confirmed that temperature accelerates the aggregation of nanoparticles dispersed within the base fluid, leading to a decrease in the stability of the nanofluid. Hence, an increase in temperature will raise the kinetic energy of the molecules of the nanoparticles and base fluid, which in turn promotes aggregation.

12.3.2.4 Influence of nanoparticle size

The stability of a nanofluid depends on the velocity of sedimentation, which is directly proportional to the square of the nanoparticle's radius (Stokes law). To obtain high stability, the sedimentation velocity must be lessened [57]. Hence, for high stability, the nanoparticle size must be smaller, resulting in low sedimentation velocity. From Derjaguin-Landau-Verwey-Overbeek (DLVO) aggregation theory, the stability of nanofluids is a function of the Van der Waal forces (interactive and attractive energy) and the electrical repulsive interaction forces (interactive repulsive energy). Nanoparticles with a repulsive force higher than the Van der Waal force tend to be more stable irrespective of their size [58]. Nithiyanantham et al. [59] studied the influence of nanoparticle size on the stability of SiO_2-Molten salt nanofluids. SiO_2 with particle sizes between 27 and 800 nm was investigated, and they observed that smaller nanoparticle sizes (27 nm) had low stability, while those with large sizes (450 and 800 nm) had higher stability.

12.3.2.5 Influence of nanoparticle shape

The nanoparticle shape effect on stability has been extensively studied in the literature [42,60–65]. Kim et al. [60] experimentally and theoretically explained the influence of particle shape on the stability of Al_2O_3 nanofluids using the DLVO

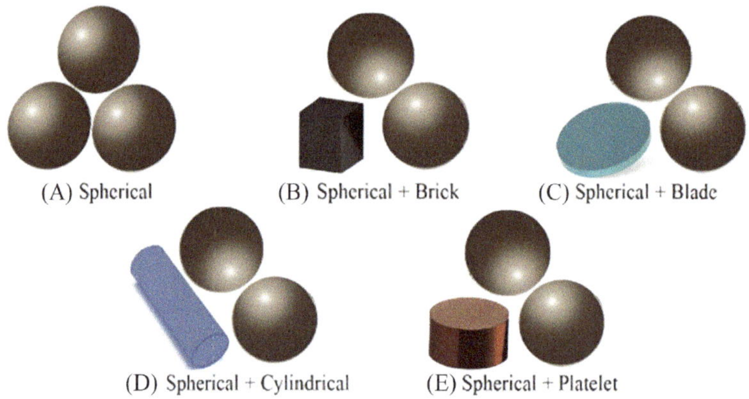

FIGURE 12.3

Some nanoparticles shapes examined in nanofluid-jet impingement cooling studies [62].

Reprinted with permissions from B. Boudraa, R. Bessaïh, Numerical investigation of jet impingement cooling an isothermal surface using extended jet holes with various binary hybrid nanofluids, Int. Commun. Heat Mass Transf. *127 (2021) 105560.*

aggregation theory. The repulsive interaction energy was proportional to the particle's surface area. Hence, different shapes (surface area) of the same nanoparticle will have further stability as the repulsive force will vary. However, some literature suggests that DLVO aggregation theory is only valid for spherical-shaped nanoparticles [42,61,66]. Other studies have stated that different shape complicates the interactive, attractive energy and the interactive repulsive energy prediction as presented by DLVO theory [42,67]. Typical shapes of nanoparticles examined in literature are (as shown in Fig. 12.3) blade, platelet, spherical, oblate spheroid, prolate spheroid, cylinder, hexahedron, laminar, tetrahedron, column, and brick [62,68,69]. Boudraa and Bessaih [62] numerically studied the impact of multishaped hybrid nanoparticles of Al_2O_3 (spherical-shaped) and MgO (platelet, blade, and brick). They recommended the combination of nanofluids of different shapes as it improved thermal performance compared to using only spherical-shaped nanoparticles. Zahmatkesh et al. [64] also highlight that natural and forced convection studies have shown that platelet-like nanoparticles enhance better, while mix-convection studies have demonstrated that the blades-like shape is superior. Also, they proposed that the control of particle shape could be a future means of thermal augmentation.

12.3.2.6 Influence of nanoparticle concentration

Increased nanofluid concentration reduces the mean particle distance. From the DLOV theory, the closer the particle, the more interactive and attractive the energy (Van de Waal force), which leads to agglomeration. Hence, increasing nanoparticle concentration yields reduces the stability of the nanofluid. Sadeghi et al. [70], Mahbubul et al. [71], and Chakraborty et al. [42] studied Al_2O_3 nanofluid, and they observed that stability reduced with particle concentration. The same outcome was observed in the study by Hong et al. [72] while examining the stability of Fe nanofluid.

12.3.2.7 Influence of preparation method

As indicated earlier, nanofluids prepared using the one-step technique are generally more stable than those prepared using the two-step technique. Also, several authors have empirically studied the influence of the stirring temperature and sonication features on the thermophysical properties and stability of nanofluid and its hybrids [73–75]. Mahbubul et al. [76] recommend that longer sonication time and high sonication amplitude for the effective dispersion of the nanoparticle molecules in the base fluid during preparation. However, this as well needs to be optimized for effective thermophysical property results.

12.3.3 Applications of nanofluids

Nanofluids have found application in various fields of study, among them are biomedical, cosmetology, military, chemistry, physics, and engineering. Data analysis from SCOPUS shows that nanofluids have found most of their applications in engineering and related fields. Recently, Xiong et al. [28] summarized the utilization of several nanofluids in solar thermal systems in the review study. Wong and Leon [77] extensively reviewed the prospects of nanofluids and their wide range of applications in biomedicine (nanodrug delivery, cancer therapy, cryosurgery sensing, and imaging) and engineering systems and subsystems (electronics, automobile fuel, and coolant). Fig. 12.4 summarizes the wide-ranging application of

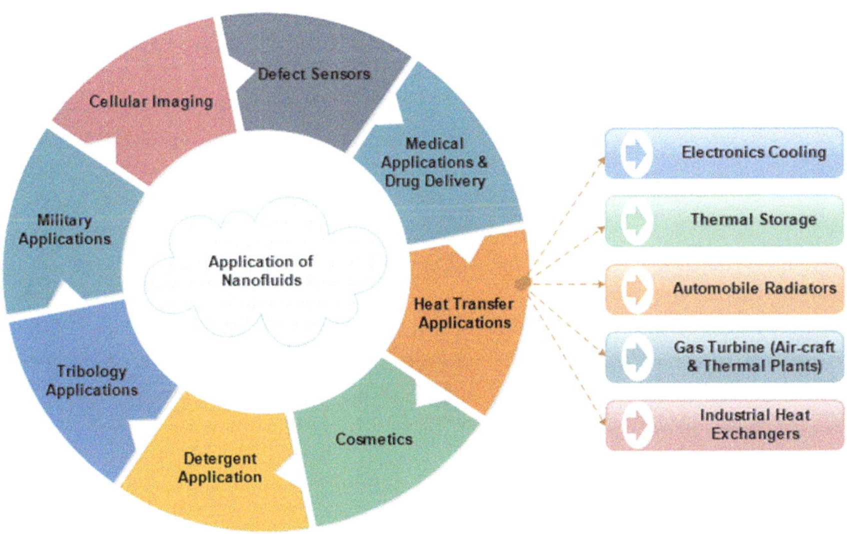

FIGURE 12.4

General application of nanofluids.

nanofluids. One of the fundamental applications of nanofluid in engineering heat transfer systems applies to JIC techniques.

12.4 Nanofluid jet impingement cooling

Generally, JIC stands as a well-established and notably efficient method for facilitating heat transfer. It involves the use of a high-speed fluid jet to extract heat energy from target hot surfaces or bodies. JIC was basically used for turbine blade and metal treatment in the earlier years of its study. However, the resourcefulness of gas turbines in aeronautical and energy applications attracted several researchers to study how they can operate optimally. The fluids used for this cooling process are generally called coolant, heat transfer fluid, heat transfer medium, or thermal transport mediums. In earlier studies, air and water were the conventional heat transfer fluids used for impingement cooling, with research emphasizing the fluid dynamics and boundary layer condition that favors enhancement of heat transfer. The unique fluid dynamics of impinging jet cooling are attributed to factors such as high turbulent intensity, substantial pressure drop, and a thin thermal boundary layer. This boundary layer exists between the fluid jet and the target surface, spanning from the point of maximum impact (stagnation point) to the periphery (edge) of the cooled surface. These characteristics collectively contribute to the improvement of heat transfer within the system. Table 12.1 presents some basic geometrical models used in JIC, while Fig. 12.5 presents a mind map of NJIC studies.

12.4.1 Application of jet impingement cooling

JIC is a significant thermal dissipation technique in domestic and industrial settings. The versatility and effectiveness of the technique have been proved from its wide range of applications in many electronics systems [88–97], heating, ventilation, and air conditioning (HVAC) systems [98–100], steel and paper industry [101–108], automobiles [109–112], and power generating systems [113–119]. Wei et al. [89] empirically investigated the impact of a 3D-shaped water jet impingement cooler on electronic chips of high-power devices. It was observed that the jet impingement cooler with a 4×4 nozzle array had the best heat transfer performance with the least pump power. Kwon et al. [91] considered the cooling performance of high-capacity gallium nitride transistors using an additively manufactured multijet air cooler with different configurations and compared it to the single jet cooler. The report shows that the multijet air cooler was more efficient than the benchmarked single jet coolers, especially during direct cooling. The performance of inline and staggered multiple jet impingement internal cooling of turbine blades under the influence of crossflow was studied by Salem et al. [94]. Their experimental and numerical study revealed that the inline multiple jet arrangement had better heat transfer enhancement than the staggered jet arrangement. This result corresponds to earlier studies by Metzger et al. [120]. Further

Table 12.1 Typical models utilized in jet impingement cooling studies.

S/n	Author(s)	Mtd	Nanofluids	Flow condition	Type/nature of jet	Remark on nozzle behavior	Remarks on nanofluid	Model geometry
1	Boudraa and Bessaih [62]	Num (FVM)	Hybrid Al_2O_3 (spherical) and MgO (platelet, blade, and brick)/ water /0% ≤ φ ≤ 5% (vol)	5000 ≤ Re ≤ 15,000/D = 0.4 mm, H/D = W/D = X/D = 5	Extended circular orifice, inline multijet (15 × 15 array)/ free-surface, confined, normal jet.	Extended jet improved heat transfer by 30%; compared to using water	Improved HT was observed when using multishaped nanoparticles compared to using only spherical-shaped nanoparticles.	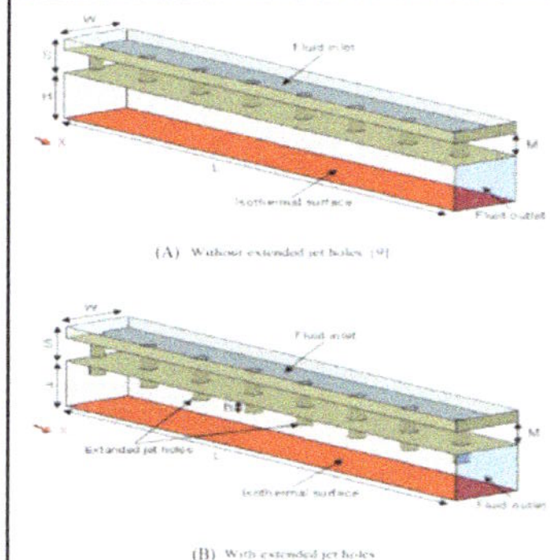
2	Tie, Li and Xuan [49]	Exp	Cu (26 nm)/ water/SDBS (0.05 wt.% to 0.1 wt. %)/0.17% ≤ φ ≤ 0.64% (vol)	4000 ≤ Re ≤ 10,000/D = 1.5 mm, H = 15 mm	Circular orifice, staggered multijet/free-surface, unconfined, normal jet.	N/A	"h" was enhanced by 6.8%, while using the nanofluid surfactant reduced the HT.	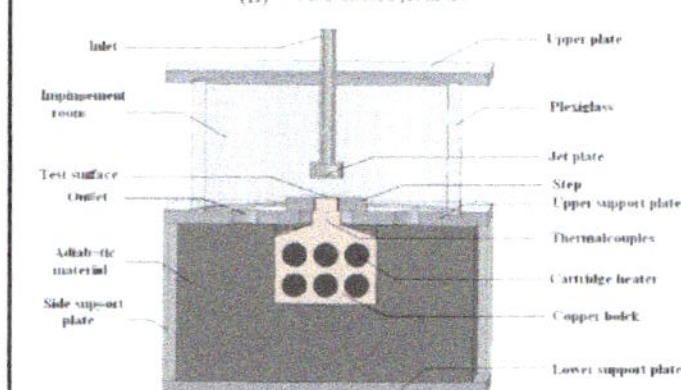
3	Selimefendigil and Öztop [78]	Num (FVM)	Al_2O_3/water/ 0% ≤ φ ≤ 6% (vol)	100 ≤ Re ≤ 400/H/W = 4, Pulsating freq (f) 1 < f < 10 (Hz)	Rectangular single-slot jet/ free-surface, confined, normal-pulsating jet.	Nu at the stagnation point increased with increasing pulsating frequency.	18.8% enhancement was obtained for nanofluids with 6 vol.%.	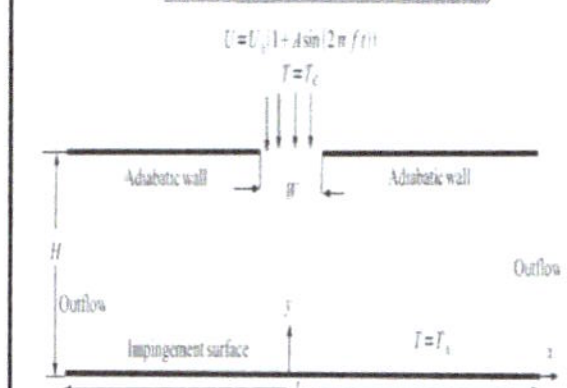

(Continued)

Table 12.1 Typical models utilized in jet impingement cooling studies. *Continued*

S/n	Author(s)	Mtd	Nanofluids	Flow condition	Type/nature of jet	Remark on nozzle behavior	Remarks on nanofluid	Model geometry
4	Santosh et al. [79]	Exp	Al_2O_3, TiO_2 (20 nm)/water/ $0.01\% \leq \varphi \leq 0.07\%$ (wt)	$P_{Jet} = 0.1863$ and 0.245 bar/D = 2 mm, $120 < H < 240$;X/D = 20	Circular orifice, inline double-jet/free-surface, unconfined, normal jet	Increasing the H/D reduced the heat transfer enhancement and the cooling rate.	Al_2O_3/water had the best thermal performance compared to TiO_2/ water and D.I. water for cooling	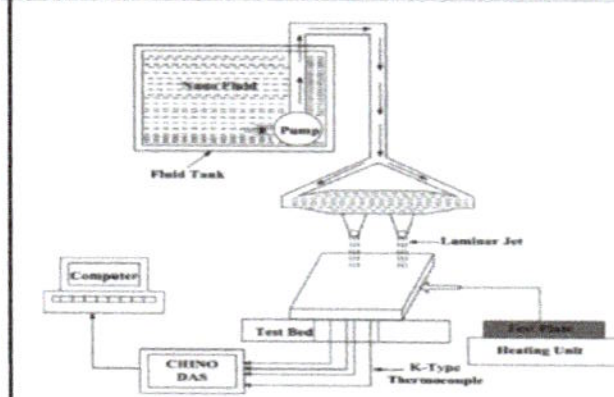
5	Zhang et al. [80]	Num	NEPCM/ Polyalphaolefin	V = 2, 3 and 6 m/s/W = 0.1 mm, H/W = 3, X/D = 4	Rectangular inline slot multijet/ submerged, confined, and normal jet.	N/A	"h" oscillated in a normal distribution pattern with increasing NEPCM concentration.	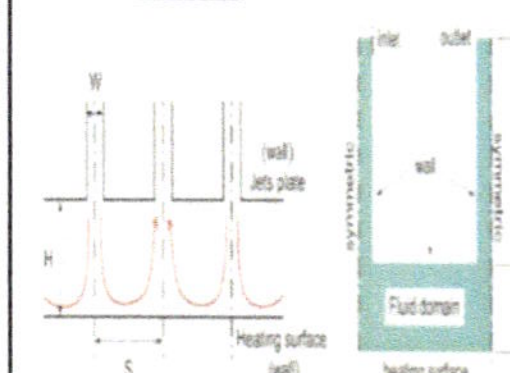
6	Sun, Qu and Yang [81]	Exp	Cu/water/SDBS/ $0.1\% \leq \varphi \leq 0.5\%$ (wt)	$2000 < Re < 7000$/D = 1, 3 and 6 mm; $1 \leq H/D \leq 2$	Circular orifice, single or inclined free-surface jet, unconfined, normal jet.	H/D = 2 @ D = 3 mm and jet angle of 90 degrees was the optimum jet parameter relative to best cooling effect.	HT increased with φ of nanofluid at fixed flow conditions	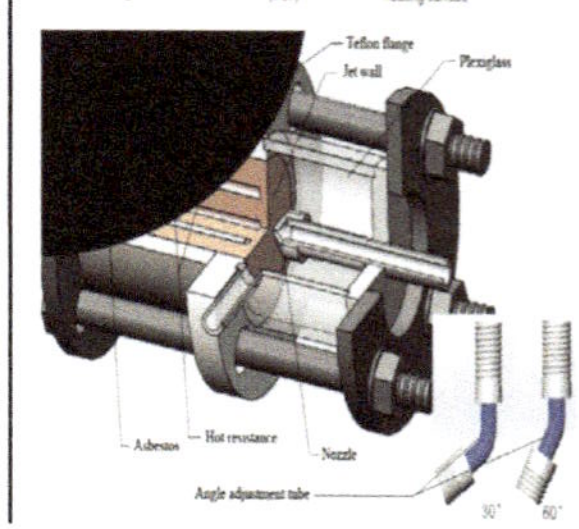

7	Pratap et al. [82]	Num (FVM) and Exp	Al_2O_3/water/3% (vol)	Re = 20,000/D = 2 mm, 2 ≤ H/D ≤ 10	Circular orifice single jet/free-surface, confined, normal jet.	Nu at the stagnation zone increased with an increase in H/D at fixed Re.	Nanofluids improved thermal performance by 8.03% @3% (vol) compared to water.	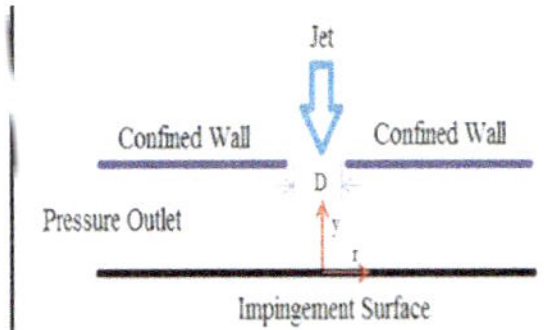
8	Datta and Halder [83]	Num (FVM)	Hybrid Ag–Al_2O_3/ SiC–Al_2O_3/water/ 1% ≤ φ ≤ 2% (vol)	10,000 ≤ Re ≤ 30,000/2 ≤ H/D ≤ 6	Circular orifice, single and straight jet/free-surface, confined, normal jet.	As H/D decreases, heat transfer improves but the friction factor rises; H/D = 4 is the optimal jet height.	42% and17% increase in HTC and pumping power was observed resp. For the 0.5% conc. Of Al_2O_3-SiC hybrid nanofluid.	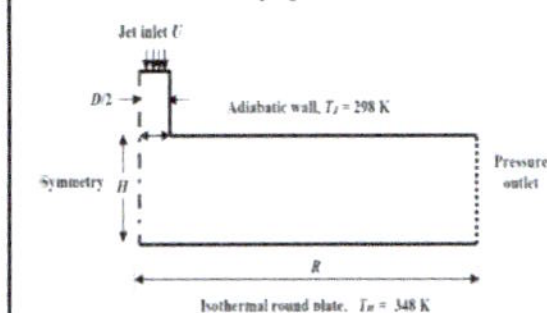
9	Li, Xuan and Yu [84]	Exp	Cu (25 and 100 nm)/water/ 1.5% ≤ φ ≤ 3.0% (vol)	(2000 < Re < 16,000)/D = 2 mm, 2 < H/D < 6	Circular orifice, single and straight jet/ submerged, confined and normal jet.	N/A	The 25 nm nanofluid outperformed the 100 nm nanofluid, enhancing the base fluid's "h."	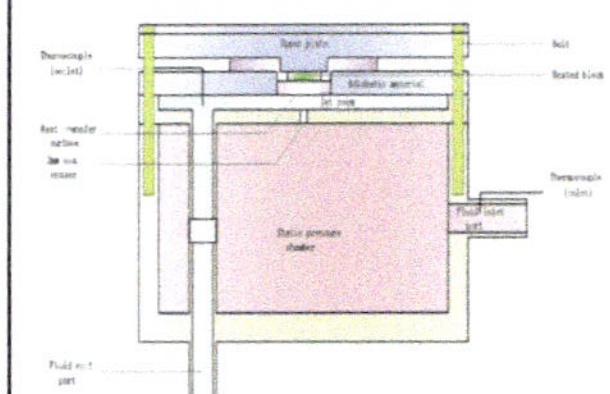
10	Nimmagadda et al. [85]	Num (FVM)	Single and Hybird {Cu, AL, TiO_2, and (Cu + TiO_2)} (20–80) nm/water/ 1% ≤ φ ≤ 3% (vol)	200 ≤ Re ≤ 600/H/W = 4	Single rectangular slot jet/free-surface, confined, normal jet with crossflow, under magnetic field (MF) turbulent promoter.	Crossflow effect promotes higher Nu_{avg} than direct jet impingement. Nonuniform MF was insignificant for the transverse/ crossflow impingement.	At 3 vol.% TiO_2, Cu, AL, and CU + TiO_2 with magnetic field (Ha = 40), similar Nu_{avg} values resulted from consistent thermo-hydrodynamic conditions.	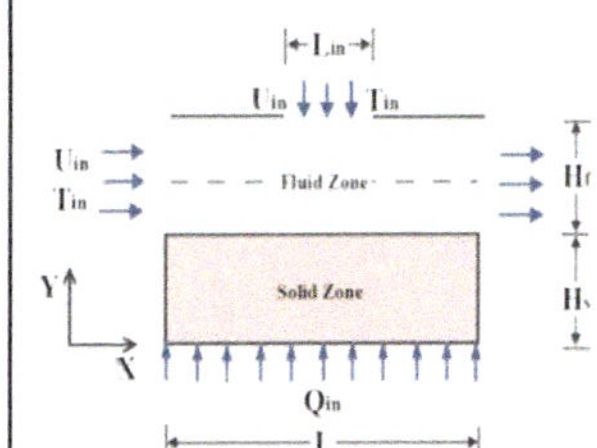

(*Continued*)

Table 12.1 Typical models utilized in jet impingement cooling studies. *Continued*

S/n	Author(s)	Mtd	Nanofluids	Flow condition	Type/nature of jet	Remark on nozzle behavior	Remarks on nanofluid	Model geometry
11	Naphon et al. [86]	Exp	TiO_2 (21 nm)/ water/ $0.005\% \leq \varphi \leq 0.015\%$ (vol)	8.75–11.75 g/s/ $0.5 \leq D \leq 1.5$ mm, $1.0 \leq H \leq 2.0$ mm	Circular orifice single jet.	As D increases and H reduces the heat transfer rate increases.	Heat transfer rate is by 18.56% increased at ($\varphi = 0.015\%$).	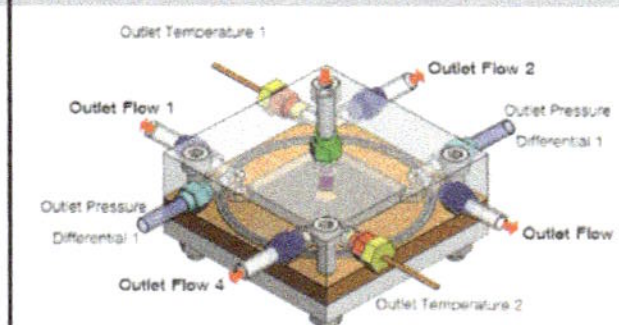
12	Peng et al. [87]	Num	Cu_2O (25 nm)/ water/ $1.5\% \leq \varphi \leq 3.0\%$ (vol)	$2000 \leq Re \leq 16{,}000$	Circular orifice single jet/free-surface, confined normal jet.	N/A	Multiphase models with dispersed nanoparticles improved heat transfer more than single-phase models.	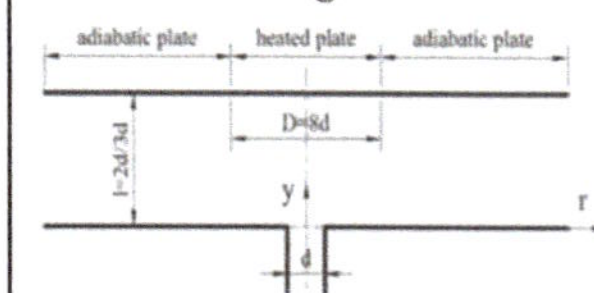

Table 12.1.1 Reprinted with permissions from B. Boudraa, R. Bessaïh, Numerical investigation of jet impingement cooling an isothermal surface using extended jet holes with various binary hybrid nanofluids, Int. Commun. Heat Mass Transf. *127 (2021) 105560. Table 12.1.2 Reprinted with permissions from P. Tie, Q. Li, Y. Xuan, Heat transfer performance of Cu–water nanofluids in the jet arrays impingement cooling system,* Int. J. Therm. Sci. *77 (2014) 199–205. Table 12.1.3 Reprinted with permissions from F. Selimefendigil, H.F. Öztop, Pulsating nanofluids jet impingement cooling of a heated horizontal surface,* Int. J. Heat Mass Transf. *69 (2014) 54–65. Table 12.1.4 Reprinted from K. Santosh, P. Nayak, M. Chandra, S.K.S. Parashar, S. Kumar Nayak, C. Mishra, Enhancement of heat transfer by water–Al_2O_3 and water–TiO_2 nanofluids jet impingement in cooling hot steel surface, 11(16) (2016) 1253–1273, (Open Access). Table 12.1.5 Reprinted with permissions from C.Y. Zhang, T. Wang, D.H. Chen, F.J. Hong, P. Cheng, Confined jet array impingement cooling with spent flow distraction using NEPCM slurry,* Int. Commun. Heat Mass Transf. *77 (2016) 140–147. Table 12.1.6 Reprinted with permissions from B. Sun, Y. Qu, D. Yang, Heat transfer of single impinging jet with Cu nanofluids,* Appl. Therm. Eng. *102 (2016) 701–707. Table 12.1.7 Reprinted with permissions from A. Pratap, Y.K. Baghel, V.K. Patel, Effect of impingement height on the enhancement of heat transfer with circular confined jet impingement using nanofluids,* Mater. Today Proc. *28 (2020) 1656–1661. Table 12.1.8 Reprinted from A. Datta, P. Halder, Thermal efficiency and hydraulic performance evaluation on Ag-Al_2O_3 and SiC-Al_2O_3 hybrid nanofluid for circular jet impingement,* Arch. Thermodyn. *42 (1) (2021) 163–182, (Open Access). Table 12.1.9 Reprinted with permissions from Q. Li, Y. Xuan, F. Yu, Experimental investigation of submerged single jet impingement using Cu–water nanofluid,* Appl. Therm. Eng. *36 (1) (2012) 426–433. Table 12.1.10 Reprinted with permissions from R. Nimmagadda, H.D. Haustein, L. Godson Asirvatham, S. Wongwises, Effect of uniform/non-uniform magnetic field and jet impingement on the hydrodynamic and heat transfer performance of nanofluids,* J. Magn. Magn. Mater. *479 (2019) 268–281. Table 12.1.11 Reprinted with permissions from P. Naphon, L. Nakharintr, S. Wiriyasart, Continuous nanofluids jet impingement heat transfer and flow in a micro-channel heat sink,* Int. J. Heat Mass Transf. *126 (2018) 924–932. Table 12.1.12 Reprinted from W. Peng, L. Jizu, B. Minli, Y. Wang, H. Chengzhi, A numerical investigation of impinging jet cooling with nanofluids,* Nanoscale Microscale Thermophys. Eng. *18 (4) (2014) 329–353, (Full Access).*

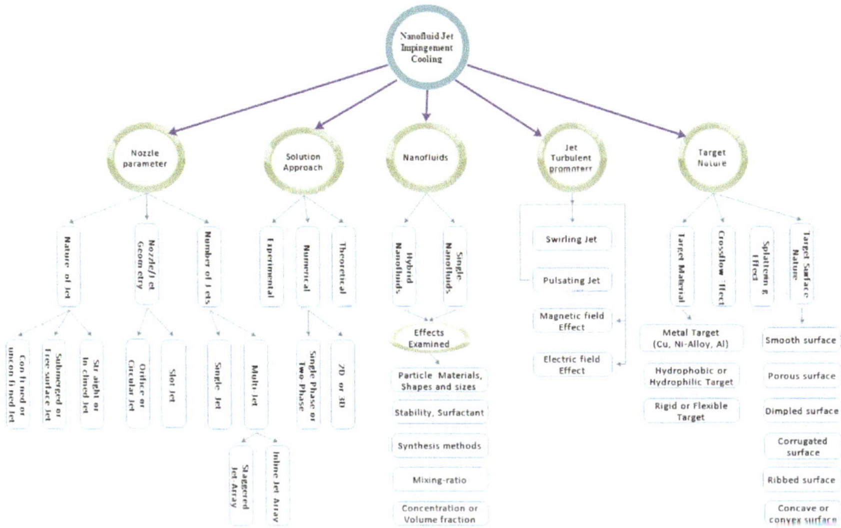

FIGURE 12.5

Mind map on nanofluid-jet impingement cooling studies.

studies on the optimized arrangement for multiple jets as applicable to microchannel heat sinks was numerically carried out by Zunaid et al. [121]. Their work examines the inclined jet arrays while varying the jet diameter and interjet spacing. They concluded that a heat sink with 13 zigzag (staggered) inclined micro jets had the best heat transfer potential with the slightest pressure drop. While Salem et al. concluded that inline multiple jet arrangement had the best thermal performance for jets generally positioned to the target, Zunaid et al. present that staggered jet arrangement had better performance for incline jets.

In recent decades, several studies have been carried out on the enhancement of thermal performance using impinging jet cooling. Impingement cooling studies cooled be categories based on the submergence of the jet, nozzle-jet parameter, nature of the target, phase of coolant, the effect of crossflow, the effect of confinement, splattering phenomenon, and its application [94]. These studies focused on different categories or aspects of JIC.

12.4.1.1 Submerged and free-surface jets

Impinging jet involves the discharge of liquid, gaseous, or nanofluids on a target surface or body. The typical classification of impinging jets based on the fluid and its ambient or surrounding fluid is discussed here. Submerged jets are fluid jets impinged or discharged into regions where their fluid is the same as the ambient jet (i.e., fluids of the same density). An example of submerged jet is the impingement of a cold air jet to an air-bounded flat plate. The right section of Fig. 12.6 illustrates the flow pattern of a submerged jet. The submerged jet

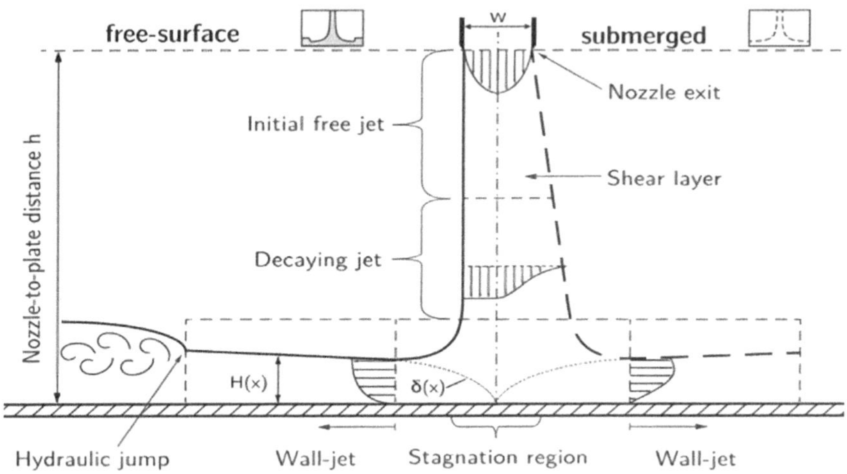

FIGURE 12.6

Free-surface jet and submerged jet in a typical flat plate impinging flow field [122].

Reprinted with permissions from M. Bieber, R. Kneer, W. Rohlfs, Self-similarity of heat transfer characteristics in laminar submerged and free-surface slot jet impingement, Int. J. Heat Mass Transf. *104 (2017) 1341–1352.*

immediately after exiting the nozzle looks like a free jet, after which it becomes a decaying jet. The flow path shows slight deflection of the jet from the axial center of the nozzle. The velocity profile of the jet flow observes the Gaussian distribution, where the normal velocity reduces and the streamwise tangential velocity increases. The stagnation zone/region is the region noted for high stress, the smallest thermal boundary layer thickness, and high convective heat transport. The thermal boundary layer increases while convective heat transfer decreases as the fluid moves away from the stagnation zone along the wall jet [122,123].

Several researchers have studied the effect of submerged jets on heat transfer enhancement. Unsubmerged or free-surface jets are jets exuded into regions where jet fluid is different from the ambient jet (i.e., different density). An example is the impingement of a cold liquid jet on an air-bounded flat plate. Free-surface jet impingement is illustrated in the left section of Fig. 12.6. Free-surface jets exhibit relaxation on exiting the jet nozzle due to the absence of shear between the fluid jet and the less-dense ambient fluid [124,125].

For a circular or tube nozzle, a parabolic velocity profile is observed upstream immediately after the jet exits from the nozzle, the profile flattens (i.e., has constant velocity) downstream until impingement. The boundary layer thickness of the jet commences from the jet core and increases in the streamwise flow direction until the hydraulic jump, where recirculation is observed. Most NJIC studies in the literature typically considered free-surface jets. However, in some cases, confined jets submerge the jet after continuous impingement.

12.4.2 Confinement effect

Impinging a jet can either be confined or unconfined. Confined jets are such jets where the flow dynamics of the jet is retrained with a barrier parallel to the target surface, placed at the tip of the jet outlet, as shown in Fig. 12.7. The barrier or confinement tends to influence the dynamics of the wall jet after impingement at the stagnation zone. The influence of confined and unconfined jets has been critically evaluated in several studies by Behnia et al. [126], Choo and Kim [127], Colucci and Viskanta [128], Markal [129] and Qiu et al. [130]. The numerical investigation on the thermal performance of confined and unconfined air jets by Behnia et al. [126] shows that the confinement has very low significance to the turbulent heat transferred during jet impingement, except at very small target-to-nozzle distance. Experimental studies by Choo and Kim [127], using air and water jets, respectively, with a jet diameter of 4 mm, target-to-nozzle distance ($0.125 < H/d < 1.0$), and pumping power ($1.35 \times 10^{10} < P < 4 \times 10^{13}$) compared the thermal effectiveness of both confined and unconfined impingement jets. The results revealed that thermal efficiencies of both confined and unconfined jets a nonvariable pumping conditions are quite similar. However, the thermal performance for a confined jet reduces by 20%–30% for a fixed Reynolds number value, compared to the unconfined jet. Colucci and Viskanta's [128] finding revealed that the thermal efficiency of a confined jet is sensitive to changes in flow rate and target-to-nozzle distance when compared to unconfined jets.

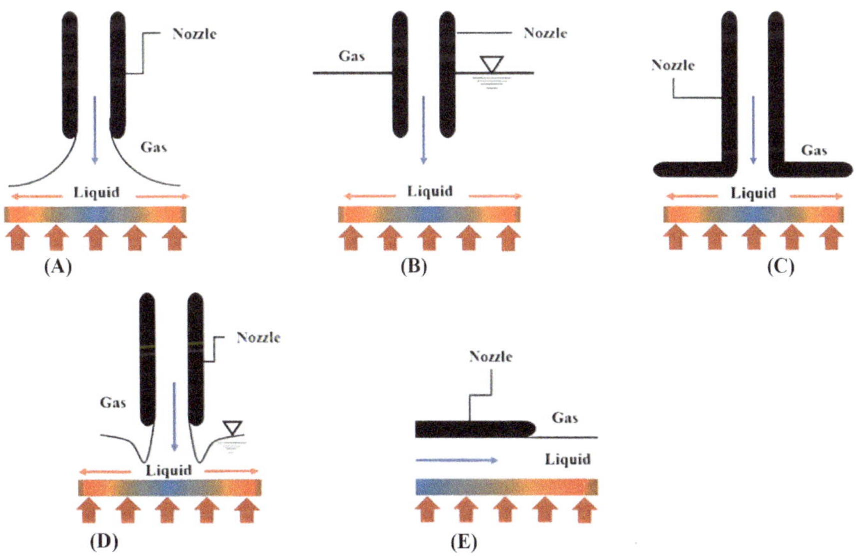

FIGURE 12.7

Impinging jet configurations (A) unconfined free-surface jet, (B) unconfined submerged jet, (C) confined submerged jet, (D) plunging jet, and (E) wall jet [131].

Reprinted from O.J. Wai, P. Gunnasegaran, H. Hasini, A review on experimental and numerical investigations of jet impingement cooling performance with nanofluids, Micromachines *13 (12) (2022) 2059. (Open Access).*

12.5 Studies on parametric factor affecting thermal enhancement of nanofluid-jet impingement

In this section, a detailed description of JIC and the parametric factors affecting its heat transfer performance will be considered.

12.5.1 Nozzle-jet parameters

The nozzle is the outlet of a flow channel (pipe, tube, or hose) used to regulate the fluid jet's pressure and velocity on exit. The controlled fluid from a nozzle is called the jet. Jet could be liquid, gaseous, or a mixture of micro/nanoscale solid with either liquid or gas. According to Zuckerman and Lior [123], major nozzle parameters that influence the nature and effectiveness of a fluid jet for heat transfer purposes include nozzle geometry, jet aperture shape, number of nozzles, and nozzle-to-target orientation.

12.5.1.1 Influence of nozzle geometry

The influence of several nozzle geometries on thermal augmentation has been studied in previous studies. Circular jet apertures or orifices [49,132–135] and rectangular jet apertures or slots [85,136,137] are the most common jet aperture in literature. Li et al. [138] experimentally studied the thermal performance of Cu-H_2O nanofluids in a submerged impinging jet cooling setup. He considered the impact of four different nanoparticle concentrations between the range (1.5 vol.% $\leq \varphi \leq$ 3.0 vol.%), at Reynold number range (2000 $\leq$ Re $\leq$ 16,000), and nozzle parameters (D = 2 mm, 1 < H/D < 3) for a circular orifice single jet. The result shows that enhanced heat transfer improved with nanoparticle concentration, and the 3.0 vol.% concentration and H/D = 2 had the best heat transfer coefficient. Subsequent investigations on the same nanofluid were conducted by Sun et al. [81], using nanoparticle concentration range (0.1 wt.% $\leq \varphi \leq$ 0.5 wt.%), SDBS as surfactant, flow Reynold number (2000 < Re < 7000), and nozzle parameter (D = 1, 3 and 6 mm; 1 $\leq$ H/D $\leq$ 2) considering free-surface single circular orifice in straight and inclined jet orientations, respectively. The outcome revealed that thermal enhancement increased with the nanofluid concentration at fixed flow conditions, with 0.4 wt.% being the optimum mass fraction. According to experimental studies by Naphon et al. [86], as the jet diameter increases and the target-to-nozzle gap reduces, the heat transfer rate increases. Optimum thermal performance was recorded at their highest nanofluids volume concentration (0.015%), and the heat transfer rate increased by 18.56%.

Selimenfendigil and Oztop [137] numerically studied the merged influence of porous mediums and single and multislot alumina-oxide nanofluid-jet heat transfer considering free-surface rectangular jets in normal (90 degrees) orientation. The study outcome revealed that the highest nanoparticle volume fraction improved the Nuavg by 11% for both single jet and multijet. The nanoparticle size effect was not significant. Multijet and single jet impingement had approximately 119% and 84.5% improvement of Nu_{avg}, respectively, comparing the least and largest permeability of the lower layer porous plates. Boudraa and Bessaih [62] also studied the impingement cooling of an isothermal surface with different nanoparticle shape combinations of Al_2O_3-MgO/water hybrid nanofluid, volume concentration ($0\% \leq \varphi \leq 5\%$) using finite volume method. The study used an extended circular orifice multijet (15×15 array) arranged inline, with nozzle parameter (D = 0.4 mm, H/D = W/D = X/D = 5). The result shows that the extended/confined jet improved heat transfer by 30%, while the stagnation zone had the highest Nu value.

12.5.2 Influence of number of nozzles

Impinging jets could be single jet, double-jet, or array/multi/multiple jet depending on the number of nozzles. Multijet can also be classified as either an inline array or a staggered array. Single jet and multijet have been extensively investigated individually in many kinds of literature using air, water, or nanofluids as heat transfer mediums for the impinging jet(s). However, few studies are comparing the single jet with the multijet nanofluid impingement. However, Selimenfendigil and Oztop [137] have shown that multijet impingement has better thermal enhancement than single jet impingement.

12.5.2.1 Influence of nozzle orientation

The orientation of the nozzle concerning the target plate considerably affects the heat transfer performance. Sun et al.[81] studied the influence of jet orientation/angle on the heat transfer, he examined the effect of three nozzle orientations (30, 60, and 90 degrees) under different flow conditions. The optimum jet parameter for effective cooling was observed at a dimensionless target-to-plate distance (H/D) of 2, jet diameter of 3 mm, and jet orientation of 90 degrees.

12.5.3 Target surface/object nature parameters

The thermal boundary condition of the target surface considered in JIC studies could be an isothermal or constant heat flux surface. Most studies on isothermal surfaces are numerical, Boudraa et al. [62] and Selimefendigil et al. [139–141] among others. Meanwhile, most experimental studies consider constant heat flux surface conditions [134,142,143]. Also, the target surface could either be flexible [141] or rigid. Some common target surface parameters examined in the literature are target surface geometry, surface roughness, and target inclination. Table 12.2 shows a summary of some of the target surfaces examined by some authors and their respective impact on the performance of NJIC.

Table 12.2 Summary of target surface parameter's influence on nanofluid-jet impingement cooling performance.

S/n	Author (s)	Approach	Nanofluid	Target	Remark on target parameter	Remark on nanofluids	Model geometry
1	Ahmadi et al. [144]	Num	Al_2O_3/H_2O	Concave surface	Optimum jet-to-plate distance (H/D) for both average and stagnation NU was H/D = 5 for the concave surface cooled with the nanofluid mentioned.	Nanofluid thermal performance increased with nanoparticle concentration. At 5% conc., 13.4% and 11.2% enhancement of Nu_{avg} and Nu_{stag} were obtained.	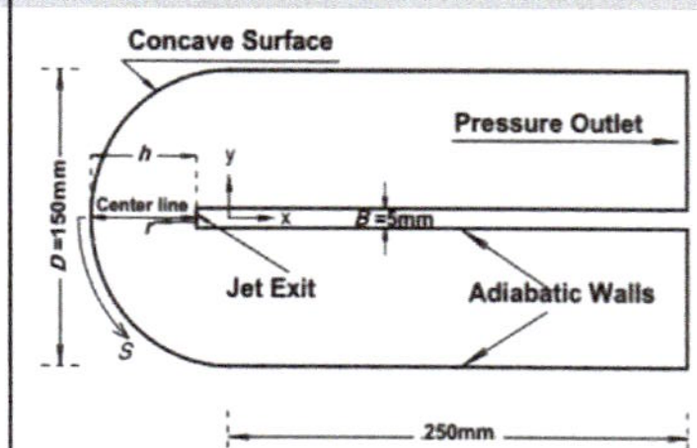
2	Jaiswal et al. [145]	Num	Al_2O_3/H_2O	Concave surface	At constant surface diameter, as the H/W is increased from 4 to 6, enhancement of HT is significant. Also, temperature increment was observed at the region of impingement plate and declined at the increased edge of the plate.	N/A	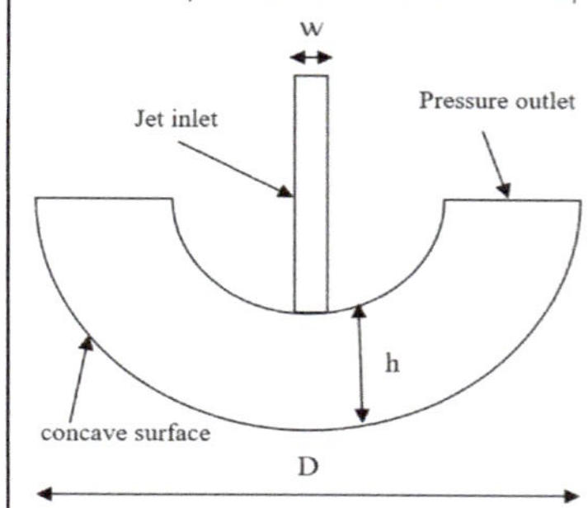
3	Selimefendigil and Oztop [146]	Num (FEM and COBYLA)	CuO_2/H_2O	Convex (Semielliptical surface) and flat surface	Higher heat transfer is observed for the curved wall than for the flat wall.	A peak thermal enhancement of 20% was reported when the nanofluids had the highest volume fraction.	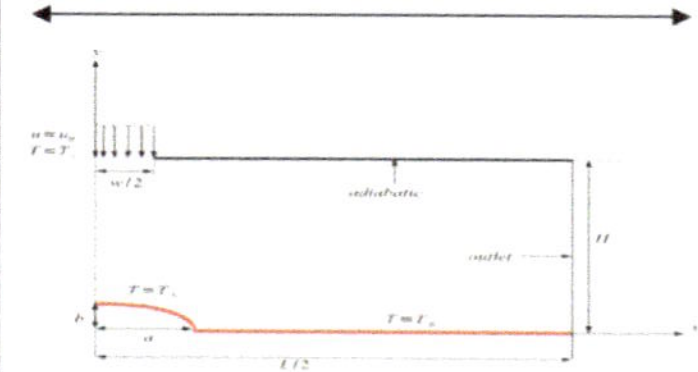

4	Ersayın and Selimefendigil [147]	Num (FVM)	Al_2O_3/H_2O	Flat surface (Linearly moving)	The plate velocity is significant to the thermal and flow performance. low plate velocity aided the increase in thermal enhancement.	The nanofluid had better thermal enhancement compared to the water.	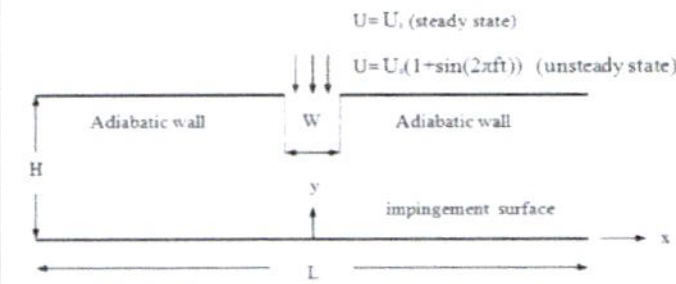
5	Selimefendigil et al. [148]	Num	Al_2O_3/H_2O (with non-Newtonian behavior)	Partially-curved (convex) surface	The convex curve's aspect ratio is slightly significant with a heat transfer enhancement of 5% at optimum value.	The Nu_{avg} increased as particle volume fraction and particle size increased.	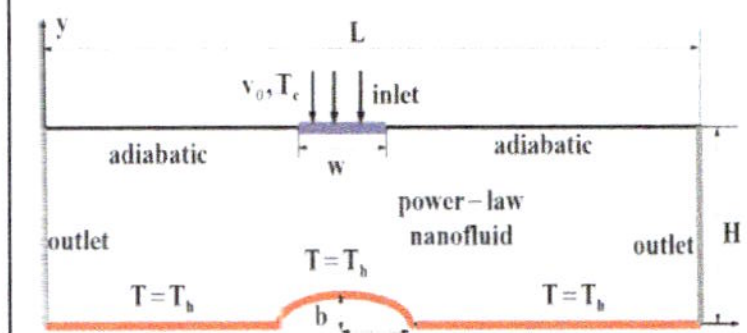
6	Mahdavi, Sharifpur and Meyer [149]	Num	Al_2O_3/H_2O	Flat surface (Rotating-disk)	High torque is required for the plate to have thermal enhancement for a high concentration of particles in the nanofluids.	The nanofluids are significant for the enhancement of heat transfer.	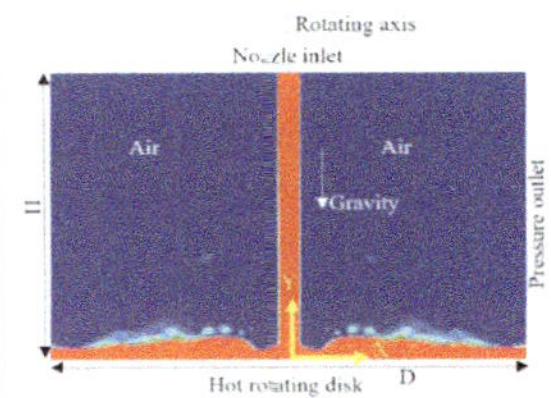

Table 12.2.1 Reprinted with permissions from H. Ahmadi, R.M. Moghari, K. Esmailpour, A.S. Mujumdar, Numerical investigation of semi-confined turbulent slot jet impingement on a concave surface using an Al_2O_3−waternanofluid, Appl. Math. Model. *40 (2) (2016) 1110−1125. Table 12.2.2 Reprinted from A. Jaiswal, A. Datta, P. Halder, Heat transfer augmenation using nanofluid of a confined slot impingement jet on concave surface,* IOP Conf. Ser. Mater. Sci. Eng. *377 (1) (2018), 012075, (Open Access). Table 12.2.3 Reprinted with permissions F. Selimefendigil and H.F. Öztop, Jet impingement cooling and optimization study for a partly curved isothermal surface with CuO-water nanofluid,* Int. Commun. Heat Mass Transf. *89 (2017) 211−218. Table 12.2.4 Reprinted from E. Ersayin, F. Selimefendigil, Numerical investigation of impinging jets with nanofluids on a moving plate,* Math. Comput. Appl. *18 (3) (2013) 428−437, (Open Access). Table 12.2.5 Reprinted from F. Selimefendigil, H.F. Oztop, A.J. Chamkha, Jet impingement heat transfer of confined single and double jets with non-newtonian power law nanofluid under the inclined magnetic field effects for a partly curved heated wall,* Sustainability *13 (9) (2021) 5086, (Open Access). Table 12.2.6 Reprinted with M. Mahdavi, M. Sharifpur, J.P. Meyer, L. Chen, Thermal analysis of a nanofluid free jet impingement on a rotating disk using volume of fluid in combination with discrete modelling,* Int. J. Therm. Sci. *158 (2020).*

12.5.3.1 Influence of target surface geometry

Different rigid target surface geometries have been studied in previous works, flat, convex, and concave surface targets are the most investigated. Tie et al. [49] and Amjadian et al. [134] experimentally conducted an experimental study to investigate the thermal enhancement achieved through nanofluid impingement jet for cooling a constant heat flux on a flat plate. Datta and Halder [150] numerically studied the implication of surface curvature on the enhanced heat transfer from a convex, concave, and flat plate. The result shows that surface curvature significantly affects the heat transfer rate for smaller H/D. The convex surface had higher local Nu than the concave and flat plate at the stagnation region.

12.5.3.2 Influence of target surface roughness

Surface roughness significantly affects friction factor and pressure drop during fluid flow in the wall jet zone. Some of the investigated surface roughness in literature are; micro/nanostructures roughness [132,151,152], dimples [153], grooves [154], macro/micro pin-fins [135,155–157], ribs [158,159], transverse perforation [160], wavy [161], and ceramic matrix composite [162]. Very few studies have focused.

(Hydrophobic/hydrophilic) significantly affects the nanofluids' contact angle on the silicon surface. In the case of a hydrophobic-coated surface, after impingent, the nanofluid droplet spreads, retracts, swirls, and retracts. Also, nanofluid enhanced the heat transfer by 33% and 214% for the uncoated and hydrophobic surfaces, respectively.

12.5.4 Studies on nanofluid phase change effect

Different authors have studied phase change effects in JIC, considering boiling occurrence in conventional fluids and nanofluids. Other works considered the phase change of a microencapsulated phase change material or nanoencapsulated phase change material (NEPCM). Mitra et al. [163] experimentally explored MWCNT-water and TiO_2-water nanofluid-jet boiling heat transfer for a cooled steel surface. The study outcome revealed that nanofluid promoted a prompt shift of the boiling regime from film boiling to transition boiling. Also, the critical heat flux (CHF) was marginally varied because of using nanofluids instead of water. Singh et al. [164], in furtherance to Mitra et al.'s work, investigated the influence of SiO_2-, Al_2O_3-, and TiO_2-water nanofluid concentration on boiling heat transfer while cooling a steel plate. The result revealed that the CHF increase was proportional to the increase in TiO_2-water nanofluid concentration. Chang et al. [165] also studied Al_2O_3-DI water nanofluid-jet impingement flow-boiling. They observed that applying slight acoustic excitation to the heated surface eliminated nanosorption layers, which enhanced thermal performance while using nanofluid instead of DI water.

Zhang et al. studied the thermal enhancement of arrays of NEPCM-slurry jet impingement cooled heated surface with "spent flow distraction." The result shows that the addition of NEPCM was insignificant to the localized heat transfer.

12.5.5 Studies on splattering phenomenon

During the impingement, the high-velocity fluid jet impacts the surface, spreads, and splashes with different droplets depending on the nature of the surface and fluid flow conditions. Liquid jet impingement on the target surface could be bulk or atomized fluid droplets. For intensive comprehension of the splattering phenomenon, several researchers have considered the spread-splash of different fluid droplets with varied flow properties and target surface nature [25,166–170]. The spreading and splashing of impinging Al_2O_3/glycerol nanofluid droplets have been experimentally studied by Aksoy et al. [170] using high-speed photography. They observed that the droplet dynamics are a function of the droplet impact velocity and that slashing is enhanced by increasing nanoparticle concentration at a low Reynolds number. Both Rioboo et al. [166] and Aksoy et al. [170] have classified impinging fluid droplet dynamics into spreading, prompt splash, corona splash, partial rebound, and complete rebound. Some of these classifications are presented in Fig. 12.8. Tyagi et al. [27] and Aksoy et al. [25,170] in their recent review and investigational studies, emphasize the relevance of droplet dynamics for boosting heat transfer from the substrate or target surface.

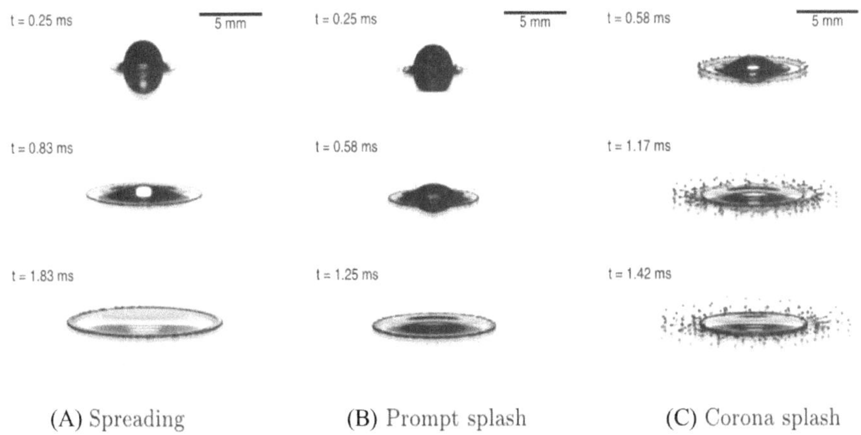

FIGURE 12.8

Typical spread-splash phenomenon of Al_2O_3/glycerol nanofluid on a smooth sapphire target [170].

Reprinted by permission from Y.T. Aksoy, P. Eneren, E. Koos, M.R. Vetrano, Spreading-splashing transition of nanofluid droplets on a smooth flat surface, J. Colloid Interface Sci. *606 (2022) 434–443, https://doi.org/10.1016/J.JCIS.2021.07.157.*

12.5.6 Studies on turbulence promoters in nanofluid-jet impingement cooling

Several studies have investigated implementing turbulence-promoting nozzle flow for enhanced thermal performance in JIC. Fluid jets obtained from such setups are commonly called synthetic jets. Most studies on the turbulence-promoted nozzle flow can either be "swirling jets" (using helical inserts in the nozzle), "pulsating jets," or "magnetic-field effected" flow. Both numerical and experimental studies have been conducted on swirling jet and pulsating jet heat transfer.

12.5.6.1 Swirling-jet impingement cooling

Wongcharee et al. experimentally studied the thermal enhancement of swirling (using twist tape inserts) nanofluid impingement cooling jets using TiO_2-water [171] and CuO-water [172–174] nanofluid. Their studies concluded that the swirling jets had better enhancement than the conventional jet for all nanofluid concentrations examined. In further studies by Wongcharee et al. [173] to determine the optimum twist ratio of the twisted tape for the swirled CuO-water nanofluid impinging jet. For nanoparticle concentration range of 2–4 vol.%, jet-to-target distance 2, 3, and 4, twist ratios of 1.43, 2.86, and 4.28, and Reynolds number between 1600 and 9400. They reported that the thermal performance was slightly augmented while reducing the twist ratio of the twisted tape at the optimum jet-to-target gap. Monfared et al. also studied swirling nanofluid impinging jet and concluded that both nanofluid and swirling contribute to heat transfer enhancement. The average Nusselt number for a swirling nanofluid jet was improved by 82% compared to a plain tube. Xu et al. examined the thermal performance of swirling air-impingement cooling jets using a novel threaded nozzle instead of conventional helical inserts. In the study, they examined the effect of Reynold number range (6000–30,000), twist angle (0–45 degrees), and dimensionless jet-to-target distance range (1–8) on the thermal performance. From their smoke flow visualization results, they concluded that the 45 degrees twist angle had the best thermal performance due to the absence of a hyperbolic recirculation zone. The thermal performance was enhanced by 11.4% compared to the conventional (nonswirling) impingement jet.

12.5.6.2 Pulsating jet impingement cooling

Intermediate fluid flow or flow pulsation is another technique of enhancing thermal performance in jet impingement heat transfer by promoting turbulence. Pulsating/oscillating/intermittent jet impingement has found wide application in heat transfer management in aerospace designs and metal treatment and machining workshops [175,176]. Several experimental studies on active and passive PJIC using conventional fluids (air and water) have been investigated in the literature [176–182]. Generally, active pulse generation requires an external energy source, while passive pulse generation does not require an external energy source. Active pulse generation had been studied in previous experimental works using crank-wheel mechanism (consisting of an electric motor, gearbox, pulley-belt, and flywheel) [156,183], variable speed electric motor and ball valve [184], blower (with cycling suction) [185], or solenoid valve and function/signal generator [177,179,180,186–190].

Most of the earliest investigations on pulsating jet impingement using water or air as cooling fluid were conducted by Zumbrunnen et al. [182,184,191–193]. In all their studies, the result revealed that the pulsating jets improved the thermal performance compared to steady jet. The enhanced thermal performance was attributed to continuous thermal and hydraulic boundary layer renewal. Sheriff and Zumbrunnen [191] investigated the impact of the Strouhal number (a function of the oscillating frequency of pulsation) on the thermal performance of pulsating water jet impingement for the pulse frequency range of 5–280 Hz. The study observations were explained regarding the thermal and hydrodynamic boundary layer, boundary layer renewal, and bulge formation on the free surface. Thermal performance enhancement for the pulsating jet flow condition was attributed to boundary layer renewal on the surface.

Yang et al. [183] experimentally examined the heat transfer performance of pulsating water jet on a confined concave surface using a crank-wheel mechanism (electric motor, gearbox, pulley-belt, and flywheel) for fluid flow Reynold number ($17{,}500 < Re < 43{,}200$) and pulsating frequency ($0 < F < 2.08$ Hz). The study outcome revealed that the stagnation Nusselt number improved by a factor of 3.2 because of the pulsating water jet cooling compared with nonreciprocating or steady jet cooling. Enhanced thermal performance of 240% was reported in the study by Chang et al. [156], where pulsating or reciprocating impingement jet heat transfer on ribbed horizontal surfaces was investigated. In the study by Travnicek and Vit [185], enhanced heat transfer of 18% was obtained while using a hybrid synthetic air jet (combination of the fluid diode to the synthetic jet actuator) or reversible pulsating air jet generated by the means of a blower with cycling suction.

Utturkar et al. [186] utilized a more advanced means to generate pulsating air jet, which consists of a function generator (time-harmonic signal control) and an oscilloscope. The study revealed that the enhanced heat transfer is proportional to the pulsating frequency and amplitude. Arik and Icoz [187] later advanced this pulse generation method using a function generator, laser vibrometer, oscilloscope, and piezoelectric disks. Arik et al. [188,189] and Ghaffari et al. [190] utilized this method of pulse generation, and their outcomes show that better cooling performance can be obtained from synthetic jets compared to steady jets due to vortex shedding generated during pulsation, jet-to-target distance effect on heat transfer was insignificant for synthetic jets and further recommend synthetic (pulsating) jet as a solution to heat transfer challenges in miniaturized and highly slim electronic devices such as tablets, smartphone, laptop, and LED/LCD TVs. Recently, Abishek and Narayanaswamy [177] modified and simplified Arik et al.'s pulse generation method using a solenoid valve, electric relay, and function generator for water jet. However, the pulsating jet effect was insignificant within the frequency and amplitude range ($0.25 < F < 2.5$ Hz; $0.5 < A < 1$) examined in the study.

Sabelberg et al. [194] developed and optimized a passively controlled pulsating water jet experimental test rig for general-purpose impingement cooling. In the study, pulsation amplitude, Strouhal number, and pressure fluctuation were monitored for three different designs of pulse-generation modules (axial wheel, swirl body, and rotational sphere). The result showed that the modified axial rotation recorded the best Strouhal number (0.19), which implies the maximum heat transfer rate. Fig. 12.9 shows some of the pulse-generation modules nozzle designs used and their corresponding flow fields recorded using high-speed camera. Saliba et al. [176] experimentally studied the heat transfer in a passively controlled pulsating air jet. The work revealed that the passive fluidic oscillator significantly improved the thermal performance at a higher Reynolds number and smaller jet-to-target gap compared to the steady jet.

Selimenfendigil and Oztop [78] numerically studied and optimized the thermal performance of pulsating NJIC of a heated surface. In their work, the pulsation frequency, amplitude, and behavior were modeled using the time-dependent velocity equation $\{V = V_o(1 + A\sin(2\pi ft))\}$. The study outcome revealed that the stagnation Nusselt number increased because of increasing pulsating jet frequency, Reynolds number, and nanofluid volume fraction, respectively. Li et al. [183] and Zhang et al. [182] reported other recent numerical works on pulsating NJIC. To date, there are no experimental studies on pulsating nanofluid-jet impingement.

12.5.6.3 Influence of magneto-hydrodynamics jet

Magneto-hydrodynamics involves the study of the flow dynamics of magnetic fluids or ferrofluids under the influence of magnetic fields. Studies on the thermal performance of magneto-hydrodynamic jet flow are very scarce in the literature. This could be attributed to the difficulty of effectively building a controlled magnetic field around jet impingement test sections. However, Nakharintr and Naphon [195] presented the first experimental study on this topic. Their work investigated the combination of four different thermal enhancement techniques (impinging jet, magnetic field, mini-fin heat sink, and nanofluids) while significantly examining the influence of magnetic field. The results obtained revealed that the effect of the magnetic field was significant to the thermal dissipation from the target surface. Also, the thermal enhancement improved proportionally to the magnetic field strength. Recently, Selimenfendigil et al. [148] and Nimmagadda et al. [85] conducted numerical studies considering the effect of magnetic field on NJIC systems. The numerical study by Selimefendigil et al. examined the effect of a magnetic field on the heat transfer performance while impinging power law nanofluid jet(s) on a convex surface. The outcome revealed that increasing magnetic field strength improved the average Nusselt number for shear-thinning fluids or pseudoplastics; an enhancement of 12.58% was observed compared to the unmagnetized flow. However, the mean Nusselt number exhibited a reduction as the magnetic field strength increased for shear-thickening fluids or dilatants.

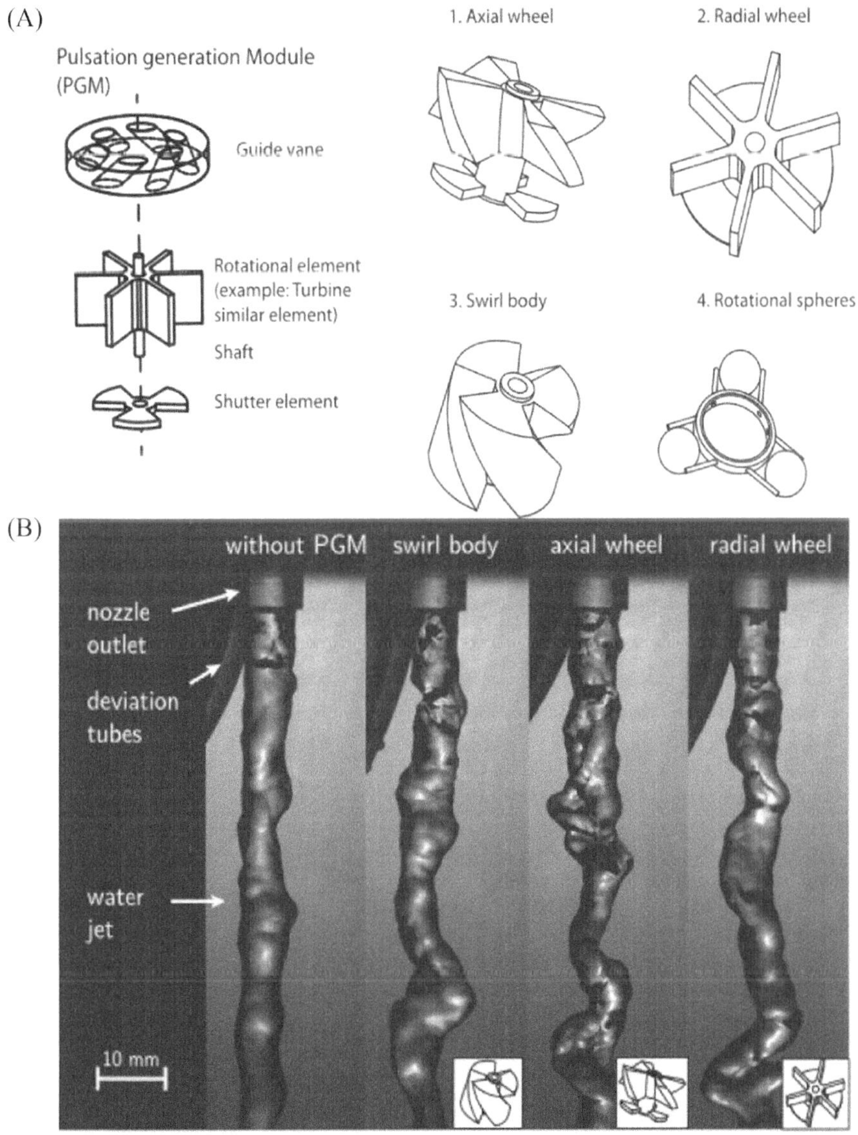

FIGURE 12.9

Passive pulse-generation modules (A) some nozzle designs and (B) their recorded flow field [194].

Reprinted from E. Sabelberg, M. Cardenas, R. Kneer, W. Rohlfs, Design, development, and validation of concepts for generating passive pulsation in cooling nozzles, Case Stud. Therm. Eng. *7 (2016) 103–108. (Open Access).*

12.6 Correlations for nanofluid-jet impingement cooling in literature

Some empirical correlations that approximately predict the heat transfer in nanofluid-jet impingement and their accuracy are presented [84,196–203]. Webb and Ma [196], in their studies on a single circular submerged air jet, obtained a correlation for Nusselt number as a function of Reynolds number, Prandtl number, and jet-to-target distance. However, the accuracy of their correlation was not reported. In the extensive experimental work of Li et al. [84], a more comprehensive accurate (2% error) heat transfer correlation was proposed. Generally, most of the correlation presents the Nusselt number as a function of nanofluid concentration, Reynolds number, Peclet number, Prandtl number, and the dimensionless height of the nozzle from the target (H/W or H/D), as well as the number and jet-to-target distance for a single submerged confined jet on a horizontal flat surface, within the Reynold number range (2000–14,000). Datta et al. had a similar correlation but with lower accuracy ($\pm$ 15%). Comprehensive details of some Nusselt number correlations for NJIC are presented in Table 12.3.

12.7 Contemporary and future application of nanofluid-jet impingement cooling

Several studies have shown that NJIC can be applied in most engineering systems where liquids (water) are used as coolants. Some of these areas include gas turbines used in aircraft and power stations, metallurgical heat treatment, power electronic cooling, solar system management, HVAC ventilation, and automobiles.

12.7.1 Gas turbine

Gas turbine remains a notable prime mover with vase applications ranging from power plants to aerospace drives. However, due to the enormous heat emitted in the turbine, there is a need for in-operation internal cooling of some of its vital parts (blade, vanes, end wall, cascades, engine after-burner, combustor liner backside, etc.) to improve its energy efficiency and durability [206]. Internal cooling of gas and steam turbine parts has been the most adopted, versatile, and effective cooling method. The relative performance of internal JIC and external film cooling technique was tested by Mensch and Thole [207], they observed that internal jet impingement had better cooling effectiveness. Several authors have studied air JIC of gas turbine parts due to the density and abundance of air [94,180,184]. However, to the best knowledge of this author, there are no studies in open-access articles concerning the potential of nanofluids in cooling gas turbines.

Table 12.3 Correlation for nanofluid-jet impingement cooling.

S/n	NJIC correlation	Impingement condition	Applicable range	Error	Author and year
1	$\mathrm{Nu} = c\mathrm{Re}^m \mathrm{Pr}^n [1.0 + \alpha_1 (H/D) + \alpha_2 (H/D)^2 + \alpha_3 (H/D)^3]$	Single circular submerged jet.	N/A	N/A	Webb and Ma [196]
2	$Nu_{nf} = 0.264(1.0 + 2.2061\varphi^{0.864}\mathrm{Pe}_d^{0.2715})\mathrm{Re}_{nf}^{0.5375}0.5375$ $\mathrm{Pr}_{nf}^{1/3} \times [1.0 + 0.3923(H/D) + 0.0086(H/D)^2 - 0.0259(H/D)^3$	Single submerged confined jet on a horizontal flat surface.	$2000 < \mathrm{Re} < 14{,}000$	2%	Li, Xuan, and Yu [138]
3	$Nu_{nf} = 51.29 \times (\mathrm{Re})^{0.60} \times (\mathrm{Pr})^{-2.93} \times (H/D)^{-0.25}$	Single circular Free-surface unconfined jet on a vertical flat plate.	$5000 < \mathrm{Re} < 12{,}000$, $7.03 < \mathrm{Pr} < 7.53$, $6 < l/d < 12$.	± 30%	Modak et al. [197]
4	$Nu_{endwall} = 0.846(\mathrm{Re})^{0.563} \times (H/W)^{-0.0653}$	Single-slot free-surface confined jet on a flexible curved wall.	$25 < \mathrm{Re} < 200$, $H/W = 8$	± 5%	Selimefendigil and Oztop [141]
5	$Nu_{wall(\mathrm{Re}_m,\varphi_m)} = a + b^{*}\mathrm{Re}_m + c^{*}\varphi_m + d^{*}\mathrm{Re}_m^2 + e^{*}\mathrm{Re}_m^{*}\varphi_m$ Where $\mathrm{Re}_m = \frac{\mathrm{Re}-300}{130}$, $\varphi_m = \frac{\varphi - 0.02189}{0.01255}$	Single-slot free-surface confined jet on the flat and curved wall.	$100 < \mathrm{Re} < 500$, $0.0001 < \varphi < 0.0400$	± 25%	Selimefendigil and Oztop [146]
6	$Nu_{avg} = (634 \times 10^{-4} Pe) + \frac{654.5}{13.61 + \frac{r}{d} + (1-\varphi)^2} + \frac{283 \times 10^{-6}}{\frac{r}{d} \times \frac{z}{d}} - \frac{r}{d} - (146 \times 10^{-7} \frac{r}{d} Pe)$	Single free-surface jet on the flat horizontal plate.	$0 \le \varphi \le 8.5\%$, $0.5 \le z/d \le 8$, $2.5 \le r/d \le 12.5$	± 10%	Sorour et al. [198]

(Continued)

Table 12.3 Correlation for nanofluid-jet impingement cooling. *Continued*

S/n	NJIC correlation	Impingement condition	Applicable range	Error	Author and year
7	$\frac{\overline{Nu_{nf}}}{\overline{Nu_{bf}}} = \begin{cases} 1+\varnothing\left(-0.169+44.966\varnothing+0.650307\varnothing^2\right) & (single\ ph) \\ 1+\varnothing\left(44.0049-236.4944\varnothing+3.437\varnothing^2\right) & (Multi\ ph) \end{cases}$ Where $\overline{Nu_{bf}} = 0.462624$ $Re^{0.4633}\ H/W^{-0.43711}$	Single- and two-phase single-slot jet on horizontal flat surface.	$0\% \leq \varnothing \leq 4\%$ $100 \leq Re \leq 400$ $0.5 \leq H/W \leq 4$	N/A	Abdelrehim et al. [200]
8	$\overline{Nu} = 0.77(1-0.7534V_p+0.85V_p^2)Re^{0.768}$ $\overline{Nu} = 0.125Re^{0.76}Pr^{0.34}(1+\varnothing)^{0.58}$	Single free-surface confined jet on flat rotating horizontal flat surface.	$0 \leq V_p \leq 1$ $20{,}000 \leq Re \leq 30{,}000$ $0\% \leq \varnothing \leq 5\ \%$	± 5%	Datta et al. [204]
9	$Nu = 3.953\left(1-0.53\varnothing^{1.411}Pe^{2.4\times10^{-4}}\right)Re^{0.134}Pr^{\left(\frac{1}{3}\right)}\left[\frac{H}{D}\right]^{0.161}$	Single free-surface jet on horizontal flat surface.	$0.02\% \leq \varnothing \leq 0.1\ \%$ $2000 < Re < 10{,}000$, $3 < Pr < 8$, $2 < H/D < 7.5$	± 15%	Barewar et al. [201]
10	$Nu_{nf} = 0.00078\left\{Re^{0.58999}Pr^{\frac{1}{3}} \times \left[1-0.47033\left(\frac{H}{D}\right)+0.08819\left(\frac{H}{D}\right)^2+0.01762\left(\frac{H}{D}\right)^3-0.00398\left(\frac{H}{D}\right)^4\right\}\right.$	Single circular free-surface jet on horizontal flat surface.	$8000 \leq Re \leq 13{,}000$ $1\% \leq \varnothing \leq 3\%2$ $\leq H/D \leq 5$	± 10%	Lv et al. [202]
11.	$Nu_{nf} = [1.4510\times10^{-19}+0.00746\left(\varnothing^{0.06242}Pe^{-2.4849}\right)$ $(Re^{-3.7224}Pr^{\frac{1}{3}} \times \left[1+6.8732\left(\frac{H}{D}\right)-4.3599\left(\frac{H}{D}\right)^2\right]+1.2045\left(\frac{H}{D}\right)^3-0.1134\left(\frac{H}{D}\right))^4$	Single circular free-surface jet on horizontal flat surface.	$4000 \leq Re \leq 12{,}000$ $0.5\% \leq \varnothing \leq 2\%2$ $\leq H/D \leq 5$	± 12%	Lv et al. [205]
12	$Nu_0 = 0.846Re^{0.5633}(H/W)^{-0.0653}$ {stagnation point} $Nu_{av} = 0.02785Re^{0.7941}(H/W)^{0.0405}$ {Average}$\frac{Nu_{0,nf}}{Nu_{bf}} = 1.000+\varphi(3.129-172.8\varphi+3103\varphi^2)$ {stagnation point}$\frac{Nu_{avg,nf}}{Nu_{avg,bf}} = 1.000+\varphi(3.455-196.1\varphi+3529\varphi^2)$ {Average}	Single-slot confined free-surface on horizontal flat surface.	$100 \leq Re \leq 400$, $4 \leq H/W \leq 10$ & $0\% \leq \varphi \leq 5\%$	N/A	Manca et al. [203]
13	$Nu_0 = Pr^{0.42}\left[\frac{1.53Re^{\left\{0.695-\left(\frac{w_h}{2D}+\frac{H}{D}^{0.333}+3.06\right)^{-1}\right\}}}{1.39+\frac{w_h}{2D}+\frac{H}{D}}\right]$	Single confined & submerged slot jet.	$3000 \leq Re \leq 90{,}000$, $2 \leq W/2D \leq 25$ $2 \leq H/D \leq 10$	7.38%	Abishek and Narayanaswamy [177]

12.7.2 Metallurgical heat treatment

Due to the vital role of heat treatment in the improvement of the mechanical properties of metals (Steel, Titanium, etc.), several authors have experimentally studied the application of fluid jet impingement in metallurgical processes [208–212]. Santosh et al. [213] studied the application of nanofluids (Al_2O_3-DIW and TiO_2-DIW) as a coolant for hot steel strips. When compared with conventional fluid (water). The result from Santosh et al. study revealed that the thermal dissipation rate of the steel strips improved substantially while using nanofluids. Also, they reported that the thermal performance improves with increasing concentration of the nanofluids until a certain limit of the concentration. Sarkar et al. report similar cooling performance while using Cu-Al double-layered hydroxide (Cu-Al LDH), they concluded that this enhancement could improve the quality of steel production. Further studies on industrial scale application of NJIC were conducted by Oliveira et al. [108] and Kashyap et al. [104]. Their studies showed that industrial-scale experiments unveil vital cooling data that could not be observed in laboratory-scale experiments. They observed that the rate of heat dissipated at a point far away from the impingement zone is proportional to the flow Reynolds number.

12.7.3 Automobile/power electronic cooling

According to FreedomCAR goal in 2020 [214], to ensure the competitiveness of hybrid electric vehicles, low cost, highly efficient, small in size and weight, and reliable components and systems must be involved. To achieve this goal, studies have been carried out for optimized thermal management in the Insulated Gate Bipolar Transistors present in modern inverters/converters. Jones-Jackson et al. [92] summarized recent research outcomes on JIC to alleviate the thermal stress imposed on the modules and increase their durability and reliability. Wu et al. [215], Wei et al. [90], Sabato et al. [216], Klinkhamer et al. [217], and others, have shown experimentally and numerically that JIC could improve the performance of power electronic systems. We recommend further study on applying liquid nanofluids as coolants in this field.

12.7.4 Solar systems

Photovoltaic (PV) cell modules and concentrated solar power (CSP) systems have been reported to perform best at certain temperature ranges, which deteriorate the energy efficiency and lifespan of their components if exceeded. A systematic review by Eng-Ewe et al. [218] revealed that 3% of open-access literature on "solar technology JIC application" topics involves using nanofluid-jet for cooling. Among these studies are Javidan and Moghadam [219], Rostami et al. [220], Ebaid et al. [221], and Hasan et al. [222]. Their studies revealed that the electrical and heat energy was enhanced concerning the nanofluid concentration at the expense of the hydrodynamics requirement. A novel study by Rostami [220] used ultrasound waves to atomize CuO-DIW nanofluid on PV modules, the result of the study shows a 57.27% improvement in cooling efficiency compared to uncooled PV modules. Ebaid et al. conducted a similar study using TiO_2 and Al_2O_3 nanoparticles in water-cetyltrimethylammonium bromide mixture and water-polyethylene glycol mixture, respectively. They also obtain enhanced cooling efficiency while using the nanofluids, respectively. There is a need for more studies on the industrial-scale application of nanofluid as heat transfer fluid in CSP systems and PV modules.

12.7.5 Heating, ventilation, and air conditioning ventilation

As the need for better indoor air quality increases in this age, the relevance of efficient, low cost, and energy-consuming HVAC systems cannot be overemphasized. Recently, the application of nanoparticle-enhanced phase change materials (NEPCMs) as refrigerants is a novel approach. Milanese et al. [223] conducted an experimental campaign at the University of Salento, comparing the COP of a conventional refrigerant with an NEPCM in two similar HVAC systems. The result revealed that the NEPCM-operated HVAC had an average improvement of 9.8% and 8.9% in winter and summer, respectively. Among other works, this has shown the edge of using nanofluids in HVAC and impinging jet ventilation systems.

12.8 Conclusion and outlook

In summary, this comprehensive exploration of NJIC has shed light on its intricate heat transfer dynamics, which are intricately linked to a range of key parameters. Throughout this review, we have delved into various factors, including the characteristics of the target surface, nozzle type, the influence of turbulence promoters, and the specific nanofluids employed.

Our examination has encompassed a broad spectrum of experimental and numerical investigations into NJIC. These studies have encompassed scenarios involving single or multiple jet impingements of single or hybrid nanofluids onto surfaces with diverse geometries, such as flat, concave, convex, corrugated, and pin-fin structures.

While NJIC exhibits remarkable potential and versatility in diverse applications such as gas turbines, HVAC systems, solar-thermal installations, automobiles, and metallurgical processes, the review underscores a persistent challenge: the absence of conclusive optimal parameters. Despite the wealth of research and applications, the precise configuration for achieving peak heat transfer performance remains elusive.

To overcome this challenge and unlock the full potential of NJIC, it is imperative that future endeavors focus on application-specific investigations. These studies should be tailored to the unique demands of different industrial contexts, ultimately paving the way for the optimization of NJIC parameters for maximum efficiency and efficacy in respective applications.

References

[1] J.H. Rosenfeld, M.T. North, Porous media heat exchangers for cooling of high-power optical components, Opt. Eng. 34 (2) (1995) 335–341. Available from: https://doi.org/10.1117/12.194100.

[2] A. Izadi, M. Siavashi, Q. Xiong, Impingement jet hydrogen, air and CuH_2O nanofluid cooling of a hot surface covered by porous media with non-uniform input jet velocity, Int. J. Hydrog. Energy 44 (30) (2019) 15933–15948. Available from: https://doi.org/10.1016/J.IJHYDENE.2018.12.176.

[3] J. Lee, I. Mudawar, Assessment of the effectiveness of nanofluids for single-phase and two-phase heat transfer in micro-channels, Int. J. Heat. Mass. Transf. 50 (3–4) (2007) 452–463. Available from: https://doi.org/10.1016/J.IJHEATMASSTRANSFER.2006.08.001.

[4] L. Cheng, G. Ribatski, J.R. Thome, Analysis of supercritical CO_2 cooling in macro- and micro-channels, Int. J. Refrig. 31 (8) (2008) 1301–1316. Available from: https://doi.org/10.1016/J.IJREFRIG.2008.01.010.

[5] J.S. Coursey, Enhancement of Spray Cooling Heat Transfer Using Extended Surfaces and Nanofluids (Dissertation), University of Maryland, College Park, 2007.

[6] J.K. Calautit, P.W. Tien, S. Wei, K. Calautit, B. Hughes, Numerical and experimental investigation of the indoor air quality and thermal comfort performance of a low energy cooling windcatcher with heat pipes and extended surfaces, Renew. Energy 145 (2020) 744–756. Available from: https://doi.org/10.1016/J.RENENE.2019.06.040.

[7] O. Aydin, W.J. Yang, Natural convection in enclosures with localized heating from below and symmetrical cooling from sides, Int. J. Numer. Methods Heat Fluid Flow 10 (5) (2000) 518–529. Available from: https://doi.org/10.1108/09615530010338196/FULL/PDF.

[8] M. Sharifpur, S.O. Giwa, K.Y. Lee, H. Ghodsinezhad, J.P. Meyer, Experimental investigation into natural convection of zinc oxide/water nanofluids in a square cavity, Heat Transf. Eng. 42 (19–20) (2020) 1675–1687. Available from: https://doi.org/10.1080/01457632.2020.1818384.

[9] R. Hannemann, J. Marsala, M. Pitasi, Pumped liquid multiphase cooling, Am. Soc. Mech. Eng. Electron. Photonic Packag. EPP. 4 (2008) 469–473. Available from: https://doi.org/10.1115/IMECE2004-60669.

[10] R.E. Chupp, H.E. Helms, P.W. McFadden, T.R. Brown, Evaluation of internal heat-transfer coefficients for impingement-cooled turbine airfoils, J. Aircraft 6 (3) (1969) 203–208. Available from: https://doi.org/10.2514/3.44036.

[11] D.E. Metzger, Spot Cooling and Heating of Surfaces with High Velocity Impinging Air Jets Part 1—Slot Jets on Plane Surfaces, 1962.

[12] D.R. Walz, Spot cooling and heating of surfaces with high velocity impinging air jets. Part 2. Circular jets on plane and curved surfaces, slot jets on curved surfaces, 1964. [Online]. Available: https://apps.dtic.mil/sti/citations/AD0607727 (accessed February 2, 2022).

[13] D.E. Metzger, T. Yamashita, C.W. Jenkins, Impingement cooling of concave surfaces with lines of circular air jets, J. Eng. Power 91 (3) (1969) 149–155. Available from: https://doi.org/10.1115/1.3574713.

[14] J.N.B. Livingood, P. Hryeak, Impingement heat transfer from turbulent air jets to flat plates: a literature survey, Security X–2778 (1973) 43 [Online]. Available: http://hdl.handle.net/2060/19730016200 (accessed February 2, 2022).

[15] M. Muneeshwaran, G. Srinivasan, P. Muthukumar, C.C. Wang, Role of hybrid-nanofluid in heat transfer enhancement – a review, Int. Commun. Heat. Mass.

Transf. 125 (2021) 105341. Available from: https://doi.org/10.1016/J.ICHEATMASSTRANSFER.2021.105341.

[16] O. Mahian, A. Kianifar, S.A. Kalogirou, I. Pop, S. Wongwises, A review of the applications of nanofluids in solar energy, Int. J. Heat. Mass. Transf. 57 (2) (2013) 582–594. Available from: https://doi.org/10.1016/J.IJHEATMASSTRANSFER.2012.10.037.

[17] M. Sharifpur, A.B. Solomon, J.P. Meyer, J.S. Ibrahim, B. Immanuel, Thermal conductivity and viscosity of Mango Bark/water nanofluids, in: 13th International Conference on Heat Transfer, Fluid Mechanics and Thermodynamics, July 2017.

[18] S.U.S. Choi, J.A. Eastman, Enhancing thermal conductivity of fluids with nanoparticles, ASME Int. Mech. Eng. Fluids Eng. Div. FED. 8 (231) (1995) 99–105. Available from: https://doi.org/10.1021/je60018a001.

[19] B. Wang, X. Wang, W. Lou, J. Hao, Rheological and tribological properties of ionic liquid-based nanofluids containing functionalized multi-walled carbon nanotubes, J. Phys. Chem. C. 114 (19) (2010) 8749–8754. Available from: https://doi.org/10.1021/JP1005346.

[20] T.X. Phuoc, M. Massoudi, R.H. Chen, Viscosity and thermal conductivity of nanofluids containing multi-walled carbon nanotubes stabilized by chitosan, Int. J. Therm. Sci. 50 (1) (2011) 12–18. Available from: https://doi.org/10.1016/J.IJTHERMALSCI.2010.09.008.

[21] K. Anoop, R. Sadr, M. Al-Jubouri, M. Amani, Rheology of mineral oil-SiO_2 nanofluids at high pressure and high temperatures, Int. J. Therm. Sci. 77 (2014) 108–115. Available from: https://doi.org/10.1016/J.IJTHERMALSCI.2013.10.016.

[22] A.K. Sharma, A.K. Tiwari, A.R. Dixit, Rheological behaviour of nanofluids: a review, Renew. Sustain. Energy Rev. 53 (2016) 779–791. Available from: https://doi.org/10.1016/J.RSER.2015.09.033.

[23] S.V. Ravikumar, et al., Heat transfer enhancement using air-atomized spray cooling with water-Al_2O_3 nanofluid, Int. J. Therm. Sci. 96 (2015) 85–93. Available from: https://doi.org/10.1016/J.IJTHERMALSCI.2015.04.012.

[24] B. Sun, Y. Zhang, D. Yang, H. Li, Experimental study on heat transfer characteristics of hybrid nanofluid impinging jets, Appl. Therm. Eng. 151 (2019) 556–566. Available from: https://doi.org/10.1016/J.APPLTHERMALENG.2019.01.111.

[25] Y.T. Aksoy, Y. Zhu, P. Eneren, E. Koos, M.R. Vetrano, The impact of nanofluids on droplet/spray cooling of a heated surface: a critical review, Energies 14 (1) (2021) 80. Available from: https://doi.org/10.3390/EN14010080. 2020.

[26] G. Liang, I. Mudawar, Review of spray cooling – part 1: single-phase and nucleate boiling regimes, and critical heat flux, Int. J. Heat. Mass. Transf. 115 (2017) 1174–1205. Available from: https://doi.org/10.1016/J.IJHEATMASSTRANSFER.2017.06.029.

[27] P.K. Tyagi, R. Kumar, P.K. Mondal, A review of the state-of-the-art nanofluid spray and jet impingement cooling, Phys. Fluids 32 (12) (2020) 121301. Available from: https://doi.org/10.1063/5.0033503.

[28] Q. Xiong, A. Hajjar, B. Alshuraiaan, M. Izadi, S. Altnji, S.A. Shehzad, State-of-the-art review of nanofluids in solar collectors: a review based on the type of the dispersed nanoparticles, J. Clean. Prod. 310 (2021) 127528. Available from: https://doi.org/10.1016/J.JCLEPRO.2021.127528.

[29] Y. Li, J. Zhou, S. Tung, E. Schneider, S. Xi, A review on development of nanofluid preparation and characterization, Powder Technol. 196 (2) (2009) 89–101. Available from: https://doi.org/10.1016/j.powtec.2009.07.025.

[30] M. Theingi, K.T. Tun, N.N. Aung, Preparation, characterization and optical property of $LaFeO_3$ nanoparticles via sol-gel combustion method, SciMedicine J. 1 (3) (2019) 151–157. Available from: https://doi.org/10.28991/SCIMEDJ-2019-0103-5.

[31] S.M. Abdel-Samad, A.A. Fahmy, A.A. Massoud, A.M. Elbedwehy, Experimental investigation of TiO_2-water nanofluids thermal conductivity synthesized by sol-gel technique, Curr. Nanosci. 13 (6) (2017). Available from: https://doi.org/10.2174/1573413713666170619124221.

[32] M.S. Liu, M.C.C. Lin, C.Y. Tsai, C.C. Wang, Enhancement of thermal conductivity with Cu for nanofluids using chemical reduction method, Int. J. Heat. Mass. Transf. 49 (17–18) (2006) 3028–3033. Available from: https://doi.org/10.1016/J.IJHEATMASSTRANSFER.2006.02.012.

[33] J.N. Solanki, R. Sengupta, Z.V.P. Murthy, Synthesis of copper sulphide and copper nanoparticles with microemulsion method, Solid. State Sci. 12 (9) (2010) 1560–1566. Available from: https://doi.org/10.1016/J.SOLIDSTATESCIENCES.2010.06.021.

[34] A. Alyan, S. Abdel-Samad, A. Massoud, S.A. Waly, Characterization and thermal conductivity investigation of copper-polyaniline nano composite synthesized by gamma radiolysis method, Heat. Mass. Transf. und Stoffuebertragung 55 (9) (2019) 2409–2417. Available from: https://doi.org/10.1007/S00231-019-02588-Z/FIGURES/9.

[35] L. Marinescu, et al., Optimized synthesis approaches of metal nanoparticles with antimicrobial applications, J. Nanomater. 2020 (2020). Available from: https://doi.org/10.1155/2020/6651207.

[36] H.J. Kim, I.C. Bang, J. Onoe, Characteristic stability of bare Au-water nanofluids fabricated by pulsed laser ablation in liquids, Opt. Lasers Eng. 47 (5) (2009) 532–538. Available from: https://doi.org/10.1016/J.OPTLASENG.2008.10.011.

[37] T.X. Phuoc, Y. Soong, M.K. Chyu, Synthesis of Ag-deionized water nanofluids using multi-beam laser ablation in liquids, Opt. Lasers Eng. 45 (12) (2007) 1099–1106. Available from: https://doi.org/10.1016/J.OPTLASENG.2007.06.005.

[38] H.S. Kim, F. Yilmaz, P. Dharmaiah, D.J. Lee, T.H. Lee, S.J. Hong, Characterization of Cu and Ni nano-fluids synthesized by pulsed wire evaporation method, Arch. Metall. Mater. 62 (2B) (2017) 999–1004. Available from: https://doi.org/10.1515/AMM-2017-0141. 2.

[39] G.J. Lee, C.K. Kim, M.K. Lee, C.K. Rhee, S. Kim, C. Kim, Thermal conductivity enhancement of ZnO nanofluid using a one-step physical method, Thermochim. Acta 542 (2012) 24–27. Available from: https://doi.org/10.1016/J.TCA.2012.01.010.

[40] C.H. Lo, T.T. Tsung, L.C. Chen, Shape-controlled synthesis of Cu-based nanofluid using submerged arc nanoparticle synthesis system (SANSS), J. Cryst. Growth 277 (1–4) (2005) 636–642. Available from: https://doi.org/10.1016/J.JCRYSGRO.2005.01.067.

[41] A. Choukourov, et al., Residual- and linker-free metal/polymer nanofluids prepared by direct deposition of magnetron-sputtered Cu nanoparticles into liquid PEG, J. Mol. Liq. 336 (2021) 116319. Available from: https://doi.org/10.1016/J.MOLLIQ.2021.116319.

[42] S. Chakraborty, P.K. Panigrahi, Stability of nanofluid: a review, Appl. Therm. Eng. 174 (2020) 115259. Available from: https://doi.org/10.1016/J.APPLTHERMALENG.2020.115259.

[43] M.A. Khairul, K. Shah, E. Doroodchi, R. Azizian, B. Moghtaderi, Effects of surfactant on stability and thermo-physical properties of metal oxide nanofluids, Int. J. Heat. Mass. Transf. 98 (2016) 778–787. Available from: https://doi.org/10.1016/J.IJHEATMASSTRANSFER.2016.03.079.

[44] P.K. Das, N. Islam, A.K. Santra, R. Ganguly, Experimental investigation of thermophysical properties of Al_2O_3−water nanofluid: role of surfactants, J. Mol. Liq. 237 (2017) 304−312. Available from: https://doi.org/10.1016/J.MOLLIQ.2017.04.099.

[45] S. Chakraborty, I. Sengupta, I. Sarkar, S.K. Pal, S. Chakraborty, Effect of surfactant on thermo-physical properties and spray cooling heat transfer performance of Cu-Zn-Al LDH nanofluid, Appl. Clay Sci. 168 (2019) 43−55. Available from: https://doi.org/10.1016/J.CLAY.2018.10.018.

[46] H.W. Xian, N.A.C. Sidik, R. Saidur, Impact of different surfactants and ultrasonication time on the stability and thermophysical properties of hybrid nanofluids, Int. Commun. Heat. Mass. Transf. 110 (2020) 104389. Available from: https://doi.org/10.1016/J.ICHEATMASSTRANSFER.2019.104389.

[47] A.O. Borode, N.A. Ahmed, P.A. Olubambi, M. Sharifpur, J.P. Meyer, Effect of various surfactants on the viscosity, thermal and electrical conductivity of graphene nanoplatelets nanofluid, Int. J. Thermophys. 42 (11) (2021) 1−15. Available from: https://doi.org/10.1007/S10765-021-02914-W/FIGURES/12.

[48] J. Wang, G. Li, T. Li, M. Zeng, B. Sundén, Effect of various surfactants on stability and thermophysical properties of nanofluids, J. Therm. Anal. Calorim. 143 (6) (2021) 4057−4070. Available from: https://doi.org/10.1007/S10973-020-09381-9/TABLES/3.

[49] P. Tie, Q. Li, Y. Xuan, Heat transfer performance of Cu−water nanofluids in the jet arrays impingement cooling system, Int. J. Therm. Sci. 77 (2014) 199−205. Available from: https://doi.org/10.1016/J.IJTHERMALSCI.2013.11.007.

[50] S. Umar, F. Sulaiman, N. Abdullah, S.N. Mohamad, Investigation of the effect of pH adjustment on the stability of nanofluid, AIP Conf. Proc. 2031 (1) (2018) 020031. Available from: https://doi.org/10.1063/1.5066987.

[51] P.I.P. Soares, et al., Iron oxide nanoparticles stabilized with a bilayer of oleic acid for magnetic hyperthermia and MRI applications, Appl. Surf. Sci. 383 (2016) 240−247. Available from: https://doi.org/10.1016/J.APSUSC.2016.04.181.

[52] Xju Wang, D. sheng Zhu, S. yang, Investigation of pH and SDBS on enhancement of thermal conductivity in nanofluids, Chem. Phys. Lett. 470 (1−3) (2009) 107−111. Available from: https://doi.org/10.1016/J.CPLETT.2009.01.035.

[53] J. Huang, X. Wang, Q. Long, X. Wen, Y. Zhou, L. Li, Influence of pH on the stability characteristics of nanofluids, in: 2009 Symp. Photonics Optoelectron. SOPO 2009, 2009. doi: 10.1109/SOPO.2009.5230102.

[54] S.O. Giwa, M. Sharifpur, M.H. Ahmadi, S.M. Sohel Murshed, J.P. Meyer, Experimental investigation on stability, viscosity, and electrical conductivity of water-based hybrid nanofluid of MWCNT-Fe_2O_3, Nanomater 11 (1) (2021) 136. Available from: https://doi.org/10.3390/NANO11010136.

[55] S.A. Adam, X. Ju, Z. Zhang, J. Lin, M.M. Abd El-Samie, C. Xu, Effect of temperature on the stability and optical properties of SiO_2-water nanofluids for hybrid photovoltaic/thermal applications, Appl. Therm. Eng. 175 (2020) 115394. Available from: https://doi.org/10.1016/J.APPLTHERMALENG.2020.115394.

[56] H. Chang, et al., Temperature effect on the stability of CuO nanofluids based on measured particle distribution, Key Eng. Mater. 295−296 (2005) 51−56. Available from: https://doi.org/10.4028/http://WWW.SCIENTIFIC.NET/KEM.295-296.51.

[57] D. Dey, P. Kumar, S. Samantaray, A review of nanofluid preparation, stability, and thermo-physical properties, Heat. Transf. Res. 46 (8) (2017) 1413−1442. Available from: https://doi.org/10.1002/HTJ.21282.

[58] I. Popa, G. Gillies, G. Papastavrou, M. Borkovec, Attractive and repulsive electrostatic forces between positively charged latex particles in the presence of anionic linear polyelectrolytes, J. Phys. Chem. B 114 (9) (2010) 3170–3177. Available from: https://doi.org/10.1021/JP911482A/ASSET/IMAGES/JP911482A.SOCIAL.JPEG_V03.

[59] U. Nithiyanantham, et al., Effect of silica nanoparticle size on the stability and thermophysical properties of molten salts based nanofluids for thermal energy storage applications at concentrated solar power plants, J. Energy Storage 51 (2022) 104276. Available from: https://doi.org/10.1016/J.EST.2022.104276.

[60] H.J. Kim, S.H. Lee, J.H. Lee, S.P. Jang, Effect of particle shape on suspension stability and thermal conductivities of water-based bohemite alumina nanofluids, Energy 90 (2015) 1290–1297. Available from: https://doi.org/10.1016/J.ENERGY.2015.06.084.

[61] C. Rodríguez-Abreu, Nanocolloids: some basic concepts and principles of their stabilization, Nanocolloids A Meet. Point Sci. Technol. (2016) 1–36. Available from: https://doi.org/10.1016/B978-0-12-801578-0.00001-1.

[62] B. Boudraa, R. Bessaïh, Numerical investigation of jet impingement cooling an isothermal surface using extended jet holes with various binary hybrid nanofluids, Int. Commun. Heat Mass. Transf. 127 (2021) 105560. Available from: https://doi.org/10.1016/J.ICHEATMASSTRANSFER.2021.105560.

[63] A.K. Alanazi, Y. Khetib, H.M. Abo-Dief, M. Rawa, G. Cheraghian, M. Sharifpur, The effect of nanoparticle shape on alumina/EG-water (50:50) nanofluids flow within a solar collector: entropy and exergy investigation, Case Stud. Therm. Eng. 28 (2021) 101510. Available from: https://doi.org/10.1016/J.CSITE.2021.101510.

[64] I. Zahmatkesh, et al., Effect of nanoparticle shape on the performance of thermal systems utilizing nanofluids: a critical review, J. Mol. Liq. 321 (2021) 114430. Available from: https://doi.org/10.1016/J.MOLLIQ.2020.114430.

[65] Y. Khetib, K. Sedraoui, A.A. Melaibari, A. Alzaied, R. Alsulami, M. Sharifpur, Heat transfer and pressure drop in turbulent nanofluid flow in a pin-fin heat sink: fin and nanoparticles shape effects, Case Stud. Therm. Eng. 28 (2021) 101378. Available from: https://doi.org/10.1016/J.CSITE.2021.101378.

[66] E.M. Hotze, T. Phenrat, G.V. Lowry, Nanoparticle aggregation: challenges to understanding transport and reactivity in the environment, J. Environ. Qual. 39 (6) (2010) 1909–1924. Available from: https://doi.org/10.2134/JEQ2009.0462.

[67] S.N.do Santos, E.S. Bernardes, R. Santos-Oliveira, Nanoradiopharmaceuticals in current molecular medicine, Fundam. Nanopart. Classif. Synth. Methods Prop. Charact. (2018) 553–569. Available from: https://doi.org/10.1016/B978-0-323-51255-8.00018-5.

[68] Y. Lin, B. Li, L. Zheng, G. Chen, Particle shape and radiation effects on Marangoni boundary layer flow and heat transfer of copper-water nanofluid driven by an exponential temperature, Powder Technol. 301 (2016) 379–386. Available from: https://doi.org/10.1016/J.POWTEC.2016.06.029.

[69] S.S. Ghadikolaei, M. Gholinia, Terrific effect of H_2 on 3D free convection MHD flow of $C_2H_6O_2H_2O$ hybrid base fluid to dissolve Cu nanoparticles in a porous space considering the thermal radiation and nanoparticle shapes effects, Int. J. Hydrog. Energy 44 (31) (2019) 17072–17083. Available from: https://doi.org/10.1016/J.IJHYDENE.2019.04.171.

[70] R. Sadeghi, S.G. Etemad, E. Keshavarzi, M. Haghshenasfard, Investigation of alumina nanofluid stability by UV–vis spectrum, Microfluid. Nanofluidics 18 (5–6) (2015) 1023–1030. Available from: https://doi.org/10.1007/S10404-014-1491-Y.

[71] I.M. Mahbubul, et al., Effect of ultrasonication duration on colloidal structure and viscosity of alumina–water nanofluid, Ind. Eng. Chem. Res. 53 (16) (2014) 6677–6684. Available from: https://doi.org/10.1021/IE500705J.
[72] K.S. Hong, T.K. Hong, H.S. Yang, Thermal conductivity of Fe nanofluids depending on the cluster size of nanoparticles, Appl. Phys. Lett. 88 (3) (2006) 031901. Available from: https://doi.org/10.1063/1.2166199.
[73] A. Asadi, et al., Effect of sonication characteristics on stability, thermophysical properties, and heat transfer of nanofluids: a comprehensive review, Ultrason. Sonochem. 58 (2019) 104701. Available from: https://doi.org/10.1016/J.ULTSONCH.2019.104701.
[74] R.A. Farade, et al., Investigation of the effect of sonication time on dispersion stability, dielectric properties, and heat transfer of graphene based green nanofluids, IEEE Access. 9 (2021) 50607–50623. Available from: https://doi.org/10.1109/ACCESS.2021.3069282.
[75] A.K. Tiwari, N.S. Pandya, Z. Said, S.H. Chhatbar, Y.A. Al-Turki, A.R. Patel, 3S (Sonication, surfactant, stability) impact on the viscosity of hybrid nanofluid with different base fluids: an experimental study, J. Mol. Liq. 329 (2021) 115455. Available from: https://doi.org/10.1016/J.MOLLIQ.2021.115455.
[76] I.M. Mahbubul, R. Saidur, M.A. Amalina, E.B. Elcioglu, T. Okutucu-Ozyurt, Effective ultrasonication process for better colloidal dispersion of nanofluid, Ultrason. Sonochem. 26 (2015) 361–369. Available from: https://doi.org/10.1016/J.ULTSONCH.2015.01.005.
[77] K.V. Wong, O. De Leon, Applications of nanofluids: current and future, Adv. Mech. Eng. 2010 (2010) 11. http://doi.org/10.1155/2010/519659.
[78] F. Selimefendigil, H.F. Öztop, Pulsating nanofluids jet impingement cooling of a heated horizontal surface, Int. J. Heat. Mass. Transf. 69 (2014) 54–65. Available from: https://doi.org/10.1016/J.IJHEATMASSTRANSFER.2013.10.010.
[79] S.K. Nayak, C.P. Mishra, S.K.S. Parashar, Enhancement of heat transfer by water–Al_2O_3 and water–TiO_2 nanofluids jet impingement in cooling hot steel surface, J. Experiment. Nanosci. (2016).
[80] C.Y. Zhang, T. Wang, D.H. Chen, F.J. Hong, P. Cheng, Confined jet array impingement cooling with spent flow distraction using NEPCM slurry, Int. Commun. Heat. Mass. Transf. 77 (2016) 140–147. Available from: https://doi.org/10.1016/J.ICHEATMASSTRANSFER.2016.08.013.
[81] B. Sun, Y. Qu, D. Yang, Heat transfer of single impinging jet with Cu nanofluids, Appl. Therm. Eng. 102 (2016) 701–707. Available from: https://doi.org/10.1016/J.APPLTHERMALENG.2016.03.166.
[82] A. Pratap, Y.K. Baghel, V.K. Patel, Effect of impingement height on the enhancement of heat transfer with circular confined jet impingement using nanofluids, Mater. Today Proc. 28 (2020) 1656–1661. Available from: https://doi.org/10.1016/J.MATPR.2020.05.016.
[83] A. Datta, P. Halder, Thermal efficiency and hydraulic performance evaluation on Ag-Al_2O_3 and SiC-Al_2O_3 hybrid nanofluid for circular jet impingement, Arch. Thermodyn. 42 (1) (2021) 163–182. Available from: https://doi.org/10.24425/ATHER.2021.136953.
[84] Q. Li, Y. Xuan, F. Yu, Experimental investigation of submerged single jet impingement using Cu–water nanofluid, Appl. Therm. Eng. 36 (1) (2012) 426–433. Available from: https://doi.org/10.1016/J.APPLTHERMALENG.2011.10.059.
[85] R. Nimmagadda, H.D. Haustein, L. Godson Asirvatham, S. Wongwises, Effect of uniform/non-uniform magnetic field and jet impingement on the hydrodynamic and

heat transfer performance of nanofluids, J. Magn. Magn. Mater. 479 (2019) 268–281. Available from: https://doi.org/10.1016/J.JMMM.2019.02.019.

[86] P. Naphon, L. Nakharintr, S. Wiriyasart, Continuous nanofluids jet impingement heat transfer and flow in a micro-channel heat sink, Int. J. Heat. Mass. Transf. 126 (2018) 924–932. Available from: https://doi.org/10.1016/J.IJHEATMASSTRANSFER.2018.05.101.

[87] W. Peng, L. Jizu, B. Minli, Y. Wang, H. Chengzhi, A numerical investigation of impinging jet cooling with nanofluids, Nanoscale Microscale Thermophys. Eng. 18 (4) (2014) 329–353. Available from: https://doi.org/10.1080/15567265.2014.921749.

[88] A. Pavlova, M. Amitay, Electronic cooling using synthetic jet impingement, J. Heat. Transf. 128 (9) (2006) 897–907. Available from: https://doi.org/10.1115/1.2241889.

[89] T. Wei, et al., High-efficiency polymer-based direct multi-jet impingement cooling solution for high-power devices, IEEE Trans. Power Electron. 34 (7) (2019) 6601–6612. Available from: https://doi.org/10.1109/TPEL.2018.2872904.

[90] T.W. Wei, et al., Experimental characterization and model validation of liquid jet impingement cooling using a high spatial resolution and programmable thermal test chip, Appl. Therm. Eng. 152 (2019) 308–318. Available from: https://doi.org/10.1016/J.APPLTHERMALENG.2019.02.075.

[91] B. Kwon, T. Foulkes, T. Yang, N. Miljkovic, W.P. King, Air jet impingement cooling of electronic devices using additively manufactured nozzles, IEEE Trans. Components Packag. Manuf. Technol. 10 (2) (2020) 220–229. Available from: https://doi.org/10.1109/TCPMT.2019.2936852.

[92] S. Jones-Jackson, R. Rodriguez, A. Emadi, Jet impingement cooling in power electronics for electrified automotive transportation: current status and future trends, IEEE Trans. Power Electron. 36 (9) (2021) 10420–10435. Available from: https://doi.org/10.1109/TPEL.2021.3059558.

[93] M.V. Pukhovoy, K.A. Kunts, S.E. Spesivtsev, O.A. Kabov, Maximum heat fluxes and features of heat transfer mechanisms with boiling during jet impingement cooling of electronics, J. Phys. Conf. Ser. 1867 (1) (2021) 012036. Available from: https://doi.org/10.1088/1742-6596/1867/1/012036.

[94] A.R. Salem, F.N. Nourin, M. Abousabae, R.S. Amano, Experimental and numerical study of jet impingement cooling or improved gas turbine blade internal cooling with in-line and staggered nozzle arrays, J. Energy Resour. Technol. Trans. ASME 143 (1) (2021). Available from: https://doi.org/10.1115/1.4047600/1084723.

[95] M.S. Zaman, et al., Multiphysics optimization of thermal management designs for power electronics employing impingement cooling and stereolithographic printing, IEEE Trans. Power Electron. 36 (11) (2021) 12769–12780. Available from: https://doi.org/10.1109/TPEL.2021.3076532.

[96] J. Fu, Y. Li, Z. Cao, B. Sundén, J. Bao, G. Xie, Effect of an impinging jet on the flow characteristics and thermal performance of mainstream in battery cooling of hybrid electric vehicles, Int. J. Heat. Mass. Transf. 183 (2022) 122206. Available from: https://doi.org/10.1016/J.IJHEATMASSTRANSFER.2021.122206.

[97] J.N. Berg, R.C. Allen, S. Sobhansarbandi, A novel method of cooling a semiconductor device through a jet impingement thermal management system: CFD modeling and experimental evaluation, Int. J. Therm. Sci. 172 (2022) 107254. Available from: https://doi.org/10.1016/J.IJTHERMALSCI.2021.107254.

[98] X. Ye, H. Zhu, Y. Kang, K. Zhong, Heating energy consumption of impinging jet ventilation and mixing ventilation in large-height spaces: a comparison study, Energy Build. 130 (2016) 697–708. Available from: https://doi.org/10.1016/J.ENBUILD.2016.08.055.

[99] H.J. Chen, B. Moshfegh, M. Cehlin, Investigation on the flow and thermal behavior of impinging jet ventilation systems in an office with different heat loads, Build. Environ. 59 (2013) 127–144. Available from: https://doi.org/10.1016/J.BUILDENV.2012.08.014.
[100] H. Yamasawa, T. Kobayashi, T. Yamanaka, N. Choi, M. Cehlin, A. Ameen, Effect of supply velocity and heat generation density on cooling and ventilation effectiveness in room with impinging jet ventilation system, Build. Environ. 205 (2021) 108299. Available from: https://doi.org/10.1016/J.BUILDENV.2021.108299.
[101] N. Karwa, P. Stephan, Experimental investigation of free-surface jet impingement quenching process, Int. J. Heat. Mass. Transf. 64 (2013) 1118–1126. Available from: https://doi.org/10.1016/J.IJHEATMASSTRANSFER.2013.05.014.
[102] F. Montevecchi, G. Venturini, N. Grossi, A. Scippa, G. Campatelli, Heat accumulation prevention in Wire-Arc-Additive-Manufacturing using air jet impingement, Manuf. Lett. 17 (2018) 14–18. Available from: https://doi.org/10.1016/J.MFGLET.2018.06.004.
[103] S. Chakraborty, I. Sarkar, A. Roshan, S.K. Pal, S. Chakraborty, Spray cooling of hot steel plate using aqueous solution of surfactant and polymer, Therm. Sci. Eng. Prog. 10 (2019) 217–231. Available from: https://doi.org/10.1016/J.TSEP.2019.02.003.
[104] D. Kashyap, V. Prodanovic, M. Militzer, Transient bottom jet impingement cooling of steel, ISIJ Int. 60 (8) (2020) 1743–1751. Available from: https://doi.org/10.2355/ISIJINTERNATIONAL.ISIJINT-2019-691.
[105] G. Liu, F. Zhang, X. Zhou, D. Zhu, B. Wang, Experimental research on spray characteristics and heat transfer of cast-rolled strip subjected to impinging air-atomized cooling, Ironmaking & Steelmaking (2020).
[106] D. Mei, W. Duan, Y. Zhu, Y. Li, F. Xing, W. Chen, Heat transfer enhancement and application of multiple stepped jet cooling in a vertical alloying furnace, Int. J. Therm. Sci. 170 (2021) 107183. Available from: https://doi.org/10.1016/J.IJTHERMALSCI.2021.107183.
[107] N.H. Saeid, Thermal stress analysis of jet impingement cooling of a solid block, Heat Transf. Eng. (2022).
[108] A.V.S. Oliveira, et al., Experimental study of the heat transfer of single-jet impingement cooling onto a large heated plate near industrial conditions, Int. J. Heat. Mass. Transf. 184 (2022) 121998. Available from: https://doi.org/10.1016/J.IJHEATMASSTRANSFER.2021.121998.
[109] A. Bhunia, C.L. Chen, Jet Impingement cooling of an inverter module in the harsh environment of a hybrid vehicle, Proc. ASME Summer Heat. Transf. Conf. 4 (2009) 561–567. Available from: https://doi.org/10.1115/HT2005-72574.
[110] P.R. Parida, S.V. Ekkad, K. Ngo, Impingement-based high performance cooling configurations for automotive power converters, Int. J. Heat. Mass. Transf. 55 (4) (2012) 834–847. Available from: https://doi.org/10.1016/J.IJHEATMASSTRANSFER.2011.10.024.
[111] H. Park, Numerical assessment of liquid cooling system for power electronics in fuel cell electric vehicles, Int. J. Heat. Mass. Transf. 73 (2014) 511–520. Available from: https://doi.org/10.1016/J.IJHEATMASSTRANSFER.2014.02.038.
[112] X. Feng, E. Cousineau, K. Bennion, G. Moreno, B. Kekelia, S. Narumanchi, Experimental and numerical study of heat transfer characteristics of single-phase free-surface fan jet impingement with automatic transmission fluid, Int. J. Heat. Mass. Transf. 166 (2021) 120731. Available from: https://doi.org/10.1016/J.IJHEATMASSTRANSFER.2020.120731.

[113] M. Forster, B. Weigand, Experimental and numerical investigation of jet impingement cooling onto a concave leading edge of a generic gas turbine blade, Int. J. Therm. Sci. 164 (2021) 106862. Available from: https://doi.org/10.1016/J.IJTHERMALSCI.2021.106862.

[114] A.Ü. Tepe, Numerical investigation of a novel jet hole design for staggered array jet impingement cooling on a semicircular concave surface, Int. J. Therm. Sci. 162 (2021) 106792. Available from: https://doi.org/10.1016/J.IJTHERMALSCI.2020.106792.

[115] S. Abraham, R.P. Vedula, Distribution of effectiveness and Nusselt number over a corrugated surface impinged by a row of circular jets, IOP Conf. Ser. Mater. Sci. Eng. 1132 (1) (2021) 012007. Available from: https://doi.org/10.1088/1757-899X/1132/1/012007.

[116] M. Salman, R. Chauhan, S.C. Kim, Exergy analysis of solar heat collector with air jet impingement on dimple-shape-roughened absorber surface, Renew. Energy 179 (2021) 918–928. Available from: https://doi.org/10.1016/J.RENENE.2021.07.116.

[117] M. Salman, M.H. Park, R. Chauhan, S.C. Kim, Experimental analysis of single loop solar heat collector with jet impingement over indented dimples, Renew. Energy 169 (2021) 618–628. Available from: https://doi.org/10.1016/J.RENENE.2021.01.043.

[118] M. Salman, R. Chauhan, G. kumar Poongavanam, M.H. Park, S.C. Kim, Utilizing jet impingement on protrusion/dimple heated plate to improve the performance of double pass solar heat collector, Renew. Energy 181 (2022) 653–665. Available from: https://doi.org/10.1016/J.RENENE.2021.09.082.

[119] M.P. Sargunaraj, M. Otto, L. Vesely, E. Fernandez, J.S. Kapat, V. Viti, Use of supercritical CO_2 impingement cooling for a hypersonic leading edge, in: AIAA SCITECH 2022 Forum, January 2022. doi: 10.2514/6.2022-0266.

[120] D.E. Metzger, L.W. Florschuetz, D.I. Takeuchi, R.D. Behee, R.A. Berry, Heat transfer characteristics for inline and staggered arrays of circular jets with crossflow of spent air, J. Heat. Transf. 101 (3) (1979) 526–531. Available from: https://doi.org/10.1115/1.3451022.

[121] M. Zunaid, A. Husain, B. Singh Chauhan, R. Sahu, Numerical analysis of inclined jet impingement heat transfer in microchannel, Mater. Today Proc. 43 (2021) 557–563. Available from: https://doi.org/10.1016/J.MATPR.2020.12.048.

[122] M. Bieber, R. Kneer, W. Rohlfs, Self-similarity of heat transfer characteristics in laminar submerged and free-surface slot jet impingement, Int. J. Heat. Mass. Transf. 104 (2017) 1341–1352. Available from: https://doi.org/10.1016/J.IJHEATMASSTRANSFER.2016.08.104.

[123] N. Zuckerman, N. Lior, Jet impingement heat transfer: physics, correlations, and numerical modeling, Adv. Heat. Transf. 39 (C) (2006) 565–631. Available from: https://doi.org/10.1016/S0065-2717(06)39006-5.

[124] M. Molana, S. Banooni, Investigation of heat transfer processes involved liquid impingement jets: a review, Braz. J. Chem. Eng. 30 (3) (2013) 413–435. Available from: https://doi.org/10.1590/S0104-66322013000300001.

[125] N.J. Wilken, Experimentalinvestigation of Free-Surface Jet-Impingement Cooling by Means of TiO_2-Water Nanofluids, University of Pretoria, Pretoria, 2019.

[126] M. Behnia, S. Parneix, Y. Shabany, P.A. Durbin, Numerical study of turbulent heat transfer in confined and unconfined impinging jets, Int. J. Heat. Fluid Flow. 20 (1) (1999) 1–9. Available from: https://doi.org/10.1016/S0142-727X(98)10040-1.

[127] K.S. Choo, S.J. Kim, Comparison of thermal characteristics of confined and unconfined impinging jets, Int. J. Heat. Mass. Transf. 53 (15–16) (2010) 3366–3371. Available from: https://doi.org/10.1016/J.IJHEATMASSTRANSFER.2010.02.023.
[128] D.W. Colucci, R. Viskanta, Effect of nozzle geometry on local convective heat transfer to a confined impinging air jet, Exp. Therm. Fluid Sci. 13 (1) (1996) 71–80. Available from: https://doi.org/10.1016/0894-1777(96)00015-5.
[129] B. Markal, Experimental investigation of heat transfer characteristics and wall pressure distribution of swirling coaxial confined impinging air jets, Int. J. Heat. Mass. Transf. 124 (2018) 517–532. Available from: https://doi.org/10.1016/J.IJHEATMASSTRANSFER.2018.03.101.
[130] D. Qiu, L. Luo, S. Wang, B.A. Sunden, X. Zhang, Analysis of heat transfer and fluid flow of a slot jet impinging on a confined concave surface with various curvature and small jet to target spacing, Int. J. Numer. Methods Heat. Fluid Flow. 29 (8) (2019) 2885–2910. Available from: https://doi.org/10.1108/HFF-07-2018-0354/FULL/XML.
[131] O.J. Wai, P. Gunnasegaran, H. Hasini, A review on experimental and numerical investigations of jet impingement cooling performance with nanofluids, Micromachines 13 (12) (2022) 2059. Available from: https://doi.org/10.3390/MI13122059.
[132] E. Demir, et al., Multiphase submerged jet impingement cooling utilizing nanostructured plates, in: ASME 2012 10th International Conference on Nanochannels, Microchannels, and Minichannels Collocated with the ASME 2012 Heat Transfer Summer Conference and the ASME 2012 Fluids Engineering Division Sum, ICNMM 2012, July 2013, pp. 49–56. doi: 10.1115/ICNMM2012-73062.
[133] C. Senkal, S. Torii, Thermal fluid flow transport phenomena in nanofluid jet array impingement, J. Flow. Vis. Image Process. 22 (1–3) (2015) 59–79. Available from: https://doi.org/10.1615/JFLOWVISIMAGEPROC.2015015238.
[134] M. Amjadian, H. Safarzadeh, M. Bahiraei, S. Nazari, B. Jaberi, Heat transfer characteristics of impinging jet on a hot surface with constant heat flux using Cu_2O–water nanofluid: an experimental study, Int. Commun. Heat. Mass. Transf. 112 (2020) 104509. Available from: https://doi.org/10.1016/J.ICHEATMASSTRANSFER.2020.104509.
[135] U. Allauddin, et al., Heat transfer enhancement caused by impinging jets of Al_2O_3-water nanofluid on a micro-pin fin roughened surface under crossflow conditions—a numerical study, J. Enhanc. Heat. Transf. 27 (4) (2020) 367–387. Available from: https://doi.org/10.1615/JENHHEATTRANSF.2020033413.
[136] F.J. Hong, C.Y. Zhang, D.H. Chen, G. Chen, Confined jet array impingement cooling using NEPCM nanofluids, in: ASME 2016 5th International Conference on Micro/Nanoscale Heat and Mass Transfer MNHMT 2016, 1, March 2016. doi: 10.1115/MNHMT2016-6531.
[137] F. Selimefendigil, H.F. Öztop, Combined effects of double porous layers and nanofluids on the performance of confined single and multi-jet impingement heat transfer, (2021). https://doi.org/10.1080/00986445.2021.1928650
[138] Q. Li, Y. Xuan, F. Yu, J. Tan, Experimental investigation of submerged impinging jet using Cu-water nanofluid, 2010 14th Int. Heat Transf. Conf. IHTC 14 (5) (2011) 469–473. Available from: https://doi.org/10.1115/IHTC14-22040.
[139] F. Selimefendigil, A.J. Chamkha, Cooling of an isothermal surface having a cavity component by using CuO-water nano-jet, Int. J. Numer. Methods Heat. Fluid Flow. 30 (4) (2020) 2169–2191. Available from: https://doi.org/10.1108/HFF-12-2018-0724.

[140] F. Selimefendigil, M. Çoğan, H.F. Öztop, Combined effects of local curvature and elasticity of an isothermal wall for jet impingement cooling under magnetic field effects, J. Cent. South. Univ. 28 (11) (2021) 3534–3544. Available from: https://doi.org/10.1007/S11771-021-4873-9.

[141] F. Selimefendigil, H.F. Öztop, Cooling of a partially elastic isothermal surface by nanofluids jet impingement, ASME J. Heat. Mass. Transf. 140 (4) (2018). Available from: https://doi.org/10.1115/1.4038422/367817.

[142] S. Wiriyasart, P. Naphon, Liquid impingement cooling of cold plate heat sink with different fin configurations: high heat flux applications, Int. J. Heat. Mass. Transf. 140 (2019) 281–292. Available from: https://doi.org/10.1016/J.IJHEATMASSTRANSFER.2019.06.020.

[143] R. Ekiciler, M.S.A. Çetinkaya, K. Arslan, Effect of shape of nanoparticle on heat transfer and entropy generation of nanofluid-jet impingement cooling, Int. J. Green. Energy Taylor Fr. Online 17 (10) (2020) 555–567. Available from: https://doi.org/10.1080/15435075.2020.1739692.

[144] H. Ahmadi, R.M. Moghari, K. Esmailpour, A.S. Mujumdar, Numerical investigation of semi-confined turbulent slot jet impingement on a concave surface using an Al_2O_3–water nanofluid, Appl. Math. Model. 40 (2) (Jan 2016) 1110–1125. Available from: https://doi.org/10.1016/J.APM.2015.06.021.

[145] A. Jaiswal, A. Datta, P. Halder, Heat transfer augmenation using nanofluid of a confined slot impingement jet on concave surface, IOP Conf. Ser. Mater. Sci. Eng. 377 (1) (2018) 012075. Available from: https://doi.org/10.1088/1757-899X/377/1/012075.

[146] F. Selimefendigil, H.F. Öztop, Jet impingement cooling and optimization study for a partly curved isothermal surface with CuO-water nanofluid, Int. Commun. Heat. Mass. Transf. 89 (2017) 211–218. Available from: https://doi.org/10.1016/J.ICHEATMASSTRANSFER.2017.10.007.

[147] E. Ersayin, F. Selimefendigil, Numerical investigation of impinging jets with nanofluids on a moving plate, Math. Comput. Appl. 18 (3) (2013) 428–437. Available from: https://doi.org/10.3390/MCA18030428.

[148] F. Selimefendigil, H.F. Oztop, A.J. Chamkha, Jet impingement heat transfer of confined single and double jets with non-newtonian power law nanofluid under the inclined magnetic field effects for a partly curved heated wall, Sustainability 13 (9) (2021) 5086. Available from: https://doi.org/10.3390/SU13095086.

[149] M. Mahdavi, M. Sharifpur, J.P. Meyer, L. Chen, Thermal analysis of a nanofluid free jet impingement on a rotating disk using volume of fluid in combination with discrete modelling, Int. J. Therm. Sci. 158 (2020). Available from: https://doi.org/10.1016/J.IJTHERMALSCI.2020.106532.

[150] A. Datta, P. Halder, Curved surface and thermal performance factor effect using SiC-Al_2O_3 hybrid nanofluid jet impingement, Interfacial Phenom. Heat. Transf. 8 (4) (2020) 321–326,. Available from: https://doi.org/10.1615/INTERFACPHENOMHEATTRANSFER.2020036357.

[151] M. Kahani, R.G. Jackson, G. Rosengarten, Experimental investigation of TiO_2/water nanofluid droplet impingement on nanostructured surfaces, Ind. Eng. Chem. Res. 55 (7) (2016) 2230–2241. Available from: https://doi.org/10.1021/ACS.IECR.5B04465.

[152] L.A. El-Gabry, D.A. Kaminski, Experimental investigation of local heat transfer distribution on smooth and roughened surfaces under an array of angled impinging jets, J. Turbomach. 127 (3) (2005) 532–544. Available from: https://doi.org/10.1115/1.1861918.

[153] S.M. Kim, A. Afzal, K.Y. Kim, Optimization of a staggered jet-convex dimple array cooling system, Int. J. Therm. Sci. 99 (2016) 161–169. Available from: https://doi.org/10.1016/J.IJTHERMALSCI.2015.08.013.

[154] Y.H. Liu, S.J. Song, Y.H. Lo, Jet impingement heat transfer on target surfaces with longitudinal and transverse grooves, Int. J. Heat. Mass. Transf. 58 (1–2) (2013) 292–299. Available from: https://doi.org/10.1016/J.IJHEATMASSTRANSFER.2012.11.042.

[155] C. Wan, Y. Rao, P. Chen, Numerical predictions of jet impingement heat transfer on square pin-fin roughened plates, Appl. Therm. Eng. 80 (2015) 301–309. Available from: https://doi.org/10.1016/J.APPLTHERMALENG.2015.01.053.

[156] S.W. Chang, L.M. Su, Y. Zheng, Reciprocating impingement jet heat transfer with surface ribs, A J. Therm. Energy Gener. Transp. Storage Convers. 13 (4) (2000) 275–297. Available from: https://doi.org/10.1080/08916150050175462.

[157] L.G. Hansen, B.W. Webb, Air jet impingement heat transfer from modified surfaces, Int. J. Heat. Mass. Transf. 36 (4) (1993) 989–997. Available from: https://doi.org/10.1016/S0017-9310(05)80283-2.

[158] S. Caliskan, S. Baskaya, Experimental investigation of impinging jet array heat transfer from a surface with V-shaped and convergent-divergent ribs, Int. J. Therm. Sci. 59 (2012) 234–246. Available from: https://doi.org/10.1016/J.IJTHERMALSCI.2012.04.013.

[159] Y. Rao, P. Chen, C. Wan, Experimental and numerical investigation of impingement heat transfer on the surface with micro W-shaped ribs, Int. J. Heat. Mass. Transf. 93 (2016) 683–694. Available from: https://doi.org/10.1016/J.IJHEATMASSTRANSFER.2015.10.022.

[160] S. Caliskan, Flow and heat transfer characteristics of transverse perforated ribs under impingement jets, Int. J. Heat. Mass. Transf. 66 (2013) 244–260. Available from: https://doi.org/10.1016/J.IJHEATMASSTRANSFER.2013.07.027.

[161] P. Xu, A.P. Sasmito, S. Qiu, A.S. Mujumdar, L. Xu, L. Geng, Heat transfer and entropy generation in air jet impingement on a model rough surface, Int. Commun. Heat. Mass. Transf. 72 (2016) 48–56. Available from: https://doi.org/10.1016/J.ICHEATMASSTRANSFER.2016.01.007.

[162] K. Krishna, M. Ricklick, Heat transfer analysis of jet impingement cooling on a simulated ceramic matrix composite surface, Proc. ASME Turbo Expo. 5A-2017 (2017). Available from: https://doi.org/10.1115/GT2017-64991.

[163] S. Mitra, S.K. Saha, S. Chakraborty, S. Das, Study on boiling heat transfer of water–TiO_2 and water–MWCNT nanofluids based laminar jet impingement on heated steel surface, Appl. Therm. Eng. 37 (2012) 353–359. Available from: https://doi.org/10.1016/J.APPLTHERMALENG.2011.11.048.

[164] M.K. Singh, D. Yadav, S. Arpit, S. Mitra, S.K. Saha, Effect of nanofluid concentration and composition on laminar jet impinged cooling of heated steel plate, Appl. Therm. Eng. 100 (2016) 237–246. Available from: https://doi.org/10.1016/J.APPLTHERMALENG.2016.01.032.

[165] T.B. Chang, Y.K. Yang, Heat transfer performance of jet impingement flow boiling using Al_2O_3-water nanofluid, J. Mech. Sci. Technol. 28 (4) (2014) 1559–1566. Available from: https://doi.org/10.1007/S12206-013-1143-2.

[166] R. Rioboo, C. Tropea, M. Marengo, Outcomes from a drop impact on solid surfaces, At. Sprays 11 (2) (2001) 155–165. Available from: https://doi.org/10.1615/ATOMIZSPR.V11.I2.40.

[167] C. Josserand, S.T. Thoroddsen, Drop impact on a solid surface, Annu. Rev. Fluid Mech. (2016).

[168] M.A. Quetzeri-Santiago, K. Yokoi, A.A. Castrejón-Pita, J.R. Castrejón-Pita, Role of the dynamic contact angle on splashing, Phys. Rev. Lett. 122 (22) (2019). Available from: https://doi.org/10.1103/PHYSREVLETT.122.228001.

[169] A.L. Yarin, Drop impact dynamics: splashing, spreading, receding, bouncing..., Annu. Rev. Fluid Mech. (2005).

[170] Y.T. Aksoy, P. Eneren, E. Koos, M.R. Vetrano, Spreading-splashing transition of nanofluid droplets on a smooth flat surface, J. Colloid Interface Sci. 606 (2022) 434–443. Available from: https://doi.org/10.1016/J.JCIS.2021.07.157.

[171] K. Wongcharee, V. Chuwattanakul, S. Eiamsa-ard, Heat transfer of swirling impinging jets with TiO_2-water nanofluids, Chem. Eng. Process—Process Intensif. 114 (2017) 16–23. Available from: https://doi.org/10.1016/J.CEP.2017.01.004.

[172] K. Wongcharee, S. Eiamsa-ard, Enhancement of heat transfer using CuO/water nanofluid and twisted tape with alternate axis, Int. Commun. Heat. Mass. Transf. 38 (6) (2011) 742–748. Available from: https://doi.org/10.1016/j.icheatmasstransfer.2011.03.011.

[173] K. Wongcharee, V. Chuwattanakul, S. Eiamsa-ard, Influence of CuO/water nanofluid concentration and swirling flow on jet impingement cooling, Int. Commun. Heat. Mass. Transf. 88 (2017) 277–283. Available from: https://doi.org/10.1016/J.ICHEATMASSTRANSFER.2017.08.020.

[174] K. Wongcharee, S. Eiamsa-ard, Heat transfer enhancement by using CuO/water nanofluid in corrugated tube equipped with twisted tape, Int. Commun. Heat. Mass. Transf. 39 (2) (2012) 251–257. Available from: https://doi.org/10.1016/j.icheatmasstransfer.2011.11.010.

[175] J. Doty, et al., "Dynamic thermal management for aerospace technology: review and outlook, J. Thermophys. Heat. Transf. 31 (1) (2017) 86–98. Available from: https://doi.org/10.2514/1.T4701/ASSET/IMAGES/LARGE/FIGURE12.JPEG.

[176] G. Saliba, et al., Jet impingement cooling using fluidic oscillators: an experimental study, J. Phys. Conf. Ser. 2116 (1) (2021) 012028. Available from: https://doi.org/10.1088/1742-6596/2116/1/012028.

[177] S. Abishek, R. Narayanaswamy, Low frequency pulsating jet impingement boiling and single phase heat transfer, Int. J. Heat. Mass. Transf. 159 (2020) 120052. Available from: https://doi.org/10.1016/J.IJHEATMASSTRANSFER.2020.120052.

[178] J. Wassenberg, P. Stephan, T. Gambaryan-Roisman, Heat transfer during pulsating liquid jet impingement onto a vertical wall, Heat. Mass. Transf. und Stoffuebertragung 57 (4) (2021) 617–629. Available from: https://doi.org/10.1007/S00231-020-02973-Z/FIGURES/13.

[179] N. Neumann, A. Berthold, F. Haucke, D. Peitsch, P. Stathopoulos, Pulsed impingement turbine cooling and its effect on the efficiency of gas turbines with pressure gain combustion, J. Turbomach. 143 (7) (2021). Available from: https://doi.org/10.1115/1.4050361/1101649.

[180] Z. Zhang, Q. Li, C. Bruecker, Q. Zhang, Enhanced thermal performance with high-amplitude intermittent impingement cooling, Int. J. Heat. Mass. Transf. 185 (2022) 122359. Available from: https://doi.org/10.1016/J.IJHEATMASSTRANSFER.2021.122359.

[181] P. Li, D. Guo, R. Liu, Mechanism analysis of heat transfer and flow structure of periodic pulsating nanofluids slot-jet impingement with different waveforms, Appl. Therm. Eng. 152 (2019) 937–945. Available from: https://doi.org/10.1016/J.APPLTHERMALENG.2019.01.086.

[182] D.A. Zumbrunnen, M. Aziz, Convective heat transfer enhancement due to intermittency in an impinging jet, J. Heat. Transf. 115 (1) (1993) 91–98. Available from: https://doi.org/10.1115/1.2910675.

[183] T.L. Yang, S.W. Chang, L.M. Su, C.C. Hwang, Heat transfer of confined impinging jet onto spherically concave surface with piston cooling application, JSME Int. J. Ser. B Fluids Therm. Eng. 42 (2) (1999) 238–248. Available from: https://doi.org/10.1299/JSMEB.42.238.

[184] E.C. Mladin, D.A. Zumbrunnen, Alterations to coherent flow structures and heat transfer due to pulsations in an impinging air-jet, Int. J. Therm. Sci. 39 (2) (2000) 236–248. Available from: https://doi.org/10.1016/S1290-0729(00)00242-8.

[185] Z. Trávníček, T. Vít, Impingement heat/mass transfer to hybrid synthetic jets and other reversible pulsating jets, Int. J. Heat. Mass. Transf. 85 (2015) 473–487. Available from: https://doi.org/10.1016/J.IJHEATMASSTRANSFER.2015.01.125.

[186] Y. Utturkar, M. Arik, C.E. Seeley, M. Gursoy, An experimental and computational heat transfer study of pulsating Jets, J. Heat. Transf. 130 (6) (2008). Available from: https://doi.org/10.1115/1.2891158/466852.

[187] M. Arik, T. Icoz, Predicting heat transfer from unsteady synthetic jets, J. Heat. Transf. 134 (2012) 8. Available from: https://doi.org/10.1115/1.4005740/468175.

[188] M. Arik, R. Sharma, J. Lustbader, X. He, Steady and unsteady air impingement heat transfer for electronics cooling applications, J. Heat. Transf. 135 (2013) 11. Available from: https://doi.org/10.1115/1.4024614/367257.

[189] M. Arik, R. Sharma, J. Lustbader, X. He, Comparison of synthetic and steady air jets for impingement heat transfer over vertical surfaces, Intersoc. Conf. Therm. Thermomech. Phenom. Electron. Syst. ITHERM (2012) 1354–1363. Available from: https://doi.org/10.1109/ITHERM.2012.6231578.

[190] O. Ghaffari, M.B. Dogruoz, M. Arik, An investigation into momentum and temperature fields of a meso-scale synthetic jet, Thermomech. Phenom. Electron. Syst.—Proceedings Intersoc. Conf. (2014) 889–896. Available from: https://doi.org/10.1109/ITHERM.2014.6892375.

[191] H.S. Sheriff, D.A. Zumbrunnen, Effect of flow pulsations on the cooling effectiveness of an impinging jet, J. Heat. Transf. 116 (4) (1994) 886–895. Available from: https://doi.org/10.1115/1.2911463.

[192] E.C. Mladin, D.A. Zumbrunnen, Nonlinear dynamics of laminar boundary layers in pulsatile stagnation flows, J. Thermophys. Heat. Transf. 8 (3) (1994) 514–523. Available from: https://doi.org/10.2514/3.573.

[193] E.C. Mladin, D.A. Zumbrunnen, Nonlinear dynamics of laminar boundary layers in pulsatile stagnation flows, J. Thermophys. Heat Transf. (2012).

[194] E. Sabelberg, M. Cardenas, R. Kneer, W. Rohlfs, Design, development, and validation of concepts for generating passive pulsation in cooling nozzles, Case Stud. Therm. Eng. 7 (2016) 103–108. Available from: https://doi.org/10.1016/J.CSITE.2016.03.005.

[195] L. Nakharintr, P. Naphon, Magnetic field effect on the enhancement of nanofluids heat transfer of a confined jet impingement in mini-channel heat sink, Int. J. Heat. Mass. Transf. 110 (2017) 753–759. Available from: https://doi.org/10.1016/J.IJHEATMASSTRANSFER.2017.03.078.

[196] B.W. Webb, C.F. Ma, B.W. SinWebb, C.F. Ma, Single-phase liquid jet impingement heat transfer, Adv. Heat. Transf. 26 (C) (1995) 105–217. Available from: https://doi.org/10.1016/S0065-2717(08)70296-X.

[197] M. Modak, S.S. Chougule, S.K. Sahu, An experimental investigation on heat transfer characteristics of hot surface by using CuO-water nanofluids in circular jet impingement cooling, J. Heat. Transf. 140 (2018) 1. Available from: https://doi.org/10.1115/1.4037396/375610.

[198] M.M. Sorour, W.M. El-Maghlany, M.A. Alnakeeb, A.M. Abbass, Experimental study of free single jet impingement utilizing high concentration SiO_2 nanoparticles water base nanofluid, Appl. Therm. Eng. 160 (2019) 114019. Available from: https://doi.org/10.1016/J.APPLTHERMALENG.2019.114019.

[199] F. Selimefendigil, Pulsating hybrid nanofluids double slot jets impingement onto an isothermal wall, Heat. Transf. Res. 49 (2) (2018) 173–188. Available from: https://doi.org/10.1615/HEATTRANSRES.2017015650.

[200] O. Abdelrehim, A. Khater, A.A. Mohamad, A. Radwan, Two-phase simulation of nanofluid in a confined single impinging jet, Case Stud. Therm. Eng. 14 (2019) 100423. Available from: https://doi.org/10.1016/J.CSITE.2019.100423.

[201] S.D. Barewar, S. Tawri, S.S. Chougule, Heat transfer characteristics of free nanofluid impinging jet on flat surface with different jet to plate distance: an experimental investigation, Chem. Eng. Process.—Process Intensif. 136 (2019) 1–10. Available from: https://doi.org/10.1016/J.CEP.2018.12.001.

[202] J. Lv, C. Hu, M. Bai, K. Zeng, S. Chang, D. Gao, Experimental investigation of free single jet impingement using SiO_2-water nanofluid, Exp. Therm. Fluid Sci. 84 (2017) 39–46. Available from: https://doi.org/10.1016/J.EXPTHERMFLUSCI.2017.01.010.

[203] O. Manca, D. Ricci, S. Nardini, G. Di Lorenzo, Thermal and fluid dynamic behaviors of confined laminar impinging slot jets with nanofluids, Int. Commun. Heat. Mass. Transf. 70 (2016) 15–26. Available from: https://doi.org/10.1016/j.icheatmasstransfer.2015.11.010.

[204] A. Datta, S. Kumar, P. Halder, Heat transfer and thermal characteristics effects on moving plate impinging from Cu-water nanofluid jet, J. Therm. Sci. 29 (1) (2019) 182–193. Available from: https://doi.org/10.1007/S11630-019-1107-7.

[205] J. Lv, S. Chang, C. Hu, M. Bai, P. Wang, K. Zeng, Experimental investigation of free single jet impingement using Al_2O_3-water nanofluid, Int. Commun. Heat. Mass. Transf. 88 (2017) 126–135. Available from: https://doi.org/10.1016/J.ICHEATMASSTRANSFER.2017.08.017.

[206] R. Amano, B. Sundén, Impingement Jet Cooling in Gas Turbines, WIT Press, 2014, p. 25.

[207] A. Mensch, K.A. Thole, Overall effectiveness of a blade endwall with jet impingement and film cooling, J. Eng. Gas. Turbines Power 136 (3) (2014). Available from: https://doi.org/10.1115/1.4025835/373835.

[208] M. Jena, P.C. Mishra, S.S. Sahu, Comparative performance of jet and spray impingement cooling in steel strip run-out table: experimental results, Aust. J. Mech. Eng. (2022). Available from: https://doi.org/10.1080/14484846.2022.2066854.

[209] M.R. Pais, L.C. Chow, E.T. Mahefkey, Multiple jet impingement cooling, J. Thermophys. Heat. Transf. 7 (3) (1993) 435–440. Available from: https://doi.org/10.2514/3.437.

[210] J.M. Jha, S.V. Ravikumar, I. Sarkar, S.K. Pal, S. Chakraborty, Jet impingement cooling of a hot moving steel plate: an experimental study, Exp. Heat. Transf. 29 (5) (2016) 615–631. Available from: https://doi.org/10.1080/08916152.2015.1046019.

[211] H. Leocadio, J.C. Passos, Experimental investigation of heat transfer characteristics during water jet impingement cooling of a high-temperature steel surface, Ironmaking & Steelmaking (2021).

[212] I. Sarkar, S. Chakraborty, A. Roshan, D.K. Behera, S.K. Pal, S. Chakraborty, Application of TiO_2 nanofluid-based coolant for jet impingement quenching of a hot steel plate, Exp. Heat. Transf. 32 (4) (2019) 322–336. Available from: https://doi.org/10.1080/08916152.2018.1517835/SUPPL_FILE/UEHT_A_1517835_SM6787.TIF.

[213] K. Santosh, P. Nayak, M. Chandra, S.K.S. Parashar, S. Kumar Nayak, C. Mishra, Enhancement of heat transfer by water-Al_2O_3 and water-TiO_2 nanofluids jet impingement in cooling hot steel surface, J. Exp. Nanosci. 11 (16) (2016) 1253–1273. Available from: https://doi.org/10.1080/17458080.2016.1209789.

[214] FreedomCar and Fuel Partnership, "Partnership Plan," 2006. [Online]. Available: http://www.eere.energy.gov/vehiclesandfuels/pdfs/program/fc_fuel_partnership_plan.pdf.

[215] R. Wu, T. Hong, Q. Cheng, H. Zou, Y. Fan, X. Luo, Thermal modeling and comparative analysis of jet impingement liquid cooling for high power electronics, Int. J. Heat. Mass. Transf. 137 (2019) 42–51. Available from: https://doi.org/10.1016/J.IJHEATMASSTRANSFER.2019.03.112.

[216] M. Sabato, A. Fregni, E. Stalio, F. Brusiani, M. Tranchero, T. Baritaud, Numerical study of submerged impinging jets for power electronics cooling, Int. J. Heat. Mass. Transf. 141 (2019) 707–718. Available from: https://doi.org/10.1016/J.IJHEATMASSTRANSFER.2019.06.081.

[217] C. Klinkhamer, S. Abishek, K.L.V. Iyer, R. Balachandar, R. Barron, Characterization of a jet impingement heat sink for power electronics cooling, Therm. Sci. Eng. Prog. 34 (2022) 101408. Available from: https://doi.org/10.1016/J.TSEP.2022.101408.

[218] W. Eng Ewe, et al., Jet impingement cooling applications in solar energy technologies: systematic literature review, Therm. Sci. Eng. Prog. 34 (2022) 101445. Available from: https://doi.org/10.1016/J.TSEP.2022.101445.

[219] M. Javidan, A.J. Moghadam, Effective cooling of a photovoltaic module using jet-impingement array and nanofluid coolant, Int. Commun. Heat. Mass. Transf. 137 (2022) 106310. Available from: https://doi.org/10.1016/J.ICHEATMASSTRANSFER.2022.106310.

[220] Z. Rostami, M. Rahimi, N. Azimi, Using high-frequency ultrasound waves and nanofluid for increasing the efficiency and cooling performance of a PV module, Energy Convers. Manage. 160 (2018) 141–149. Available from: https://doi.org/10.1016/J.ENCONMAN.2018.01.028.

[221] M.S.Y. Ebaid, A.M. Ghrair, M. Al-Busoul, Experimental investigation of cooling photovoltaic (PV) panels using (TiO_2) nanofluid in water -polyethylene glycol mixture and (Al2O3) nanofluid in water-cetyltrimethylammonium bromide mixture, Energy Convers. Manage. 155 (2018) 324–343. Available from: https://doi.org/10.1016/J.ENCONMAN.2017.10.074.

[222] H.A. Hasan, K. Sopian, A.H. Jaaz, A.N. Al-Shamani, Experimental investigation of jet array nanofluids impingement in photovoltaic/thermal collector, Sol. Energy 144 (2017) 321–334. Available from: https://doi.org/10.1016/J.SOLENER.2017.01.036.

[223] M. Milanese, F. Micali, G. Colangelo, A. de Risi, Experimental evaluation of a full-scale HVAC system working with nanofluid, Energies 15 (8) (2022) 2902. Available from: https://doi.org/10.3390/EN15082902.

Index

Note: Page numbers followed by "*f*" and "*t*" refer to figures and tables, respectively.

O

P

T

U

V

W

X

Y

Z